Insects

Insects

Evolutionary Success, Unrivaled Diversity, and World Domination

David B. Rivers

Department of Biology
Loyola University of Maryland

Johns Hopkins University Press   Baltimore

© 2017 Johns Hopkins University Press
All rights reserved. Published 2017
Printed in Canada on acid-free paper
9 8 7 6 5 4 3 2 1

Johns Hopkins University Press
2715 North Charles Street
Baltimore, Maryland 21218-4363
www.press.jhu.edu

Library of Congress Cataloging-in-Publication Data

Names: Rivers, David, 1966– author.
Title: Insects : evolutionary success, unrivaled diversity, and world
    domination / David B. Rivers.
Description: Baltimore, Maryland : Johns Hopkins University Press,
    2017. | Includes bibliographical references and index.
Identifiers: LCCN 2016018627 | ISBN 9781421421704 (hardcover :
    alk. paper) | ISBN 1421421704 (hardcover : alk. paper) |
    ISBN 9781421421711 (electronic) | ISBN 1421421712 (electronic)
Subjects: LCSH: Insects—Textbooks.
Classification: LCC QL463 .R58 2017 | DDC 595.7—dc23
    LC record available at https://lccn.loc.gov/2016018627

A catalog record for this book is available from the British Library.

*Special discounts are available for bulk purchases of this book. For
more information, please contact Special Sales at 410-516-6936 or
specialsales@press.jhu.edu.*

Johns Hopkins University Press uses environmentally friendly book
materials, including recycled text paper that is composed of at least
30 percent post-consumer waste, whenever possible.

Contents

Preface

Insects are extraordinary creatures. In every facet of biology imaginable, these six-legged life forms accomplish tasks uniquely or with an efficiency not realized by any other organisms, certainly not by any other animal. Telling evidence of their great adaptive success is their incredible biological diversity and sheer abundance on the planet. On any given day and at any moment in time, an estimated one million species or more are represented by two to ten quintillion individual insects residing on Earth. For those not scientifically inclined, and even for those who are, we call that a heck of a lot! It is nearly impossible to grasp the enormity of insects on the planet and the influence they have had (and continue to have) on life on Earth. For me, the words of Dr. Gilbert Waldbauer,* in his engaging text *Insects through the Seasons* (1996, xii-xiii), expresses all that needs to be said: "Insects are endlessly fascinating—wonderful in their amazing variety and marvelous in the perfection of the evolutionary adaptations that have made them the dominant animal group on the earth. When I wake up in the morning to get ready for work, I often think about how fortunate I am to be an entomologist."

Perhaps it seems intuitively obvious that the title of this textbook was derived from Dr. Waldbauer's words. A very good assumption on your part. But in actuality, his words were not the true source of inspiration for this book. That honor is bestowed on two genetically enhanced cartoon rats, Pinky and the Brain—creations of Acme Laboratories by movie mogul Steven Spielberg—who starred in two self-titled animated series that aired in the United States during the 1990s. In every episode, the two rats would begin by pondering the same theme, world domination (Ponderings of Pinky and the Brain, http://yhoo.it/1ioVDvw):

> BRAIN: We must prepare for tomorrow night.
> PINKY: Why? What are we going to do tomorrow night?
> BRAIN: The same thing we do every night, Pinky—try to take over the world![†]

This quote struck a chord with me as being very apropos for what insects do. Unlike the Brain, however, insects are innocent in their conquests, achieving domination of the planet with no conscious drive to do so. As acclaimed insect chemical ecologist

---

*Dr. Gilbert Walbauer is Professor Emeritus in the Department of Entomology at the University of Illinois.

†The cartoon series aired from 1995 to 1998 on the WB network, and then on Cartoon Network from 1998 to 2007, originating as characters in the Animaniacs cartoon series, also developed by Steven Spielberg.

Dr. Thomas Eisner* stated so clearly in his book *For Love of Insects* (2003, 1), insect dominance has been achieved through an accumulation of life history successes:

> Taken as a group, and viewed on a grand scale, insects have achieved a great deal. They metamorphose—or at least most of them do—growing up as larvae and taking to the wing as adults. They have adapted for direct sperm transmittal through insemination, thereby relinquishing forever dependence on water for spawning. And as a consequence of acquisition of an exoskeleton—an external skeleton consisting of a hard cuticle—they have attained quickness of motion, resistance to desiccation, and the capacity to achieve dominance on land. They have succeeded in one major respect where humans have failed. They are practitioners of sustainable development. Although they are the primary consumers of plants, they do not merely exploit them. They also pollinate them, thereby providing a secure future, both for themselves and for their plant partners.

The path laid out thus far presents insects as fascinating, efficient, responsible, and remarkably successful. These words reveal another facet of insect biology; that is, those who study insects often view the world through a much different lens than the majority of the populace, the latter often referred to as "normal." When most people are polled for opinions about insects, "fascination" is not the first word that comes to mind. No, most use terms like repulsion, fear, disease, destruction, and many others devoid of warmth and caring for creatures with six legs. Renowned entomologist and humorist Dr. May Berenbaum summed up normal human sentiments toward insects:

> The vast majority of people consider it a high priority to minimize the extent of their interaction with the insect world. Homes are sealed, sprayed, and kept meticulously clean so as to reduce the probability that they will be invaded by insects; similarly, bodies are bathed, hair is shampooed, and clothing regularly washed in order to eliminate any unwanted contact with six-legged life forms. In the overwhelmingly vast majority of daily conversations, insects are conspicuous in their absence; those rare conversations in which insects feature prominently are generally carried out in guarded tones, often with a touch of embarrassment. After all, no one likes to admit, even to close personal friends, to being stung, bitten, infested, invaded, or otherwise bested by the loathsome insects that manage to get around the safeguards.†

A resounding "amen" was likely just breathed by the majority. I readily admit that opinions on six-legged critters are mixed. Family, friends, and most students usually do not share my enthusiasm for insects. And that is okay; the mission in writing this book was not conversion of lost souls to entomology. The book is designed for lovers and haters and anyone who falls in between. A passion for insects is not required to understand the facts, ideas, and concepts that unfold on the pages of this text. However, an open mind certainly makes learning much easier and more enjoyable. It also will allow you to see the true world of insects, no matter what view you currently hold toward such beasts.

We know that several species of insects transmit diseases and evoke huge economic losses in terms of food production and damage to buildings, as well as just annoy us to no end. But in the grand scheme of things, very few are responsible for disturbing the human condition. The overwhelming majority of insects aid our lives in some fashion, oftentimes in ways we do not realize. In *Insects,* I aim to discuss the good and the bad sides of insects, in an open, honest, and engaging examination of insect biology. At the end of the journey, you can decide for yourself what to think of these creatures. Personally, I find them more fascinating than any other creature I have ever encountered.

Why this book?

When I was an undergraduate majoring in biology, insects were not yet part of my future. I enrolled in an entomology course, used a textbook that focused on insect morphology, learned to identify these creatures using tedious jargon-laden identification keys, experienced the joys of lab practicals where the goal was to identify

---

*Dr. Thomas Eisner was Professor Emeritus in the Department of Entomology at Cornell University.

†From Dr. Berenbaum's entertaining book *Bugs in the System* (1995, xi). She is head of the Department of Entomology at the University of Illinois.

insects to genus and species within thirty seconds, and completed the massive undertaking of a producing a large insect collection while the seasons were changing and my TA was becoming more cranky. Despite this introduction to the world of entomology, I become enthralled with insects. Not because of the approach or the book; no, I found that both squelched my interest. The instructor was unbelievably knowledgeable but lectured in a dry, monotonous manner, relying on paper notes that had jaundiced with age.

A few students in the class loved everything about the insect collections, pinning, preserving, and identifying. They had collected for years, and this course was the next step in their life's journey toward entomological nirvana. Most everyone else looked like death, as they dragged their bodies into the lecture hall three mornings a week, lulled to sleep by the waves of taxonomic information. They took entomology to complete a category requirement for the biology major; other courses they desired were already closed. Every facet of this class just confirmed that they disliked—no, hated—anything related to insects. So each person just needed to survive for fifteen weeks, and then freedom. I suppose to a degree I was of a similar mindset. The course did not stimulate my interest in insects, although I had hoped it would. Instead, my curiosity was flamed to life by a research adviser who shared with me his enthusiasm for learning and the biology of insects. In short order, I knew my calling would involve entomology.

As a college professor now entering my twenty-second year in the classroom, my goal as an instructor has been to share enthusiasm and stimulate curiosity for all topics discussed. Insects are my real passion, and I believe that, when shared in an open and genuine manner, it (passion, not insects) becomes available to everyone. I have tried to capture that philosophy in this book. Most students enrolled in a general entomology course in the United States have no intention of pursuing a career centered on insects. Since this is no secret to anyone teaching an entomology course, why would a majors' level textbook seem appropriate for the class? Simple: most instructors teach a course the way it was taught to them as undergraduate students. The approach used today in a typical entomology course is similar to what I experienced as an undergraduate. And the net effect is also still

the same: most students become uninterested, get overwhelmed with terminology, and shut the door on insects. Yes, this is a broad stroke that does not encompass all courses. But it reflects the experience of many.

Insects was designed with non-entomology students in mind as the target audience. The intent is not to compete with current texts designed for undergraduate or graduate students majoring in entomology. Rather, this book focuses on the fascinating biology of insects and how these beasts interact with humans. This gets at the underlying question, to put it concisely if crudely: Why should we care about insects? Unlike other entomology textbooks, this one focuses on pedagogical presentation of ideas, concepts, and analyses. Consistent with this approach, outlines of key concepts are located at the beginning of each chapter and then developed more fully as a chapter review. Review and synthesis questions are included throughout each chapter. The questions were developed with Bloom's Taxonomy in mind, so that the assessment of student development of learner skills, including higher-order learning, can be accomplished in addition to content goals.

Several topics are addressed with in-depth coverage not typically found in other entomology textbooks, including the impact of insects on the human condition, the history of entomology (e.g., origins of the discipline), forensic entomology, the use of insects as weapons of war, and new threats to the human condition (e.g., invasive species, changing insect populations with global warming). The book addresses other more familiar topics in nontraditional ways. For example, insect classification is discussed within the context of social media (e.g., Facebook and Twitter). The idea is to introduce students to the topic but not overwhelm them with morphology and taxonomy. Taxonomic keys are not included, which likely will be viewed as sacrilegious to the discipline. To make up for this intentional omission, the appendix provides links to several websites with identification keys.

Aspects of nutrition and diet are tackled by exploring why insects do not get fat, which allows the course to explore a whole range of related subjects, including the problem of obesity in human populations and the location of taste receptors in the context of toxins and meeting nutrient needs, as well as the concepts of hunger and satiety. The goal of each chapter is to present insect biology in a manner that captivates the attention of students who may not be enthralled with entomology at the onset. Humor, anecdotes, and connections to our everyday lives are sprinkled into the writing so that each chapter is not as dense as a typical textbook. The hope is that *Insects* will have broad appeal to college students with vastly different backgrounds, as well as to individuals with an interest in insects, because of its pedagogical approach to each topic, the unique topics covered, and a more relatable coverage of traditional subjects.

Features

The textbook is organized into seventeen chapters and an appendix, which can be explored in any order. Whereas a traditional approach relies on organization of an entomology textbook by emergent properties, or an initial introduction to the scientific method, the philosophy here is to build a foundation of student interest based on topics that keeps learning motivation high throughout a fifteen-week term. To aid in this process, the text incorporates a pedagogical theme coupled with supporting features that are common in introductory biology textbooks but usually absent from entomology books. What follows is an explanation of chapter components.

Chapter Overviews

Synopsis of the key ideas and pedagogy for each chapter. The focus is to pique student interests, like a movie trailer does, about the content to come.

Key Concepts

Familiar with Google Maps? Who isn't? You can easily get text directions to lead you from any location in the world to anywhere else, provided each location has an address. But sometimes the directions, no matter how clear, are just not where we want to start. Thus, a quick click on satellite view gives us that overview needed to put the detailed directions in proper context. And here, too, is where the Key Concepts element provides the big picture of what each chapter is all about.

Fly Spot

To consume it you must exude it. Several species of flies, especially filth and necrophagous species, are known to regurgitate portions of their diet to the outside of their bodies, to be deposited on a surface in the form of a spot. In some cases, the spot is immediately reconsumed, and at other times, reconsumption occurs at a later date. Why do this at all? There are several explanations, but one is that the flies concentrate and digest food outside the body so that once consumed the second time, it is nutrient-packed and ready for absorption. In what other course has fly vomiting been used as a metaphor to learning special topics? Fly Spots are included in each chapter, covering unique and fascinating features of insect biology, including how they interact with us. ("Time for Frass" was the alternate idea, but it completely changed the intent and view—check the glossary to learn more.)

Quick Check

Questions are placed throughout the chapters to help students gauge their understanding of key concepts or ideas that are essential components of their knowledge foundation. Students will have an immediate idea of whether they should spend more time with the topic before moving on.

Beyond the Text

These questions, also positioned throughout the chapters, focus on conceptual understanding or application of the ideas presented in the text and provide feedback regarding higher learning of the material.

Bug Bytes

Video clips of tidbits of insect behavior or natural history add an extra dimension to chapter topics. Most are YouTube videos that, once viewed, offer suggestions for several others related to the subject matter.

Chapter Reviews

The chapter reviews add meat to the bones of the key concepts introduced at the beginning of each chapter. In the chapter reviews, key facts and ideas are included with each key concept so that students can quickly review the chapter content and gauge their level of comprehension of each topic.

Mushroom Farming (aka self-testing)

What's with the name? Mushroom bodies in insects, also known as neuropils, are structures associated with olfactory processing (sense of smell) but, for our purposes, also function in learning and memory. To help you learn and retain some really cool information about the insect, we are literally cultivating learning and memory by self-testing. In other words, we are mushroom farming. The self-test questions are designed to help you determine if any of the facts, concepts, or crazy ideas have "stuck" in your large frontal ganglion.

The Entomologist Bookshelf

The bookshelf is a list of supplemental readings related to the chapter topics that provides more in-depth coverage to background or ancillary subjects. Several are included to permit comparisons to other animal groups, especially humans.

Additional Resources

Included at the end of each chapter are links to websites that provide information to expand on the topics covered in the text. This provides a starting point for students with a specific interest to explore a topic in more depth or follow a tangential idea well beyond what the textbook can cover.

Appendix

An extensive list of resources available for the study of insects is included in the appendix and divided into three categories: (1) sources of live and preserved specimens; (2) sources of study materials, especially in the form of collecting materials; and (3) the final category, which is more of a potpourri in which everything else tends to fall. This last section includes insects in art, clothing, jewelry, and much more.

Scratch-n-Sniff Entomology

Still working on this aspect of the book, although the dung beetles have been the easiest technology to develop so far. Maybe in the next edition!

REFERENCES

Berenbaum, M. R. 1995. Bugs in the System. Addison-Wesley Publishing, Reading, MA.

Eisner, T. 2003. For Love of Insects. The Belknap Press at Harvard University Press, Cambridge, MA.

Waldbauer, G. 1996. Insects through the Seasons. Harvard University Press, Cambridge, MA.

Insects

1 An Introduction to Insects

Busting the Myths, Lies, and Urban Legends

Trying to understand an insect taken out of its ecological milieu is as futile as trying to see significance in a single word lifted from its context on this page.

Dr. Gilbert Waldbauer
Insects through the Seasons (1998)

Where do you start when telling the story of the most amazing and successful creatures on Earth? The question itself likely needs explanation, because you may not share my view. Insects are the most amazing? Arguably a subjective statement, but one we will revisit throughout this book. Most successful? Here I stand on solid ground. By nearly any measure, members of the class Insecta hold no rivals, especially when the standard focuses on species abundance, efficiency, or contributions to ecosystem functioning. Such aspects may be too "biological," for the moment, so in terms of what matters most to us, insects are amazingly successful at eating our food, invading our homes, and causing disease. Their success may appear to be at our expense but, as we will see in this chapter, appearances may be deceiving. Insects do not need us, but we most certainly need them. Chapter 1 is dedicated to introducing you to the fascinating world occupied by so many (the insects) but understood and appreciated by too few (humans).

No doubt you formed an image long ago about what an insect is and what it does, and have a mental ledger of their worth to you individually and to humankind as a whole. I would wager the value is relatively low. That is fine; we need a starting point. Return to those images as you read. The goal in this first chapter is to challenge your perceptions about insects: what they look like, what they do, where they live, and several other facets of insect biology. Our goal is to identify the myths, lies, and urban legends so that we can begin to focus on what really matters: entry into an extraordinary world, in which insects have an evolutionary answer to almost any challenge posed to them. What we will find are some of the most efficient, entertaining, and, yes, beautiful creatures ever to inhabit the Earth. And if it is not clear to you yet, it will eventually become evident that these are the creatures that truly dominate the planet.

Key Concepts

- No better time than now to find out what an insect actually is
- Who cares about insects, or, why study them darn bugs?
- Insects are the most successful animals on the planet

Figure 1.1. Insects display tremendous variation in size, shape, and color. Photos courtesy (*clockwise from upper left*) of Clemson University–USDA Cooperative Extension Slide Series (German cockroach); Jessica Lawrence, Eurofins Agroscience Services (cuckoo wasp); Joseph Berger (wheel bug); Susan Ellis (scorpionfly); and Gerald J. Lenhard, Louisiana State University (hickory horn devil), all via Bugwood.org.

- Insects are bad, bad, bad
- What I have learned on my own: Insects are way cool
- How do you know it is not a spider?
- "Bugs" versus insects: The importance of knowing what you are talking about

No better time than now to find out what an insect actually is

Let me see if I got this straight. You enrolled in a course focused on insects, purchased this book, paid to have it shipped to your house, and paced anxiously for days to weeks by the mailbox until it arrived. Yet you have waited until now to find out what an insect is. One of us may have issues! We will start with a concise definition: insects are amazingly cool. And those who study them are

even cooler. In my mind, both statements are completely true. But perhaps you need a bit more substance, since you're the one who is going to be tested on this material.

Insects are a group of animals that display tremendous variety when it comes to appearance (figure 1.1). Though in comparison to vertebrate animals (insects do not have an internal skeleton, making them "invertebrates" by classification) they are small, some are nearly two feet long. Most, however, are an inch long or less. So, "small" is an appropriate term. Their bodies are divided into three distinct regions (head, thorax, and abdomen), with each region responsible for specific functions (figure 1.2). As adults, insects typically have six legs, located on the middle section (the thorax) of the body and positioned equally on either side as a reflection of their mirrored or **bilateral symmetry**.

Figure 1.2. The evolutionary process of tagmosis results in adult insects with tagmata, or three distinct body regions: head, thorax, and abdomen. Photo courtesy of Joseph Berger, Bugwood .org.

Insects are the only animals other than birds and bats that can fly, at least naturally, and thus wings (two pairs) are also present on most adult insects. The head displays large, prominent compound eyes that may be brightly colored, red, green, or yellow, and one pair of antennae, extending from the top portion of the cranium to probe the surrounding environment or to engage in battle. The body and appendages, as with all members of the phylum Arthropoda (*arthro* = jointed, *poda* = foot) that insects belong to, are segmented or jointed. In fact, aspects of segmentation are evident both externally and internally. Insects also have paired openings (termed **spiracles**) to internal tubes called **trachea**, along the thorax and abdomen, for gas exchange.

Internally, insect body design shares little with what you know about the human body. The thick blood (more correctly, **hemolymph**) bathes all the internal organs, as insects have no blood vessels. The insect gut is a long tube that varies in shape and composition based on the main food type consumed. Circulation is accomplished by a tubular heart and series of mini pumping organs. Multiple "brains" are positioned throughout the body, and if the head is lopped off, no problem, the insect can keep on walking and may even attempt to fly. The body's defenses are poised to ward off parasites as large as the insect itself, and urine formation occurs right in the digestive tract. Pretty cool, right?

The study of these cool beasts is called **entomology**. Most often, insects and other closely related terrestrial arthropods get lumped together, at least in textbooks and most undergraduate courses. This probably contributes to the lay public's confusion over whether a spider is an insect, and the use of the term "bug" to describe any small creature with more than two legs. Those who study insects are typically called "odd," "weird," and, sometimes, "unemployed." We prefer the more general "entomologist," but "way cool" is acceptable as well. It is not the goal of this book to attempt to convert you to the world of insects—not that there is anything wrong with that. Rather, the goal is to introduce you to an absolutely fascinating world. I am convinced that if you let yourself see the wonders in this unique world, you will, at the very least, come to have a new appreciation for insects. And if you're not careful, you will be truly amazed by their biology.

Who cares about insects, or, why study them darn bugs?

Insects are not what they seem, or at least, not what you have been told. You have been lied to. Time and time again. And that lying probably will not end anytime soon. Thankfully, you have arrived here before the point of no return: I will now attempt to set you on the correct path, one of truth and enlightenment. (Note: stand clear of expected lightning strike.) All right, perhaps I am laying the **frass*** a little thick. But it is safe to say that your impressions of insects likely lie with the vast majority of non-entomologists (figure 1.3), who dislike, fear, and possibly loathe beasts with six legs (or more†), wings, and a propensity to scurry. I contend, however, that the beast you think you know—that is, the dreaded multilegged, slimy, biting, stinging, out-to-get-you critter with beady little eyes, gnashing teeth, and a poor attitude—is really not so bad. Insects are merely misunderstood.

---

*Frass is a delightful term especially useful as a cipher between geeky (entomologists) friends. Frass is a combination of partially digested foods, metabolic wastes, and water, eliminated from the hindgut via the anus. In other words, it is insect poop.

†Insect adults typically have six legs, located on the middle region of the body known as the thorax, but several closely related animals can have far more than six. Regardless of anatomical differences, many-legged animals often all get piled in one group loosely referred to as "bugs" or "worms" or "whatever."

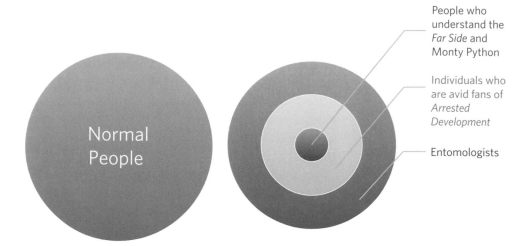

People who
understand the
Far Side and
Monty Python

Individuals who
are avid fans of
*Arrested
Development*

Entomologists

Normal
People

Figure 1.3. Relationship between entomologists and everyone else (normal people). Notice that the circles do not overlap. *The Far Side* is the brilliant cartoon strip by Gary Larson, Monty Python is the genius comedy troupe that starred in such classics as *The Meaning of Life* and *Life of Brian*, *Arrested Development* is the former Fox television series now airing on Netflix, and entomologists are the cool individuals who study insects.

How can this be true? After all, a thorough study of the World Wide Web reveals countless pages dedicated to the evil that insects do. Sites like Biodiversity Explorer (http://www.biodiversityexplorer.org/insects/why.htm) quickly get to the heart of the matter: insects negatively affect the human condition. By this logic, entomological disdain would be completely warranted—maybe.

What such websites also reveal is that it is all about being humancentric. Humans are what we consider important, which in turn shapes what we study. In other words, humans consider insects worthy of investigation only if they affect us. This statement is not meant to downplay the insect influence on the human condition. Insects can powerfully shape human existence through destructive interactions, an aspect examined in detail in chapter 4. In fact, here are the top four reasons, as identified by Biodiversity Explorer as well as most any other source consulted, that we (as in, humans as a whole) study insects:

1. Insects spread disease (to us).
2. Insects eat our crops.
3. Insects eat our stored food.
4. Insects maim and kill our livestock (as well as pets).

Maim our livestock. Wow! How often do you even use "maim" in a sentence? That sounds awful. Please do not

misconstrue the message; there is no denying the potential for destruction in the world of insects. An examination of one disease alone, malaria, clearly illustrates the potent blow that one group of insects can unleash on human populations. But the bad side of the insect world is just one part of the story. In reality, only about 2% of **extant** insects are actually responsible for the destructive effects outlined above. The remaining 98%, meaning nearly one million species, can be classified as either neutral (i.e., neither harmful nor useful*) or beneficial to humans (figure 1.4). The latter needs clarification. As a group, the class Insecta is essential to our well-being. Let's take it a step further: our very survival as a species depends on insects. Yet most of us have no clue that we need insects. Why? As discussed earlier, we are too busy despising them.

Okay, if insects are truly needed for human survival, what do insects do for us? Chapter 3 is dedicated to an in-depth discussion examining the many beneficial roles insects play in the human condition. Here, I offer just a

---

*This classification is clearly subjective, as many insects influence cultural practices, including music, religion, movies, and literature. Thus, on an aesthetic level, which perhaps is determined individually, some neutral insects are regarded as beneficial or pleasing.

Figure 1.4. This insect was the inspiration and star of the movie *Silence of the Lambs*. In the movie, murder victims were found with pupae of the death's head moth, *Acherontia lachesis*, placed in their mouths. Photo by Eric Gagnon at http://bit.ly/1FrqPVH.

Figure 1.5. The European honey bee, *Apis mellifera*, is a common source of honey in the United States. However, contrary to the implications of this photo, ear wax does not come from bees. Photo courtesy of Whitney Cranshaw, Colorado State University, Bugwood.org.

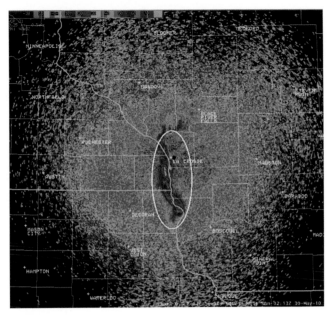

Figure 1.6. Doppler radar detection of mass mayfly (order Ephemeroptera) emergence along the Mississippi River at 3 p.m. CDT on May 29, 2010. That's right, the insects can be detected by weather radar! The mayflies show up as bright pink, purple, and white. Image available in public domain at http://bit.ly/1j6cJ29.

few enticements to illustrate the good side of the insect world:

Pollination. About 80% of all flowering plants depend on insects to transport pollen. Now, if the idea of insects serving as a plant sperm transportation system is not exciting to you, maybe realizing that this relationship results in many foods we eat, like apples, peaches, squash, almonds, and pie (I might be lying about this last one, but in fairness, I told you the lying would continue) will spawn your appetite.

Insect-derived foods. Yes, you have! You have consumed food made by insects. Just take some excreted plant fluids (nectar), regurgitated ("vomited" sounds so crude) from the mouths of bees, add some plant sperm (pollen) and maybe a few bacteria. Viola, bread spread. Okay, you call it honey (figure 1.5).

Insects recycle almost all dead things. Imagine if they did not do this; within a few months, during the warm days of summer, bodies of dead critters would start to accumulate like snow in your yard. Meaning, you would measure the corpse buildup in inches to feet. Gross but interesting (figure 1.6).

Insect anal drippings can be used to polish your shoes and furniture. That delightful thought will serve as our cliffhanger trailer (more on this later).

There is much more to the world of insects than just the rotten few that cause anguish. As this chapter unfolds, many of those stories will be told. If all goes well, your perceptions of what insects are and why they should be studied will change, for the good, and the lies that have been shared with you can be purged. To understand insects requires that you are interested in some aspect of them, and if that interest goes beyond just the negative interactions, then a true appreciation for their value to all life on Earth can develop.

Insects are the most successful animals on the planet

You still don't believe me about the most successful animals on Earth. I get it. You're mad because I lied in the last section; who doesn't like pie? Let it go. Let's return to the original premise: insects are the most evolutionary successful group of animals ever to have existed on the planet. An argument can be made to support this statement using several criteria. For now, we will zero in on just two attributes, abundance and diversity. By abundance, I mean the sheer number of insects that exist. This includes all individuals of every insect species that occur at any given time. Diversity is a smaller measure, focusing on the total number of described **species** of insects. What a species is will be discussed later in the chapter, but for now it is sufficient to say that this represents a means for grouping insects based on unique characteristics. Thus diversity is interested in how many of these distinct groups occur within the class Insecta.

Our examination of insect success will begin with abundance. How many total insects are on the planet right now? And how in the world could you measure all the insects that exist, anywhere on the planet, at any moment in time? The answer to the first question is dependent on the second. What this means is that any estimation or educated guess of the abundance of insects, or any organism for that matter, relies on the quality of our methods for counting or sampling. The "how" question represents a combination of dedicated sampling and estimation of insect populations. Dedicated sampling? Yes, and how. There is no one method for collecting insects. Rather, a range of techniques has been developed, often specific for the species or group of interest, based on the modes of locomotion (e.g., flying, walking, swimming) or habitat (e.g., ground dwelling, subterranean, aquatic) of the targeted beasts (figure 1.7). Even with an arsenal of sampling devices in hand, all methods suffer limitations and none are designed to collect

Figure 1.7. One common method for collecting flying insects is to use a Malaise trap. Insects fly under the canopy and then walk up (or away from gravity) toward a collecting funnel. Image available at http://bit.ly/27s03XZ.

more than just a few of the total numbers of individuals found in any one location.

Some elaborate approaches have been used, such as **pheromone** trapping that relies on the sex attractants produced by one gender to entice the other, or insecticide fogging of tree canopies (as has been done in tropical locales such as Central and South America), with thousands of specimens falling like rain onto canvas tarps, yielding estimates of abundance and diversity that far surpass previous estimates. Of course approaches like the latter raise questions about ecological impact and **environmental stewardship**. Once collected, the specimens are killed* and then identified, the latter process sometimes being quite tedious and time-consuming (figure 1.8). In most instances, an entomologist is capable of making broad categorical identifications or specializes in a particular group of insects; other experts often need to be consulted to make correct identifications outside the area of specialty. The data generated provides a snapshot of diversity and abundance within the narrow scope permitted by the collecting method. This information can then in turn be used to start developing estimates of abundance and diversity, at least for the specimens iden-

---

*The "killing" aspect of entomology creates an apparent paradox for an entomologist, in that to study the critters that evoke fascination and wonder, the target of interest (insects) must be killed for further investigation.

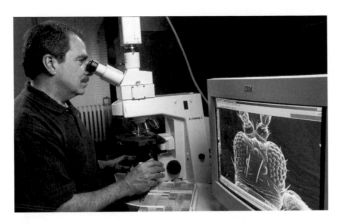

Figure 1.8. Entomologist David Nickle examines an unidentified thrip species found in a shipment of cut flowers from South Africa and compares it to the image on the screen of a known thrip species. Photo courtesy of Peggy Greb, USDA Agricultural Research Service, Bugwood.org.

tified, for a larger area of interest. Information and observations from other sources about other species located in other habitats can then be added to build the case for species richness or abundance for a particular habitat or location. The take-home message is that estimates are the best we can do and are limited by the methods of collection.

One way to address insect abundance is to simply look around you. Insects are everywhere. But "everywhere" reaches much farther than your immediate surroundings. Yes, it is true that insects are living in nearly every square inch of your yard, and have undoubtedly invaded regions of your home, spaces that will make you shudder once revealed (warning: chapters 4 and 14 may be a source of discomfort and repulsion for you). Their range by no means ends with one domicile. Insects thrive on land and in freshwater. Some live in the severe cold of the Arctic, some in opposite conditions like hot springs where temperatures reach levels that should melt biological membranes; others occur at high altitude in mountainous regions (six-legged critters have been observed at 20,000 feet above sea level in the Himalayas) where oxygen is nearly nonexistent; there are species that thrive in pools of crude petroleum, and ones that have specialized as parasites within the nostrils of walruses, diving to locations in oceans and seas that would not be possible for them to reach by any other natural means. There are even insects that prefer the warmth and texture of ma-

nure, which by comparison to the other extreme environments, seems like a life of luxury.

These examples serve as testament to the amazing evolutionary adaptations displayed by members of the class Insecta, permitting not only survival, but also dominance in nearly every ecosystem on Earth. "Nearly every" implies kinks in the armor of superiority, but not really. Some exceptions are to be expected: anywhere ice persists, like the polar ice caps or Antarctica, it is difficult for most any forms of life to exist. A bit surprising is that insects are rare in marine environments, being restricted to regions close to shore or to a parasitic lifestyle within marine hosts. However, in the context that insects evolved as creatures for a terrestrial existence, and only secondarily moved on to live in aquatic environs, their presence in fresh and salt water, no matter how abundant, should be viewed as a remarkable evolutionary achievement. Other than these few examples, insects can be found anywhere and everywhere, and in abundance.

Insect abundance varies on a daily basis, as new insects are born (hatch from eggs) and old ones die. Throw in changes in environmental conditions, and population levels will fluctuate. For example, low (in other words cold) or high temperatures for long periods, or sudden dramatic shifts in temperature, can wreak havoc with insect populations. So an estimate of insect abundance is not constant. One reasonable estimate places the total number of insects at about one quintillion (1,000,000,000,000,000,000), equal to about 2.7 billion tons of insects, outweighing the human population by nearly ten times! Insects also outnumber humans by a ratio of about two hundred million to one. Biologically speaking, there are a lot of insects. No other animal group is even within a magnitude of scale to the insects in this regard. Thus insects can be argued to be the most successful group of animals, based on outnumbering all competitors.

Quick check
When calculating the abundance of insects, why are the values presented as estimates?

The only rival to the impressive abundance of insects may be provided by insects themselves. What I mean by this is that the diversity of insects deserves even more accolades than the abundance. Our measure of diversity is species—more correctly, the total number of species. Current estimates place insect diversity somewhere between 900,000 to 1,000,000,000 species. This range reflects the total number that have been "described," meaning that they have been physically collected and given a description as an animal unique from any previous specimens ever observed. Appropriate responses to this statement include, "Wow!" Or, "wow" (insert sarcastic inflection). Or, possibly, "What is a species?"

When addressing the last question, amazing to say, there is no agreed-upon definition. The basic working definition is that a species is the largest group of organisms capable of interbreeding and producing fertile offspring. Generally, members of the same species share strong similarities in appearance (also referred to as morphology), DNA, and habitat. Insects can complicate the definition, because it has recently been discovered that some insects believed to be separate species are capable of producing viable offspring if geographical, physical, or physiological barriers are removed. One example is a group of parasitic wasps thought to represent at least two species: each supposed species becomes capable of mating with the other when treated with antibiotics. As it turns out, bacteria residing within each insect are responsible for their reproductive incompatibility; once the microorganisms are removed by antibiotic treatment, the two insects no longer meet the definition of separate species. How widespread this phenomenon is within the insects specifically and the animal kingdom as a whole is unknown. One immediate conclusion is that, based on this example, there clearly is an overestimation of the number of insect species. Perhaps—although more recent discussions actually argue just the opposite.

Beyond the text

Does estimation of human populations have the same limitations as estimation of insect populations?

A defining study by Erwin (1983) on the insect richness in South America concluded that the number of insect species had been grossly underestimated, and that a truer estimate is closer to thirty million species. That's right, thirty times the accepted number. If correct, this would also argue that our numbers for insect abundance are greatly underestimated. No need to fear that the world is about to end from insect infestation (although, again, remember, they are everywhere), because many experts have challenged this view of diversity. Erwin's estimates were based on sampling one type of insect (beetles, or members of the order Coleoptera), from tree canopies of nineteen specimens of one tree species located in Panama. In other words, the values were derived from a very narrow sample, within a remote location in one region of one country. Hardly an all-inclusive sampling approach. Yet despite obvious weaknesses in his early estimates, Erwin (1988) revamped his calculations to state that the number of insect species had been underestimated and now should be expected to exceed thirty million.

Today, more conservative estimates are used. Typically, the total number of species of any animal is predicted to be a little less than ten times the number of described species. This would suggest a maximum of between nine million and ten million extant insect species, but more conservative approaches argue for a value nearer five million. So Erwin's estimates would be pared back to a meager five million to ten million species. What does this even mean? The easiest way to understand these numbers is to compare them to other animals. Humans display an impressive abundance of nearly six billion people, but represent only one species, *Homo sapiens*. Not even a blip on the animal diversity scale. By contrast, if we simply say that the number of insect species is approximately one million, then insects would represent more than 75% of all animal life on the planet. So if the truer number is closer to five million, that means insects are even more dominant, approaching 90%-95% of the existing animal life on Earth. Is that reason enough for you to spend at least some time chatting about insect success?

Insects are bad, bad, bad

The case was made above to study insects because of their success in terms of abundance and diversity. Clearly an impressive argument was laid out. Yet perhaps you remain skeptical that insects are worthy of study. Okay, fine, it's time to get serious. "All insects are bad": I don't believe that statement, but you might. We've already touched on the fact that some insects are highly destructive. But unlike in the previous section, where diversity and abundance were highlighted as key features of insect success, numbers tell a much different story in terms of the harmful effects on the human condition. How so? The diversity of destructive species is incredibly low in comparison to the total. What this means is that though the effects of destructive insects are considerable, particularly in terms of injury and death to human populations, only a small number of six-legged beasts (approximately 2%) can be held responsible. Nonetheless, those few rotten apples* are probably enough to convince most individuals to lump all insects into one category: "bad." Regardless of how many insects are to blame, the effects of direct and indirect insect activity can be devastating, representing a macabre way of classifying insect success.

Generally, destructive insects are grouped by the effect they have on the human condition. This classification scheme yields three distinct categories of destructive insects or pests:

1. *Medically important species*, which transmit disease pathogens, or directly attack or feed on humans (or pets);
2. *Structural pests*, often referred to as urban insects, which damage or destroy dwellings or other types of physical structures;
3. *Agricultural pests*, which do damage to crops and livestock or other aspects of food production.

Each category represents highly adapted species that have evolved to take advantage of the way humans live. There is certainly variation in how dependent each is on human existence. For example, some insects have essentially adapted to the ways humans grow food, dispose of refuge, or live together in communities. Such species are referred to as **synanthropic** because they live in association with humans, either directly or indirectly. Other insects are not really dependent on our species at all. Rather, they are merely opportunistic, taking every advantage of our vulnerability. Survival for such species is mostly independent of whether *Homo sapiens* continue to inhabit the planet or not.

Figure 1.9. The triumphant young David holds the head of the giant Philistine warrior Goliath. The image of the painting by Caravaggio (1571–1610) is available in public domain at http://bit.ly/24V0o3h.

Chapter 4 explores in detail examples of insect success at the expense of humans, with stories about insect species that have the capacity to cripple peoples and societies through attacking food supplies and transmitting disease. Here, we present just a few examples of the power of the insect, which serve to demonstrate that size does not matter, as such relatively tiny creatures can dominate the mighty *Homo sapiens*. A modern yet twisted adaptation of David and Goliath,† staged in our everyday lives (figure 1.9).

One of the cruelest ways that insects torment human populations is through disrupting agriculture. Several insect species directly or indirectly attack crops or other aspects of food production and storage. Direct attack should be obvious, as the insect feeds on a food source, damaging the crop as potential human nutriment. In indirect attack, insect activity does not directly target the

---

*A reference to the proverb, "Don't let one bad apple spoil the whole bunch," in which the moral is that the few do not represent the many.

†A biblical story discussed in I Samuel, in which the giant Philistine warrior Goliath is defeated by the seemingly outmatched and much smaller David, future king of Israel.

Table 1.1. Nations facing unrest and government action due to food shortage and rising food prices

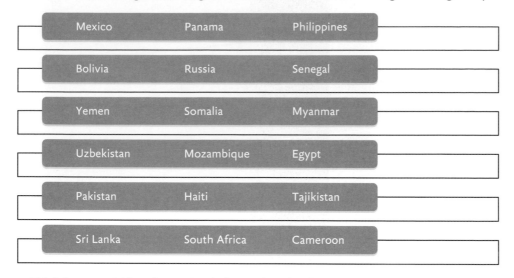

Mexico	Panama	Philippines
Bolivia	Russia	Senegal
Yemen	Somalia	Myanmar
Uzbekistan	Mozambique	Egypt
Pakistan	Haiti	Tajikistan
Sri Lanka	South Africa	Cameroon

This is just a partial list of countries dealing with civil and government unrest due to food shortages and/or rising food costs.

Rising fuel prices, increased costs associated with crop protection, particularly from insects, embargos, and government freezes are just some of the factors contributing to food shortages. Information reported by the Food and Agriculture Organization of the United Nations (www.fao.org).

food source but ultimately does lower food or nutrient production. The end result is the same regardless of ethnicity, age, health, or wealth: a food shortage that can lead to a slow wasting away. In modern times, insect damage to food production is most destructive in nations already ravaged by disease and poverty (table 1.1).

Perhaps the most destructive agricultural pest of all time is the desert locust or short horn grasshopper, *Schistocerca gregaria*. This eating machine is capable of devouring more than its own body weight in food each day. During sustained migration, large swarms consisting of thousands to millions of individuals can consume more than 20,000 tons (18,144,000 kilograms) of grain and vegetation per day. It is no wonder that *S. gregaria* is thought to be the beast referenced in the Christian Bible as the plague of locusts unleashed on the pharaoh of Egypt (Exodus 8:10) (figure 1.10). To be selected by God as a tool of destruction is quite the honor! Regardless of the forces that may have selected them, great swarms of locusts so thick that the sky darkened as their masses took flight have been reported throughout the world for centuries, including portions of Africa, Asia, and the Middle East, leaving a wake of vegetative destruction behind.

The largest locust swarm ever recorded covered an area equivalent to four hundred square miles (an area about the size of New York City) and was composed of an estimated forty billion individuals! To put those numbers in proper perspective, consider this: that one swarm represented about seven times the abundance of the entire human population, yet was composed of just one insect species found in one location. As we might imagine, nearly all vegetation was entirely consumed in the path of the locusts. Modern outbreaks are more localized, but still very damaging. The last major outbreak of the desert locust occurred in West Africa in 2004–2005, yielding economic losses estimated to exceed $2.5 billion, and though the locust damage was not considered the primary reason for the famine that followed, it was considered a contributing factor. In fact, both the desert locust and the migratory locust, *Locusta migratoria*, sharply increase the severity of established famine when swarms invade regions already faced with food shortage brought on by drought, war, or other factors. Considering that this represents but one example of insect damage to agriculture, an image starts to form of how just a few hundred or thousand species of insects can elicit enormous economic losses worldwide, totaling over $100 billion

Figure 1.10. Depiction of the plague of locusts unleashed by God toward the pharaoh of Egypt during the time of Moses. Illustration from the 1890 Holman Bible available in public domain via http://bit.ly/1TjNg2r.

annually between food loss and control efforts to suppress insect activity. Couple these estimates with human lives lost through insect-induced food shortage or famine-related disease, and the impact of agricultural insects is staggering.

Not to be outdone are medically important insects. Several blood-feeding insects serve as vectors for human diseases that today place more than 65% of the world's population at risk. Top on this list are mosquitoes that vector malaria. Malaria is a disease that at one time was broadly distributed across the globe, but today is mostly isolated to tropical and subtropical regions, particularly in Africa, Asia, and Central and South America. The disease is caused by parasitic protozoans in the genus *Plasmodium*, which reside in adult anopheline mosquitoes. Female mosquitoes seek a blood meal before producing eggs, and during the process of dining on a host—meaning you—saliva is injected into the host circulatory system to anesthetize the area where the syringe-like mouth has pierced through the skin and to prevent blood clotting and other host defenses from interfering with feeding (figure 1.11). It is during saliva injection that the *Plasmodium* parasites are deposited into human hosts. Once in the new host, the parasite invades human liver cells to reproduce, eventually producing thousands of new individuals that will take up residence in red blood

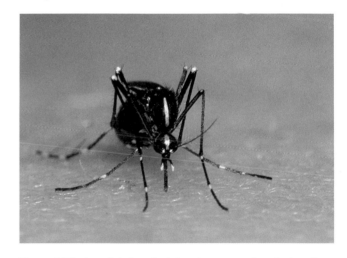

Figure 1.11. An adult female Asian tiger mosquito, *Aedes albopictus*, feeding. Just after the long mouthparts are inserted into the host, saliva is injected, permitting transfer of disease-causing microorganisms. Photo courtesy of Susan Ellis, Bugwood.org.

cells. As the disease runs its course, infected individuals initially display flu-like symptoms. But as the condition worsens, the malarial infection begins to severely compromise the body, causing an array of complications including severe fever, septicemia, convulsions, and respiratory distress and often culminating in death, particularly in children, the elderly, and individuals whose health is already compromised.

Despite the fact that malaria is one of the most intensively studied diseases of all time, and that most aspects of the parasite-host association are well understood, it remains one of the most prevalent and lethal diseases today. In 2010, the Centers for Disease Control and Prevention estimated 219 million cases of malaria worldwide, with 660,000 deaths reported in the same year. Malaria is just one example of insect-transmitted disease. Between the blood-feeding behavior of biting flies, mosquitoes, fleas, and true bugs (the last section of chapter 1 explains why all insects are not bugs), approximately 3.5 billion people live in regions at risk of debilitating insect-borne diseases. Maybe there is genuine reason to live in fear after all!

With so much death and destruction going around at the hands* of agriculturally and medically important insects, structural pests must be pretty dull? I guess that depends on whether you like sleeping under a roof that might collapse during the middle of the night, leaving you maimed or worse (I just needed to use "maim" in a sentence one more time). Yes, a tad bit of exaggeration. Yet it still gets to the point: many species of insects express their influence by boring into or chewing on natural and man-made structures. If this happens to be your home, the result can be thousands of dollars in damages, or worse, the house can become uninhabitable. And yes, there have been instances in which structural integrity has been lost, and all or portions of homes collapsed. Wood-based products are the principal targets of the majority of structural insects, with public enemy number one being the termite.

All termites are not the same and do not induce damage in an identical manner. However, since you are likely familiar with at least some aspects of termites, we will leave that discussion until chapter 4. Instead, our focus here will be on a lesser-known wood terrorist, the emerald ash borer, *Agrilus planipennis* (figure 1.12). Here is where beauty and the beast become entwined. The adult beetles are a brilliant metallic green from head to tail,

*Insects do not have hands, so the more correct appendage name would be **tarsi**, with **pulvilli**, or footpads, located as equivalents to the palm of a hand. Such terminology is not needed for our understanding of insect destruction, but you are the one who had to know, as evidenced by continuing to read this footnote.

Figure 1.12. A new structural pest to the United States, the emerald ash borer, *Agrilus planipennis*. Photo courtesy of David Cappaert, Michigan State University, Bugwood.org.

with a hint of blue splashed across the body to yield an entomological jewel that faintly resembles a precious jade stone, provided of course that stones have mandibles and wings. But underneath that beauty is a ferocious eating machine that in some ways rivals the desert locust discussed earlier. The adults are really not the problem. Since their accidental introduction into the United States around 2002, adults of this **invasive species** fly to the top of tree canopies, spending each day munching on the leaves of ash trees. If that were all the adults did, then there would be very little to the story. The problem is that emerald ash borers like to eat and have sex, with sex leading to eggs, and eggs leading to juvenile beetles, better known as larvae. Larvae cause all the problems, not unlike human children.

What do the larvae do? They eat. More specifically, the beetle juveniles feed by boring through the bark of the ash trees, leaving serpentine trails that damage the tree's vascular tissue (phloem). The effect is akin to pinching shut a human artery: nutrient circulation is suppressed. Eventually, if the damage to the tree phloem is significant, death will result. Since the emerald ash borer's introduction into the United States, an estimated 150–200 million ash trees have died. For regions in which ash trees are essential components of the economy via the nursery industry, the financial losses already amount to over $1 billion.

The beetles' effect can also be felt on city infrastructure. Baltimore, Maryland, serves as a prime example.

Baltimore has approximately 300,000 ash trees in the city proper, and approximately six million are distributed throughout the entire metropolitan area, accounting for more than 10% of the total tree population. As of 2013, *A. planipennis* was knocking on the door, having been discovered in Montgomery county, just south of the city. In a worst-case scenario, in which the beetle heavily infests Baltimore's ash tree population, economic losses are expected to exceed $225 million. Even if the beetle never invades the city, preventive steps must be taken, and prevention costs money. So a would-be pest, meaning one that might not actually ever do any damage in a suspected area, may still cost millions of dollars in prophylactic treatment.

Keep in mind the idea of *millions to billions* when discussing the numbers of people in the path of insect destruction. The phrase also provides an accurate depiction of the economic losses stemming from insect damage, no matter what category of pest we discuss. Insects have been and continue to be amazingly successful at wreaking havoc in our lives, and for that reason alone, they warrant our serious attention. Many individuals have devoted their lives to studying the class Insecta with the goal of putting an end to the types of entomological dominance discussed in the preceding examples. Some of their efforts are chronicled in chapter 14, while the success of insects at our expense will be revealed from a historical perspective in chapter 2, and the problems they evoke in modern times appear in chapter 4.

Bug bytes

Mosquito blood feeding

http://www.youtube.com/watch?v=AWt-SQNp0Xs

What I have learned on my own: Insects are way cool

While the hope is that the preceding section captivated your attention with the destructive power that can be unleashed by six-legged biological cylinders with wings, by no means was the goal to convince you that they are indeed all bad. Remember that only a small fraction of the whole participates in tormenting humans. The overwhelming majority play significantly important roles in our existence that we, and the plants and animals we love

and need, depend on. Chapter 3 is home to the insect public relations campaign—in other words, recounting all the good they do.

Now, to be totally honest, I did not enter college with a love of insects, or a hatred either. I was entomologically neutral. Insects were an occasional point of fascination for me while growing up. I did not choose to collect insects like baseball trading cards, as many of my current and former students always expect to learn. Rather, when my attention was turned away from sports, I chased after lightning bugs with my friends, for hours on end, fascinated by the their (I mean the insects', not my friends') ability to produce light and driven by the desire to sell the collected "bugs" to a midwestern chemical company to make my first million. Long story, but in brief I was duped by my older sister into spending an entire summer trapping lightning bugs, so that together we could sell them to make ten dollars for every thousand collected. Do you have any idea how long it takes to catch a thousand lightning bugs with your bare hands? Answer: longer than one summer! We made exactly ten dollars, split eleven ways: my sister's cut, mine, and nine of my friends who were recruited as the bug posse.

So, early in my life, I was neither entomologically inclined nor bright. Is there a point to my tale? Yes, there is. My love of insects did not start as a youth, as frequently occurs with those who have devoted their lives to studying them. I began the journey as a young biologist who jumped at the chance to become engaged in undergraduate research and was then transformed from someone seeking ways to kill these dreaded beasts (the goal of the research project) to someone fascinated with insect biology. I do not look into the insect world with rose-colored glasses, for I understand the destructive power of insects. However, I choose to focus on the biology of the masses, that is, the nondestructive varieties. They are adaptive wonders, amazing in their efficiency and beautiful in their diversity.

You, however, may need some convincing regarding the last statement. No problem. I plan to share in the chapters to come some of the insect wonders that have been revealed to me, through readings, research, and conversations with others who share my passion. How can organisms that can move by almost any method of locomotion known not impress you? Insects can run,

Beetle troubles for Noah

Insect success has often resulted in problems for humankind. Imagine the issues faced by Noah, the biblical figure discussed in the book of Genesis of the Old Testament, who was given the ultimate homework assignment: build a boat (an ark) large enough to house two of every kind of animal on the planet for at least forty days while the Earth is flooded. Technically, this was his final project, because if he did a poor job, a failing grade was the least of his problems. Presumably Noah had never constructed an ark before, the Internet was not yet up and running for instructions, and the building materials necessary to construct a ship mandated by God (Genesis 6:15) to be 300 cubits* long (135 meters), 50 cubits wide (22.5 meters) and 30 cubits high (13.5 meters) were not readily available (no Lowe's or Home Depot on the scene). Once the ark was constructed, it appears the job became somewhat easier in that all the animals just showed up (Genesis 6:16). So Noah did not encounter the collecting or sampling problems that would be common to those interested in biodiversity. Nonetheless, the insects had to be a major challenge. If we consider that approximately one million species of insects have been identified to date in modern times, with likely many more in existence but yet to be discovered, Noah minimally needed room for two million insects (two of each species). There is no reason to believe that there were fewer species at the time of Noah than there are today. In fact, it is far more likely that modern humans have caused extinction of insects that once lived during the Old Testament period. Since the Earth was to be flooded, presumably many of the aquatic species did not need a boat ride, which would lower the insect numbers somewhat, but not that much.

Even if the focus was on just the largest group of insects, beetles, or members of the order Coleoptera, Noah's task was still overwhelming. Beetles account for 40% of all insects and about 25% of all animal life. They have been found in nearly every habitat imaginable, including freshwater and marine environments. If the aquatic representatives were not provided with a boarding pass, at least two major issues still remained for Noah and his crew: (1) how to feed the passengers, and (2) how to stop

*A cubit is a unit of measure approximately the length of an adult man's arm from shoulder to fingertips.

jump, fly, swim, and glide. There are fly maggots that leap by grabbing their anus with their mouth, creating muscular tension like a spring, and then, releasing their grip, shooting straight up. Not impressed? Well, you try that trick! Insects can fend off parasites nearly the size of their own bodies, even after the invaders have burrowed into their body cavities through the skin, displacing organs as they move about. Some beetles escape skillful predators by releasing a knockout gas from glands near the anus, leaving the would-be attackers "sleeping" and vulnerable to being eaten themselves; this protective beetle "gas" is affectionately referred to as fatal flatulence.

When it comes to sex, insects could write their own Kama Sutra! Did I mention traumatic insemination? That's right, a few species mate by the male ramming his penis through the body wall of his mate, who actually may be simultaneously engaged in intercourse via more traditional means with a separate partner. Speaking of efficiency, with some insect species, the female needs to mate only once in her life, for she can store the sperm of her "lover" for eternity (which is usually about four to five days), releasing sperm to fertilize eggs through conscious control, thereby controlling the sex of her offspring.*

*This mechanism of fertilization, as also known as **arrhenotoky**, results in unfertilized eggs becoming males or sons, and fertilized eggs becoming daughters.

feeding the passengers. The vast majority of adult beetles residing within terrestrial habitats are **phytophagous**, meaning they consume plants as a food source. While we cannot be sure of the stage of development of the insects (nor any aspect of the story), it is reasonable to assume only adults could have made the journey to the ark. Consequently, vegetative materials of all sorts were needed to keep the beetles alive during the forty-day trip. Now of course if Noah had the capacity to induce **quiescence**, **aestivation**, **diapause**[†] or some other dormancy in the beetles, or if the adults mated, laid eggs, and then died, well then, problem solved! The Bible does not address whether Noah had knowledge of the physiological ecology of insects.

The second problem would be how to stop the feeding activity of select beetle passengers. Few coleopteran species are parasitic. Thus, issues that, say, blood-feeding insects would have caused for Noah were not relevant with beetles. However, many are predatory. These carnivorous species would have been a major problem for the ship's crew, in that ending the cruise with half the number of species they began with would probably not have impressed Noah's boss. Even more challenging is that several beetles are **xylophagous**, a term used for animals that eat wood. Approximately six thousand species of beetles from at least six families (Anobiidae, Bostrichidae, Buprestidae, Cerambycidae, Lyctidae, and Platypodidae) exist today that wreak havoc on a range of tree species. In the grand scheme of beetles, wood-eating species represent only about 1.5% of all known members of the order Coleoptera. The problem would have been that all the xylophiles would have been concentrated in one small location, the ark. It is true that the damage they do depends on a number of factors, including their stage of development, the type of wood, the temperature, and so on. All species are not fast eaters either, so damage to the wooden ark might have been slow. There is also no way of knowing what tree type(s) the ark was built from. Gopher wood is mentioned in Genesis, but scholars disagree whether this represents a reference to a tree species or a wood treatment. The effects of beetle chewing and boring would differ greatly between an ark constructed of cypress versus one of pine or oak. What can be postulated is that if Noah's boat sprung leaks during the trip, and at least twelve thousand suspects come to mind.

[†]All these terms refer to a state of physiological reduction analogous to dormancy or hibernation that allows insects to survive during periods of inhospitable environmental conditions.

Efficiency is also apropos for describing insect digestion. Most will not eat if they do not have a hunger need, meaning that insects obey the commands of a hunger center that is only active when a nutritional need must be met. If that food nutrient cannot be found, many would rather die than eat without purpose. Just how exactly can insects tell what nutrients are in their food? I am glad you asked! By using their feet. Insects have taste receptors on the bottoms of their legs (footpads), and also on the outside of the mouth as well as inside. Several species have literally thousands of taste receptors located in multiple locations on and in the body. Even more impressive is the fact that most are able to detect the nutritional suitability of a food source, including whether it is also potentially toxic or noxious, prior to ingestion. When do humans make such decisions? For me, usually after I have regurgitated the night away!

I can go on and on with examples, and I will, as the chapters proceed. My journey to the insects was not along the same path as so many before me. I did not find them as a child and stay with my entomological interest. No, my journey was probably more similar to yours. I started out trying to kill them. Only after I opened my eyes and mind to truly learning about insects did I begin to discover the wonders of insect biology. For me, insects became cool. Your journey may well have a different end result. However, if you follow along through the next several pages, you may come to see the reasoning of my treatise that insects are phat, but definitely not fat.

How do you know it is not a spider?

Before moving too deep into the book, a couple of myths need to be exposed. Everyone don you berets so that we can play our own version of MythBusters.*

Myth #1: Spiders are insects.
Myth #2: All insects are bugs.

I am angry that we even need to do this, particularly that we have to deal with the second myth at all. So for the sake of keeping my cool, I am putting some distance between the myths by providing explanations in separate sections. Let's start with the spider dilemma by seeking the most reputable source I could find, my mother.

To my mother, if an animal has more than two legs, there is a good chance it is an insect. Concise, but a wee bit too broad for a definition. Actually, her eyesight is really not that good, so for something smaller than say, a cow, she probably cannot tell how many, or if any, legs are present at all (figure 1.13). For her and many other individuals, insects and spiders and any other gross creepy creatures are lumped together when lots of legs are present. For my children, who believe that insects and spiders may attack them, particularly at night while they sleep, I have given an easy working definition: If it bites you and you die, probably a spider. If you live, clean your room. I suppose this is a decent starting point for examining spiders and insects.

Spiders are not insects, though they are close cousins. Myth busted! Both spiders and insects do have six legs, but spiders take it an extra step (actually two) by owning two extra legs (figure 1.14). Spider bodies are composed of just two regions (remember that insects have three), they cannot fly, and they have way more than two eyes. Oh yeah, almost forgot, spiders have fangs designed for piercing the bodies of their prey and produce venom that

Figure 1.13. Though members of my family might confuse this for an insect, the animal pictured possesses only four legs and lacks wings. Thus, a cow (*Bos* sp.) is not an insect. Photo courtesy of Keith Weller, USDA Agricultural Research Service, Bugwood.org.

is a cocktail of toxins and enzymes, which is injected into the unlucky food source that will eventually be sucked dry by the eight-legged conqueror. What creeps a lot of folks out is the way spiders walk. They display a slow, purposeful gait, with each hairy leg gracefully taking a delicate step, one making contact before another is lifted off the ground. The pattern of walking is surprisingly distinct between spiders and insects, as the spider relies on hydraulic movement of blood for leg movement, whereas insect legs are muscle-driven and give a zigzag appearance to the gait, since walking relies on alternating pivot legs. If I had to guess, you probably have never paid much attention to how small creepy things glide across the terrain of your kitchen floor. Maybe you should!

So if ever in doubt about whether the multilegged creature walking across the floor or your face is an insect or spider, gently pick it up, move it close to your good eye, and count the legs. If still unsure, dangle it in your mother's face and measure the effect of her scream on your eardrums: numbing, it's an insect; bleeding, a spider. Fathers are no help here because the cursing that ensues is undecipherable and at too high a decibel level. Of course, if bitten during any of the preceding steps, wait five to ten minutes, so that a trained EMT can decide for you. What was the point of this discussion? Quite simple: if you take the time to observe the two, spiders and insects

---

MythBusters is a US TV series on the Discovery Channel. The hosts, Adam Savage and Jamie Hyneman, with the aid of their ample staff, tackle the validity of myths, rumors, wives' tales, movie scenes, news stories, and anything else that suits their fancy, using aspects of the scientific method. Those that do not hold up to the testing are considered busted myths.

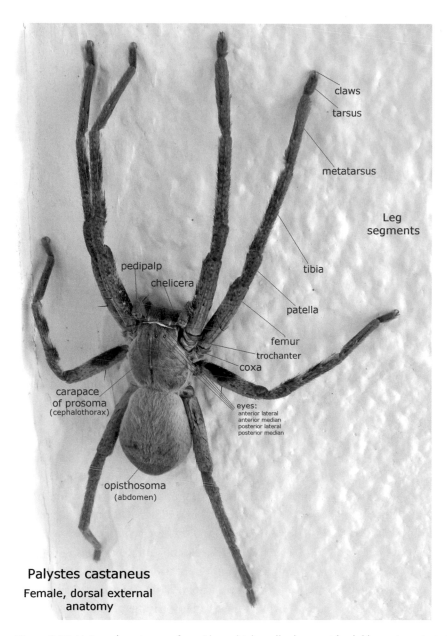

Figure 1.14. External anatomy of a spider, which really does not look like an insect at all! Image by Jon Richfield available at http://bit.ly/1Ox2IcN.

"Bugs" versus insects: The importance of knowing what you are talking about

We will now address our second myth; that is, "all insects are bugs." The phrase gives entomologists heart palpitations—or at least those who teach entomology. Obviously my tone suggests that this myth is about to get busted. Yes indeed! What's the big deal? So people refer to insects as bugs—so what? Get a life! The significance is much like that discussed in the last section: just because two things look similar does not mean they necessarily should be lumped together. A similarity in anatomy or behavior is why we have made the association (figure 1.15). I understand that connection. However, when a close inspection reveals real divergence, well then, that is why we use different names for different groups.

resemble each other only cursorily. The myth should not exist in the first place!

Figure 1.15. All insects are not bugs, but all bugs are insects. Why does it matter? The swallowtail butterfly poses no threat to the human condition, but the bed bug on the right is a blood-feeding parasite that readily feeds on humans. Photos by Daniel Schwen (http://bit.ly/1LtuwLC) and the Centers for Disease Control and Prevention (http://bit.ly/1G0m9kp).

In the case of insects, we will learn in chapter 7 about classification, including the distinctions between animals that are close cousins, such as spiders and insects, which are grouped in the same phylum (Arthropoda) but separated into unique subdivisions (subphyla). Within each form of classification, the degree of relatedness between species increases as the groupings becomes more and more specific. So members of the same phylum are related and share several but not all features, members of a subphylum share closer relationships, and species found in the same class are very similar. However, even species that share a lot in common are still not the same animal. Later, we will discover that all insects do not look the same, behave similarly, reproduce by identical means, eat the same food, and so on. Yes, there are unifying features, which is what led to them being placed in the same class, but differences in anatomy or molecular profiles warrant separate groups. To distinguish among different insects, subdivisions are made within the class Insecta that include groupings such as orders, families, genus, and species. Even each of these can be further subdivided depending on the level of complexity we wish to explore. For now, suffice to say that such divisions exist because all insects are not the same.

So, returning to our original goal of debunking the insects versus bugs myth. Here goes: all members of the class Insecta are insects, but all insects are not "bugs." Technically speaking, only one group of insects has the title of bugs: those that belong to a division called the order Hemiptera (also called order Heteroptera). The class name Insecta also goes by Hexapoda, in reference to the number of legs (six), typical of adult insects, and the order name has also undergone several name changes as a result of research, discussion, debate, and argument among those who study classification—taxonomists and systematists. Thus, bugs (often called true bugs) belong to the order Hemiptera or Heteroptera, which represents but one subdivision of the class Insecta or Hexapoda. My head hurts.

Does the difference in terminology matter? Yes it does, particularly when the identity of the insect in question is critical to assessing whether a potential pest is present or not, so that proper action can be put in place. The identity is absolutely critical if the insect in question is being used as physical evidence in a homicide investigation to estimate the time of death (more correctly the time of insect colonization). The identity is also important to you in determining whether the red and black "bug" that landed on your head is simply an insect, like a harmless ladybird beetle (commonly called lady bug) or an assassin bug (a real hemipteran) with a nasty disposition and a propensity to bite. Is it critical that you understand insect classification in order to study them or appreciate them? No. And because that is true, our textbook attempts to deemphasize classification and terminology as a means to learn about insects, and instead focuses on the fascinating features of insect biology. So sit back, buckle up, and get ready to explore the most amazing world on the planet!

CHAPTER REVIEW

❧ **No better time than now to find out what an insect actually is!**

- Insects are a group of animals that display tremendous variety when it comes to appearance. They can be distinguished by their anatomy as being small, their bodies are divided into three distinct regions, adults have six legs located on the middle section (the thorax) of the body and positioned equally on either side of the body, they have wings, and two large compound eye are positioned on the head, as are a pair of antennae.

- Internally the body cavity is filled with thick blood (more correctly, hemolymph), the gut is a long tube that varies in shape and composition based on the main food type consumed, circulation is accomplished by a tubular heart and series of mini pumping organs, and multiple "brains" are positioned throughout the body.

- The study of these cool beasts is called entomology. Most often, insects and other closely related terrestrial arthropods usually get lumped together, at least in textbooks and most undergraduate courses.

Who cares about insects, or, why study them darn bugs?

- Why study insects? Insects can powerfully shape human existence through destructive interactions. In fact, the top four reasons why we study insects are because (1) insects spread disease; (2) insects eat our crops; (3) insects eat our stored food; and (4) insects maim and kill our livestock.

- Only about 2% of extant insects are actually responsible for the destructive effects outlined above. The remaining 98%, meaning nearly one million species, can be classified as either neutral or beneficial to humans.

- Insects are truly needed for human survival. Just a few examples of what humans rely on or need insects for include plant pollination, insect-derived foods, insects' tremendous recycling efforts, and a wide range of insect-derived products.

Insects are the most successful animals on the planet

- Insects are the most evolutionary successful group of animals ever to have existed on the planet. An argument can be made to support this statement based on abundance and diversity. Abundance is the sheer number of insects that exist. This includes all individuals of every insect species that occur at any given time. Diversity is a smaller measure, as it focuses on the total number of described species of insects. Diversity is interested in how many distinct types or groups of insects exist.

- Abundance is concerned with how many total insects are on the planet right now. How in the world could you measure all the insects that exist, anywhere on the planet, at any moment in time? Any estimation or educated guess of the abundance of insects, or any organism for that matter, relies on the quality of our methods for counting or sampling insects. The "how" question represents a combination of dedicated sampling and estimation of insect populations.

- One way to address insect abundance is to simply look around you. Insects are everywhere. But "everywhere" reaches much farther than your immediate surroundings. Their range by no means ends with one domicile. Insects thrive on land and in freshwater. Some exceptions do exist: anywhere ice persists, like the polar ice caps or Antarctica, it is difficult for most any forms of life to exist. Insects are also rare in marine environments, being restricted to regions close to shore or to a parasitic lifestyle within marine hosts.

- Insect abundance varies on a daily basis, as new insects are born and old ones die. One reasonable estimate places the total number of insects at about one quintillion (1,000,000,000,000,000,000), equal to about 2.7 billion tons of insects, outweighing the human population by nearly ten times!

- Perhaps the only rival to the impressive abundance of insects is provided by insects themselves. What this means is that the diversity of insects deserves even more accolades than abundance. Current estimates place insect diversity somewhere between 900,000 to 1,000,000,000 insect species.

- If we simply say that the number of insect species is approximately one million, then insects would represent more than 75% of all animal life on the planet.

Insects are bad, bad, bad

- Some insects are highly destructive. The diversity of destructive species is incredibly low by comparison to the total. What this means is that though the effects of destructive insects are considerable, particularly in terms of injury and death to human populations, only a small number of six-legged beasts (approximately 2%) can be held responsible.

- Generally, destructive insects are grouped by the effect they have on the human condition. This classification scheme yields three distinct categories of destructive insects: medically important species, structural pests, and agricultural pests.

- One of the cruelest ways that insects torment human populations is through disrupting agriculture. Several insect species directly or indirectly attack crops or other aspects of food production and storage. Direct attack should be obvious, as the insect feeds on a food source, damaging the crop as potential human nutriment. In indirect attack, insect activity does not directly target the food source but ultimately does lower food or nutrient production.

- Not to be outdone are medically important insects. Several blood-feeding insects serve as vectors for human disease that today place more than 65% of the world's population at risk. Top on this list are mosquitoes that vector malaria. Malaria is a disease that at one time was broadly distributed across the globe, but today is mostly isolated to tropical and subtropical regions, particularly in Africa, Asia, and Central and South America.

- Many species of insects express their influence by boring or chewing on natural and man-made structures. If this happens to be your home, the result can be thousands of dollars in damages, or worse, the house can become uninhabitable. Wood-based products are the principal targets of the majority of structural insects, with public enemy number one being the termite.

- Keep in mind the idea of *millions to billions* when discussing the numbers of people in the path of insect destruction. The phrase also provides an accurate depiction of the economic losses stemming from insect damage, no matter what category of pest we discuss. Insects have been

and continue to be amazingly successful at wreaking havoc in our lives, and for that reason alone, they warrant our serious attention.

✤ What I have learned on my own: Insects are way cool

- A small fraction of insects participate in tormenting humans. The overwhelming majority play significantly important roles in our existence that we, and the plants and animals we love and need, depend on.

- Insects do wondrous things that captivate the imagination. There are fly maggots that leap by grabbing their anus with their mouth, creating muscular tension like a spring, and then, releasing their grip, shooting straight up. Insects can fend off parasites nearly the size of their own bodies, even after the invaders have burrowed into their body cavities through the skin, displacing organs as they move about. Some beetles escape skillful predators by releasing a knockout gas from glands near the anus, leaving the would-be attackers "sleeping" and vulnerable to being eaten themselves. When it comes to sex, insects could write their own Kama Sutra! With some insect species, the female only needs to mate once in her life, for she can store the sperm of her "lover" for eternity, releasing sperm to fertilize eggs through conscious control, thereby controlling the sex of her offspring. Efficiency is also apropos for describing insect digestion.

- If you allow yourself to look into the world of insects with open eyes and open mind, you may very well come to see the reasoning behind the treatise that insects are phat, but definitely not fat.

✤ How do you know it is not a spider?

- Spiders are not insects, though they are close cousins. Both spiders and insects do have six legs, but spiders take it an extra step (actually two) by owning two extra legs. Spider bodies are composed of just two regions, they cannot fly, and they have way more than two eyes.

- What creeps a lot of folks out is the way spiders walk. They display a slow, purposeful gait, with each hairy leg gracefully taking a delicate step, one making contact before another is lifted off the ground. The pattern of walking is surprisingly distinct between spiders and insects, as the spider relies on hydraulic movement of blood for leg movement, whereas insect legs are muscle-driven and give a zigzag appearance to the gait, since walking relies on alternating pivot legs.

✤ "Bugs" versus insects: The importance of knowing what you are talking about

- Just because one thing looks similar to another thing does not mean they necessarily should be lumped together. Similarities in anatomy or behavior are why we make the association. In the case of spiders and insects, both are grouped in the same phylum (Arthropoda) but separated into unique subdivisions (subphyla). Within each form of classification, the degree of relatedness between species increases as the groupings become more and more specific. So members of the same phylum are related and share several but not all features, members of the same subphylum share closer relationships, and species found in the same class are very similar. However, even species that share a lot in common are still not the same animal.

- To distinguish among different insects, subdivisions are made within the class Insecta that include groupings like orders, families, genus, and species. Each of these can be further subdivided, depending on the level of complexity we wish to explore. For now, suffice to say that such divisions exist because all insects are not the same.

- All members of the class Insecta are insects, but all insects are not "bugs." Technically speaking, only one group of insects has the title of bugs, and those belong to a division called the order Hemiptera, sometimes referred to as Heteroptera. Does the difference in terminology matter? Yes it does, particularly when the identity of the insect in question is critical to assessing whether a potential pest is present or not, whether the insect in question is to be used as evidence in a criminal investigation, or whether the beast in question is likely to bite or not.

MUSHROOM FARMING (SELF-TEST)

Level 1: Knowledge/Comprehension

1. Define the following terms:
   - (a) insect
   - (b) humancentric
   - (c) synanthropic
   - (d) species
   - (e) invasive species
   - (f) entomology

2. Insects are generally studied for their negative effects on the human condition. Describe the major categories or ways in which insects harm humans.

3. Despite the negative influences of insects, only a small fraction of the whole actually represents true pests of humans. The vast majority of insects are considered useful to humans. Detail as many examples as you can of ways in which humans benefit from an association with members of the class Insecta.

Level 2: Application/Analysis

1. Insects are considered the most evolutionary successful of all animals, particularly when abundance and diversity are examined. Explain how these two parameters help illustrate insect dominance on the planet.

2. Describe how anatomical and behavioral features can be used to distinguish between different animal groups.

Level 3: Synthesis/Evaluation

1. A distinction can be made between members of the class Hexapoda and the specific group of insects known commonly as true bugs. Why does it matter that such distinction be made?

REFERENCES

Centers for Disease Control and Prevention. Malaria. http://www.cdc.gov/malaria/. Accessed March 27, 2014.

Erwin, T. L. 1983. Tropical forest canopies: The last biotic frontier. *Bulletin of the Entomological Society of America* 29:14-19.

Erwin, T. L. 1988. The tropical forest canopy: The heart of biotic diversity. In: Biodiversity (E. O. Wilson, ed.), pp. 123-229. National Academies Press, Washington, DC.

Evans, A. V., C. L. Bellamy, and L. C. Watson. 1996. An Inordinate Fondness for Beetles. Henry Holt, New York, NY.

Gaston, K. J. 1991. The magnitude of global insect species richness. *Conservation Biology* 5(3): 283–296.

Hocking, B. 1968. Six-Legged Science. Schenkman Publishing, Cambridge, MA.

Hölldobler, B., and E. O. Wilson. 1998. Journey to the Ants: A Story of Scientific Exploration. The Belknap Press at Harvard University Press, Cambridge, MA.

Maryland Department of Natural Resources. Emerald ash borer.http://www.dnr.state.md.us/dnrnews/infocus /emerald_ash_borer.asp. Accessed March 26, 2014.

May, R. M. 1988. How many species are there on earth? *Science* 241:1441–1449.

Pedigo, L. P., and M. E. Rice. 2006. Entomology and Pest Management. 5th ed. Pearson/Prentice Hall, Upper Saddle River, NJ.

Waldbauer, G. 1996. Insects through the Seasons. Harvard University Press, London, UK.

Waldbauer, G. 2000. Millions of Monarchs, Bunches of Beetles: How Bugs Find Strength in Numbers. Harvard University Press, Cambridge, MA.

THE ENTOMOLOGIST BOOKSHELF (SUPPLEMENTAL READINGS)

Berenbaum, M. R. 1996. Bugs in the System: Insects and Their Impact on Human Affairs. Helix Books, New York, NY.

Cranshaw, W., and R. Redak. 2013. Bugs Rule!: An Introduction to the World of Insects. Princeton University Press, Princeton, NJ.

Gullan, P. J., and P. S. Cranston. 2010. The Insects: An Outline of Entomology. Wiley-Blackwell, West Sussex, UK.

Johnson, N. F., and C. A. Triplehorn. 2004. Borror and DeLong's Introduction to the Study of Insects. 7th ed. Cengage Learning, New York, NY.

Whitfield, J. B., and A. H. Purcell III. 2012. Daly and Doyen's Introduction to Insect Biology and Diversity. 3rd ed. Oxford University Press, Cambridge, MA.

ADDITIONAL RESOURCES

Entomology for beginners
http://bijlmakers.com/insects/entomology-for- beginners/

Extension entomologists blogs
http://entomology.tamu.edu/extension/blogs/

Insect biology
http://wiki.bugwood.org/Insect_Biology

Insect biology and ecology: A primer
http://www.biocontrol.entomology.cornell.edu/bio.html

Insects in humans' life
http://www.efabre.net/insects-humans039-life

2   History of Entomology

A Discipline Founded on Death

The mosquito's a clever little bastard. You can track him for days and days until you really get to know him like a friend. He knows you're there, and you know he's there. It's a game of wits. You hate him, then you respect him, then you kill him.

Eric Idle
Mosquito Hunters,*
Monty Python's Flying Circus (1970)

The origins of entomology are uniquely different from those of many scientific disciplines. This should come as no surprise, because those who study insects have never been regarded as champions of the "normal" path. The field was established because of insects' negative effects on the human condition: a small number of species can profoundly reduce local and global food production; several blood-feeding insects serve as vectors for human disease that today place more than 65% of the world's population at risk; and still others invade human environs or annoy in other ways. Since recorded history, man has sought ways to reduce the negative influence of insects on humans, mostly through developing means to kill them, and thus those who studied insects generally did so (and still do) to discover new avenues for insect control. The point is that entomology developed as an applied discipline that has stayed, for the most part, true to its roots. Along the way, insect biology came into vogue for naturalists and scientists alike. In fact in many cultures, insects were held in high esteem, sometimes even worshipped, based on aspects (real or otherwise) of their biology. Singing, flying, and rolling dung captured the imagination of ancient civilizations and was often immortalized in paintings, carvings, and writings. This chapter explores key historical events and figures that shaped the formation of entomology as a discipline, as well as influenced the modern study and use of insects in academic, cultural, and terroristic pursuits.

Key Concepts
- Insects: Unwelcome guests since the beginning of the human "party"
- Insect plagues and deities in ancient and modern civilizations
- Naturalists, physicians, and the clergy: An intriguing new pinup calendar or prominent figures in entomology?
- Politicians at work: Two acts of Congress established entomology in North America

---

Mosquito Hunters. http://www.youtube.com/watch?v=BHBbJAIcnBI.

- War anyone? Insects have been the true victors of most wars
- Evolution of entomology: Insects as tools for biology, agriculture, and war

Insects: Unwelcome guests since the beginning of the human "party"

To quote Monty Python* in any venue is always a delight. But alas, their genius and wit is not without entomological errors: hunting for male mosquitoes is misguided, as they do not and cannot blood-feed. Rather, the honor of dining on human blood is restricted to the adult female, who does so to obtain the nutrients needed for provisioning eggs. Thus, females are the dangerous game to be pursued and not the tiny, inoffensive males. However, what does ring true is that mosquitoes and many other insects have triggered in humans the desire to kill, probably since the very onset of the human-entomological association. The relationship was formed by insects, uninvited, to take advantage of mankind. Humans formed dwellings in convenient locales—convenient for the insects, that is, to find and use them. As humans began their foray into agriculture, insects were there, eager to partake of the harvest before the laborers interfered. Organized communities (e.g., villages, towns, cities) formed by man offered new opportunities for insects: concentrated food sources (human populations) that were nutrient rich (individual humans). Coupled with the fact that methods of insect control were largely unknown to most peoples, stored foods were left unprotected, and hygienic conditions were awful, the advent of human civilization has been a golden time for many insect species, particularly blood-feeding varieties, to become unwelcome neighbors. In other words, synanthropy, or the condition of living in close association with mankind, developed for insects very early in human history, perhaps with the evolution of *Homo sapiens*. This section examines several examples to understand human perceptions of insect influences on everyday life, including cultural and religious beliefs of entomological origin. Insects as a whole pre-date man

by over 400 million years, plenty of time to lie in wait for the perfect host (parasitic and otherwise). Ever since, humans have sought ways to end the relationship.

Based on such examples, entomology should be an old discipline, but the reality is much different. Recognizing insects or their activity is not the same as establishing a discipline or field of study. Humans have acknowledged insects in their lives since the beginnings of recorded history. But the study of insects—from the perspective of an organized or systematic examination of the biology of these creatures—is not nearly as old. Scientific or academic study of insects, and hence the true origins of entomology, likely did not begin until around the sixteenth century. The seminal work *Historiae animalium* (*History of Animals*), a five-volume treatise (1551-1558) by Swiss naturalist Conrad Gesner, is considered the beginnings of modern **zoology**,[†] and with reference to insects in the work, earns the award for first recognized study of insects (figure 2.1). Some regard the works of William Kirby (1759-1850), an English entomologist, as the official foundation for the study of insects. Consequently, he is regarded as the father of entomology, especially among English and other Europeans (figure 2.2). As such, entomology as an academic discipline had its beginnings in Europe at the turn of the nineteenth century.

In North America, insect investigations did not gain a foothold on the continent until the early 1800s, and even then, those engaged in such activities may have been inclined to keep their fondness for insects to themselves. The public as a whole tended to view a fascination for insects as worthy only of the foppish (foolish or silly), or those with idle minds or nothing better to do with their time (figure 2.3). It would take another sixty years and intervention by the US Congress before studying insects became less fodder for ridicule or scorn and entomology became truly recognized as a scientific study and was viewed as a bona fide discipline, and to be an entomologist became a worthy profession.

Later in this chapter we examine key individuals who played prominent roles in laying the foundation for en-

*The BBC TV series *Monty Python's Flying Circus* aired from 1969 to 1974 and starred John Cleese, Michael Palin, Terry Jones, Eric Idle, Terry Gilliam, and Graham Chapman. The comedy troupe also generated such legendary films as *Monty Python and the Holy Grail*, *The Meaning of Life*, and *The Life of Brian*.

†Zoology is the study of animals as a whole. At the time of Gesner, most interest was directed toward macro-vertebrates, meaning big animals with backbones. The initial four volumes described the quadrupeds, fishes, amphibians, and birds residing in Zürich, Switzerland, and the fifth focused on snakes.

Figure 2.1. Swiss naturalist Conrad Gesner, whose works represent some of the earliest study of insects. Photo by Roland zh available at http://bit.ly/1TirptH.

Figure 2.2. Image of William Kirby, regarded by many as the father or founder of entomology. Original lithograph created by T. H. McGuire (1851) is available in public domain at http://bit.ly/1OOsJlv.

tomology. Here, we take a step back in time to examine some of the earliest influences of insects on the human condition, periods long before scientific study was postulated for any organism. These early entomological encounters pre-date human civilizations and organized agriculture. Obviously this suggests that insect synanthropy and all its negative consequences had yet to take center stage in the lives of humans. Insect participation in human activity is rooted to prehistoric times, with early cave drawings revealing the recognition of six-legged arthropods. Ancient Egyptians noted that death and insects went hand in hand: some insects were the cause of death (later learned to be vectors of disease) and others were attracted to the dead (mummification was a means to protect the dead from **necrophagous** insects, i.e., those that feed on carrion or dead animals). This

recognition was evident in paintings, carvings, and writings in the form of hieroglyphs (figure 2.4). Even tissue samples collected from Egyptian mummies confirm that disease-causing insects were present as early as pre-pharaonic times (before the pharaohs of Egypt, meaning earlier than 3000 BC) (figure 2.5). Those insect vectors included fleas harboring the causative agent of bubonic plague, mosquitoes transmitting malaria, and lice delivering typhus to human hosts. The effect of these diseases on the development of entomology is discussed in the next section, and the devastating consequences for human civilizations are examined in detail in chapter 4.

Origins of cultural appreciation of insects

Ancient peoples' interest in insects transcended disease and death. For example, certain ethnic assemblages or

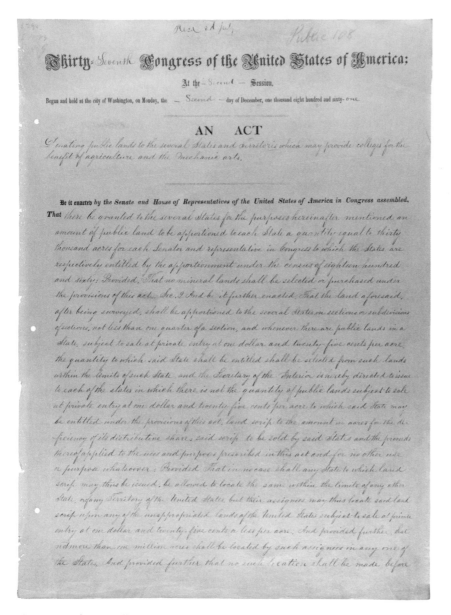

Figure 2.3. The Morrill Land Grant Act of 1862 was instrumental in establishing entomology as an academic discipline in the United States. Image available at http://bit.ly/1L3vwr8.

civilizations revered if not worshipped insects. This is perhaps best illustrated by ancient Egypt. Many insects held special meaning in association with the gods, death, or the afterlife. Beetles particularly resonated with Egyptians from the period known as predynastic (prior to circa 3000 BC) to the onset of New Kingdom Egypt (c. 1550 BC). Dung and buprestid beetles held symbolic and religious significance, features that are evident throughout hieroglyphs, carvings, jewelry, and even stone statues (figure 2.6). One example is the god Khepri, who held dominion over all insects. His likeness is depicted as hav-

ing a head in the form of an adult dung* beetle (figure 2.7). Why would a beetle that specializes in using animal

---

*The term dung is one I never assumed would require a definition. However, I have been asked on more than one occasion in class, "What do you mean by dung?" My quip response, that it's the vertebrate equivalent to frass, does not satisfy everyone. So, if curious, dung is material eliminated or removed from the digestive tract via the anal opening and is composed of food material in various states of digestion, gut fluids, mucus, bacteria, and any other object that has traveled through the alimentary canal. Also affectionately known as animal poop, but generally only if in a solid or semisolid state.

Figure 2.4. Scarab beetle in hieroglyphics from New Kingdom Egypt (1333–1279 BC). Image available at http://bit.ly/1NKb6lV.

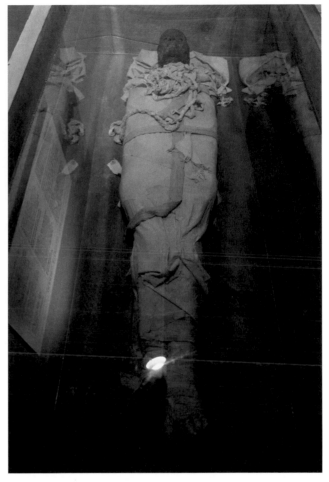

Figure 2.5. Egyptian mummy from c. 1500 BC. Preserved blood-feeding insects have been recovered from similar mummies, testing positive for agents responsible for such diseases as bubonic plague and malaria. Photo by Daderot available at http://bit.ly/1WgvLB1.

dung symbolize a god? Good question. The answer literally lies *in* dung. Well, *with* dung, but either way, balls of animal poop are the reason. (A detailed explanation is revealed in the section devoted to insect deities.)

Buprestid beetles also captured the imagination of ancient Egyptians for their apparent ability to be reborn from tree cores. Beetles in the family Elateridae (click beetles) were common subject matter for carvings and jewelry, perhaps because of the adult's ability to make a clicking sound by striking the abdomen against the ground. The movement is used as a righting reflex when the beetles are flipped on their dorsal surface ("backs"), and giving the appearance that the beetles can jump without using their legs.

Several non-beetle insects also influenced Egyptian culture. Bees and wasps are evident in carved reliefs and pottery, in part because their fierce fighting behavior was revered as well as manipulated (bee bombs were used as weapons to remove enemies from hidden fortifications and trenches). Honey bees provided protection from evil spirits and were the source of nectar for the gods, that is, honey (based on discussions from chapter 1, the gods must not have known the source of this gift). Not surprisingly, a honey pot was found in the tomb of the boy pharaoh, King Tutankhamun. At least for royalty and priests, adult flies were used to transport the soul from the mortal world to the afterlife (figure 2.8). Interestingly, mummification was used as a means to protect a corpse from being consumed by those same flies, par-

tially out of fear that the soul would be lost if the physical shell was disrupted by necrophiles.

Insects were intimately associated with the gods, mythology, and death throughout the ancient empires of the Maya and Aztecs of Central and South America. Butterflies, rather than the beetles of Egypt, dominated the insect connection to the people of ancient Mesoamerica. A variety of butterflies symbolized beauty, rebirth, and control over death. The butterfly also symbolized the human quest to achieve immortality. The Moche people of pre-Columbian Peru deliberately exposed dead bodies to the environment for several days after death to encourage fly colonization. The Moche believed that fly larvae would engulf the "anima" or spirit

Figure 2.6. Scarab amulet from the Third Intermediate-early Late Period (946–525 BC) Egypt. The sculpture depicts a scarab beetle, a revered insect in ancient Egypt. Photo available at http://bit.ly/1G0RUtP.

Figure 2.7. Wall painting of Nun, the Egyptian god of the waters of chaos, lifts the baroque of the sun god Ra (represented by both the scarab and the sun disk) into the sky at the beginning of time. The head of this deity is commonly depicted as the body of a scarab beetle, presumably the dung beetle *Scarabaeus sacer*. Image available in public domain at http://bit.ly/256WtE7, painter unknown.

of the individual as they consumed the flesh of the corpse, eventually metamorphosing into adult flies, which in turn would return to live with the people. In North America, mortuary pottery found in ruins of the Mimbreno tribes that inhabited the Mimbres River Valley of New Mexico and southeastern Arizona were adorned with insects, yet such creatures were absent from bowls used for other purposes. These examples suggest that civilizations throughout the New World recognized a connection between insects and death, which was tied to beliefs in the afterlife.

Insect recognition in ancient Asia was associated with their contributions to aesthetics and culture rather than worship or fear. For example, the Japanese valued insect depictions in art, literature, and recreational pursuits. In modern times, this appreciation has carried on in the form of Pokémon,* a video game series owned and produced by Nintendo Corporation. The game is based on insect collecting, which was a childhood hobby of game creator Satoshi Tajiri, and the characters themselves seem to be an amalgamation of Japanese mythological creatures and woodland animals, including insects. Unfortunately, the video game's popularity has not sparked a rise in junior entomologists. The Chinese expressed a

---

*Pokémon has become the second most popular video game-based media franchise in the world since its launch in 1996. Many of the 792 fictional characters introduced in the series display clear insect influences in terms of anatomy and behaviors.

Figure 2.8. Mummification was practiced by both ancient Egyptians and Peruvians, but for different purposes. In Egypt, bodies were protected from necrophagous insects by mummification, but insect feeding was encouraged in Peru. Image of the Peruvian mummy available at http://bit.ly/1OOTZ38.

usually in the form of a would-be predator. As we learn in chapter 11, singing can be accomplished by more than one mechanism, as this group of insects demonstrates.

Though a cricket serenade was pleasing to the Chinese, their true insect love affair was with a moth that produced silk. **Sericulture**, or the practice of breeding silk moths for the production of raw silk, originated in China over five thousand years ago and then spread to other parts of Asia and Europe. The silk is produced by a moth, *Bombyx mori*, often referred to as the mulberry silk moth (for its preference for mulberry leaves), or Chinese or Japanese silk moth, but it is better known as the domesticated silk moth. The latter name reflects an amazing case of human intervention: the moth was domesticated from a wild variety of silk moth, *Bombyx mandarina*. Today, the insect is not capable of surviving without human care. Silk is a natural fiber that can be woven into textiles. Since raw silk is derived from an insect that must be farmed like cattle, the ancient silk trade was quite lucrative for the Chinese, who kept secret for centuries how to produce the luxuriant material.

Any story of ancient insect influences would be incomplete without a discussion of the entomological connection to vampires. Vampires? Yes, as in reference to mythical beings that feed on living creatures to obtain blood or "life essence" for their own continuing existence. Folklore surrounding vampires can be found on almost any continent, but the origins are generally linked to eastern Europe and the Balkans. In those regions, vampires were described as appearing much like a decomposing corpse: bloated, with dark skin consistent with **necrosis** and **marbling**, the latter the result of sulfhemoglobin formation in capillaries near the skin surface. In the West African countries of Togo (Togolese Republic) and Ghana, the Ewé people describe vampires (known as Adze) as mythical creatures that can assume the form of an adult firefly (figure 2.9) or any of a number of common insects. The idea is that Adze can hide in plain sight, appearing deceptively as harmless insects, that is, until they have a taste for blood. For like other vampires, they must feed on blood for sustenance and can only do so by changing back into the form of a human. Had I known about this in my youth, my career path may have been seriously altered! (See chapter 1's story of firefly catching.) Or perhaps I had nothing to fear from

fondness for the musical talents of orthopterans, the insect group that contains crickets, grasshoppers, and katydids. Each of these insects "sings" with purpose; that is, they possess the ability to produce a distinctive song, with the aim of attracting the attention of a potential mate or sound a warning alarm that danger is present,

Figure 2.9. The eastern firefly, *Photinus pyralus*. Not to be confused with an Adze vampire. Or is it? Photo by Art Farmer available at http://bit.ly/1Jnt0Vi.

the Adze, because these African vampires prefer the blood of the innocent, namely infant children, and not the precociously stupid.

Foundations of academic entomology

Prior to the sixteenth century, there is no evidence that insects were studied in a manner akin to what we would call today a scientific study. Even the work of the Swiss zoologist Conrad Gesner cannot be regarded as a true study of insects, because these creatures were merely mentioned in *Historiae animalium*, along with hundreds of other animals, as existing. In most written records, when insects were mentioned, it was with disdain or hatred since these beasts tormented humans with as much ferocity as the devil (figure 2.10)—but there are a few notable exceptions.

Dutch biologist and microbiologist Jan Swammerdam (1637-1680) received acclaim for developing dissecting techniques involving a microscope for a wide range of animals, as well as for being the first to observe and describe red blood cells (erythrocytes). His study of insects was focused on comparing them to "higher" animals, revealing through observation and dissection that the

Figure 2.10. Painting *Satan, Sin and Death* by William Hogarth (1697-1794). Available in public domain at http://bit.ly/1KyDRh6.

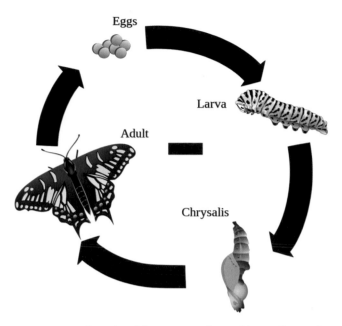

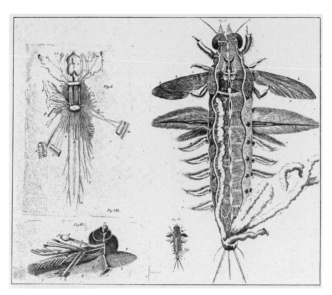

Figure 2.12. Drawings by Jan Swammerdam (1637-1680) of a mayfly (order Ephemeroptera). Image available at http://bit.ly /1KtpQDl.

Figure 2.11. Life cycle of the anise swallowtail butterfly, *Papilio zelicaon*. Jan Swammerdam was instrumental in demonstrating that insects pass through multiple life stages, as does this butterfly, which develops from egg to larva to pupa (cocoon or chrysalis) and adult. Image by Bugboy52.40 available at http:// bit.ly/1QXl03N.

differences were not as large as popularly believed. Interestingly this experience is similar for many students today when studying insects for the first time. Swammerdam was instrumental in establishing that insects develop through stages, so that an egg, larva, pupa, and adult represent the same insect (figure 2.11). In 1669, he published *History of Insects*, in which he correctly described the reproductive organs of insects and detailed the process of metamorphosis. His line drawings of insect anatomy were exceptional, even more impressive when considering the quality of microscopes available to him (figure 2.12).

Perhaps rivaling Swammerdam, the Italian physician and biologist Marcello Malpighi (1628-1694) used an early version of the microscope to see capillaries in animals for the first time and to characterize insects' internal anatomy. Malpighi used larvae of the silkworm, *Bombyx mori*, as the specimen to characterize the insect excretory organs, now called **Malpighian tubules**. He determined that insects do not have lungs, and instead, their gas exchange

relies on a series of tubes, termed tracheae, that open directly to the outside of the skin (the openings are termed **spiracles**), and he generated extraordinary drawings of insect musculature and internal anatomy that could still be used in entomology classes taught today (figure 2.13). Malpighi also dabbled in plant anatomy, earning the attention of another prominent figure in entomology, Carolus Linnaeus, who bestowed on Malpighi one of the highest honors available to a biologist: Linnaeus named the plant genus *Malpighia* after the Italian biologist.

Insect plagues and deities in ancient and modern civilizations

What is the quickest way to get a population of people to notice insects? The answer is simple: choose an insect that transmits disease. If that insect vectors a disease-causing agent that can induce an **epidemic**, or better yet, a **pandemic**,* well then, the marketing strategy will yield instantaneous results. But, for that to be true, people have to know that it is the insects transmitting the dis-

---

*An epidemic occurs when a disease reaches unexpected levels of infection, or is perhaps even new to a particular location. If that disease spreads to cover a large area, including other countries or continents, it is referred to as a pandemic.

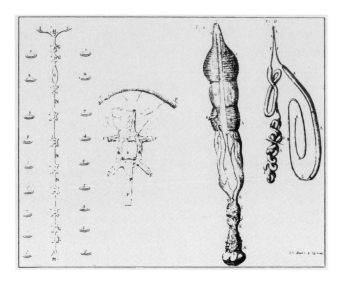

Figure 2.13. Internal anatomy of a silkworm, drawn by the Italian physician Marcello Malpighi (1628–1694). Image available at http://bit.ly/1OsTeh7.

eases. Early in human history, insect-borne diseases reached pandemic levels, yet no one had any idea what caused them. The most accepted explanations stemmed from religious beliefs, in which failing to follow the decrees of the creator, or somehow displeasing one or more gods in general, brought on illnesses. Severe deity disappointment resulted in plagues, although as we will see, not all insect plagues were necessarily associated with disease. The best-known insect plagues are directly tied to the best-known deity. In the Christian Bible, as well as in the Quran, the texts reveal that God or Yahweh unleashed ten plagues on Egypt to force the pharaoh to release the Israelites from the bondage of slavery. Three of those reported plagues involved insects as the causative agents to instill pain and suffering:

1. *Plague of lice*. Exodus 8:16–19 mentions the Hebrew noun *kinam*, which can be translated as lice (presumably plant lice), gnats, or fleas. Some biblical scholars believe that fleas were used and that they in turn transmitted bubonic plague, accounting for the sixth plague, that of boils. In chapter 4 we discuss the bubonic plague in detail, including examining the origins of the Black Death in Egypt, rather than in Europe as was traditionally believed.

2. *Plague of flies* (sometimes referred to as wild animals). In the fourth plague Yahweh sends swarms of flies (Exodus 8:20–23). The Hebrew word *arob* in the biblical text refers to biting flies,* conceivably the stable fly, *Stomoxys calcitrans*, suggesting the "swarms" were not just an annoyance but also a source of pain and, potentially, disease. Today, stable flies remain a serious pest of livestock worldwide, especially of cows and horses, and readily blood-feed on humans located in close proximity.

3. *Plague of locusts*. The eighth plague of Egypt was locusts (Exodus 10:1–20). As discussed in chapter 1, the insect was most likely the desert locust, *Schistocerca gregaria*, a beast known throughout history as capable of achieving incredible swarm densities and cleansing the land of all vegetation in their destructive paths. The locust plague described in the book of Exodus was so heavy with insects that it covered the sky, casting a shadow over Egypt.

Amazingly, the story told in Exodus indicates that it took an additional two plagues to convince the pharaoh to release the Hebrew people to Moses. If the events are assumed to be historically correct, the most plausible pharaoh referenced in Exodus is Ramesses II of the Nineteenth Dynasty, placing the time line of the insect plagues sometime between 1290 and 1213 BC.

Whether the sixth plague described in the Christian Bible was bubonic plague is left to conjecture, since there is no means to test this hypothesis historically or experimentally. However, the question of interest to us is, what connection does this devastating disease have to insects, more specifically, to fleas? If you have ever owned a dog or cat, then you are quite familiar with these bloodthirsty insects, which rely on warm-blooded animals to serve as a nutrition source, meaning blood. In the case of bubonic plague, fleas harboring the disease-causing agent (the bacterium *Yersinia pestis*) fed on black rats. Eventually, the rats died. The fleas sought out new hosts, easy to find since the rats lived in close association with humans. The fate of those infected was the same for human as for rat: death typically occurred in less than a week. Bubonic plague swept through parts of Europe on at least three occasions, leaving a wake of death. In the

---

*There is considerable debate whether the original Hebrew text refers to swarms of biting flies, wild animals, or even to beetles.

sixth century, the Roman Empire faced an epidemic of bubonic plague that left an estimated death toll of fifty million people.

Plague outbreak during the late Middle Ages (1340–1400) in Europe was perhaps the most deadly, reaching pandemic levels and decimating the total population by at least one-third. This period in the history of bubonic plague is referred to as the Black Death. The initial infection leading to the Black Death is believed to have started in China and then spread to Europe via a Mongolian attack on an Italian trading post, which in turn led to infested rats aboard trading ships transmitting the disease to Italy. We will explore other possible origins of bubonic plague in chapter 4, but it is important to understand that bubonic plague represents one of the most deadly insect-borne diseases and has had a notoriously profound influence in shaping the history of mankind throughout the world (figure 2.14).

Beyond the text

Considering that *Yersinia pestis* is a parasite, why is it odd that black rats and humans both died when infected by the bacterium?

One of the more unexpected outcomes of bubonic plague is the familiar children's nursery rhyme "Ring o' Roses" or "Ring around the Roses," in which the lyrics appear to refer to the symptomology and finality of the plague as it spread through England in 1662. The American version of the rhyme goes like this:

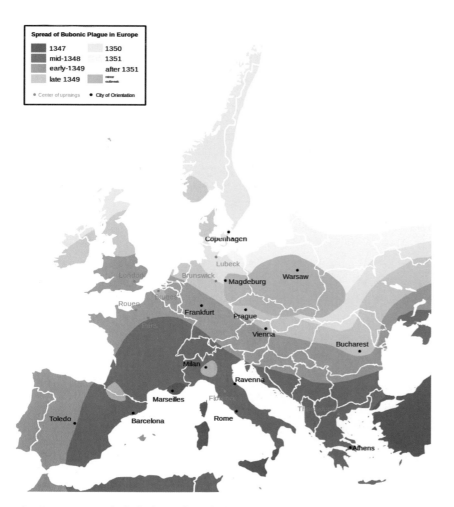

Figure 2.14. Spread of Black Death, or bubonic plague, through Europe during the Middle Ages. Image created by Andy85719 and available at http://bit.ly/1V85OGS.

Ring-a-round the rosie,
A pocket full of posies,
Ashes! Ashes!
We all fall down.

The first line is assumed to refer to the external appearance of swollen lymph nodes throughout the body that can be seen as red to purple circles, or buboes, under the skin. "Posies" refers either to snuff taken to alleviate pain from the swelling or, literally, to a "posy" (bouquet) of flowers that provides a fragrant odor to mask the distinctive scent of plague victims. And of course the final two lines detail the fate of most who succumb to the disease: death, followed by burning the bodies. The rhyme's original author is unknown, but its earliest known appearance in print was 1881 (figure 2.15). Some literary scholars are emphatic that the rhyme has no link to bubonic plague—but if it truly does describe buboes and death, it is quite disturbing how many children have unknowingly learned about Black Death as a happy sing-a-long rhyme.

Insects' linkage to deities has not always been tied to plagues or destruction. Ancient Egyptians, Chinese, Merovingians, Mayans, Aztecs, and countless other civilizations have associated insects with deities, or at least nobility. Of course the question "why" surfaces—as with the displeasing-the-gods explanations for the cause of diseases mentioned earlier, people needed reasonable explanations for the unexplainable. So the phenomenon of a creature that could fly when man cannot, or transform from an ugly being (caterpillar) into a thing of beauty (like a swallowtail butterfly), or a tree that could seemingly give life or rebirth to a beetle found in its core could have only one possible explanation: supernatural beings were involved (figure 2.16). And in many cultures, the gods only spoke to those of noble birth. So kings, emperors, pharaohs, and the like were linked to the deities, and the insects were gods, or worshipped as messengers of the gods.

Many examples illustrate the relationships between insects and deities. For example, the Maya believed that animal beings, many in the form of insects, inhabited the Earth. The wife of one ancestral hero, Xbalanque, was transformed into an array of animals, including bees and other insects. Eventually she became the moon, but only

Figure 2.15. Some believe that the children's nursery rhyme Ring o'Roses or Ring around the Roses depicts the events of the Great Plague of London during 1665. Others contend there is no such relationship. *The Great Plague 1665* is by Rita Greer and available at http://bit.ly/1YFx39m.

Origins of the Plague

Black Death is another name for the bubonic plague, which has been one of the most devastating pandemics ever encountered by human civilizations. The exact death toll cannot be calculated, but estimates place the carnage between seventy-five million and two hundred million people. The plague reached its zenith in Europe during a period extending from 1346 to 1353. A central question that arises when examining this potent disease is, where did it originate before becoming a pandemic? The accepted view is that the plague initiated in the arid plains of central Asia, transmitted along the Silk Road to Crimea. Mongolian invaders and their associated caravans are believed to have been the vehicle that propagated the disease, via transportation of the Oriental rat flea, *Xenopsylla cheopis*. The fleas eventually infected black rat stowaways on merchant ships, and upon arrival at European ports, the rats and fleas, and hence the bubonic plague, arrived to a new continent. Challenging this accepted view is evidence from pharaonic Egypt: mummies and fossilized rat fleas from excavations in Amarna from pharaonic Egypt confirm the presence of plague as an endemic disease earlier than the European pandemics. The working hypothesis is that the causative agent of plague, *Yersinia pestis*, coevolved with the Nile rat, *Arvicanthis niloticus*, and that the rat flea *X. cheopis* transmitted the bacterium to a new host, the black rat that stowed away on ships engaged with trade between Egypt and various other ports, including Europe. This theory does not preclude the possibility that disease reached Asia and then followed the Silk Road, but it does offer an alternative hypothesis that the European pandemics originated in Egypt and not Asia.

temporarily. Aztecs identified insect life cycles, particularly those of butterflies, with numerous aspects of their lives, including death and reincarnation. Two Aztec goddesses, Xochiquetzal and Izpapalotl (figure 2.17), are depicted in many paintings as having wings of butterflies, and the goddess Izpapalotl was claimed to control death since she could swallow darkness. It is entirely possible that this belief spawned the superstition in Mexico that if a black butterfly, *Ascalapha odorata*, stops at your door, somebody will die. Perhaps one way the superstition gets debunked is because *A. odorata* is actually a moth and not a butterfly.

In dynastic Egypt, buprestid or jewel beetles symbolized rebirth and thus were held in high esteem. This relationship stems from the ancient myth involving Osiris, a primeval king of Egypt, who later became lord of the underworld and afterlife (figure 2.18). As the story goes, Osiris was murdered by his brother Set, resulting in his entrapment inside a tamarisk tree. His ultimate escape, and hence his return to life, occurred when his queen, Isis—the goddess of motherhood, nature, and magic—split open the tree to release him. The story is assumed tied to the beetles via a scenario in which carpenters splitting trees to make wood for coffins came upon the buprestids on the inside; larvae fed on the wood and transformed into adults, which would have appeared like a reincarnation to those unfamiliar with insect life cycles.

Another beetle, the dung beetle, held a vaunted place among Egyptians. Dung beetles are a type of scarab (family Scarabaeidae), intimately linked to the god Khepri, a sun or solar deity. The association was made because of an unusual behavior displayed by the beetles. A pair of adult beetles, composed of a male and female *Scarabaeus sacer*, form large balls out of elephant dung that they roll from one location to another (figure 2.19). To Egyptians of long ago, the movements of the dung balls represented the physical movements of the sun across the sky. Since the god Khepri held dominion over insects and the sun, dung beetles were thought to control the movement of the sun each day.

Egyptian mythology also links these scarab beetles to rebirth or reincarnation because (unbeknownst to the

Figure 2.16. Butterflies represent the transformation from ugly (relatively speaking) to beautiful, in this case an adult swallowtail butterfly. Photo of caterpillar by Michael Gimelfarb (http://bit.ly/1QwwRVw) and of an adult butterfly by SFAJane (http://bit.ly/1V8J0kL).

Egyptians) adult females lay eggs in the dung balls, where the larvae feed and develop, and in turn, the dung or dead matter appears to give rise to the new beetles. This latter concept, known as **abiogenesis,** serves as an excellent segue to seventeenth-century Europe, where the idea that insects arise spontaneously from the dead spawned the beginnings of scientific investigation using insects.

Bug bytes
African dung beetles
http://bit.ly/1iLSy9S

Naturalists, physicians, and the clergy: An intriguing new pinup calendar or prominent figures in entomology?

Abiogenesis is the idea that living organisms can form from inanimate objects. While the term may not strike a chord with most people, the phrase "spontaneous generation" usually does, a term synonymous with abiogenesis. The theory stated that living organisms could arise from something other than a similar organism. For example, fleas were believed to originate from dust, maggots (e.g., fly larvae) from rotting meat, and, by extension, dung beetles from manure. How could such views become accepted theories? The same way that unsubstantiated claims still arise today: anecdotal observations and statements never challenged can become "fact" or urban legend. Here is where the **scientific method** proves its importance to any scientific discipline: ideas, concepts, and theories are tested rigorously and repeatedly through formation of hypotheses (plausible explanations), observation, and experimentation. Only after such deliberate scrutiny do ideas become accepted as facts. In more practical terms, the whole basis for the popular television show *MythBusters* is to use the scientific method to test a myth, urban legend, or popular belief.

Italian physician and naturalist Francesco Redi (1626–1697) used aspects of the scientific method to test the validity of the theory of spontaneous generation. In 1667, Redi designed a simple yet ingenious series of experiments to test the theory that fly maggots spontaneously formed on meat (figure 2.20). His classic experiments relied on placing meat in jars, some covered with

Figure 2.17. The Aztec goddesses Xochiquetzal (*left*) and Itzpapalotl (*right*) from pre-Columbian drawings. Images available at http://bit.ly/1WBibM1 (Xochiquetzal) and http://bit.ly/1MH5tWU (Itzpapalotl).

Figure 2.18. Frieze of Osiris in the tomb of Nefertari (c. 1295–1256 BC). Image available at http://bit.ly/1LugPw6.

parchment, some with fine wire screen, and some not covered at all. As would be expected today, the meat left uncovered attracted flies, which laid eggs, and hence gave rise to maggots, otherwise known as **larvae**. No eggs were laid on meat in the jars covered with parchment. Those jars covered with wire screen attracted flies, but since the adults could not reach the meat, they laid eggs on the screen. Redi's experiments demonstrated conclusively that flies and their juveniles do not form from meat but rather that the adults were attracted to the spoiling meat (if they could detect the odor) and laid eggs on or near the food.

Redi was instrumental in reestablishing what the ancient Chinese had previously determined in the eighth century: that certain fly species were attracted to dead animals and that fly reproduction relied on utilization of the carcass. For avid fans of television crime shows like *CSI* or *NCIS*, these flies are quite familiar visitors, feeding on corpses and providing entomological clues into time-of-death investigations (topics explored in chapter 15). Redi's interest in insects also represents an important trend in the early history of entomology: those who would be classified as insect naturalists frequently were employed as physicians or clergy. No doubt holding positions held in high esteem by the public was influential in establishing the side hobby of studying insects as "acceptable" behavior.

Figure 2.19. A flightless dung beetle, *Circellium bachus*, rolling a dung ball. In ancient Egypt, it was another species, *Scarabaeus sacer*, that was revered for pushing the sun across the sky. Photo by Kay-africa available at http://bit.ly/1QXnoav.

Figure 2.20. Portrait of the Italian physician and naturalist Francesco Redi (1626–1697). Redi's famed experiments helped disprove the theory of spontaneous generation. Image available at http://bit.ly/1MH68HV.

The eighteenth century gave rise to the Swedish naturalist Carolus (Carl) Linnaeus (also known as Carl von Linné, 1707-1778). Linnaeus was fascinated by the morphology (anatomy) of plants and animals, and recognized the need for a means to classify organisms more efficiently than was practiced at the time. For example, the concept of a species was well established by the mid-eighteenth century, but names of organisms were cumbersome phrases or entire descriptions, which also varied depending on the scientist or country. Linnaeus developed a **binomial system of nomenclature** for classifying plants and animals that unified and organized botanical and zoological specimens. The method greatly truncated the length of scientific names, and was much simpler than other protocols. More importantly, the binomial system allowed botanists, zoologists, and naturalists across the world to communicate much more effectively about the same organisms, because the names were unified (or stabilized) in approach. His approach to organismal systematics led to Linnaeus being regarded as the father of **taxonomy** (figure 2.21).

Linnaeus was also instrumental in collecting and identifying thousands of animals, and, among them, about two thousand insects were initially described by his efforts. He is one of only two individuals who are recognized by just a single initial when giving authorship to a species. In the case of the common house fly, *Musca domestica* L. (order Diptera, family Muscidae), which he described in 1758, the "L." refers to Linnaeus. The other individual who shares this distinction is Johan Christian Fabricius (1745-1808), a Danish student of Linnaeus and the first individual dedicated to insect taxonomy.

Fabricius built on the foundation laid by Linnaeus in terms of naming insects and improving the classification of all animals (figure 2.22). He was devoted almost entirely to studying insects, which is evident in his lifetime total of describing 9,776 species from across the world. Fabricius also contributed to enhancing insect classification in general by using artificial and natural characteristics. Artificial features allowed for identifying an insect species, while natural characteristics were used to compare a particular species with other genera. He used mouthparts as the central feature for separating insects into orders (his wording was "classes"), which are large groupings of insects that generally share simi-

Figure 2.21. Portrait of Carolus Linnaeus (1707–1778), the Swedish naturalist who developed the binomial system of nomenclature for classifying plants and animals. Linnaeus identified about 2,000 insect species during his lifetime. Image available at http://bit.ly/1V9ay9G.

lar morphological characteristics, like types of wings, mouthparts, antennae, and body plan. Fabricius was one of the first individuals to attempt a worldwide classification of insects, but he soon realized that a single individual could not accomplish the feat. Today, entomologists generally specialize on a single group (order or even family) of insects.

The British naturalist and country priest William Kirby (1759–1850) was generally regarded as an entomologist by his fellow naturalists, including Fabricius. Kirby is often called the founder of entomology for his seminal work, *Introduction to Entomology*, coauthored with friend William Spence, a four-volume treatise published between 1815 and 1826, and for laying the groundwork for the study of insects. He helped found the Entomological

Society of London in 1833 and served as the president of the Ipswich Museum (1847–1850), an institution that Kirby worked on developing for more than forty years.

The North American counterpart to Kirby was Thomas Say (1787–1834). Say was a member of a group of naturalists that helped form the Academy of Natural Sciences in Philadelphia in 1812 (figure 2.23). He was the entomologist of the group and devoted himself to studying the natural history of insects (as well as molluscs) and classifying them. Say braved the untamed frontiers of the United States to collect insect specimens, eventually settling in New Harmony, Indiana, to begin his monumental work, *American Entomology, or Descriptions of Insects in North America*, a three-volume set completed in 1828. In all, Say described 1,400 insect species in North America, the vast majority of them beetles. His collective and tireless dedication to studying insects led to him being proclaimed the father of entomology in North America (often refined as "descriptive" entomology). Interestingly, Say referred to his mentor Frederick Valentine Melsheimer as the "Parent of American Entomology," largely based on his text *A Catalogue of Insects in Pennsylvania*, in which he described 1,363 species of insects collected in the keystone state.

The late nineteenth century was characterized by advances in general entomology as a whole, led by the excellent descriptive works of Jean-Henri Casimir Fabre (1823–1915) (figure 2.24). Fabre was a true renaissance man, trained as a physicist, chemist, botanist, teacher, and writer. Today he is probably best known for his remarkable writings, which meticulously describe the natural history, behavior, and morphology of a wide range of insects. Fabre maintained a dedication to scientific truth, reflected in his writings, but not with dry, detached prose, as might be expected; rather, his writing engaged the reader, despite the background or interest level for the subject, a true testament that passion and excitement can be associated with scientific writing. Over the course of his career, Fabre wrote a series of papers and texts on insects that collectively are referred to as *Souvenirs Entomologiques* (*Souvenirs of Insect Life*). In his works, Fabre provided observations on numerous insects and arachnids that were very influential for numerous scientists, including Charles Darwin. Darwin did little to impress Fabre, however, as Fabre was a skeptic

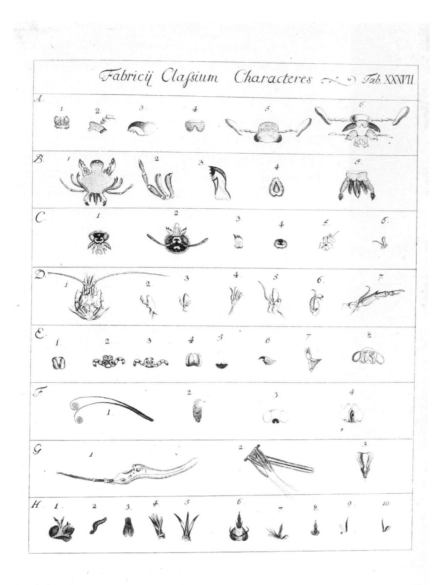

Figure 2.22. Illustration by Johan Christian Fabricius (1745-1808) depicting the anatomical features used as the basis for his classification system of insects. The classification scheme was presented in *Genera insectorum Linnaei et Fabricii iconibus illustrata* (1789) with Carl von Linné (Linnaeus), J. J. Römer, Johann Rudolf Schellenberg, and J. H. Sulzer as co-authors. Image available at http://bit.ly /1sBRhXM.

of Darwin's theory of evolution (as well as many other theories of the time).

Not surprisingly, Charles Darwin (1809-1882) was an influential character in the early history of entomology, but not necessarily for his theory of evolution (figure 2.25). Yes, evolution is today considered the foundation for all life sciences, but Darwin's ideas were mostly subject to scorn and ridicule during his lifetime. His *On the Origin of Species* (1859) was part of the foun-

dation that led to the theory of evolution becoming accepted as fact, but his views did not become a mainstay until around the 1930s. In addition to being an accomplished geologist and naturalist, he trained as a medical student for two years, and as a clergyman and a taxidermist, and was an "amateur" entomologist, specializing in beetles. Insects have more than fifty references in *Origin of Species*, and are used as examples to support his theories on evolution and sexual selection and to

Figure 2.23. Illustration of Thomas Say that appeared in *Popular Science Monthly* (volume 21) in 1882. Say (1787–1834) is considered the father of North American entomology. Image available at http://bit.ly/1qqAo07.

Figure 2.24. Jean-Henri Fabre (1823–1915) was not only an extraordinary naturalist, writer, teacher, chemist, and botanist, he was also able to recognize great talent as well. This illustration of the dung beetle *Scarabaeus sacer* was one of many beautiful plates created by E. J. Detmold for Fabre's books. Image available at http://bit.ly/1ssxl8Y.

make comparisons between insects observed in Brazil and his homeland of England.

Darwin also discussed the significance of insects as a whole to pollination of flowers and plants in his book *On the Various Contrivances by which British and Foreign Orchids Are Fertilized by Insects,* and used numerous insect examples in his outstanding text *The Descent of Man*. In more practical terms, Darwin described insect collecting, beetles specifically, as an activity that gave him some of his greatest pleasure while a student at Cambridge University. He even claims to have invented two novel methods for collecting beetles: scraping moss from trees during the winter and collecting debris from underneath barges. To his joy, Darwin was able to harvest some very rare beetle species using these methods.

Politicians at work: Two acts of Congress established entomology in North America

By the mid-nineteenth century, studying insects had become an acceptable (notice I refrain from saying "normal") practice. However, entomology was not recognized as an academic discipline. A few universities in Europe and North America established professorships in which a single individual was allowed to serve as the insect naturalist or insect curator for the institutions. But a formal education in entomology did not become available until sometime later. In North America, the formation of entomology into an academic discipline, and hence a scholarly pursuit at a college or university, is rooted in two acts of the US Congress: the Morrill Land Grant Act of 1862, and the United States Entomological Commission of 1876.

The Morrill Land Grant Act of 1862 was realized by the hard-fought efforts of Jonathan Baldwin Turner, a professor at Illinois College. In the original bill, Turner proposed setting aside 30,000 acres of land per senator and representative in each state for the construction of an agricultural college. The land would be sold to raise capital for the venture and placed in an endowment. Turner had envisioned that a college education would be made available for everyone, regardless of social class,

Figure 2.25. Charles Darwin as he appeared in 1869, long after his voyages on the HMS *Beagle*. Photo by J. Cameron available at http://bit.ly/1FdWgmy.

and that the newly formed universities would offer degrees in agriculture, mechanical arts, and home economics (hence the A&M in the title of several of the original land grant universities). The act was rejected by then President James Buchanan. However, the bill was tweaked to include a provision that each university would teach military tactics, a feature considered very important at the time since the nation was on the verge of southern secession and civil war. The supporters of the bill believed it was essential for the modified version of the bill to be introduced by a congressman from the east, since this region was more populous, thus the majority of senators and representatives represented this region, which would result in more land for their respective states. Consequently, representative Justin Smith Morrill of Vermont introduced the bill to Congress, which was subsequently signed by President Abraham Lincoln in 1862. Entomology was not necessarily an official academic program at any of these new colleges, at least not

initially, but the subject was taught for the first time in North America, raising awareness of insects. Of course, the focus of the teaching was on the fact that all insects were bad, since they inflicted disease and targeted agriculture.

Perhaps equally important to legitimizing entomology as a profession and academic discipline was the formation of the United States Entomological Commission in 1876. The commission came about because of increasing pressure on the US Congress to do something about the Rocky Mountain locust plague. From 1874 to 1876, the migratory locust (actually the Rocky Mountain locust, *Melanoplus spretus*) invaded a number of important grain-producing states—generally those states located west of the Mississippi River in the Great Plains (figure 2.26). Earlier in this chapter, and in chapter 1, we have noted the massive destruction that large swarms of locusts can inflict. Reports to Congress were that the locusts had reached such incredible densities that their dead bodies were blocking trains on the tracks. I am not sure if a formal study has ever been done to determine how many locusts are necessary to block a train, but the qualitative answer is, a lot! Prior to the formation of the commission, no federal entity was charged with the task of monitoring populations of destructive insects or developing plans of action to combat them. States were left to fend for themselves.

The commission itself was short term, lasting only three years. During that short window, however, commission members were extraordinarily successful in investigating numerous economically important insects, detailed in an array of reports and bulletins. The fact that the United States government formed the commission was a key element in establishing entomology, particularly applied entomology, as a bona fide profession and one that should be studied in an academic setting to protect the citizenry. Coincidently—and never mentioned as important to popularizing entomology during this time—was the rapid demise of the Rocky Mountain locust. Sightings of this species in the western part of United States during the 1870s suggested that *M. spretus* could reach densities that far surpassed the swarms of any other locust species, yet in less than thirty years it was considered extinct. The work of the entomology commission had nothing to do with the locust decline, but

Figure 2.26. Range of the Rocky Mountain locust as determined by the United States Entomological Commission in 1877. Image available at http://bit.ly/1V9byL0.

the public did not know that those entomologists were not responsible for saving them from future outbreaks.

Beyond the text

If the United States Entomological Commission was not responsible, what could have caused the rapid extinction of the Rocky Mountain locust?

War anyone? Insects have been the true victors of most wars

The origins of entomology in the ancient world have a common linkage to study of insects in modern times: war. What do insects have to do with war? Plenty! Insects have been instrumental in shaping the outcome of nearly every major military conflict prior to the 1940s. The most obvious influence has been insect-borne disease: the death toll of soldiers succumbing to disease vectored by insects far exceeds that of all deaths attributable to war-related injuries in every war prior to World War II. But the role of insects during human conflict has not been limited to disease transmission. Almost since the beginnings of recorded history, insects have been recruited into man's conquests by being forged into weapons and agents of torture. In the twentieth century, human ingenuity led to the full-fledged development of biological terrorism, with entomological terrorists serving as some of the most creative and destructive agents

of war. The influence of insects in war can be classified based on their specific roles as either intentional or opportunistic soldiers: insects as weapon of war or insects as opportunists.

Insects as weapons of war

Insects have been drafted into military campaigns for thousands of years. Arguably the oldest known description of insects used as weapons was discussed earlier: the Christian Bible states that God used three (possibly four) insect-related plagues to inflict pain and suffering on Egypt as a means to prompt the pharaoh (Ramesses II) to release the Israelites (Exodus 8:10) to Moses. This undoubtedly represents the first written record detailing the concept of biological, and certainly entomological, terrorism.

Ancient Egyptian, Mayan, and Aztec civilizations were known to use bees and wasps as bombs. Obviously they recognized that some insects, in this case stinging Hymenoptera (Hymenoptera is another example of an insect order), sting and produce toxins that inflict minimal pain but often are capable of inducing lasting effects, including death. Entire bee or wasp nests were placed in porcelain jars, dropped or thrown into entrenchments and fortifications, with the intent of driving the enemy out of strongholds. Such weapons had the disadvantage of their handlers not being able to control in which direction the insects traveled. Once bees or wasps were released, generally by smashing the bomb casing, they were in a foul mood and were driven to attack most any large vertebrate in their path, including the soldiers that designed the bombs.

Stinging and biting insects also provided opportunities in methods of interrogation or torture. Entomological interrogation techniques were ingenious yet diabolical. Examples include staking or burying individuals next to anthills, tying naked captives to trees in mosquito- and biting fly-infested areas, force-feeding milk and honey to prisoners to promote anal **myiasis**, and entrapping prisoners with blood-sucking bed bugs, or placing them into earthen pits filled with ticks and assassin bugs. In many instances, the mere threat of the insect was sufficient to extract the desired information.

Using insects to kill and torment has not been restricted to small-scale techniques like torturing prisoners. Six-legged soldiers have been fashioned into biological terrorism agents, threatening human populations through disease transmission, mimicking the effects of insect bombs, or targeting food production. Chapter 16 explores topics related to past and modern uses of insects in warfare, focusing on ento-terrorism as one the most potent and realistic new threats to civilizations.

Insects as opportunists

Generally speaking, insects did not start any of the wars between nations, civilizations, or peoples, but they have been the victors in most battles. The conditions that arise from conventional warfare are ideal for certain species of insects to capitalize on the misfortunes of war. Poor hygiene; bodies stressed from lack of food, insufficient sleep, and mental terror; living in close proximity, particularly during trench warfare; reusing clothing from injured or dead soldiers; and residing in new environs (e.g., tropics) were ideal avenues to attract and retain several types of blood-feeding insects. Insect-borne diseases such as typhus, yellow fever, malaria, and bubonic plague went hand in hand with the military conflicts and dictated the outcomes of many battles. If that statement appears to overreach the true significance of insects, let me assure you, it is definitely on the mark. Let us use one example to illustrate the point: typhus.

Typhus represents a cluster of closely related diseases induced by bacteria of the genus *Rickettsia*. Epidemic ty-

Figure 2.27. Male body louse, *Pediculus humanus* var. *corporis*. Body lice are parasites of humans that can be transmitted via clothing, bedding, or close contact. Photo by Janice Harney Carr (2006) available at http://bit.ly/1j82WbS.

phus is transmitted via body lice that easily migrate from one human host to another (figure 2.27). Living in tents, trenches, or makeshift barracks placed soldiers in close contact, and sharing clothing or bedding with infested individuals created access for lice movement. Without treatment, mortality occurs in 10%–60% of infected individuals. However, soldiers stressed and weakened by combat are at greater risk of death. In fact, the Emperor Napoleon Bonaparte found this out firsthand during the French invasion of Russia (or the Patriotic War of 1812).

Napoleon marched across the Neman River with his Grand Armée, nearly half a million strong, with the goal of defeating the Russian army (figure 2.28). The French forces were superior in every respect and, if engaged in a traditional ground battle, should have easily defeated the tsar's army within days. But things did not go as Na-

Figure 2.28. Napoleon Bonaparte, first consul and emperor following the French Revolution. Painting by Andrea Appiani (1805) available at http://bit.ly/1OPOQry.

poleon had planned. The Russian army retreated deeper and deeper into the country, burning buildings and food supplies as they traveled and pulling the Grand Armée farther into Russia. Napoleon's troops became cut off from their own supply routes, were ill prepared for the harsh winter conditions of northern Russia, and became exposed to lice harboring the bacterium *Rickettsia prowazekii*. Cold and starving French soldiers scavenged for food and clothing, frequently donning articles removed from the dead, thereby inoculating themselves with lice. After months of chasing the Russian army and with no clear victory in sight, Napoleon returned home with just remnants of the French army: only 27,000 fit soldiers remained; 380,000 had died, the majority from typhus, and another 100,000 were captured. Within months of their return to France, thousands more soldiers died from epidemic typhus. A tiny blood-feeding insect (lice) was capable of doing what no human army had done before:

decisively and completely defeat the mighty General and Emperor Napoleon Bonaparte.

The story of insects' involvement in war contains several important take-home messages. First, the high insect-induced death toll would have continued in subsequent wars if not for the discovery and development of new drugs (e.g., antimalarials) and insecticides (e.g., DDT, formally known as dichlorodiphenyltrichloroethane; see chapter 14), and deciphering the life histories of the insects involved. Second, disease transmission by insects during military campaigns still occurs, evident during the US occupation of Afghanistan (2001–2014), and the war (2003–2011) involving Iraq and the United States. Collectively, war has reinforced the image that insects are bad for the human condition. The message that is lost, however, is that insects were either involuntarily drafted into human conflict or simply took advantage of mass assemblages of soldiers. Had peace been maintained, the role of insects would have been much different.

Evolution of entomology: Insects as tools for biology, agriculture, and war

Entomology has evolved significantly by the twenty-first century. Public perception of insects is no longer centered exclusively on the negative aspects of insect-human interactions. Some insects are actually regarded as good, especially those that display obviously beneficial products or activities (figure 2.29). In fact, public concern has been expressed over the rapid and unexplained demise of honey bees (a common name that generally refers to one species, *Apis mellifera*) throughout North America and parts of Europe, and **colony collapse disorder*** has been the subject of major media coverage over the past few years. Entomology has become a standard academic discipline offered at most state universities (the land grant colleges and universities) in the United States, and

---

*Colony collapse disorder (CCD) is a recent condition associated with the European honey bee, *Apis mellifera*, in which workers leave the hive to forage but do not return. A few individuals always disappear from the hive, but with CCD, as many as one-third or more of all individuals in a hive are lost. The reasons for the disappearance are unknown, although much speculation has been made on a number of man-made influences. The demise of these honey bees has an enormous economic and ecological effect, as many flowering plants are pollinated by the European honey bee.

Figure 2.29. Adult honey bees (*Apis mellifera*) pollinating a dandelion flower. Photo by Guérin Nicolas available at http://bit.ly /1gLhxbr.

remains at the forefront in addressing issues related to agriculture, ecology, and insect-borne diseases. Perhaps what has not changed so much is a parent's reaction to learning that their beloved child has selected entomology as his or her course of study in college, over such choices as physician, biomedical researcher, lawyer, or mime (cue Pharrell Williams's "Happy" in the background. http:// www.youtube.com/watch?v=y6Sxv-sUYtM&feature=kp).

Profound leaps in our knowledge of insects have occurred over the last hundred years. With the advent of advanced molecular tools such as genomic mapping and next-generation sequencing, the interest in and use of insects is transforming, from being confined to entomology to investigators using insects as model organisms in an array of disciplines. It is fair to estimate that more non-entomologist scientists use insects as test subjects today than those trained in entomology. How can that be possible? For years, such insects as *Drosophila* (so-called fruit flies) and *Nasonia* (a parasitic wasp) have served as tools for genetic studies not based on insect-specific questions (figure 2.30). Species from both genera, as well as others, have had their entire genomes sequenced, opening avenues of investigation for those interested in development, cell regulation, neurobiology, reproduction, human disease, and many other areas of the life sci-

Figure 2.30. The entire genome of *Drosophila melanogaster* has been sequenced, elevating the fruit fly to even greater utility as a model organism. Photo by Sanjay Acharya available at http:// bit.ly/1OD3HIp.

ences. Several insects have served as investigative leads in pharmaceutical studies aimed at discovery of new agents to treat cancer, suppress microorganisms, or as anti-inflammatory compounds. Many species synthesize insecticidal chemicals used in their own defense or to aid reproduction, so research is being conducted to determine if those same compounds can be used in crop, livestock, or human protection from harmful insects. Efforts are even afoot to capitalize on the behaviors of certain species for use in national security, such as using flying insects—or just their cells—to detect bombs, or equipping bees with recording devices for surveillance and spying (see chapter 16 for more details). Insects are even aiding law enforcement by revealing information relevant to time-of-death determinations, presence of narcotics in a corpse, or detecting gunshot residues not visible to the naked eye on bodies. The last example serves as the basis for the subfield forensic entomology (the subject of chapter 15).

Perhaps most satisfying is a simpler evolutionary change: the goal of entomology is no longer to kill or eradicate insects. Of course, some entomologists still have the primary task of reducing the effects of insects on the human condition. They do so, however, with an arsenal of tools that permits managing rather than attempting to control insect populations (a subject addressed in chapter 14). Humans are beginning to accept coexistence with insects. The new attitude makes sense, as no organism has ever defeated the class Insecta. Maybe it is time to spend the next millennium taking a few history lessons from them: world domination has only successfully been achieved by one group of animals, and all its members have six legs.

CHAPTER REVIEW

❋ **Insects: Unwelcome guests since the beginning of the human "party"**

- Many insects have triggered in humans the desire to kill, probably since the very onset of the human-entomological association. The relationship was formed by insects, uninvited, to take advantage of mankind. Humans formed dwellings in convenient locales; convenient, that is, for the insects to find and use them. As humans began their foray into agriculture, insects were there, eager to partake of the harvest before the laborers interfered. Humans forming organized communities offered new opportunities for insects: concentrated food sources that were nutrient-rich. Because methods of insect control were largely unknown to most peoples, stored foods were left unprotected, and hygienic conditions were awful, the advent of human

civilization was a golden time for many insect species, particularly blood-feeding varieties, to become unwelcome neighbors.

- Given humans' long association with insects, entomology should be an old discipline, but the reality is much different. Recognizing insects or their activity is not the same as establishing a discipline or field of study. Humans have acknowledged insects in their lives since the beginnings of recorded history. But the study of insects is not nearly as old, at least not from the perspective of an organized or systematic examination of insect biology. Scientific or academic study of insects, and hence the true origins of entomology, likely did not begin until around the sixteenth century.

- Ancient peoples' interest in insects transcended disease and death. For example, certain ethnic assemblages or civilizations revered if not worshipped insects. Beetles particularly resonated with Egyptians from the period known as predynastic or prehistoric Egypt to the onset of New Kingdom Egypt. Insects were intimately associated with the gods, mythology, and death throughout the ancient empires of the Maya and Aztecs of Central and South America. Butterflies, rather than the beetles of Egypt, dominated the insect connection for the people of ancient Mesoamerica. Insect recognition in ancient Asia was associated with insects' contributions to aesthetics and culture rather than worship or fear.

- Prior to the sixteenth century, there is no evidence that insects were studied in a manner akin to what we would call today a scientific study. The works of Swammerdam and Malpighi were instrumental in establishing scientific study of insects; both were highly acclaimed for other academic achievements and thus brought credibility to using insects as test subjects.

❋ **Insect plagues and deities in ancient and modern civilizations**

- Early in human history, insect-borne diseases reached pandemic levels, yet no one had any idea what caused the illnesses. The most accepted explanations stemmed from religious beliefs, in which failing to follow the decrees of the creator or somehow displeasing one or more gods in general brought on illnesses. Severe deity disappointment resulted in plagues, although, as we will see, not all the insect plagues were necessarily associated with disease. The best-known insect plagues are directly tied to the best-known deity: In the Christian Bible, as well as in the Quran, the text reveals that God or Yahweh unleashed ten plagues on Egypt to convince the pharaoh to release the Israelites from the bondage of slavery. Three of those reported plagues involved insects as the causative agents to instill pain and suffering.

- In the case of bubonic plague, fleas harboring the causative agent of the disease (the bacterium *Yersinia pestis*) fed on black rats. Eventually the rats died. The fleas sought out new hosts, easy to find since the rats lived in close association with humans. The fate of the infected was the same for humans as for rats: death typically occurred in less than a week. Bubonic plague swept through parts of Europe on at least three occasions, leaving a wake of death.

- The insect linkage to deities has not always been tied to plagues or destruction. Ancient Egyptians, Chinese, Merovingians, Mayans, Aztecs, and countless other civilizations have associated insects with deities or, at least, nobility. Of course the question "why" surfaces—as with the displeasing-the-gods explanations for the cause of diseases mentioned earlier, people needed reasonable explanations for the unexplainable. So the phenomenon of a creature that could fly when man cannot, or transform from an ugly being into a thing of beauty, or a tree that could seemingly giving life or rebirth to a beetle found in its core, could have only one possible explanation: supernatural beings were involved.

✦ **Naturalists, physicians, and the clergy: An intriguing new pinup calendar or prominent figures in entomology?**

- The Italian physician and naturalist Francesco Redi (1626-1697) used aspects of the scientific method to test the validity of the theory of spontaneous generation. In 1667, Redi designed a simple yet ingenious series of experiments to test the theory that fly maggots spontaneously formed on meat.

- The eighteenth century gave rise to the Swedish naturalist Carolus (Carl) Linnaeus (1707-1778). Linnaeus developed the binomial system of nomenclature for classifying plants and animals that unified and organized botanical and zoological specimens, which greatly simplified the process of naming organisms. He was also an avid insect collector, identifying over 2,000 species in his lifetime.

- The British naturalist and country priest William Kirby (1759-1850) was generally regarded as an entomologist by his fellow naturalists. Kirby is often called the founder of entomology for his seminal work, *Introduction to Entomology*, coauthored with friend William Spence, a four-volume treatise published between 1815 and 1826, and for laying the groundwork for the study of insects.

- The North American counterpart to Kirby was Thomas Say (1787-1834). Say was a member of a group of naturalists that helped form the Academy of Natural Sciences in Philadelphia in 1812. He devoted himself to studying the natural history of insects (as well as molluscs) and classifying them. Say braved the untamed frontiers of the United States to collect insect specimens, eventually settling in New Harmony, Indiana, to begin his monumental work, *American Entomology, or Descriptions of Insects in North America*, a three-volume set completed in 1828. In all, Say described 1,400 insect species in North America, the vast majority of them beetles.

- The late nineteenth century was characterized by advances in general entomology as a whole, led by the excellent descriptive works of Jean Henri Casimir Fabre (1823-1915). He is probably best known for his remarkable writings, which meticulously describe the natural history, behavior, and morphology of a wide range of insects. Over the course of his career, Fabre wrote a series of papers and texts on insects that collectively are referred to as *Souvenirs Entomologiques*.

- Charles Darwin (1809-1882) was an influential character in the early history of entomology, but not necessarily for his theory of evolution. In addition to being an accomplished geologist and naturalist, he trained as a medical student for two years, and as a clergyman and a taxidermist, and was an "amateur" entomologist, specializing in beetles. Insects have more than fifty references in *Origin of Species*, and are used as examples to support his theories on evolution and sexual selection and to make comparisons between insects observed in Brazil and his homeland of England.

Politicians at work: Two acts of Congress established entomology in North America

- By the mid-nineteenth century, studying insects had become an acceptable practice. However, entomology was not recognized as an academic discipline. In North America, the formation of entomology into an academic discipline, and hence a scholarly pursuit at a college or university, is rooted in two acts of the US Congress: the Morrill Land Grant Act of 1862 and the United States Entomological Commission of 1876.

- The Morrill Land Grant Act proposed setting aside 30,000 acres of land per senator and representative in each state for the construction of an agricultural college. Entomology was not necessarily an official academic program at any of these new colleges, at least not initially, but the subject was taught for the first time in North America, raising awareness of insects.

- Equally important to legitimizing entomology as a profession and academic discipline was the formation of the United States Entomological Commission in 1876. The commission came about because of increasing pressure on the US Congress to do something about the Rocky Mountain locust plague (1874-1876). Prior to the formation of the commission, no federal entity was charged with the task of monitoring populations of destructive insects or developing plans of action to combat them. States were left to fend for themselves. The commission itself was short term, lasting only three years.

War anyone? Insects have been the true victors of most wars

- What do insects have to do with war? Plenty! Insects have been instrumental in shaping the outcome of every major military conflict prior to the 1940s. The most obvious influence has been insect-borne disease: the death toll of soldiers succumbing to disease vectored by insects far exceeds that of all deaths attributable to war-related injuries in every war prior to World War II. But the role of insects during human conflict has not been limited to disease transmission. Almost since the beginnings of recorded history, insects have been recruited into humans' conquests by being forged into weapons and agents of torture.

- Ancient Egyptian, Mayan, and Aztec civilizations were known to use bees and wasps as bombs. Obviously they recognized that some insects, in this case stinging Hymenoptera, sting and produce toxins that inflict minimal pain but often are capable of inducing lasting effects, including death.

- Stinging and biting insects also provided opportunities in methods of interrogation or torture. Examples include staking or burying individuals next to anthills, tying naked captives to trees in mosquito- and biting fly-infested areas, force-feeding milk and honey to prisoners to promote anal myiasis, and entrapping with blood-sucking bed bugs, or placing them into earthen pits filled with ticks and assassin bugs.

- Insect-borne diseases such as typhus, yellow fever, malaria, and bubonic plague went hand in hand with the military conflicts and dictated the outcomes of many battles.

- War has reinforced the image that insects are bad for the human condition. What is lost, however, is that insects were either involuntarily drafted into human conflict or simply took advantage of mass assemblages of soldiers.

✤ **Evolution of entomology: Insects as tools for biology, agriculture, and war**

■ Entomology has evolved significantly by the twenty-first century. Public perception of insects is no longer centered exclusively on the negative aspects of insect-human interactions. Some insects are actually regarded as good, especially those that display obviously beneficial products or activities.

■ Entomology has become a standard academic discipline offered at most state universities in the United States, and remains at the forefront in addressing issues related to agriculture and insect-borne diseases.

■ Profound leaps in our knowledge of insects have occurred over the last hundred years. With the advent of advanced molecular tools such as genomic mapping and next-generation sequencing, the interest in and use of insects is transforming, from being confined to entomology to investigators using insects as model organisms in an array of disciplines. For years, such insects as *Drosophila* and *Nasonia* have served as tools for genetic studies not based on insect-specific questions. Species from both genera, as well as others, have had their entire genomes sequenced, opening avenues of investigation for those interested in development, cell regulation, neurobiology, reproduction, human disease, and many other areas of the life sciences.

MUSHROOM FARMING (SELF-TEST)

Level 1: Knowledge/Comprehension

1. Define the following terms:
   (a) synanthropy   (d) binomial system of nomenclature
   (b) Khepri   (e) myiasis
   (c) sericulture   (f) Black Death

2. Explain the significance of Carolus Linnaeus's new binomial system of nomenclature in promoting entomology.

3. Identify two major insect-borne diseases that have shaped the outcomes of military battles.

4. Describe the connection between dung beetles and the Egyptian deity Khepri.

Level 2: Application/Analysis

1. Prior to World War II, more soldiers suffered from insect-transmitted disease than actual injuries incurred from battle. Explain what changed during this period that led to the decreased influence of insects during war.

2. Discuss the significance of the Morrill Land Grant Act of 1862 on establishing entomology as a bona fide discipline in North America.

3. How has entomology changed in the twenty-first century in terms of who studies insects and why?

Level 3: Synthesis/Evaluation

1. Explain who has had a greater influence on public acknowledgement of insects: Charles Darwin, the Christian god Yahweh, Carolus Linnaeus.

2. Speculate on whether entomology as a discipline would have evolved at the same pace and with similar importance in the absence of World Wars I and II.

REFERENCES

Beutelspacher, C. R. 1988. Las mariposas entre los antiguos Mexicanos. Fondo de Cultura Economica, Avenida de la Universidad, Mexico.

Bishop, F. 1913. The stable fly (*Stomoxys calcitrans* L.): An important livestock pest. *Journal of Economic Entomology* 6:112-126.

Bowsky, W. 1971. The black death: A turning point in history. Holt, Rinehart, and Winston, New York, NY.

Bunson, M. 2000. The Vampire Encyclopedia. 2nd ed. Gramercy Press, New York, NY.

Capinera, J. L. 1993. Insects in art and religion: The American Southwest. *American Entomologist* 39(4): 221-229.

Carpenter, M. M. 1953. Bibliography of biographies of entomologists (supplement) *American Midland Naturalist* 50(2): 257-348.

Darwin, C., and N. Barlow. 1993. The Autobiography of Charles Darwin: 1809-1882. Reissue edition. W. W. Norton, New York, NY.

Gigal Research. Archaeological and historical researches. http://gigalresearch.com/uk/publications-pharaohs.php. Accessed April 27, 2014.

Hogue, C. 1987. Cultural entomology. *Annual Review of Entomology* 32:181-199.

Huchet, J.-B., and B. Greenberg. 2010. Flies, mochicas and burial practices: A case study from Huaca de la Luna, Peru. *Journal of Archaeological Science* 37:2846-2856.

Kritsky, G., and R. Cherry. 2000. Insect Mythology. Writers Club Press, Lincoln, NE.

Lockwood, J. A. 2004. Locust: The Devastating Rise and Mysterious Disappearance of the Insect that Shaped the American Frontier. Basic Books, New York, NY.

Lockwood, J. A. 2009. Six-Legged Soldiers: Using Insects as Weapons of War. Oxford University Press, New York, NY.

McNeill, W. H. 1977. Plagues and Peoples. Anchor Books, New York, NY.

Panagiotaopulu, E. 2004. Dipterous remains and archaeological interpretation. *Journal of Archaeological Science* 31:1675-1684.

Ponce-Ulloa, H. 1997. Butterflies of Ancient Mexico. *Cultural Entomology Digest* 4. http://www.insects.org.

Remington, J. E., and C. L. Remington. 1961. Darwin's contributions to entomology. *Annual Review of Entomology* 6:1-12.

Ross, H. H., C. A. Ross, and J. R. Ross. 1982. A Textbook of Entomology. 4th ed. John Wiley and Sons, New York, NY.

Tuxen, S. L. 1967. The entomologist J. C. Fabricius. *Annual Review of Entomology* 12:1-15.

THE ENTOMOLOGIST BOOKSHELF (SUPPLEMENTAL READINGS)

Cushing, E. C. 1957. The History of Entomology in World War II. Smithsonian Institution, Washington, DC.

Fabre, J. H. 1998. Fabre's Book of Insects. Dover Publications, Mineola, NY.

Gullan, P. J., and P. S. Cranston. 2010. The Insects: An Outline of Entomology. Wiley-Blackwell, West Sussex, UK.

Johnson, N. F., and C. A. Triplehorn. 2004. Borror and DeLong's Introduction to the Study of Insects. 7th ed. Cengage Learning, New York, NY.

Kelly, J. 2006. The Great Mortality: An Intimate History of the Black Death, the Most Devastating Plague of All Time. Harper Perennial, New York, NY.

Rasnitsyn, A. P., and D. L. Quicke. 2010. History of Insects. Springer, Berlin, Germany.

Stewart, A. 2011. Wicked Bugs: The Louse that Conquered Napoleon's Army and Other Diabolical Insects. Algonquin Books, Chapel Hill, NC.

Stroud, P. T. 1992. Thomas Say: New World Naturalist. University of Philadelphia Press, Philadelphia, PA.

Whitfield, J. B., and A. H. Purcell III. 2012. Daly and Doyen's Introduction to Insect Biology and Diversity. 3rd ed. Oxford University Press, Cambridge, MA.

ADDITIONAL RESOURCES

Darwin's beetles
http://friendsofdarwin.com/2009/07/20090718/

History of malaria during wars
http://www.malariasite.com/malaria/history_wars.htm

Insects and the US Civil War
http://entomology.montana.edu/historybug/civilwar2/civilwar.htm

Insects of ancient Egypt
http://www.reshafim.org.il/ad/egypt/bestiary/insects.htm

Ipswich Museum
http://www.cimuseums.org.uk/

Linnaeus Museum and home
http://www.linnaeus.se/eng/link3.html

Thomas Say
http://faculty.evansville.edu/ck6/bstud/nh.html

Yellow fever and the Mexican-American War
http://entomology.montana.edu/historybug/mexwar/mexwar.htm

3 Insects Are Not All Bad

Beneficial Aspects of Insect–Human Interactions

Humanity would probably not survive if all or only certain critically important insects were to disappear from the earth.

Dr. Gilbert Waldbauer
Millions of Monarchs, Bunches of Beetles (2000)

People can be cruel. Take human attitudes toward insects. Most individuals hold the view that all insects are bad, making this claim because some destroy our food, inflict pain, and transmit agents of disease. In reality, very few actually pose a threat to the human condition. In fact, many more species convey benefits to mankind rather than cause maladies. This statement is no doubt surprising, if not baffling, because the average person really has no idea of what insects actually do, short of stinging, biting, and annoying. But this lack of understanding has not stopped the masses from maintaining a negative bias toward an entire class of animals, the much-maligned Insecta. Can we go so far as to say that insects are our friends? Yes. Well, a definite maybe.

Chapter 3 will provide the information for you to decide by exploring the many ways that insects benefit the human condition, focusing on the products and activities derived from hexapods that positively affect human societies. Admittedly, some of the benefits are subjective. For example, not everyone enjoys bee regurgitate spread on toast. Many benefits, however, are not debatable: insects are critical to sustaining nearly every ecosystem on the planet. Other topics fall under the category of appreciating insects, in which the examples range from cultural influences on mankind (e.g., insects as symbols and in art) to purely entertainment, such as entomological stars of movies, television, and music. Still other areas of insect influence include civil and criminal law, in which beasts with six legs are the subject of litigation or, in other instances, aid law enforcement. Oh yes, we will also examine entomophagy, the idea of incorporating insects into your diet. From that vantage point, the question of friendship has already been answered, since most of us do not *ordinarily* eat our friends.

Key Concepts
- Insects as our "friends"
- From bee puke to scale poop: Useful insect products
- Oh the wonderful things that insects do!
- Insects got class: Cultural influences of six-legged creatures
- A religious experience
- Laws, litigation, and insects as evidence

Figure 3.1. Swarm of locusts near Satrokala, Madagascar, en route to strip the vegetation clear in its path. Photo by Iwoelbern available at http://bit.ly/1Ouz46r.

Insects as our "friends"

"If insects are my friends, I don't need any enemies!" A statement expressed (shouted, actually) in my household on more than one occasion. Yet I stand by the premise that insects are closer to being our friend than foe. "Are you insane?" (also frequently overheard emanating from my home). There is ample evidence to conclude yes! Nonetheless, I remain steadfast in touting the indispensable influence that insects have on the human lives (a claim I can't always make for some humans). Granted, defining insects as beneficial is inordinately more challenging than calling them destructive. Why? To nearly anyone, it is clear when an insect is more than just a mere annoyance. Destroying crops or other aspects of the human food supply can be measured tangibly, as can the toll of insect-borne disease or the effects of wood-boring insects, aspects that will be examined in detail in chapter four (figure 3.1).

However, when the task is to convince others who have little understanding of insects that insects are indeed useful, or to persuade those who hold the position that insects have no utility to life on Earth, period, that hexapods are absolutely necessary, well the word "challenging" is an understatement. The main reason for the difficulty is that many of the positive influences are subjective. Consequently, not everyone will agree that a particular insect-derived product or activity is necessarily

beneficial, and even if they did, there generally is no means to quantify the benefits reaped by humans. A case in point is the influences of insects on cultural aspects of the human condition. Insects have appeared in art, music, literature, and movies, and though some humans may find these cultural devices pleasing, none are technically necessary to human existence. Are they beneficial? Your response is neither correct nor incorrect. Art, music, and the like are almost entirely subjective and dependent on individual tastes. So while you and I may find Van Gogh's painting *Death's Head Moth* stunning (figure 3.2), Napoleon Bonaparte's symbolic use of bees fascinating (figure 3.3), and the music of Papa Roach* heart-pounding (figure 3.4), none are needed to maintain our existence or even influence our individual **homeostasis**.

Speaking of **gustation**[†] (i.e., taste), the idea of eating anything emitted from an insect's body is not a turn-on

---

*Papa Roach, a hard rock band out of Vacaville, California, has sold over twenty million album copies and is known for such songs as Last Resort, Between Angels and Insects, and Hollywood Whores.

[†]What a clever segue! Gustation is the act of tasting, a process involving sensory neurons in the form of taste receptors. For humans, taste receptors are concentrated on the tongue. So if eating an insect, the flavors are detected once placed in the mouth and usually during the act of swallowing. The insect, however, started using taste receptors when its "feet" (**pulvilli** or footpads) made contact with your hand and continued with similar sensory neurons located on other parts of the body, including the various structures constituting the mouth.

Figure 3.2. Vincent Van Gogh's painting of a death's head moth, *Giant Peacock Moth* (1889). Photograph by RAE9hGgEou6OhQ available at http://bit.ly/1MIVFMh.

Figure 3.3. Throne (1804) for Napoleon Bonaparte, displaying a bee in each corner of the backing embroidery. Honey bees were the imperial emblem during Napoleon's reign as emperor of France. Photo by Starus available in public domain at http://bit.ly/1PsU6iY.

for all and certainly cannot be considered essential. However, many people have and still do use a range of insect-derived goods, including for medicinal purposes. Here we will examine tempting treats generated from honey bees (honey, bee pollen, royal jelly) and other useful insect secretions and explore insects from the vantage point of drug discovery and as sources of antibiotics and insecticides, the latter staying true to the history of entomology (see chapter 2), in which insect death is still a driving force for the discipline. Later in the chapter, beneficial and essential insect activities will be considered through the lens of life on Earth as a whole.

Aspects of our lives that have been influenced by insects but really do not fall under the headings of beneficial or essential are broadly defined as "appreciating" insects: we will examine cultural entomology (insect influences on art, literature, music), insects that entertain

us (through television, movies, and video games), and insects that shape the law. Interestingly, in the examples of cultural influences, humans have actually sought out insects, a reversal from what was discussed in chapter 2 and what is to come in chapter 4.

From bee puke to scale poop: Useful insect products

It is time to acknowledge that you like bee puke! Or perhaps the idea of consuming materials that previously had been sucked into the mouth of a bee, carried in its gut, and regurgitated into a communal pot (honeycomb) is too gross for you? I feel your vibe; honey is apparently too disgusting for human consumption. Of course, if you consider the path that other common human foods travel to reach your table, those manufactured by the insect "food industry" do not seem too bad, or at least, not out

Figure 3.4. The hard rock band Papa Roach in concert. Photo by Ciell available at http://bit.ly/1KzRWtM.

of the ordinary (figure 3.5). Just close your eyes next time you're confronted with a jar of mouth secretions, take a bite, and let the deliciously thick sugary treat slide down your throat (figure 3.6). After all, thousands of bees have already done it before you. (Maybe that image didn't help?)

Regardless of your position on eating insect produce, a wide range of insect products have been used by humans from ancient civilizations to modern man. Only a small number of these goods are food-based. Many are tied more directly to household uses, and an increasing number are being developed for use in allopathic and homeopathic* medicinal applications. We will categorize useful insect products by uses rather than by insect group or species.

1. Food-based products
2. Uses in clothing (textile) manufacturing
3. Household uses
4. Medicinal uses

As there are hundreds of potential examples, this section will simply whet the appetite by drawing attention to some of the most common and/or interesting products.

---

*Allopathic or mainstream medicine uses pharmaceutical agents (i.e., drugs) and physical interventions to treat symptoms or physiological disorders of the human body. In contrast, homeopathic medicine relies on treatments alternative to mainstream medicine, many of which are considered controversial or nothing more than placebos.

Food-based products
BEE-BASED FOOD

It should come as no surprise to learn that honey bees are the source of the tastiest examples of insect-derived products eaten by humans. In North America, "honey bees" refers specifically to members of the genus *Apis*; the European honey bee, *Apis mellifera*, is the species most recognized and important to the honey industry. Later we will also examine several products derived from bees or their hives that have a variety of uses in our households. Here, three foods or ingredients from honey-producing bees are considered.

1. Honey. This is the standard to which all other insect-derived foods and products are compared. Honey is made through the modification of plant nectar† via regurgitation from forager bees' mouths and the evaporation of water; the evaporation typically occurs on the bee tongue (better known as the **hypoglossa**) (figure 3.7). The processes concentrate the sugars fructose and glucose, yielding an overall sweetness that approximates granulated sugar. The exact composition of honey varies by location and depends on the type of bees and the flowers used for nectar. Consequently, the color, texture, and taste of honey also varies (figure 3.8). Its high sugar concentration (65%–70%) causes honey to typically be viscous and dense.

At one time, honey was the principal sugar source for most of the world, bee and human alike. However, the discovery of sugar in cane (approximately 2,500 years ago) and in beets (at the turn of the nineteenth century) led to honey's decreased importance. That said, the sugary secretion is used today in a multitude of food products. Honey is consumed as a sweetener for food and drinks, as a flavor additive or substrate for production of some alcoholic beverages, and is commonly used as a spread for bread-based foods (e.g., toast, biscuits, scones). In reality, honey is added to and spread on almost any type of food consumed by people. More than one million tons of honey is produced worldwide annually. It is popular in the United States as well as in Europe: as domesticated bees in the United States produced 38.2 million

---

†Plant nectar is a sugar-rich liquid produced by flowering plants to attract (and reward) pollinating insects or other insects that aid the plant in some way.

Figure 3.5. Head cheese makes most insect products seem scrumptious. Photo by Rainer Zenz available at http://bit.ly/1QylzjD.

pounds of honey in 2013, valued at 212.1 cents/pound, which equates to $81 million. That is some valuable bee puke (table 3.1).

2. Royal jelly. As is honey, royal jelly is regurgitated from honey bees and used as a source of food within the hive. Normally, royal jelly is fed to developing immatures (larvae) and plays a role in determining position within the caste system of the hive: all larvae are fed the substance for three days after egg hatch, after which time, only females destined to become queens continue to receive meals of the mandibular gland secretions. Nutritionally, royal jelly is mostly water (~60%–70%), with about equal percentages of protein and monosaccharides (10% and 15%, respectively), 3%–6% lipid in the form of fatty acids, and trace amounts of vitamins and minerals. The exact composition varies, for the same reasons that honey does: variations in geography and types of bees and flowers. Royal jelly is consumed by people throughout the world, but probably more for its supposed health benefits than its taste. The milky secretion has been purported to remedy ulcers, asthma, liver disease, high cholesterol, and even premenstrual syndrome (PMS). Some applications call for applying royal jelly directly to

the skin. In the 1959 cult movie classic *The Wasp Woman* (discussed later in this chapter), the same skin treatment was prescribed to restore youthfulness. Unfortunately, the test subject turned into a wasp and went on a killing rampage. So the take-home message is be leery of the potential side effects of royal jelly.

3. Bee pollen. Bee pollen, or ambrosia, is derived from a bee colony but, technically, is not made by the bees themselves. Rather, it represents a mass of plant pollen that has been packed by worker bees into granules using honey or nectar. For hive-forming honey bees, the pollen mass represents the primary source of protein for the workers. For ground-nesting bees, the granules are typically placed into brood cells. Upon complete assembly of the bee pollen mass, a female lays an egg on top of but within the brood cell, and then seals the cell. In terms of human consumption, bee pollen adds flavor to shakes and fruit drinks, particularly those sold as healthier choices. Most human intake of bee pollen relates to claims that the substance is incredibly nutritious, high in protein and amino acids, and can treat a number of ailments. There is no empirical scientific evidence to support any of the health claims that have been made for bee

Figure 3.6. Honey: the delicious, sugary treat produced from honey bee regurgitation. Photo by Scott Bauer, USDA-ARS, available at http://bit.ly/1iwD88w.

pollen, and taste is almost never advertised, suggesting that the focus should remain on the purported nutritional benefits and not on pollen as a source for gustatory euphoria (figure 3.9).

HONEYDEW

Some websites declare that bees (meaning honey bees) are the only insects that produce food humans eat. Shockingly, these sites are maintained by retailers of all things honey bee–related. It is true that when most people are asked to name a product made by insects, the first (and usually only one) mentioned is honey, and honey bees as the source. But bees are not the only insects that produce sugary treats consumed by humans and other animals.

Several types of insects (plant hoppers, butterflies, mealybugs, aphids, and others) secrete honeydew as a sugary substance shared with other animals. Honeydew is a sugar-rich substance derived from plants in much the same way that honey bees collect nectar. A few notable differences are that sugar-laden fluids can be extracted from multiple locations on or in the plant (remember that flowers are the principal source for bees), including sucking them directly from plant tissues (phloem) and that, in most cases, honeydew is not regurgitated. Good—so I'll stop referencing bee puke. Honeydew is **excreta** from most of these insects. Which is? Kind of like a combination of urine and feces. Oh delightful.

As with honey, the exact composition of honeydew varies based on location, plant source, and insect, and it shares the similarity of being predominantly made of monosaccharides (~80%), though not quite as sweet. Honeydew is viscous and sticky, the latter quality used by some insects as a means for defense: when applied to the mouthparts of a would-be attacker, the secretion can act like glue to seal the mouth shut. In many instances, honeydew is used as a tribute or payment to other species that offer protection for otherwise defenseless insects (so insects also originated the concept of La Cosa Nostra, a.k.a. the Mafia!). In terms of human consumption, honeydew is more a food of mythology and past civilizations. For example, some biblical scholars argue that in the biblical Book of Exodus, the Israelites ate honeydew as they wandered the desert in search of the Promised Land. In Greek mythology, the supreme god Zeus was nursed as an infant with honeydew, and in Norse folklore, bees consumed "dew" (honeydew) that fell from ash trees. Interestingly, some honey bees do in fact use honeydew produced by other insects as a substitute for nectar. The resulting honey tends to have a rich amber color and is generally less sweet than honey derived from plant nectar. Most importantly, this darker honey is prized for its medicinal value, particularly in parts of Europe and Asia, though there is no evidence that honeydew-based honey has any health benefits at all.

FOOD DYES

Some insects contribute to the human diet in a nonnutritive way: they produce a compound used to make red

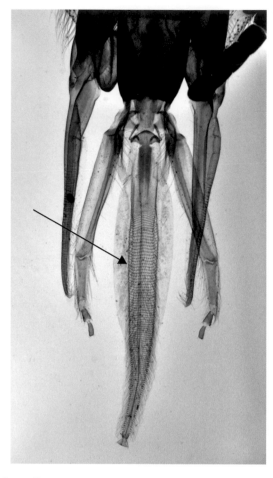

Figure 3.7. The tongue, or hypoglossa, of a honey bee is used for collecting nectar. Photos by Dominique Diaz (http://bit.ly /1ORGmAi) and dw_ross (http://bit.ly/1iM3xjF).

Figure 3.8. Honey can have different color, texture, and taste based on the flowers nectar is collected from. Photo by Andreas Praefcke available at http://bit.ly/1iM3LqN.

or crimson food coloring. This honor falls to a group of small, parasitic scale insects belonging to the genus *Dactylopius* (order Hemiptera or Heteroptera*). These insects feed on cactus pads, using their piercing mouthparts to suck nutrients from the plant (figure 3.10). Adults are dimorphic, that is, males have wings and possess long cau-

---

*We have not yet spent much time discussing taxonomic details of insects, such as classification, but here lies one of the controversies that can sometimes occur. The group of which scale insects are members is called an order, and, as can be seen in the text, two alternative names are given. Why? In short, because the experts who study these insects cannot agree whether to place them in the order formerly known for "bugs" (Hemiptera) or to create a new order name (Heteroptera) that combines bugs with hoppers, scale insects, and cicadas. Chapter 7 will explore insect classification in more detail, so that the name "order" makes more sense.

dal filaments (the appearance of tails), and females are wingless, remaining on the cactus host until death. This difference in biology has made it much easier to harvest the sex of interest, which is the females. Females produce

Table 3.1. Global natural honey production in 2012. Values are given in metric tons, and the countries listed are the top producers as reported by the United Nations Food and Agricultural Organization (http://faostat.fao.org/site/339/default.aspx.).

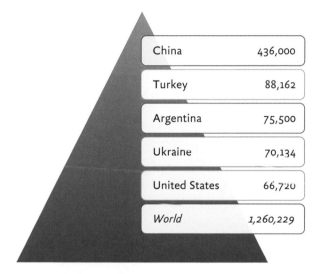

China	436,000
Turkey	88,162
Argentina	75,500
Ukraine	70,134
United States	66,720
World	1,260,229

carminic acid as a defense compound to ward off predators. They are quite good at producing the compound, as it typically represents up to 25% of the female's dry weight. Ancient Maya and Aztec civilizations discovered that the acid could be extracted from female scales insects rather easily, and when mixed with either aluminum or calcium salt, a deep red dye was produced. At that time, and through the nineteenth century, the dye, known as carmine dye or cochineal, was used to color textiles, pottery, artwork, and even tattoos. Today it is used mostly as a red coloring in food and cosmetic products, particularly lipstick. Cochineal has seen a recent resurgence as a dye used in fruit drinks, yogurt, and ice cream, largely over concerns that other types of dyes contribute to food allergies.

Quick check
What types of insect products do humans consume?

Uses in clothing (textile) manufacturing

The entomological influence on the production of clothing and other textiles dates back thousands of years to

Figure 3.9. Bee pollen, or ambrosia, in a honey bee colony. Photo by Cabajar available at http://bit.ly/1KzToME.

Figure 3.10. Scale insects from the genus *Dactylopius* in their white coccoons on a cactus plant. Red food coloring is derived from carminic acid that is produced as a defense compound by these insects. Photo by Leyo available at http://bit.ly/1V9pFFF.

the ancient civilizations of the Chinese, Aztec, and Maya. Insect-derived dyes were used for staining a range of fabric types, and protein fibers from a moth led to the creation of the most luxuriant fabric known to man, silk. In both instances, synthetic versions have almost entirely replaced the labor-intensive practices of insect farming. Nonetheless, natural insect dyes and silk are still produced (and still expensive), because there really is no substitute for the genuine article.

FABRIC DYES

Dyes derived from insects have properties for staining animal fibers that could not be matched, particularly in generating deep and intense crimson, red, and purple colors on clothing and rugs. One type of dye, carmine (the same one discussed above, as red food coloring, from scale insects in the genus *Dactylopius*), which is native to the Americas but with some species restricted to parts of Central and South America. Consequently, for centuries only the Maya and Aztecs held the secrets of when, where and how to harvest the dye. Spanish colonization of the New World brought an end to the secrecy. Within a short period of time, it was discovered

that carminic acid dyes were superior to other types in yielding intense reds and purples in textiles produced from animal fibers (as opposed to those from plants like cotton). Scale insect dyes became the preferred choice of monarchies, nobles, and the clergy throughout Europe, in some instances was prohibited from being worn by peasants or those of lesser descent (figure 3.11). The demise of imperialism, authoritarian monarchies, and caste systems throughout Europe largely ended the most exclusive period of using carmine-based dyes for clothing.

Similar in color to cochineal dyes are those found in the resinous secretions of an array of scale insects located throughout Southeast Asia. The most commonly used species is *Kerria lacca* (order Hemiptera, family Kerriidae). Processed secretions from this insect yield a dye, called lac, with color patterns similar to carmine dye. The major difference between lac and carminic acid is how the dyes are collected: the latter requires grinding up the female scale insect, while the former is secreted onto the branches of host trees. Lac is harvested as "sticklac" by cutting tree branches, and then processed to remove impurities. Lac is used principally to dye oriental rugs

Figure 3.11. Ceremonial robe of Napoleon Bonaparte and most likely dyed with insect-derived cochineal. Painting (1805) is by Jacques-Louis David (1748–1825) available at http://bit.ly/1R 6QOR4.

and wool but is frequently a component of cosmetics and, like many insect products, is purported to offer "healing" properties to humans. While lac and carmine dyes are still used today, synthetic dyes dominate the textile industry.

SILK

Silk is a protein fiber produced by several species of insects. Only one, however, is responsible for the material used in making textiles. As we learned in chapter 2, the silk moth *Bombyx mandarina* was cultivated by the Chinese during the practice of sericulture, yielding the domesticated *B. mori*,* the insect responsible for the entire

silk trade that extended from Asia into Europe, Africa, and the Middle East. Like the Maya and Aztecs with carmine dyes, only the Chinese knew the source of silk and the process for harvesting this precious fiber. Even once the secrets were revealed, not many have engaged in the very labor-demanding practice of farming silk moths. Why not? Or, worded more correctly, how difficult is it to get silk? Very! We have already discussed (in chapter 1) the fact that human intervention is required to feed this insect and ensure that mating takes place. Silk is produced by the silk moth as a last-stage caterpillar prepares for the cocoon or pupal stage.

For many insects, development from immature to adult is through a transition stage, termed pupa. Here is where all the adult structures like legs, wings, and antennae are formed. The result is an adult insect that does not resemble the juvenile at all. (In chapter 6, we'll have a more in-depth examination of how insects grow and develop.) For *B. mori*, the pupal stage occurs inside a silken cocoon. Many insects can form a cocoon, and several use silk to construct the pupal chamber, but mulberry or domesticated silk moths are unique: the cocoon is constructed from a single, unbroken strand of silk fiber. It is this property that makes the silk moth so useful for making textiles, as well as the strength of the fiber, which remains intact so long as the fibers do not get wet. Silk is collected by first placing thousands of pupae in boiling water. This process kills the developing insect and helps dissolve the glue (**sericin**) used by the caterpillar to adhere the strand together and waterproof the inside of the cocoon. Obviously, all of the moths cannot be used for silk harvesting, since the process is lethal. Silk farmers then must unwind the fiber of each cocoon as a continuous thread. Individual fibers are too weak to use to make textiles, so usually three to ten strands are wound together to form a single silk thread. In ancient China, hundreds of thousands of pupae were commonly treated in this manner in a few days' time, requiring

---

*Notice that the genus name *Bombyx* was abbreviated to just *B.* when referring to *Bombyx mori*. The reason is to avoid redundancy in scientific literature. When a genus name has already been used in an article or chapter, it is generally abbreviated from then on. Some exceptions do occur, such as

when genus names that begin with the same letter are used, in which case the first two letters of the second genus name to be introduced are used for the abbreviation. In instances when the scientific name has been used very infrequently, such as in certain types of books, the full spelling of the genus name may be used each time presented.

hundreds of individual workers involved in various aspects of sericulture.

The time-consuming and labor-intensive methods involved in producing silk would seem to discourage the continued practice of sericulture. In reality, natural silk has an attractive luster that synthetic fibers cannot replicate. Silk is also quite absorbent, allowing air to move through the fabric (similar to wicking athletic wear), yet the material traps heat at the skin surface during cool weather conditions. The fiber strength makes silk very durable and suitable for use in upholstery, wall coverings, rugs, and bedding materials. Silk production remains a multimillion-dollar business in China and India.

Household uses

A variety of insect products share a common theme: protection (for dramatic sound emphasis, see http://www.youtube.com/watch?v=JDxSnQqAYeY). Okay, not the kind of protection referenced with honeydew. No, this form is more utilitarian: insect secretions that protect our furniture, shoes, and cars. Wax, shellac, and propolis are the chief insect-derived constituents in a variety of polishes, varnishes, and sealants. **Wax** is the principal component of honeycomb produced by bee hives. The wax is collected from melting the honeycomb. Pure wax can be used as a wood protectant and sealant, foundation for shoe polish, component of cosmetics and lip balm, and to make natural candles. As with so many of the commercialized insect products, synthetic waxes are more commonly used today because of easier production and lower cost compared with collecting the naturally produced wax from honey bees.

SHELLAC

Shellac is a resinous insect secretion that rivals the utility of wax for protecting floors. The material is derived from lac-producing scale insects. Shellac is harvested from trees coated in the resin; they are then processed into dry flakes that, when dissolved in ethanol, yield a thick liquid. The liquid shellac is applied as a furniture and floor varnish. Shellac dominated the market until around the 1920s, when plant-based protectants replaced the insect product. The insect varnishes are still produced and have recently experienced a small but significant resurgence in uses to protect antique furniture. Until

Figure 3.12. Modern and classic cellos and other stringed instruments are treated with propolis to protect the finish. Photo by Andrew Plumb available at http://bit.ly/1LOk9xV.

the 1950s, shellac was important in the production of 78-rpm phonograph records. Then 78-rpm shellac records were replaced with vinyl LPs, or long-playing records, which have largely been replaced by CDs (actually LPs lost out to eight-tracks, then cassettes, and then CDs). That said, albums are being produced again but are composed of vinyl, not shellac.

Another component of honey bee hives, **propolis**, is used to make car wax. Bees do not make propolis; rather, the material is collected from tree sap or other plant sources and brought back for use in sealing small gaps in the hive. As with shellac, propolis is used as a wood varnish, but in very specific applications: to treat stringed musical instruments such as cellos, violins, and upright basses (figure 3.12). It is believed that propolis was used by Antonio Stradivari and other luthiers* of his time. Perhaps it is the type of bee employed by Stradivari that is the real secret to his unmatched excellence in producing the finest musical instruments the world has ever known!

SLUMGUM

What is left to talk about? Slumgum. The name almost reveals where it comes from. Slumgum is leftovers. Once a honeycomb has been rendered to yield clean or pure

---

*A luthier is one who makes or repairs stringed instruments.

wax, what remains—excrement, pupal linings, debris from other insects, and remnants of wax—is slumgum. What good is it? Well, it can be used to attract bees to a particular location, but as a more practical use, slumgum is used as fertilizer for ornamental plants, owing to its high content of zinc, calcium, and organic matter. As it turns out, Slumgum (http://slumgum.com/) is also the name of a jazz quartet from Los Angeles, California, that characterizes itself as risk-taking and modern; really nothing like the business-as-usual bees that gave rise to the band's name.

Medicinal uses

The preceding section shed some light on insect uses in the health care arena. Several insect-based products consumed or applied to the human body are believed to confer medicinal benefits. Unfortunately, a strong link to experimental evidence does not exist for most. Regardless, that has not been a deterrent in the advertising of some of these products. Does it border on false advertising? Here's the tricky part: it all depends on the type of product and claims made. For instance, if an item is sold as a food product or health remedy in the United States, then the Food and Drug Administration (FDA) holds jurisdiction, and the product in question must adhere to strict regulations, including providing evidence for any medicinal claims. However, if the product is sold as a dietary supplement, the FDA oversees with a different set of regulations (under the Dietary Supplement Health and Education Act of 1994). In that case, the safety of a product and/or its purported claims for a healthier life is only investigated if a complaint is filed with the FDA. The product can be produced and sold without prior FDA approval, and the manufacturer has the responsibility of making sure the labeling and advertising are truthful and not misleading. As long as the product labeling doesn't claim that the supplement prevents, cures, or treats a specific disease, there are no issues. However, if an insect secretion is proclaimed as the modern-day cure for cancer, the product will draw immediate attention and be shut down.

Several of the more common (at least to entomologists) insect-derived products with potential health benefits to humans are listed in table 3.2. This textbook would become prohibitively long (and expensive) if all were dis-

cussed in detail, so we will examine only a few of the most interesting.

APITHERAPY

Apitherapy is the practice of using bees (honey bees) and/or their products for medicinal purposes. We have already discussed a few of the medical uses for bee pollen and royal jelly, but health benefits from consuming bee products also extends to honey (energy builder, source of vitamins and minerals) and propolis (anti-inflammatory, antifungal, antibacterial, and tumor suppressant). In fact propolis has been shown in laboratory tests to possess antioxidant activity and to suppress certain types of tumors associated with a variety of cancers. Clinical studies with humans have yet to be performed to determine whether propolis will become an agent to treat or prevent this disease.

CANTHARIDIN

Cantharidin is a terpenoid compound secreted by many blister beetles belonging to the family Meloidae (figure 3.13). The chemical is used for defense by the beetle, serving as a severe irritant of the skin and mucosal membranes. In terms of medical uses, cantharidin has two "lives": one as a mild-mannered wart remover and the other as an exotic aphrodisiac. Let's focus on the boring life first. Okay, in fairness, removal of warts is not boring, but it does not sound nearly as sexy as an aphrodisiac. That said, diluted cantharidin is used as a topical ointment to stimulate serine proteases in epidermal cells to be released, aiding in the removal of not only warts, but also tattoos and the dome-shaped papules associated with the skin condition induced by the poxvirus *Molluscum contagiosum*.

The alter ego of wart removal is Spanish fly, a name that refers to the emerald green blister beetle, *Lytta vesicatoria* (figure 3.14), and that also is the common name for the aphrodisiac-like properties of cantharidin. As the story goes, ingesting the terpenoid induces erections in men, a precursor, if you will, to Viagra or Cialis.* The

---

*Viagra (sildenafil) and Cialis (tadalafil) are both prescription medications for men who suffer from erectile dysfunction, a condition in which an individual has difficulty having an erection.

Table 3.2. Insect Products with Medicinal Benefits

Insect Product	Health Benefits
honey (honey bees)	boost energy, antioxidant, antimicrobial, tumor suppressant
propolis (honey bees)	lowers blood pressure, antibacterial, histamie blocker, tumor suppresant, wart treatment
royal jelly (honey bees)	antioxidant, insulin-like activity, inhibits inflammation
cantharidin (blister beetles)	topical ointment for warts, tattoos, papules
venom (bees and ants)	tumor suppressant, histamine blocker, antimicrobial
cochineal (scale insects)	treatment for whooping cough, pain remedy
cockroach extract (cockroaches)	induction of clotting
exuvia (cicadas)	treatment of ulcers
egg shell (blister beetles and praying mantis)	treatment of kidney stones

reality is that Spanish fly works nothing like the prescription medications. The compound is an irritant of the urogenital lining, which can lead to inflammation that may result in sustained erections. Of course no good deed goes unpunished, as the ejections may last for *hours to days*, often with bloody discharge, and, if truly unfortunate, lesions can form in the epithelial lining of the urogenital tract. Ingesting cantharidin frequently induces

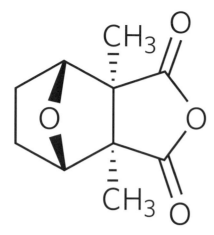

Figure 3.13. Chemical structure of cantharidin, a terpenoid compound produced by blister beetles (family Meloidae). Image by Edgar181 available at http://bit.ly/1iMolaA.

lesions anywhere along the length of the alimentary canal, with death potentially the final outcome. Just imagine the disclaimer given in any television advertisement for Spanish fly!

PHARMACEUTICALS

In the 1992 movie *Medicine Man*, Sean Connery stars as Dr. Robert Campbell, a biochemist camped in a remote locale somewhere in the Amazonian rainforest. Funded by a pharmaceutical company in the United States, he is in search of a cure for cancer. Apparently he finds it, in the form of an exotic flower, but somehow cannot synthesize the botanical compound in the flower using laboratory equipment in a hut, located in a remote location in the jungle, powered by a generator. Imagine that! Spoiler alert: Turns out the source was not the flower but a beetle residing hundreds of feet above ground in the tree canopy. Sure, this is just movie magic, but the idea of exploring exotic (and not so exotic) locations for compounds from plants and animals is very real. Insects have been the test subjects in the search for new pharmaceutical compounds, with a wide range of secretions offering potential new sources of tumor suppressants, antimicrobials, histamine blockers, and many other potential uses. Somewhat surprising is the recent discovery that venoms from tiny parasitic wasps contain promising new antibacterial and antifungal compounds, and one may pro-

Figure 3.14. The blister beetle, *Lytta vesicatoria* (family Meloidae), source of cantharidin commonly used as Spanish fly. Photo by Siga available at http://bit.ly/1G3aEJ4.

duce a peptide useful in controlling obesity in humans. The research is still in its infancy, but who knows, in a few years' time, perhaps many of our prescription medications will be farmed from animals with six legs.

Oh the wonderful things that insects do!

When discussing the utility of insects to the human condition, the idea of insects as food, or of products from multilegged creatures being essential to human health, comes off as somewhat gimmicky, a little weird, and sometimes unbelievable. Well, just wait until we discuss entomophagy, later in this section. The subjectivity of defining insects' benefits to our existence is dropped and a substantive foundation of facts lies before us when the discussion shifts to "what insects do." The roles that hexapods play in nature are critical to the functioning of every ecosystem they reside in. Humans benefit enormously from insects pollinating flowering plants, decomposing dead plant and animal matter, regulating the populations of an array of organisms, serving as a food source for many organisms (including humans), and even providing medicinal uses through their activities. Generally speaking, the human race receives far greater benefit from insects by "letting" them do what they do naturally, rather than forcing them to meet specific needs or desires, such as being cultivated for products (e.g., silk, honey). In this section, we briefly explore the activities of insects that positively affect mankind.

Insects as pollinators

Insects have an intimate relationship with flowering plants. Do not get weirded out. There is nothing creepy about that statement. It simply means that insects and plants have coevolved, and this is evident in a number of adaptations. Among them is the dependence of most flowering plants on insects for pollination. Pollination is part of sexual reproduction in the plants. Wait, plants have sex? And how! A vegetable garden is literally a botanical orgy in your backyard, and one that you encouraged. The key element to plant reproduction is fertilization; that is, male and female gametes come in contact to initiate the formation of an embryo(s). Male flowers release their gametes in a protective structure, collectively termed pollen. The pollen must reach the receiving structure (stigma) of the female flower. For plants that are not self-pollinating (some are) this generally occurs in one of two ways: nonspecific transmission (by wind) or animal-aided. Insects are the dominant animals that aid plant reproduction, and they offer several advantages over wind pollination.

1. Wind pollination is essentially a random event, whereas insects pollination is much more specific; that is, insects will visit flowers when offered sugary treats like nectar, and frequency of fertilization will be higher.
2. Insects will visit flowers under a range of climatic conditions. Wind pollination can occur only, well, when the wind is blowing.
3. Plant diversity increases in areas where coevolution has adapted plants to rely on specific insect pollination; certain insects specialize on specific flowers,

which are then no longer dependent on the randomness of wind-transmitted pollen.

Why are insects so good at pollination? Well, for one, insects are incredibly opportunistic. So if an excellent food source is available to them, such as carbohydrate in the form of nectar and protein in the form of pollen, they will take advantage of it. Plants produce so much of both that there is no issue with all of the nectar and plant sperm being consumed and none left for fertilization. The fact that insects can fly—again, remember that flight is unique to bird, bats, and insects—provides a locomotor advantage over almost everything else on the planet for visiting flowering plants. With that said, flight is not a requirement for pollination. From an anatomical perspective, insects have hairs (they are actually bristles), termed **setae,** that can trap or adhere pollen. So the insect may have no interest in the protein-rich source when visiting the flower but picks it up during feeding on nectar to mechanically transfer it to a receiving female flower. When considering that an individual foraging bee may visit several flowers in twenty-four hours, multiplied by the trillion-plus pollinating insects on the planet on any given day, one can begin to realize the vast scale of insect pollination.

Quick check

What are the physical characteristics of insects that help them excel as pollinators?

So how does this activity of insects benefit mankind? More than 90% of all flowering plants on the planet depend on insects for pollination, including those that yield the many fruits, vegetables, and nuts that we consume, and the flowers that bring aesthetic pleasure. The majority of food production for the world therefore depends on insect pollination. Flowering plants make up a large portion of nearly every ecosystem on the planet (the major exception are the oceans), and they are essential to the health and stability of each. Their survival depends on sexual reproduction, which relies on insect surrogates to intercede in gamete union. If plants stumble, then so do ecosystems, which in turn stresses our ability to survive.

Of course monetary rewards are also associated with the insect-pollen-transportation industry. For example,

Figure 3.15. Parasitic *Varroa* mite on the thorax of an adult male (drone) bee. Photo by Waugsberg available at http://bit .ly/1KBC0Ie.

in the United States, domesticated honey bees pollinate a wide range of commercial crops, yielding over $15 billion in crops (food and flowers) annually. The services of other types of bees are also recruited for pollination, such as in orchards in which leafcutter bees (*Osmia cornifrons* and *O. lignaria*) are more efficient than honey bees (only 600 females of *O. lignaria* are needed to do the same work that 50,000 *A. mellifera* accomplish). The leafcutter bees are not declining in numbers like honey bees, which are facing **colony collapse disorder** (CCD) (figure 3.15). The Fly Spot that follows addresses CCD and the effect it is having on food production in the United States and throughout other regions of the world.

Insects as decomposers

What happens to the bodies of animals that die? Or to the tissues of plants, like an enormous redwood tree (plant subfamily Sequoioideae), once all or part of the tree is no longer living? This is not meant to be an exercise in philosophy or theology, in which we speculate on existential questions of life after death or whether organisms have a soul, as the ancient Greeks once believed. Rather, the question speaks to the biology of death. Organisms decompose after death, often with the aid of other creatures. Insects' role in the decomposition processes of plants, animals, and other organic life is one of their most important contributions to ecosystem functioning. Several insect species consume or degrade other organisms

that function as producers and consumers in particular ecosystems. Once these organisms die, or begin to decay, their remains must be broken down into smaller parts, or elements, so that complex organic molecules will be fragmented. Insects perform two essential functions in this regard.

1. Their feeding activity increases the surface area of the dead material so that decomposition proceeds more rapidly by abiotic and biotic influences (figure 3.16).
2. They help release or liberate elements or molecules back into the ecosystem, for use by other organisms (figure 3.17).

Generally speaking, any animal that functions in the decomposition of dead or decaying matter is referred to as a **saprophage** (feeding on dead matter is considered **saprophagous**), and in the case of insects, they are classified by the organic material targeted: decomposers of plant material, decomposers of animal remains, and decomposers of animal excrement. The value of saprophagous activity is perhaps not as easy to visualize as with insect pollination, but it is no less important. For example, in terrestrial ecosystems, nearly 90% of the organic matter produced by green plants is left unconsumed by the entire entourage of consumers (i.e., animals, fungi, bacteria, other plants, etc.). Thus, the material must be broken down, quickly, so that other organisms can use the nutrients. Failure to do so leads to massive accumulation of organic matter in a form that is not useable to most other organisms. Likewise, the remains of animals (referred to as **carrion**) must be decomposed quickly to prevent a rapid buildup of carcasses throughout all ecosystems. The need to do so goes beyond just recycling; carrion emits noxious gases that humans find unpleasant to breathe, and a slowly decaying corpse can serve as an excellent resource for disease-causing microorganisms. Fortunately, several species of flies (order Diptera) and beetles (Coleoptera) rapidly and efficiently reduce animal remains to bone, cartilage, and hair in a matter of days (assuming the environmental conditions are favorable). As we will see in chapter 15, the activity of these same beetles and flies is the basis for the field of forensic entomology, in which saprophagous insects can provide clues about the length of time that a body has

**Two bee or not two bee:
The declining honey bee**

Colony collapse disorder, or CCD, is a condition or phenomenon associated with the European honey bee, *Apis mellifera*, in which unexpectedly high numbers of adult bees are absent from the colony. Ordinarily, some bees will die during the winter months. With CCD, however, colony losses between 30% and 60% of the total hive have been reported. Hives that have succumbed to CCD generally lack adult bees yet the queen is still alive, food in the form of honey and pollen is present, and capped brood cells are present. Generally, bees would never abandon a hive until all bees have emerged from brood cells, and they never leave food stores behind. When adult bees die, they normally do so in the colony, so dead adult bees are expected to be in trash or morgue piles within the hive. Not so with CCD—there is no trace of the adults. It is as if they just vanished.

Just how big a problem is CCD? In terms of bees, more than 50% of domesticated bee colonies have been affected in North America and parts of Europe. In 1980, an estimated 4.5 million to 5 million domesticated bee colonies existed in the United States, but today those numbers have dropped to closer to two million. As discussed, insects, primarily honey bees, are instrumental in pollinating the majority of flowering plants on the planet. Thus, the decline of the honey bee means reduced food production and a corresponding increase in food prices. Part of the reason for the hike in food costs is the increased demand for the services of a declining pool of domesticated bees, meaning the remaining bees hired to pollinate commercial crops. It is a matter of simple economics: fewer bees means the cost of renting those that remain will go up, and the grower passes the increased expenses on to the consumer. The European honey bee is not native to North America, so there are other insect and native bee species that pollinate flowers. However, none of them can be cultured as effectively as *A. mellifera* for commercial use in crop pollination.

What is the cause of CCD? Contrary to reports circulated on the Internet and reported by various media outlets, the cause of CCD has not been determined. Numerous agencies and organizations in the United States and around the globe are investigating the phenomenon. Thus far, the research has revealed that several factors may contribute to the current demise of honey bees, including pesticide use, the presence of cell phone towers, parasites, and pathogens (the *Varroa* mite always seems to be present in hives that have experienced CCD) (**Figure 3.15**), loss of habitat, changes in beekeeping practices, and environmental factors. With so many individuals intensively studying CCD, it is somewhat disconcerting that so much is still unknown about causes of the disorder and how to prevent it from continuing. With that said, there is evidence that similar colony disorders have occurred throughout history, on multiple occasions, and in each instance, the honey bees appear to have recovered. The hope is for a similar end result with CCD. Of course, the human population continues to rise at an alarming rate, meaning that even short-term reductions in food supply caused by CCD have a much larger effect on the human condition today than even ten years ago.

been dead, where the body was located, and if it had been moved.

Insects that decompose excrement provide a service that humans greatly undervalue. Anyone who owns a dog as a pet knows there is no end to the joy that comes from scooping poop from the lawn. Hopefully, for your neighbor's sake, this occurs more than once a year. Beetles and flies once again are the key contributors in this form of saprophagous activity. Of particular importance is the liberation of nitrogen—an element critical for a wide range of organisms, but particularly for phototrophic producers—into soil and water (figure 3.18).

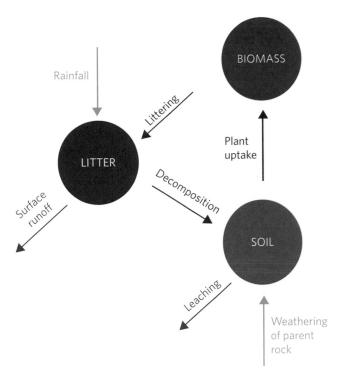

Figure 3.16. Schematic of nutrient cycling in a terrestrial ecosystem. Insects as decomposers contribute to the recycling of nutrients by helping decompose organic material in litter. Illustration by Kayau available at http://bit.ly/1NU3jU7.

Insects as population regulators

The best control methods for pestiferous insects are other insects. Talk about peer influences! In chapters 1 and 2 we discussed to some extent the negative effects of insects that transmit disease to humans and other animals, as well as the effects of those targeting our food and invading our homes. The effects can be and have been devastating to local populations. Insects are not the only creatures with such tendencies. In fact, preying on the human condition is not restricted to membership in the phylum Arthropoda: a wide range of beasts does so, and the majority are lacking backbones.* As we will learn in chapter 14, many insects that create havoc for humans do so only sometimes. At other times, we don't even consider them pests. What happens that changes a mild-mannered insect into an obnoxious pest? No single answer can account for every scenario (again, remember

that there are over one million extant insect species). One frequently altered feature is the natural regulation of insect populations. That is to say, an ecosystem has many components, living and nonliving, responsible for keeping populations in check or in balance. When any one of those is disrupted from the normal range, even slightly, the ripple effect can be quite large. Notice I refrained from naming specific set points, because environmental features are not static and show tremendous variation; a range of values is used to represent the normal condition. In the case of insect populations, many are maintained at non-pest levels (from the human perspective) by other organisms, including insects, that function as predators and parasites.

Predation involves one organism (the predator), in this case an insect, hunting or stalking a food source, which is another organism (the prey), for the purpose of consumption. For the sake of clarity, this is not quite the same as you driving to McDonald's† (hunting) and deliberating which succulent value meal to purchase (stalking), with the intent of consumption to ensure your immediate survival. Predators must eat several individual prey to survive. Approximately 25% of all insect species are considered predators, with the majority consuming other insect species as prey. Some predatory species are generalist, showing little preference for what they eat, other than size (generally feeding on those smaller than themselves), while others display a more delicate palate: that is, they are more specific in what enters their bodies.

Consuming animals, either in part or as whole, is carnivory; a predator is **carnivorous. Parasitism** is also an act of carnivory, only in this case, the carnivorous organism (the parasite) in question feeds and develops exclusively on one individual, referred to as a host. The host is typically not consumed like prey; rather the parasite feeds just enough to ensure its own survival. For insects, usually only one life stage (most commonly the juveniles or larvae) is parasitic, and the other stages are free-living. Nearly every order of insects includes

---

*Backbones, or vertebrae, are absent in invertebrate animals. The phylum Arthropoda represents just one of approximately thirty-four invertebrate animal phyla.

---

†McDonald's is a fast food chain restaurant located throughout the world, with a clown (Ronald McDonald) as the organization's face. For the record, most clowns are not considered predators but usually are carnivores.

Figure 3.17. Adult blow flies (family Calliphoridae) engaged in the decomposition of a fish head by feeding and depositing eggs, which, upon hatch, will greatly accelerate the breakdown of tissues. Photo by Laszlo Ilyes available at http://bit.ly/1QyHAyE.

Figure 3.18. Adult blow flies feeding on animal feces. Photo by Allen Watkin available at http://bit.ly/1G3byFE.

examples of parasitic species, and some groups are almost entirely reliant on parasitism. Insects also display a unique adaption to the parasitic lifestyle termed **parasitoidism**. In this condition, certain groups of insect parasites (most common with some flies and wasps) end the parasitic relationship with host death. As pointed out earlier, true parasites do not kill their hosts directly, so this is a unique outcome to the parasitic association.

Why should you care that insects eat each other? Or other animals? Not only should you care about carnivores, but your friends and family circle should expand to include **herbivorous** insects too; that is, insects that feed on vegetative, or plant, material. Wait just a minute! Didn't we discuss how bad it is when insects eat our food, aka plant stuff? Yes indeed. But we do not consume everything that insects feed on, and in many instances, herbivorous and carnivorous insects feed on plants, animals, and other organisms that have the potential to be, or always are, pests or harmful to the human condition. In other instances, the feeding activity of insects may have nothing to do with directly keeping potential pests in check but still play a significant role in regulating population densities within a specific ecosystem. Under most conditions, predators and parasites suppress insect populations by 10%–25%. While this range may not seem like much, it really is critical to maintaining the delicate balance between keeping insects in check and a population explosion.

In much the same way that insects are opportunistic at our expense, humans often take advantage of opportunity by manipulating predatory and parasitic insects for use as natural or biological control agents of pests. Such insects are typically referred to as **natural enemies** or beneficial insects (the name most commonly used to search for such insects through a Google or Yahoo web search). The way this process typically works is that the beneficial insect is one that targets a particularly important pest insect, say, a caterpillar pest of corn or a fly species on cattle feedlots. One example of the latter is the parasitic wasp *Nasonia vitripennis* (order Hymenoptera) (figure 3.19). The wasp is mass-cultured in a laboratory for later sale to farmers or other consumers (like poultry, hog, or beef producers, or those raising horses). The wasps are released and parasitize fly pupae (in this example, *N. vitripennis* is a parasitoid, so "parasitoidize" is more correct, yet awkward to say), which permits wasp reproduction to occur; the wasps become established on the feedlots without the need for additional releases, provided some hosts, albeit reduced in number, remain available. Several predatory and parasitic insect species are available commercially as biological control agents, though opinion is mixed as to whether they are truly effective at helping manage pest populations (table 3.3).

Figure 3.19. The parasitic wasp *Nasonia vitripennis* (order Hymenoptera, family Pteromalidae) using the ovipositor (apparent in the center under the wasp) to drill an access hole to the host. Photo by M. E. Clark available at http://bit.ly/1jblHes.

Beyond the text

When trying to identify insects that would serve as useful biological control agents, what characteristics should be looked for in the insects?

Insects as food

Insects as food? We have already focused on insect products that humans eat. What is left to talk about? Eating the whole enchilada. That's right, head, legs, wings, and butt. One gulp, and then savoring the taste in your mouth. I know what you're thinking: there's no way this topic is seriously being listed as an activity beneficial to mankind. Well, it is. Using insects as food is not relevant to just humans either; a wide assembly of animals, plants, and other organisms dine principally on entomological cuisine. Above we addressed some ecological implications of insects as food for predators and parasites. Viewed another way, a diet composed of six-legged entrees provides needed nutriment for growth and development for consumers. In many ecosystems, being a food source is one of the crucial roles that insects play. Turning this concept to humans, though, always seems to evoke more disgust than just about any other topic associated with insect biology. A puzzling response, because insect cousins such as shrimp, lobster, and crab are considered a delicacy that people are willing to spend a great deal of money for. Maybe the repulsion is that insects are land-based. If caterpillars were a foot long and came from the ocean, perhaps they would be perceived

Table 3.3. Commerically Available Beneficial Insects

Insect	Target
ectoparasitic wasps *Nasonia vitripennis, Muscidifurax* spp., and *Spalangia* spp.	fly pupae on feedlots, associated with horses, swine, poultry, and /horses/cattle
egg parasitoids *Trichogramma* spp.	eggs of moths associated with a variety of crops
green lacewing, *Chrysoperla rufilabris*	aphids feeding on crops
parasitic wasp, *Aphidius ervi*	adult aphids
minute pirate bug, *Orius insidiosus*	general predator of stored product pests
ladybug beetle, *Hippodamia convergens*	general predator, but especially good with aphids
mealybug destroyer, *Cryptolaemus montrouzieri*	mealybugs
whitefly parasite, *Encarsia formosa*	whiteflies

This is just a partial list of insects available for use in controlling pest insects. Several commerical suppliers can be found on-line and in the Additional Resources section of this chapter.

differently? Who knows. What we can be sure of is that insects are part of the diets of people throughout the world, and not in the form of the gimmicky foods, like beetles in lollipops, sold in the United States.

Entomophagy is the practice of consuming insects as part of the diet. All stages of insect development are com-mon dietary staples in Africa, Asia, New Zealand, Australia, and North, Central, and South America. However, entomophagy is rare in developed nations, other than as a fad or gimmick. That said, eating insects was com-mon with most tribes and civilizations pre-dating hunting and organized agriculture. Today, people eat insects

Figure 3.20. Insects (fried giant water bugs) as cuisine on display in a Thailand market. Photo by Takoradee available at http://bit.ly /1PFEw8C.

for a variety of reasons, including the high nutrient content (amino acids levels in insects are often five to ten times higher than in other animal tissues), for celebrations and rituals, and to supplement other food sources. The last is especially important in regions of the world that face food shortages caused by drought, poor soil conditions, war, and poverty. Insects are generally abundant everywhere, so they are readily available as excellent sources of protein and carbohydrates (figure 3.20). Some companies are trying to introduce insect cuisine into markets otherwise disinterested (e.g., United States and Europe), promoting the concepts of sustainability and nutrition. It is an uphill battle to gain a foothold in Western diets, but one worth trying. In a crazy twist of irony, some of the very insects being promoted as food in the United States are the same ones regulated by the Food and Drug Administration as food contaminants. We explore these regulations, better known as **defect action levels**, in chapter 14.

Insects as medicinal tools

We know from earlier in this chapter that some products from certain insects provide, or are purported to provide, medical and health benefits to humans. We also derive health benefits directly from insect activity. Two examples, **apitherapy** and **maggot therapy**, fall near the extreme end of the spectrum in terms of what humans will do for the sake of survival.

APITHERAPY

We explored apitherapy when examining potential medicinal products synthesized by honey bees or found in their hives. Apitherapy generally refers to medicinal uses of honey bee products. However, entire bees have practical medicinal uses as well. Since at least the time of Charlemagne (742–814 AD), conditions such as rheumatism and various forms of arthritis have been treated with bee stings; individuals are deliberately stung to achieve temporary relief from pain and stiffness associated with inflammation. Similar practices are still used throughout the world with other stinging Hymenoptera. For example, stings from the red imported fire ant *Solenopsis invicta* are used as a short-term remedy to treat the aches and pains associated with inflammation associated with arthritis. The main and most prevalent component of honey bee venom, a small peptide called melittin, is responsible for the anti-inflammatory action of the sting. Treatment involves holding a live bee with tweezers to allow stinging in the targeted area. Depending on the severity of the condition, this sting therapy may need to be repeated a few times a week, or as often as three to five times a week, repeated over several months. The technique has been improved to the point that isolated venom can now be injected with a syringe instead of stinging, which also permits regulating the dosage of melittin. Melittin appears to possess other medicinal properties: laboratory studies have shown the peptide is capable of targeting and destroying tumor cells and also displays anti-microbial activity, with a high specificity for inhibiting the causative agent of Lyme disease, the spirochete bacterium *Borrelia burgdorferi*.

MAGGOT THERAPY

So one would guess, from the definition of apitherapy, that maggot therapy similarly involves stinging by flies. Not quite. Maggots or fly larvae are deliberately placed on the body so that they can eat human flesh. Maybe zombie therapy would be a better name for it. In any case, live fly larvae are selected because they are saprophagous, which we learned above are insects that decompose dead organic material. Insects that feed specifically on dead animals or carrion are referred to as **necrophagous**. The ones used in maggot therapy are necrophagous fly maggots, specifically, the larvae from flies belonging to the family Calliphoridae, better known as blow flies or bottle flies. So how can the saprophagous activity of fly larvae be used in medicine? The key is that the flies are attracted to and feed on dead tissue, but a corpse is not a prerequisite for feeding; an individual with a severe enough wound can attract adult blow flies, which in turn can lay eggs in or near the wound site, and once the eggs hatch, the larvae will begin to feed, but only on the dead tissue.

What has been described so far is not maggot therapy, but instead a parasitic condition called **myiasis**. Myiasis is literally the infestation of tissue by fly larvae. It is not rare but also not common, particularly for individuals who practice good hygiene and can take of themselves. However, for those who are incapacitated, for whatever reason, myiasis is a potential threat, depending on circumstances. One historically significant example occurred during the US Civil War, when soldiers left badly wounded on the battlefield frequently had their wounds infested with fly larvae. Confederate physicians discov-

ered that such injuries were often more likely to heal following fly feeding than when left untreated. In some cases, even wounds that would normally be treated by amputation could instead recover if fly larvae fed on the dead tissue. Today, we understand that the necrophagous larvae actually **debride** the wound; that is, the feeding activity removes dead, injured, and infected tissues. At the same time, the larvae release antimicrobial components in their excrement that further "treat" the wound like an antibiotic wash.

Maggot therapy is using specific necrophagous fly larvae to treat wounds that have been otherwise unresponsive to allopathic medical therapies. It is often viewed as a last resort, particularly in situations in which the tissue has become gangrenous and amputation is a very real possibility. Depending on the size of the wound, ten to twenty sterile larvae are placed in the wound and covered with sterile gauze. The larvae are checked daily and replaced with younger ones if the treatment continues for days to weeks. Once debridement is complete, maggot therapy can no longer be used because the fly larvae will not feed on living tissue, although a few recent studies have challenged this view, claiming that the longer the flies remain on the wound, the more nonselective they become in tissue choice (living versus dead). This observation is not a reason to abandon maggot therapy, but does reinforce the need to change the larvae on the wound frequently.

Bug bytes
Medicinal maggots
http://www.youtube.com/watch?v=6Xt6NWkgydM

Insects got class: Cultural influences of six-legged creatures

A transition from maggot therapy to any other topic is not easy. Put your "oohs" away, and stop saying "gross." We now turn our attention to the ways that insects have helped shape the cultural identity of humans. The influences we are about to discuss do not clearly fall under the heading "beneficial" or "necessary," but they can definitely be entertaining, at least for some, and are more to be appreciated than anything else. Hence the phrase "appreciating insects" is more commonly used for the influence of insects in art, music, and multimedia (e.g.,

Figure 3.21. Stone sculpture of a scarab beetle located in Egyptian ruins at Karnak. Photo by JonasS available at http://bit.ly /1L6vO0o.

television, movies, and Internet). The subdiscipline of cultural entomology is devoted to the study of such influences, and also includes symbolic uses of insects. Here we explore just a few examples of the entomological presence in human culture, including the "finer" things of life, such as classical music, sophisticated art, and sports (well, that last is "high class" in my world).

The "art" of insects

Insects have a long history with art, serving as the subject matter for numerous paintings and sculptures and being immortalized in amulets and precious jewelry worn by emperors, kings, queens, priests, nobles, and many other figures of extreme or supposed importance to past civilizations. In the ancient world (Mesopotamia and Greece), sculptors would accept the tedious challenge of creating works of art depicting intricate figures of insects, sculpting the detailed bodies in stone, including the elaborate network of wing veins. A chariot pulled by reined flies is a Greek classic that highlighted the enormous technical challenge of such statues. Artists trying to separate themselves from others would include insect sculpture in their portfolios as a means to climb the ladder of success. Arguably the biggest sculptures of insects are found in Egypt, as several stone statues of scarab beetles are located in Karnak and other locales to honor the insect god Khepri (figure 3.21).

The presence of insects in paintings often reflected the technical skills of those engaged in realism. Insect body design, particularly wings, has inspired the creation of imaginary creatures and those of folklore and fantasy. Why are the wings especially significant? Close examination of the mesothoracic or forewings of insects from orders such as Odonata (dragonflies and damselflies), Mecoptera (scorpionflies) and Neuroptera (owlflies, lacewings) reveals an extensive array of veins not evident in many of the more advanced insects.* Drawings or painting of fairies, for instance, most often show wings that parallel those of a dragonfly or damselfly (figure 3.22).

In other works of art, insects are used to convey a message. The Italian Renaissance painter Carlo Crivelli (1435-1495) used a fly in his painting *Madonna with Child* as a representation of Satan, with the eyes of the Christ child appearing to stare directly at the insect (figure 3.23). This interaction is interpreted as a foreshadowing of the temptations of the adult Jesus for forty days by Satan when alone in the Judean Desert. Finding the message

---

*Why are there differences in the amounts of wing veining among insects? A common theme in animal evolution is a reduction in numbers: as an insect become more efficient at a particular task, fewer structures are needed to do the task. Thus, in advanced insects, fewer wing veins are needed than in some of the more primitive orders of insects, like Odonata and Mecoptera.

Figure 3.22. Painting *Lily Fairy* by Luis Ricardo Falero (1851-1896), clearly displaying insect influences in wing design. Photo by [1] available at http://bit.ly/1R0gcKW.

Figure 3.23. Carlo Crivelli's *Madonna with Child* (c. 1473), depicting the baby Jesus being distracted by a fly. Photo by Eugene available at http://bit.ly/1OSl8lL.

in an impressionist or surrealist use of insects is much more challenging. The Spanish Catalan surrealist Salvador Dali (1904-1989) included insects in many of his paintings, including a self-portrait. It would be safe to say that he went through a period of insect fixation, especially with butterflies—which is why it's interesting that his last painting, *The Swallow's Tail* (1983), has nothing to do with a swallowtail butterfly or insects at all.

Modern artists also take inspiration from an array of insects, as evidenced in paintings, drawings, graphical creations, and sculptures using stone, wood, iron, and produce (yes, food) as the media, jewelry made from a range of materials; insects depicted in or created from tile, shards of glass and stained glass; spray-painted images on buildings or other structures, and portrayals in clay, wax, and even ice. Preserved and living insects are commonly used to evoke sensations, conjure images, or speak universal truths (figure 3.24). Artists have always been creative in what they use to express themselves, so the wide assemblage of media used today is really not surprising, but one major difference from the past is where we find insect works of art today. Classic paintings and sculpture are housed in museums; today, many if not most works can be found online rather than on dis-

Figure 3.24. Photo by Christopher Marley of his stunning prism *Limited Aesthetica*. The prism is available at ©www.pheromone design.com.

Influences on music

play. Of course, in that respect, images of the entire inventories of many museums are now available on the web, so maybe there is nothing new about modern art.

Insects have an ear for music. Actually, they do not have ears at all, but we will deal with that in chapter 11. The entomological influence in music has been evident for hundreds of years. Song titles, lyrics, musical themes, band names, and even album—sorry—CD cover art bears the mark of insect influences. No genre of music has been untouched. In the world of classical music, such greats as Beethoven, Handel, Chopin, and Schubert have included insects in their works. *Flight of the Bumblebee* by Nikolai Rimsky-Korsakov (1844–1908) is perhaps the most recognizable of the classical works; the orchestral arrangement captures the sound and even image of a bee in flight, probably more a honey bee than bumble bee, but that is a minor point (http://www.youtube.com/watch ?v=6QV1RGMLUKE).

From classical music to modern rock and roll, insects have been serenaded in over three hundred song titles or lyrics. The songs titles capture taxonomic or classification interest as well, in that at least fourteen orders of insects have been used in song titles. Who would have thought it possible to incorporate sucking lice from the order Anoplura ("Crab Louse" by the band Lords of Acid) into a song? Even more impressive is that many of the lyrics focus on the biology of the crab louse; it's up to you to determine whether that is a good thing or not. The fascination with six-legged creatures extends to the naming of musical groups, as no less than fifty-seven bands have adopted the common names (or attributes) of insects. Some of the more recognizable include Buddy Holly and the Crickets, Iron Butterfly, Sting (I know, not a group), Papa Roach, the Bee Gees, and of course the Beatles. As for the fab four (not the Marvel Comics superheroes, but the Beatles), considering their unmatched success, one has to wonder if the acclaimed geneticist J. B. S. Haldane was referring to them when he stated that the creator must have an "inordinate fondness for beetles."

Insects and the big screen

Since the advent of cinematography in the late nineteenth century, insects have been depicted in films. Modern media has evolved to include television, web-based productions, and gaming (video, computer, smart phones), and the list is expanding at a rapid rate. As media has evolved, so too have the roles of insects, with many becoming stars. On the big screen, insects probably reached their pinnacle during the heyday of B movies, a period from the mid-1950s to the early 1960s. B movies were developed during the golden age of Hollywood, during which double features were offered; the second feature, or B movie, was less publicized and generally developed with a low budget. Many B movies focused on specific genres, such as westerns and horror films, and the latter is where insects reigned supreme. Since the average person knew nothing about insects, movie producers were free to exploit, exaggerate, and make up the biology of the creatures shown on film. So six-story-tall ants or grasshoppers and flesh-eating insects were not out of the question. Nor was the idea that insects possessed exotic or mysterious chemicals that, if introduced to the human body, could do such miraculous things as restore youth (*Wasp Woman*, 1959) or transform people into maniacal, half-insect, half-human creatures (*The Fly*, 1958; *Invasion of the Bee Girls*, 1973). Insects even learned to read from a newspaper, leading to them sharing their newly acquired knowledge by texting with their bodies such congenial phrases as "Get out" in the 1975 classic, *The Bug*. Today, insects still make appearances in horror and sci-fi adventures.

Science fiction movies, particularly those introducing a host of fierce or evil alien creatures, frequently seek inspiration from the world of insects. Insect influences are clearly evident in aliens introduced in *Enemy Mine* (1985),

Tremors (1990), *The Abyss* (1989), *Starship Troopers* (1997), the original Alien (1979), and *Men in Black* I (1997), II (2002), and III (2012). In several movies, insects make cameo appearances or are present to convey specific messages. For example, in the 1991 film *Silence of the Lambs*, a serial killer places pupae of the death head moth (genus *Acherontia*) in the mouths of victims, which backfires because the clue eventually leads to his capture; in the movie *Dragonfly* (2002), the husband of a recently deceased physician is contacted by his dead wife using dragonfly clues. The capstone of entomological cinematography was perhaps the movie *Mimic* (1997), which starred not one but two types of insects (cockroaches and a genetically engineered beast called the Judas Breed); the heroine of the movie was an entomologist who saved the world using her own blood. Lighter fare includes numerous animated films featuring insects, including *Pinocchio* (1940), *A Bug's Life* (1998), *Antz* (1998), *James and the Giant Peach* (1996), and *Bee Movie* (2007). It should also be noted that entomophagy was endorsed in Disney's *The Lion King* (1994): "slimy yet satisfying" (okay, I realize the phrase was said by a lion cub, warthog, and meerkat—back off!) (table 3.4).

Insects have made a multitude of appearances on television throughout the world, including in cartoons, sitcoms, advertisements, documentaries, and music videos. They have interfaced media sources by appearing in production broadcasts on both television and the web and have been the subject matter for millions of websites and related media. Numerous smart phone and tablet apps are dedicated to hexapods, as are video games designed for just about any gaming platform. In the video game format, insects have perhaps rivaled their B movie glory days as, in most cases, video games feature insects in horror or science fiction contexts. Of course, challenging this view is Pokémon, the enormously successful media franchise developed by Nintendo that uses insect-like creatures in relatively benign ways compared with many modern video games.

The future of insects in multimedia looks secure considering the strong foothold they've already established in just about every available media outlet. The continued development of computer-generated imaging and animation, coupled with the ever-increasing markets in gaming and handheld devices, should ensure that insects continue as media stars.

Table 3.4. Examples of Insects in the Movies

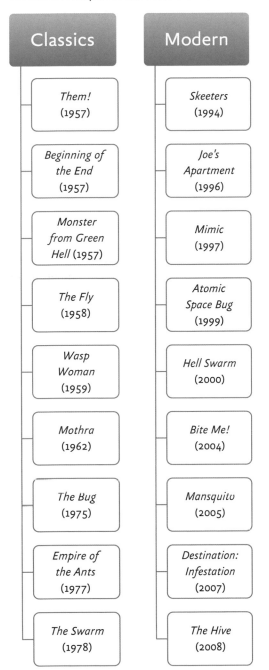

Classics	Modern
Them! (1957)	*Skeeters* (1994)
Beginning of the End (1957)	*Joe's Apartment* (1996)
Monster from Green Hell (1957)	*Mimic* (1997)
The Fly (1958)	*Atomic Space Bug* (1999)
Wasp Woman (1959)	*Hell Swarm* (2000)
Mothra (1962)	*Bite Me!* (2004)
The Bug (1975)	*Mansquito* (2005)
Empire of the Ants (1977)	*Destination: Infestation* (2007)
The Swarm (1978)	*The Hive* (2008)

Symbolic uses of insects

In several examples discussed above, insects were used to convey a message or as a representation of an idea with larger meaning than the mere anatomy of an adult insect. Those uses run parallel with symbolism, in which insects are used to physically represent an idea, concept, person, or people. In chapter 2, we examined a few examples of symbolic uses of insects in ancient Egypt (dung beetles representing the god Khepri) and Mesoamerica.

Totemism is yet another way in which insects are used in symbolism, in this case as representations of ancestors, tribes, or clans. Totem use was common with civilizations of North, Central, and South America, and throughout Africa and the aboriginal tribes in Australia (figure 3.25). Eating a totem insect was tantamount to cannibalism, as the spirit or anime of the ancestor or relative was reincarnated in the smaller life form. Likewise, harming a totem animal was to be avoided at all costs. Totemism is still practiced in many cultures throughout the world.

Heraldry is a form of symbolism in which insects and other animals were used to adorn soldiers—specifically, so that soldiers in one army could be distinguished from another. This became necessary when heavy metal armor was introduced to Europe around the eleventh century. Armor unfortunately was not originally ornate, with a coat of arms; thus, there was no way to tell friend from foe. Heraldry was instantly adopted by as many armies as possible, because of the protection it afforded soldiers during combat, with fierce or noble animals selected to adorn the armor. Lions, eagles, and unicorns were common. So too were stinging Hymenoptera: bees, wasps, and hornets (figure 3.26). Even Napoleon Bonaparte, as emperor of France, chose the honey bee as his official emblem, using the insect as ornament on royal garments and objects throughout his palaces.

Today, heraldry still exists in the form of mascots associated with sports teams. Insects, most notably stinging bees, wasps, and hornets, are mascots for an array of professional and amateur athletic teams. The NBA's Charlotte Hornets is perhaps the most recognizable, followed by Georgia Tech's Yellow Jackets (figure 3.27).

A religious experience

Entomological connections to various religions have been revealed in earlier examinations of insects and deities in chapter 2. Gods and goddesses have been depicted in paintings, carvings, engravings, and sculpture, to name just some of the known media used throughout the ancient and even modern world. The Greeks believed the soul of all creatures was a butterfly, released upon death. Ancient Egyptians considered buprestid beetles

Figure 3.25. Totem poles in Kispiox, British Columbia. Photo by Hans-Jürgen Hübner available at http://bit.ly/1WiZmd9.

Figure 3.26. Shield of arms of the Province of Salamanca, Castilla y León. The image was created by Heralder and is available at http://bit.ly/1Xk4CPp.

to be representations of reincarnation. In fact, many civilizations in the Old and New Worlds believed that release of the "soul" and aspects of life after death were linked to an array of insects. For example, the Moche

people of pre-Columbian Peru encouraged consumption of the dead by necrophagous flies so that the anime of loved ones could return to the village in the form of an insect. The Aztec and Maya worshipped gods, often in the form of butterflies, and at least one butterfly goddess controlled death by her ability to swallow darkness. As we learned in chapter 2, the God of the Christian faith employed insects in three, possibly four, plagues cast against Egypt to grab the attention of the pharaoh. Ultimately, the pharaoh relented to allow the Hebrew slaves to be freed, and as the text of the Bible states, as they wandered the desert in search of the Promised Land, many types of insects (e.g., manna in the form of honeydew, locusts) were brought to them by God to eat.

In the early Christian church, honey bees were used to symbolize frailty and virginity. The link to virginity was tied to the belief that bees do not have sex, since no one had apparently ever observed copulation between them. Like the Virgin Mary, honey bees were thought to give rise to progeny while still pure, that is, unmated. The basis for this idea was simply that many species of bees engage in sexual intercourse during flight, several meters above the ground. Consequently, most people had never witnessed the deed being done in midair, an honest

Figure 3.27. Nola, the official mascot of the NBA's New Orleans Hornets (now Pelicans) franchise. In 2014, the name "Hornets" returned to the franchise in Charlotte, North Carolina. Photo by Infrogmation available at http://bit.ly/1LwX9rm.

mistake by early church leaders. One would have to suppose that if the clergy at the time had had any sense that sex did occur, or that it was incestuous (since queens mate with their sons) or that males were considered inconsequential following copulation and were frequently chased from the hive or even killed, that they would have picked another insect as a religious symbol.

Laws, litigation, and insects as evidence

Our last example of the cultural influence of insects is a somewhat mixed bag, encompassing ideas that might be difficult to appreciate: I am speaking of the relationship between insects and the law. In some instances, laws have been written that specifically identify some insects, going so far as protecting some hexapods. For example, in the state of California, pollinating bees are protected by a statute that forbids spraying insecticides in certain areas when it is pollination season, that is, at a time of year when growers are aware that bees are actively visiting specific crops. Violators will and have faced substantial monetary fines. Failure to understand insect biology is not a viable defense.

In other cases, insects or their activity are the reason for the litigation. Without a doubt, this is where the majority of examples of insects and the law reside. Civil or criminal cases involving pest control companies are the most frequent associated with insects. As you can imagine, plaintiffs accuse companies of not honoring contracts, such as not providing adequate control or that pest insects are back, or more severe charges are filed, such as injury or death caused by negligence associated with pesticide application(s) or improper treatment of a building leading to structural failure. Similarly, discovery of insects or their body parts in food products, including items purchased at a market or restaurant, can result in litigation depending on the circumstances. Administrative law is the part of the judicial system that governs the rules, regulations, and policies that establish acceptable insect interactions during food production and processing. A violation of administrative law is required before a civil or criminal case can be made in terms of insect contamination of food.

Insects can also come to the aid of legal investigations by serving as evidence or providing clues about a particular case. For example, certain species of insects that function as decomposers of dead animals are valuable indicators of time elapsed since death, in instances of suspicious death. This is particularly true of blow flies and bottle flies (family Calliphoridae), the adults of which can detect a body within minutes following death. Females will seek out the corpse and then lay eggs, usually as the first colonizers of the body. As the corpse is fed upon and undergoes a series of physical and chemical changes, different species of insects and other organisms become attracted to it. By understanding the natural succession of these saprophagous insects under different environmental conditions, and knowing how long each stage of insect development requires in the conditions to which the body has been exposed, investigators can use flies and other insects to estimate the length of time the body was available for colonization, otherwise known as the minimum **postmortem interval**. Other useful information can also be determined from the necrophagous insects, such as whether the body has been moved from one geographic area to another, or whether the corpse had been previously submerged in water or was hidden in a location not accessible to insects, at least not initially. These topics form the basis for forensic entomology, specifically the subfield known as **medicocriminal entomology**. Chapter 15 explores forensic entomology in more depth, including an examination of whether television or movie versions of crime scenes provide an accurate portrayal of insect use in criminal investigations.

CHAPTER REVIEW

❋ **Insects as our "friends"**

- The task of trying to convince others who have little understanding of insects that insects are indeed useful, or to persuade those whose position is that insects have no utility to life on Earth—period—that in fact hexapods are absolutely necessary, well, the word "challenging" is an understatement. The main reason for the difficulty is that many of insects' positive influences are subjective. Consequently, not everyone will agree that any particular insect-derived product or activity is necessarily beneficial, and even if they did, there generally is no means to quantify the benefits reaped by humans.

- Insects provide beneficial products to humans: many that can be consumed, others that are used as household items, and still more that offer medicinal utility. Humans also reap benefits from the activities of many insect species, such as pollination, decomposition, and regulation of other organisms. Hexapods' contributions to cultural aspects of the human condition are evident in art, music, film, television, and even in religion and the law.

❋ **From bee puke to scale poop: Useful insect products**

- Insects produce a wide range of products that have been used by humans from ancient civilizations to modern man. Some are food-based, many are tied more directly to household uses, and an increasing number are being developed for use in allopathic and homeopathic medicinal applications.

- Honey bees are the source of the tastiest examples of insect-derived products eaten by humans. In North America, "honey bees" refers specifically to members of the genus *Apis*; the European honey bee, *Apis mellifera*, is the species most recognized and important to the honey industry.

- The entomological influence on the production of clothing and other textiles dates back thousands of years to the ancient civilizations of the Chinese, Aztec, and Maya. Insect-derived dyes were used for staining a range of fabric types, and protein fibers from a moth led to the creation of the most luxuriant fabric known to man, silk. In both instances, synthetic versions have almost entirely replaced the labor-intensive practices of insect farming. Nonetheless, natural insect dyes and silk are still produced (and still expensive), because there really is no substitute for the genuine article.

- A variety of insect products share a common theme: protection. Insect secretions that protect our furniture, shoes, and cars include wax, shellac, and propolis, the chief insect-derived constituents in a variety of polishes, varnishes, and sealants. Wax is the principal component of honeycomb produced by bee hives. The wax is collected from melting the honeycomb. Pure wax can be used as a wood protectant and sealant, foundation for shoe polish, component of cosmetics and lip balm, and to make natural candles. As with so many of the commercialized insect products, synthetic waxes are more commonly used today because of easier production and lower cost compared with collecting the naturally produced wax from honey bees.

- Several insect-based products consumed or applied to the human body are believed to confer medicinal benefits. Unfortunately, a strong linkage to experimental evidence does not exist for most.

❋ **Oh the wonderful things that insects do!**

- The roles that hexapods play in nature are critical to the functioning of every ecosystem they reside in. Humans benefit enormously from insects pollinating flowering plants, decomposing dead plant and animal matter, regulating the populations of an array of organisms, serving as a food source for many organisms (including humans), and even providing medicinal uses through their activities.

- Insects are the dominant animals that aid plant reproduction and offer several advantages over wind pollination. More than 90% of all flowering plants depend on insects for pollination, including those that yield the many fruits, vegetables, and nuts that we consume and the flowers that bring aesthetic pleasure. The majority of food production for the world therefore depends on insect pollination. Flowering plants make up a large portion of nearly every ecosystem on the planet (the major exception are the oceans), and they are essential to the health and stability of each. Their survival depends on sexual reproduction, which relies on insect surrogates to intercede in gamete union. If plants stumble, then so do ecosystems, which in turn stresses our ability to survive.

- Insects' role in the decomposition processes of plants, animals, and other organic life is one of their most important contributions to ecosystem functioning. Several insect species consume or degrade other organisms that function as producers and consumers in a particular ecosystem. Once these organisms die, or begin to decay, their remains must be broken down into smaller parts or elements, so that complex organic molecules will be fragmented.

- Herbivorous and carnivorous insects feed on plants, animals, and other organisms that have the potential to be, or always are, pests or harmful to the human condition. In other instances, insects' feeding activity may have nothing directly to do with keeping potential pests in check, but still plays a significant role in regulating population densities within specific ecosystems. Under most conditions, predators and parasites suppress insect populations by 10%-25%. While this range may not seem like much, it really is critical to maintaining the delicate balance between keeping insects in check and a population explosion.

- Entomophagy is the practice of consuming insects as part of the diet. All stages of insect development are common dietary staples in Africa, Asia, New Zealand, Australia, and North, Central and South America. However, entomophagy is rare in developed nations, other than as a fad or gimmick. Today, people eat insects for a variety of reasons, including the high nutrient content (amino acids levels in insects are often five to ten times higher than in other animal tissues), for celebrations and rituals, and also to supplement other food sources.

- Humans derive health benefits directly from insect activity. Two examples, apitherapy and maggot therapy, fall near the extreme end of the spectrum in terms of what humans will do for the sake of survival.

Insects got class: Cultural influences of six-legged creatures

- The phrase "appreciating insects" is more commonly used for the influence of insects in art, music, and multimedia. The subdiscipline of cultural entomology is devoted to the study of such influences, and also includes symbolic uses of insects.

- Insects have a long history with art, serving as the subject matter for numerous paintings and sculptures and being immortalized in amulets and precious jewelry worn by emperors, kings, queens, priests, nobles, and many other figures of extreme or supposed importance to past civilizations. In the ancient world, sculptors would accept the tedious challenge of creating works of art depicting intricate figures of insects, sculpting the detailed bodies in stone, including the elaborate network of wing veins. The presence of insects in paintings often reflected the technical skills of those engaged in realism, at other times was meant to convey a message, and in yet other instances interpretation has been left in the eye of the beholder.

- The entomological influence in music has been evident for hundreds of years. Song titles, lyrics, musical themes, band names, and even album and CD cover art bear the mark of insect influences. No genre of music has been untouched.

- Since the advent of cinematography in the late nineteenth century, insects have been depicted in films. Modern media has evolved to include television, web-based productions, and gaming, and the list is expanding at a rapid rate. As media has evolved, so too have the roles of insects, with many becoming stars.
- Insects have been used by many civilizations to symbolically express ideas, concepts, and people. Past cultures have relied on totemism and heraldry, and in modern times, symbolic uses of insects are frequently associated with professional and amateur athletics via the mascot.

❊ A religious experience

- Entomological connections to various religions have been evident since the formation of organized religion. Gods and goddesses have been depicted in paintings, carvings, engravings, and sculpture, to name just some of the known media used throughout the ancient and even modern world. The Greeks believed the soul of all creatures was a butterfly, released upon death. Ancient Egyptians considered buprestid beetles to be representations of reincarnation. In fact, many civilizations in the Old and New Worlds believed that release of the "soul" and aspects of life after death were linked to an array of insects. The Aztec and Maya worshipped gods, often in the form of butterflies, and at least one butterfly goddess controlled death by her ability to swallow darkness. The God of the Christian faith employed insects in three, possibly four, plagues cast against Egypt to grab the attention of the pharaoh.
- In the early Christian church, honey bees were used to symbolize frailty and virginity. The link to virginity was tied to the belief that bees do not have sex, since no one had apparently ever observed copulation between them.

❊ Laws, litigation, and insects as evidence

- The relationship between insects and the law frequently shows little evidence of a beneficial side of hexapods. In some instances, however, laws have been written that specifically identify some insects, going so far as to protecting some hexapods.
- Insects or their activities are frequent reasons for the litigation. Without a doubt, this is where the majority of the examples of insects and the law reside. Civil or criminal cases involving pest control companies are the most frequent associated with insects. Plaintiffs accuse companies of not honoring contracts, such as not providing adequate control or that pest insects are back, or more severe charges are filed, such as injury or death caused by negligence associated with pesticide application(s) or improper treatment of a building leading to structural failure.
- Insects can come to the aid of legal investigations by serving as evidence or providing clues about a particular case. For example, certain species of insects that function as decomposers of dead animals are valuable indicators of time elapsed since death in instances of suspicious death. This is particularly true of blow flies and bottle flies (family Calliphoridae), the adults of which can detect a body within minutes following death. Females will seek out the corpse and then lay eggs, usually as the first colonizers of the body. By understanding the natural succession of these saprophagous insects under different environmental conditions, and knowing how long each stage of insect development requires in the conditions to which the body has been exposed, investigators can use flies and other insects to estimate the length of time the body was available for colonization, otherwise known as the minimum postmortem interval.

MUSHROOM FARMING (SELF-TEST)

Level 1: Knowledge/Comprehension

1. Define the following terms:
   - (a) entomophagy
   - (b) gustation
   - (c) cultural entomology
   - (d) heraldry
   - (e) totemism
   - (f) maggot therapy

2. Identify products made by insects that are used (a) for food, (b) in the clothing industry, and (c) for household uses.

3. Describe how insect activity can be beneficial to humans.

4. What are some examples of insect influences on cultural aspects of the human condition?

Level 2: Application/Analysis

1. Explain the relationship between the condition of myiasis and the medical treatment called maggot therapy.

2. Honey bees may be considered the most beneficial insects to mankind. Provide evidence to support this statement.

3. Describe the relationship between insects and the law.

4. Identify some of the reasons that insects are considered ideal subject matter for works of art.

Level 3: Synthesis/Evaluation

1. Explain why insects, as opposed to other animals or plants, are being promoted for entomophagy in regions of the world facing food shortages.

2. Speculate on the reasons humans find disgust in the idea of consuming insects yet relish consuming other arthropods like crabs, lobster, and shrimp.

3. Flies that induce myiasis are used in maggot therapy. However, all flies that induce myiasis are not suitable for use in maggot therapy. Speculate on some of the attributes of that make one type of fly better suited than another for use in maggot therapy.

REFERENCES

Berenbaum, M. R. 1996. Bugs in the System: Insects and Their Impact on Human Affairs. Helix Books, New York, NY.

Danneels, E. L., D. B. Rivers, and D. C. de Graaf. 2010. Venom proteins of the parasitoid wasp *Nasonia vitripennis*: Recent discovery of an untapped pharmacopee. *Toxins* 2:494-516.

Dossey, A. T. 2013. Why insects should be in your diet. *The Scientist*. http://www.the-scientist.com/?articles.view /articleNo/34172/title/Why-Insects-Should-Be-in-Your -Diet/. Accessed July 8, 2014.

Eisner, T. 2003. For the Love of Insects. Belknap Press, Cambridge, MA.

Food and Drug Administration. 2014. Dietary supplements. http://www.fda.gov/Food/DietarySupplements/default .htm. Accessed June 23, 2014.

Gillott, C. 1995. Entomology. 2nd ed. Plenum Press, New York, NY.

Graham, J. 1992. The Hive and the Honey Bee. Rev. ed. Dadant and Sons, Hamilton, IL.

Irwin, M. E., and G. E. Kampmeier. 2002. Commercial Products, from Insects. In: Encyclopedia of Insects (V. H. Resh and R. Carde, eds.). Academic Press, San Diego, CA.

Lubke, L. L., and C. F. Garron. 1997. The antimicrobial agent melittin exhibits powerful in vitro inhibitory effects on the Lyme disease spirochete. *Clinical and Infectious Diseases* 25 (Supp. 1): S48-S51.

Mathes, E. F. D., and I. J. Frieden. 2010. Treatment of Molluscum contagiosum with cantharidin: A practical approach. *Pediatric Annals* 39(3): 124-130.

Morales-Corts, M. R., M. Á. Gómez-Sánchez, and R. Pérez-Sánchez. 2014. Evaluation of green/pruning wastes compost and vermicompost, slumgum compost and their mixes as growing media for horticultural production. *Scientia Horticulturae* 172(9): 155-160.

Mott, M. 2004. Bugs as food: Humans bite back. *National Geographic News*. http://news.nationalgeographic.com /news/pf/76388694.html. Accessed July 8, 2014.

Rivers, D. B., and G. A. Dahlem. 2014. The Science of Forensic Entomology. Wiley-Blackwell Publishing, West Sussex, UK.

Sherman, R. A., M. J. Hall, and S. Thomas. 2000. Medicinal maggots: An ancient remedy for some contemporary conflicts. *Annual Review of Entomology* 45:55-81.

Sutherland, T. D., J. H. Young, S. Weisman, C. Y. Hayashi, and D. J. Merritt. 2010. Insect silk: One name, many materials. *Annual Review of Entomology* 55:171-188.

US Department of Agriculture. 2014. Honey. ISSN: 1949-1492. http://www.nass.usda.gov/Publications /Todays_Reports/reports/hony0312.pdf. Accessed June 19, 2014.

Waldbauer, G. 1996. Insects through the Seasons. Harvard University Press, London, UK.

Waldbauer, G. 2000. Millions of Monarchs, Bunches of Beetles: How Bugs Find Strength in Numbers. Harvard University Press, Cambridge, MA.

Wouters, J., and A. Verhecken. 1989. The coccid insect dyes: HPLC and computerized diode-array analysis of dyed yarns. *Studies in Conservation* 34(4): 189-200.

Yen, A. 2009. Edible insects: Traditional knowledge or western phobia? *Entomological Research* 39(5): 289-298.

THE ENTOMOLOGIST BOOKSHELF (SUPPLEMENTAL READINGS)

Gennard, D. 2012. Forensic Entomology: An Introduction. Wiley-Blackwell Publishing, West Sussex, UK.

Gordon, D. G. 1998. The Eat-a-Bug Cookbook. Ten Speed Press, Berkeley, CA.

Hull, R., R. Katete, and M. Ntwasa. 2012. Therapeutic potential of antimicrobial peptides from insects. *Biotechnology and Molecular Biology Review* 7(2): 31–47.

Pavillard, E. R., and W. Wright. 1957. An antibiotic from maggots. *Nature* 180:916–917.

Sammataro, D., and A. Avitabile. 2011. The Beekeeper's Handbook. Cornell University Press, Ithaca, NY.

Sleigh, C. 2007. Six Legs Better: A Cultural History of Myrmecology (Animals, History, Culture). Johns Hopkins University Press, Baltimore, MD.

ADDITIONAL RESOURCES

American Apitherapy Society
http://www.apitherapy.org/

Beneficial insects
http://www.insectary.com/

Benefits of honey
http://www.benefits-of-honey.com/honey-bees.html

The bug chef
http://davidgeorgegordon.com/

Forensic entomology
http://www.forensic-entomology.com/

Insect decomposers
http://www.cals.ncsu.edu/course/ent425/library/tutorials/ecology/decomposers.html

Insect fear film festival
https://www.facebook.com/IFFFatUofI

Insects are food
http://www.insectsarefood.com/index.html

Insects in the movies
http://what-when-how.com/insects/movies-insects-in-insects/

Modern insect artwork
https://www.pheromonegallery.com/

Monarch labs living medicine
http://www.monarchlabs.com/

The resource on edible insects
http://www.entomophagy.com/

4   Insects Shaping Human Civilization

Turns Out They Might Be Bad After All

Keep your friends close, but your enemies closer.

Michael Corleone
The Godfather Part II (1974)*

Let's face it; insects are not always good neighbors. Heaven knows you've tried your best to keep the relationship going. But those pesky critters just won't cooperate. Eating your food, sharing your bed, and blood-feeding on your butt. Perhaps it's time to unfriend them? If only so simple. The reality is that those species that cause damage to the human condition do it very well. Considering the enormous success they have had by maintaining a relationship with mankind, pest insects will not go away easily.

This chapter delves into the destructive side of the class Insecta, examining insect pests of agriculture, stored products, and households, as well as those that vector disease or just annoy us. Some of the very insects discussed in chapter 3 as beneficial resurface here as occasional, or sometimes serious, pests. Such betrayal! To understand how and when entomological allegiances change, we will explore what it means to be a "pest" from an aesthetic and economic perspective. This chapter also attempts to shed light on the key features of insect biology that permit some species to be extraordinarily efficient at being pests of humans, and to decipher what man does to become a prime target.

Key Concepts

- ❋ You're making it tough to be friends!
- ❋ What does it mean to be an insect pest?
- ❋ Why are they so good at being bad?
- ❋ Where has all the food gone? Agricultural pests
- ❋ Let's live together: Household pests
- ❋ They just "bug" me: Annoying insects
- ❋ Insects, disease, and human civilizations: Medically important pests
- ❋ Implications for modern societies

---

*The acclaimed movie *The Godfather Part II* (1974), directed by Francis Ford Coppola and starring Al Pacino as Michael Corleone, in which the famous line is shared by Michael as words of wisdom learned from his father, Vito, the godfather (http://www.youtube.com/watch?v=DfHJDLoGInM).

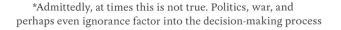

| Always | Usually | Never |

Figure 4.1. Friend status of some insects to humans. Lady bird beetles, or lady bugs, function as predators of a number of pest insects, honey bees are beneficial until they sting us, and pubic or crab lice never enter the friends circle (unless they crawl!). Photo of lady bug (by Harald Hoyer) available at http://bit.ly/1OcvvTZ, honey bee (by John Severns) at http://bit.ly/1jccBho, and crab or pubic louse (Josef Reischig) at http://bit.ly/1NUGjEF.

You're making it tough to be friends!

"Keep your friends close, but your enemies closer"; these words of wisdom may work when dealing with the Mafia, but are not as effective for handling destructive insects. Unless of course the goal is to blast 'em (with insecticide) once within range. Chapter 3 laid a foundation for an insect-mankind friendship, detailing the many ways that most insects are beneficial, or no worse than benign (including those we simply appreciate), to the human condition. Of the roughly one million described species of insects, the vast majority, nearly 990,000 species, fall into the categories of essential, entertaining, or no threat to man. It is the remaining few, less than 1% of the whole, that cast a shadow of negativity on the entire class (figure 4.1). This small minority can elicit an array of emotions—anger, fear, anxiety, repulsion, and distress—and at times create only modest physical discomfort, whereas on other occasions the outcomes can be far more severe: starvation, injury, disease, and death. These latter realties tend to trump all the good we find in insects. Not helping the situation—for insects that is—is that, as a species, we define good and bad completely from a human-centric viewpoint. We value human life above all else. There are no levels of life loss or human injury that are tolerable. When insects threaten even one us, action is usually taken,* and swiftly.

Figure 4.2. An example of a person who ate an insect but did not die. At least, he was still alive at the time of this photo. Photo by Terri Cancila, Loyola University Maryland.

It really is a no-win situation for insects. As discussed in the last chapter, it is far tougher to convince someone that insects are useful or beneficial to man than it is that they are harmful. This is partly a result of conditioning by our parents, media, and others that many, if not all, insects offer nothing but trouble for humans. Memories from my own childhood are vivid with lessons that bees are poisonous and will chase you, any "bug" in your home is a sign of uncleanliness (I believe "pigsty" was the term heard most often), and food in contact with one of these revolting creatures should never be eaten—if consumed, you most likely will die (figure 4.2). Turns out that

---

*Admittedly, at times this is not true. Politics, war, and perhaps even ignorance factor into the decision-making process

in some regions or nations, which can result in inaction against a pest insect.

most of those life lessons were factually inaccurate. Nonetheless, for me, as for so many others, an imprint was left that cast a negative impression of insects throughout my childhood. Obviously I overcame these early influences, but the vast majority of people do not. The biggest obstacle in the way of insects making a good impression on us is that they are just trying to survive. Huh? Insects are like us, in that they need to eat and prefer an abode that offers refuge from the harshness of an unfriendly natural environment. Humans provide everything insects need in one setting. We also are easier targets for insects to utilize, relatively speaking, than other organisms.

How are we easy? It's simple when you look at what humans can do that other animals cannot or simply do not do. For example, we engage in agriculture to feed ourselves, so large volumes of food become available at predictable times during the year. We store food in our homes, constantly, so insects can find food year-round if they hang out with us. Our dwellings provide protection against adverse weather and a safe house from potential predators and parasites. Humans live in large communities, making us easy to locate. And when it comes to food, humans offer a nutritious and great-tasting meal. Well, I actually do not know if the latter is true, or whether insects might actually consider us their lima beans or brussels sprouts.* Because insects take advantage of us, we classify them as pests. More specifically, humans define insects as pests based on our own subjective measures. In other words, our view of what is or is not a pest is based completely on prioritizing the human condition above all else, even over environmental concerns and ecosystem functioning. Such attitudes essentially mean

that almost any context in which an insect comes in contact with humans can be considered a pest situation. In practical terms, we tend to classify insects as pests based on what they specifically do to humans or our stuff. So, insects are pests to us because

1. they injure or feed on plants and animals associated with our food production;
2. they feed on or infest stored food or structural products;
3. they transmit disease to us, our pets, or other domesticated animals;
4. they simply annoy the heck out of us!

As we will learn in the next section, entomologists are more pragmatic in defining when and whether an insect is a pest. Be that as it may, insect pests are classified based on what they target. For example, insects that attack food in the form of crops are *agricultural pests*, whereas those that target food postharvest are *stored product pests*. *Livestock pests* or *medically important pests* frequently are **hematophagous** (blood-feeders) on domesticated cattle, sheep, swine, poultry, pets, and us. Insects that invade our homes or other dwellings are broadly termed *urban pests*, but can be further distinguished as *household* or *annoyance pests* if invading human habitation (dwellings and/or the surrounding environment), and *structural pests* if feeding on and/or burrowing into wood-based products used in building construction, as well as overlapping into stored product and medically important pests. In this chapter we examine the effects on man's existence of each type of pest situation, while chapter 14 discusses what we do to remedy problems with pests.

What does it mean to be an insect pest?

The take-home message from the last section is that, for most individuals, insects are considered pests 24/7. That could be refined somewhat. Actually, a lot! The reality is that comparatively few insect species are considered pests of the human condition in any context, and even fewer are classified as serious pests on a regular or consistent basis. Are any ever pests all the time? Subjectively speaking, yes, some species never deviate from being a problem for humans. For instance, medically important insects always represent a threat when living in close

---

*More than one child has sat at the dinner table, alone, faced with the insurmountable requirement of consuming cold lima beans or brussels sprouts before resuming their normal lives. Hiding these tools of the devil under a napkin or feeding to the family dog when no one is looking sometimes works, but usually not, especially if man's supposed best friend barfs it out onto the floor. Strategically placing under the tongue to dispose of in the toilet is tried and true, unless mom or dad wants to chat before you make it out of the kitchen. I seriously doubt that any self-respecting insect would be herbivorous on either plant product! Okay, it turns out that some insects do eat these vegetables but I for one do not agree with their decision.

Figure 4.3. Typical farming practice in the United States, in which monoculture (of, in this case, corn) occurs on hundreds of acres on a given farm or given region. Photo by Jarek Tuszynski available at http://bit.ly/1FgkfBA.

proximity to man. Of course, pest status can change, as was the case in North America with the mosquito *Anopheles quadrimaculatus* (Diptera: Culicidae*), a potent vector of malaria on the continent until around the 1950s, a period during which the causative agent of malaria—protozoa in the genus *Plasmodium*—were eradicated, thus downgrading the threat of the mosquito.

The question is, when and how are insects determined to truly be pests? Truthfully, even for entomologists, an air of subjectivity surrounds defining pests—but not nearly as much as in the stance, "They exist, therefore they are pests." Broadly defined, insects are pests whenever they affect human welfare or aesthetics. Of course, both "welfare" and "aesthetics" lie in the eye of the beholder, so a more tangible measure is needed for establishing a baseline. What this means is that it is not economically practical or physically feasible to react every time a perceived insect pest turns up in a particular area.

*Up to this point in the textbook, order and family names have been preceded by the terms "order" or "family" as appropriate. Ordinarily, when an insect is identified by genus species, the names are followed by order and family parenthetically as given here. The first name given is the order, then the family, with each separated by a colon. Both terms are capitalized and neither are underlined or italicized as a genus and species names are. In the example here, *Anopheles quadrimaculatus* belongs to the order Diptera and family Culicidae.

A baseline or threshold must be established for making decisions on when to get involved, say, by treating with insecticide or another type of artificial control method.

In an agricultural situation, pest status is influenced or defined by economics. Let's use a cornfield as an example. In the United States, corn is typically grown in monoculture on large parcels of land (figure 4.3). If a scout discovers an adult corn borer moth (*Ostrinia nubilalis*) while patrolling the field, is that insect a pest? Certainly corn borers are classified as serious pests of corn. However, the key factors are the stage of development in which it is discovered, the numbers of individuals witnessed, and the activity of the insect at that time and anticipated in the future. In short what I'm saying is, the biology of the insect must be understood to know whether the insect is even infesting the crop at all, and if so, whether it is causing injury or damage to the crop. What is the difference between these two named outcomes of insect activity? Entomologists define "injury" as physical harm to a valued commodity, corn in this case, caused by the presence or activities of the insect in question. "Damage" is a notch up in terms of intensity and represents monetary loss in the commodity as a result of the injury caused by the insect. Practically speaking, injury can occur with almost any level of insect occurrence or infestation, but damage does not always

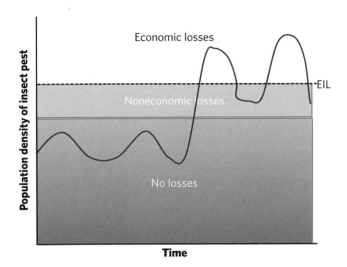

Figure 4.4. Relationship between the population density of a potential insect pest and the economic injury level (EIL). Economic losses are dependent on the value of the commodity and cost of controlling the insect. The figure is based on Pedigo and Rice (2006).

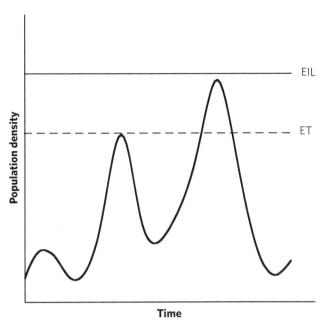

Figure 4.5. Threshold used to establish when an insect is an economic pest and when to apply artificial control. EIL = economic injury level, ET = economic threshold. The figure is based on concepts from Pedigo and Rice (2006).

occur. This also means that an insect might be considered a pest because it evokes injury, but still nothing is done about the pest because insufficient or no damage has occurred or will occur. Here is where the economic influence is most evident. A threshold level is established for a particular commodity from which to make decisions on whether injury or damage will occur or has occurred.

One such threshold is the **economic injury level** (EIL), which is defined as the lowest population density of a specific insect that will cause economic damage, or the amount of pest injury that justifies using some type of control strategy for suppressing the pest (figure 4.4). This measure is a function of both the cost of controlling the pest and the value of the crop or commodity. With this in mind, it is important to recognize that some commodities have so little overall value that it is not worth the cost of trying to save or rescue them from insect damage. Correspondingly, the value of some products is so great that any level of insect damage justifies the use of control measures. This is also true for commodities in which any injury is considered damage, as in the case of fruits and vegetables sold as produce. The average consumer in the United States will not purchase produce with any form of cosmetic injury, and especially not if it is assumed to be the result of insect activity, therefore would-be insect pests must be targeted prior to any injury occurring at

all. So simply existing can relegate some insects to the status of pest!

In some instances it is more practical to use control measures before the pest in question achieves a population density reaching or exceeding the EIL. For example, if sampling a particular area indicates that the population density is below the EIL but is on the rise, and conditions favor continued growth, then the obvious conclusion is that a pest outbreak is inevitable. The grower will usually decide to go on the offensive by using some type of control measures to prevent the pest population from reaching the EIL. This approach certainly makes sense, because it is easier to control a smaller population than a larger, and the crop will incur, at least theoretically, less damage if treatment is applied before the EIL is achieved. Treating before the EIL is reached is a common strategy in pest management. It relies on establishing a threshold level, termed an **action threshold** (AT) or **economic threshold** (ET)—a population density below the EIL—from which to make the decision whether to treat or not (figure 4.5). Now this might seem like a contradiction to our definitions, because the EIL already is derived from the economics of the situation, meaning it would seem most profitable to wait until the pest density elevates to the minimum point of the EIL. The logic is sound, but in reality, the EIL is based on the value of the commodity, which will vary with market demand, and on the cost of control. Here is a key point to the strategy of the ET: it is usually cheaper to treat when the population is lower, prior to a pest outbreak. Consequently, profit can be maximized for some insects by treating when pest density reaches the economic threshold. Later in this chapter we will examine how the concepts of economic thresholds and economic injury levels are applied to different pest situations.

Economic considerations are not relevant to determining pest status in some contexts. This is most obvious with many of the insects that invade our homes. I am referring to insects such as most species of cockroaches or flies, which really pose no economic or medical threat to us. These insects simply exist and have made the poor decision to move into our homes or places of work, restaurants, and so on. Pest status is defined by the individual, which is influenced by a whole array of factors. The parental and cultural conditioning mentioned earlier is probably the biggest influence on how we define a pest. For the average person living in the United States, no insect in the home is the acceptable level of entomological roommates. I have a higher tolerance than most people, which is irrelevant because my wife is intolerant of any six-legged creature in our (her) home. The point is that pest status is completely defined by the individual, based on personal reasoning. Any insect in question maybe a pest to one person but not to another. Entomologists define these pest determinations as **aesthetic injury levels** (AIL). Pest status is entirely subjective and individual, independent of cost analyses or other typical considerations used for determining an EIL or ET. Aesthetic injury levels certainly should not be dismissed, because they are a major factor underlying why the pest control industry in the United States is a multibillion-dollar annual entity. We will also explore in later sections the circumstances in which the AIL is the prevailing determinant of pest status.

Quick check
Is it possible to be an agricultural pest without causing monetary losses in a crop?

Why are they so good at being bad?

Now that we have an understanding of what an insect pest is, from the economic and aesthetic perspectives, the next question that may have crossed your mind is, "Why are they so good at it?" If this question has never formed, well consider this: Can you name any other invertebrate animals? If you can, congratulations! Most people cannot. And if you can, you usually know them for one of two reasons: either you eat them, or they eat something of yours. The fact is that the average person is aware of insects, can even refer to several species by common names (e.g., mosquito, termite, ant), because these beasts are viewed as pests, but most any other invertebrate, pest or not, is anonymous to us. Why? One simple explanation is that there are just so many insects (remember our discussions from chapter 1), so if even just a few of the whole are "bad," that is still a lot of species in comparison with everything else. But that does not fully explain the story. Insects tend to be extraordinarily good at any endeavor they put their little minds to. Being a pest is no exception.

The attributes that make insects exceptional at achieving pest status, by any convention (i.e., whether using EIL, ET, AIL, or ALF*), are basically the same as those that contribute to their enormous success overall. We will delve into biological explanations for why they have been so successful as a group in chapter 5 (and, really, throughout the remainder of the text). What follows is just a brief overview of some of the key features that allow insects to earn the title BDPOP (Best Darn Pests on the Planet). Okay, well, no one else says that but me, so maybe we should just stick to the biology of insects. Insects that are pests share some important features.

1. *Wings*. No other invertebrate animal has wings, yet most insects that are pests are winged during the adult stages. The increased range of locomotion is obvious, as is the unique mechanism for escaping any type of potential danger. All insects are not

*ALF is of no value to this conversation since it refers to a television sitcom that aired on NBC from 1986 to 1990 in which the title character was a friendly, yet *annoying* alien. Ah, the latter does tie in to our definitions of pest! http://www.youtube.com/watch?v=AEz_FlzLtZc.

Figure 4.6. Animal Facebook. The vast majority of the animals shown are common invertebrates that inhabit terrestrial and aquatic habitats. Composite image by Medeis available at http://bit.ly/1Tfoe7T.

equal when it comes to the efficiency of flight, which contributes to some being better pests than others.

2. *Acute olfaction.* Insects have an amazing ability to detect chemicals in the environment, otherwise known as the sense of smell or olfaction. This is an incredibly important adaption for this class of animals because they rely on chemical communication perhaps more than any other group of animals. Enhanced chemoreception (smell) allows them to detect potential food sources, and then they can rely on their wings to transport them to their next meal.

3. *Well-developed chemical communication.* As indicated with olfaction, a well-defined chemical communica-

tion system goes hand in hand with an acute sense of smell. Chapter 11 is devoted to understanding the basics of insect communication. Within the context of aiding their abilities to be pests, chemical communication is used to alert members of the same species that food or a host has been located. Scouts are commonly used to go on food reconnaissance missions, and if successful, **pheromones** are released to recruit others so that the wealth can be shared. Ants are well known for laying down trail pheromones from discovered food back to the colony, allowing any other forager to follow the same path (ant pheromone trail: http://www.youtube.com /watch?v=5HKl8Luuotw). If ants have ever invaded your home, then undoubtedly you have witnessed the mobilized ants walking "the trail" from the pantry, trashcan, or pet bowl to a hidden location behind a wall or door.

4. *Efficient and adaptable reproduction.* The remarkable world of insect reproduction is a book unto itself. Unfortunately, in this text it is limited to a single chapter (chapter 9). Their reproduction strategies aid pest insects in multiple ways. Sexually active species tend be incredibly efficient in producing offspring, and with the gender desired. What? The latter implies, correctly, that several species have the ability to influence or control the sex of their progeny, thereby maximizing reproductive effort. This is especially true for those insects that produce female-biased **clutches** and that do not require **outbreeding**, meaning that mothers can mate with male relatives without fearing deleterious consequences from inbreeding. Some insects, like aphids, actually display changes in reproductive mechanisms: early in the growing season, when adults first arrive to a field, asexual reproduction is employed to rapidly colonize the crop. Later in the season, sex becomes the norm.

5. *Body defenses capable of detoxification and resistance.* Insects are very adept at overcoming the defenses that plants or animals offer to prevent or retard attack. This is not meant to imply a one-sided relationship. Organisms that have coevolved over the course of thousands to millions of years engage in a tug-of-war in the struggle for survival. At any given point in the history of the association, one may have a slight edge over the other. However, if the relationship is relatively "new," one member of the relationship may indeed have the upper hand. This is true whether the insect in question is an herbivore, predator, or parasite; insects often have the edge. The same is true when examining an insect's own defenses against attack. For example, many plant species rely on chemical compounds to ward off herbivores, only to be circumvented by detoxification systems located within the body of the feeding insect. Individuals that survive the chemical onslaught and reproduce are likely to yield offspring with similar resistance to plant defenses (or man-

made chemicals). After just a few generations, an entire population of pest insects can essentially feed unencumbered on the plant or animal in question. They have become resistant to a particular compound or mechanism of defense. In many instances, developing resistance to one compound also confers resistance to others, even those to which the insects have never been exposed. This condition is termed **cross-resistance** and is a common occurrence found with man-made chemical insecticides.

6. *Eating machines, with extraordinary efficiency in digesting and converting nutrients to biomass.* Very few organisms on the planet can compete with insects in terms of eating. Once they find a food source, they can eat 24/7, with no need to take a break for sleep or even to defecate; they can do so while eating. Impressive! (figure 4.7), Ingested food moves rapidly from one location to the next in the digestive tract, processed according to the physical and chemical composition of the food, and utilized so that nutrient absorption is maximized along the entire length of the tube. What eventually exits the body is not useable by the insect. Almost everything that is retained is converted to biomass. The efficiency is simply astounding! Some estimates place the efficiency of assimilation—that is, the conversion of ingested food into biomass—at greater than 70% or 75%, with some species approaching 90%. By comparison, domestic livestock, whose sole existence is to become a food source for man, generally never exceed 50% assimilation efficiency. Perhaps some comfort can be gained in realizing that the bed bug sucking on your leg is amazingly efficient at what it is doing. Perhaps not.

Where has all the food gone? Agricultural pests

The answer to the question above is easy for most people: insects. Well, they cause some of the problems, but certainly not all. Their effect is global and is measured in terms of economics as well as human life. Estimating the monetary value with regard to the latter is nearly impossible. The World Health Organization has attempted to do so when considering loss of human life attributed to insect activity (the summative effect of all forms of insect pests): minimally in the hundreds of billions of dollars

Figure 4.7. Caterpillars and their frass. Sounds like a seventeenth-century painting, but here it is just a reference to the fact that some insects will continue to eat while sitting in defecate or while in the act! Photo by Fastily available at http://bit.ly/1QzczL9.

annually. How accurate is this estimate? There is no way of knowing, because there are no reliable means with which to determine the precise effect of insect activity leading toward human death. Multiple factors are generally involved, not only that insect herbivores contribute to hunger and malnutrition via food shortages or that insects vector debilitating and deadly diseases. Obviously, if someone contracts an insect-borne illness and death results, a direct cause-and-effect relationship is established. But rarely is the entomological effect on human survivorship so clear-cut. So we are left to speculation rather than empirical data to ascertain the total influence of insects on human death rates. We will revisit this discussion when examining medically important insects later in this chapter. For now, let's turn our attention to the economics of insect damage to food.

Insects that attack our food supply are thought of as agricultural pests. In reality, however, we generally distinguish between these insects based on when and what they attack. For example, the traditional category of agricultural pests includes those species that make their mark during the growth phase, that is, the preharvest phase, of crops. Insect injury and damage to crops can be direct or indirect, yielding a host of negative outcomes (see below). Insects drawn to plant-based foods postharvest are categorized as stored product pests. Meat-based

food, meaning from livestock and poultry, falls under the umbrella of livestock entomology, and hence those insects are livestock pests. Depending on their interaction with livestock, these same pests can also be medically important insects. It is also important to note that not all agricultural products are intended for human consumption, but the pests of those crops are still agricultural. For our purposes, we will broadly examine agriculturally important insects, regardless of whether the end-use product is directly intended for food consumption or not.

So is it any easier to measure the effects of agricultural insect damage in terms of economic costs? Perhaps marginally. What we know is that more than 50% of all known insects are herbivores, and that each and every type of plant humans deliberately grow (as well as those we don't) generally has some type of insect that feeds on it. The majority of the estimated 10,000 insect pest species that inhabit the globe are herbivorous and agriculturally significant and, in terms of sheer numbers, likely represent billions of individuals feeding on plants on any given day. Needless to say, the potential for extensive damage is great. That said, it is still difficult to fully calculate the effect of insect injury and damage to agriculture because so many variables are at play. Do not misunderstand, the values (for estimated damage) put forth by agricultural economists and entomologists are

Figure 4.8. The recently introduced brown marmorated stink bug, *Halyomorpha halys*, has quickly established itself as an agricultural pest of a number of crops in the eastern United States. Photo by David Lance available at bugwood.org.

Figure 4.9. An adult of the western corn rootworm, *Diabrotica virgifera*, searching for pollen in corn silk. Photo by Bemoeial available at http://bit.ly/1MqJnoJ.

based on sound models and years of observations and experience.

Nonetheless, biological systems of any kind are incredibly complex, difficult to study precisely, and therefore not fully understood. Assessing insect damage in cropping systems is no different. Why? For one, the same insect pest does not elicit identical damage from year to year. The reference to "damage" here is similar to our earlier discussion of EIL, in that the value of the commodity and cost of control vary, even during the same season. It is therefore challenging to accurately estimate the economic effect of any given insect species. Insects also cause damage in nonmeasurable ways that frequently go unnoticed or are not attributed to a specific pest, yet the end result reduces crop yield, suppresses food quality, and increases costs of harvesting or processing, as well as financial investments in control measures. So the next time you see a "bug" expert interviewed on television or in the newspaper, notice how many times "predictions" and "estimates" are used in answers to questions about a specific insect pest, such as, "What will be the effect of stink bugs (a reference to the brown marmorated stink bug, *Halyomorpha halys*) this year to farmers in Maryland?" The variability discussed above prevents any specific statements from being made (figure 4.8).

What can we say about insect damage in terms of monetary value? We will use a few examples to show what is known about the insect effect on economics and,

ultimately, you the consumer. In the United States, close to 350 million acres of land are planted in crops each year. Of those, just shy of 100 million acres, or approximately 28%, is dedicated to corn production. Corn, or maize, is used in literally thousands of corn and maize-based products, making it one of the most important agricultural crops in the United States and worldwide. With so much acreage dedicated to raising corn, it is clearly one of the crops most targeted by insect pests. Perhaps the most sinister of the corn pests are beetles in the genus *Diabrotica* (Coleoptera: Chrysomelidae), more notoriously known as corn rootworms or cucumber beetles (figure 4.9). Their destructive efforts are realized preharvest, during their larval stages, in which the immatures feed on the corn roots. Affected plants are essentially destroyed: corn plants are stunted, food quality drops, and plants topple over to rot on the ground. More than 35% of all maize in the United States is attacked by corn rootworms each year, resulting in nearly $2 billion in economic losses, and about the same amount of money is spent in control efforts, amounting to a nearly $4 billion measurable economic effect. Corn that survives to harvest then faces an onslaught of insects functioning as stored product pests, generally yielding additional losses of $1 billion–$2 billion. All told, more than 20% to 25% of all corn produced in the United States is destroyed, or at least rendered unsuitable for its intended use, as a result of insect activity.

Agriculture's Most Wanted

Figure 4.10. Some of the most destructive insect pests in the United States. Clockwise from the top left, southern corn rootworm (Pollinator, http://bit.ly/1KAEhmd), gypsy moth larva (Didier Descouens, http://bit.ly/1THyIXX), sawtoothed grain beetle (author unknown, http://bit.ly/1G3FPUN), Asian citrus psyllid (David Hall, http://bit.ly/1Ovj8kn), and the Colorado potato beetle (Fritz Geller-Grimm, http://bit.ly/1V9Z23c).

The story of steep economic losses associated with agriculture is repeated with numerous pests. Take the Asian citrus psyllid, *Diaphorina citri* (Hemiptera: Psyllidae), as an example. Since its relatively recent introduction into North America from Asia in 1998, it has caused nearly $10 billion in damage to orange and grapefruit crops in Florida and California, feeding on new shoots and sucking sap, as well as transmitting disease to citrus. The Khapra beetle, *Trogoderma granarium*, is one of the most serious invasive species of grain products and seeds worldwide, but is most prevalent in the Middle East, Africa, Asia, and the Mediterranean. Beetle larvae feed on damaged and whole grains, resulting in grain weight loss of 5%–30% but sometimes approaching 70% during severe pest outbreaks. These grain pests are capable of causing complete destruction of stored grains in a short period, yielding monetary losses exceeding $5 billion annually (figure 4.10).

In the mid-Atlantic region of the United States, the accidentally introduced brown marmorated stink bug, *Halyomorpha halys* (Hemiptera: Pentatomidae) has rapidly established itself as one of the most destructive pests of orchard fruits, and well, just about any other crop it encounters (table 4.1). As with other hemipterans, **nymphs** (juveniles) and adults use their syringe-like mouthparts to pierce the skin of targeted plants and suck out the nutritious fluids. This feeding action causes cosmetic injury as well as damage to any portion of the plant fed upon. Annual losses to apple production alone exceed $40 million in the eastern United States, with the total economic effect estimated to exceed $200 million each year. The accidental introduction of the Asian longhorned beetle, *Anoplophora glabripennis* (Coleoptera: Cerambycidae) into the United States sometime in the mid-1990s has led to this insect being one of the most severe nonnative pests of trees, causing an estimated $3.5 billion in damage each year. Severely infested trees die rapidly and lose all economic value, even for purposes of firewood and mulch. Nearly $4 billion in damage is quite significant but pales in comparison to one of the most destructive tree pests in North America, the European gypsy moth, *Lymantria dispar* (Lepidop-

Table 4.1. The Brown maramorated stink bug is a ubiquitous pest, attacking a wide range of crops and ornamental trees. As of 2013, it has been discovered in 40 states through the United States.

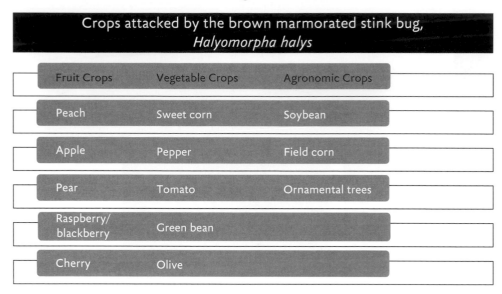

Crops attacked by the brown marmorated stink bug, *Halyomorpha halys*		
Fruit Crops	Vegetable Crops	Agronomic Crops
Peach	Sweet corn	Soybean
Apple	Pepper	Field corn
Pear	Tomato	Ornamental trees
Raspberry/ blackberry	Green bean	
Cherry	Olive	

tera: Erebidae). Gypsy moth larvae feed on the leaves of over five hundred species of trees, shrubs, and plants. Between 1970 and 2010, it is estimated that larvae of this insect caused total defoliation in over 80.4 million acres in the United States. The annual economic losses associated with the gypsy moth are approximately $14 billion, surpassing the monetary damage caused by nearly every other insect pest in North America!

This discussion could go on and on, as there is apparently no end to the examples of agriculture pests that can be provided. In reality, the number of severe pests in the United States is relatively minor, as only 150 to 200 insect species are crowned with this title. Most agricultural pests cause losses that can be calculated in millions rather than billions (either way, it's a lot of money, but a big difference when determining whether you can buy the beach house or the beach!). In many regions of the world, the damage is more severe than would occur in the United States or Europe; many countries lack the necessary resources to prevent or manage insect outbreaks. Therefore population density for a given pest can quickly exceed the EIL unchallenged (meaning no artificial control is applied) to yield extreme damage to a particular crop. The point is that, in some instances, steep economic losses are incurred not because there is no viable means to manage the pest population but because resources are

not available evenly to all nations. In chapter 14 we will examine ways that man attempts to manage populations of agricultural and other pests.

Quick check

When is an insect considered a pest?

Let's live together: Household pests

Household insects are a difficult subject for some individuals: for some, it is difficult to admit that insects do in fact live in our homes; for others, it is difficult to deal with the fact that they live among us at all. What is the difference between the two states of reality? The former is simply denial, probably associated with the whole cleanliness issue mentioned earlier, and the latter more aligned with fear. **Entomophobia** is the fear of insects, and afflicted individuals often deal with anxiety, panic attacks, or a host of irrational behaviors. In extreme cases, the fear is debilitating and can even cause periods of unconsciousness. Now, the significance of either situation identified above is that it shapes an individual's definition of a pest, meaning that the aesthetic injury level comes back into play, but not exclusively. Considerations associated with economic thresholds and economic injury levels are also relevant here, as in agricultural contexts, but not all of the time. This section examines how an insect reaches pest status within a household environment. However, before exploring that subject further, we need to broadly examine the household pest by providing a point of clarification: Household pests are a subcategory. Of what? Entomologists examine insects associated with human habitation and the human environment under the moniker of **urban entomology**. Within this area, insects can be pests

1. in association with our homes, buildings, and yards, that is, the household pests;
2. of building materials, principally wood-based products, and therefore called structural pests;
3. associated with agriculture but that secondarily invade human spaces. These pests are still considered primarily agricultural pests, usually livestock pests in the form of flies.

As you can see, "urban" does not refer exclusively to insects in cities or even specifically in homes. This is especially true when rural insects, such as flies associated with dairy farms, poultry or hog facilities, and cattle feedlots, invade nearby homes and businesses. Usually the invasion is temporary, but not always.

It is clear from our definition of urban entomology that insects can achieve pest status through measures that at times are entirely different than those discussed for agricultural pests, meaning that both the AIL and EIL factor in to the determinations. For example, for many insects that invade human habitation, the beast is a pest simply because it is present. Thus, the arbitrary measure of the aesthetic injury level is the determining factor for attaining pest status, and the threshold for pest versus no pest is completely based on individual tolerances. Non-disease-causing insects that attempt to cohabit fall into this category and range from benign moths that fly in, attracted to light escaping past a door ajar, to a disgusting cockroach scurrying across the kitchen floor. Certainly you may take exception to the cockroach example, because sometimes they transmit disease or infest food, or you might subscribe to the notion that "if you see one, then there are hundred more in hiding." "Sometimes" is the operative word. On occasion, they simply slip into a residence for a visit. In that scenario, pest status is subjective, and no real monetary loss has occurred. Of course, as stated earlier, some economic cost is associated with pests that are defined by aesthetic injury levels, since homeowners frequently apply control measures or outsource the activity (i.e. hire a pest control company) (figure 4.11).

Pest status can also be defined by traditional mathematical modeling, albeit in modified form, as discussed in agricultural contexts. Does this mean that economic thresholds and economic injury levels are established for household pests? Not exactly. The idea of the ET and EIL in a field situation relies on the ability to sample the insect of interest so that the population density can be monitored over time. Such sampling is nearly impossible in a home or business setting. Have you ever tried to find a cockroach or ant that disappears behind a wall? If so, then you know the likelihood of success in trying to model the population. Usually, too, the insect is not present all the time, and when it may occur in your home again (if at all) is unpredictable. This is in contrast to, say, the corn rootworms mentioned in the last section, which are fairly predictable as pests of corn each year, at least in those regions that have experienced previous infestations. Even keeping these limitations in mind does not rule out the possibility of establishing some type of threshold limit before acting. Obviously this is inherently more arbitrary than for agriculture pests, and depends on the insect involved. For example, most people are willing to tolerate sugar ants in the home in higher numbers and for longer than flies, and certainly cockroaches, before the decision to use artificial control is made. Even if nothing is done about the household pest, rarely is the injury or damage inflicted significant.

The situation changes radically when our attention turns to medically important insects or structural pests. On the surface, there would appear to be nothing in common among a bed bug, a frothing-at-the-mouth cockroach,* and a termite, other than the obvious physical attributes

---

*Cockroaches do not froth at the mouth, but if they did (they don't), it would be more correct to say "at the mandible." However, since we have not covered insect anatomy yet (chapter 5), let's pretend like I never said mandible. Many species of cockroaches can be mechanical vectors of disease: they pass on the causative agents as they come in contact with our food, utensils, countertops, and so on, which may eventually transfer the microorganisms to us.

Urban Entomology's Most Wanted

Figure 4.11. Some of the most despised insects in an urban environment. Clockwise from the top left, American cockroach (USGS Bee Inventory and Monitoring Lab, http://bit.ly/1Lxwvyu), bed bug (Jacopo Werther, http://bit.ly/1iMVrqL), house fly (USDA, http://bit.ly/1Kw5v0qg), eastern subterranean termite (Magnus Manske, http://bit.ly/1LORlQl), and carpenter ant (Adam Lazarus, http://bit.ly/1MJTyrm).

that classify them as insects. In reality, each achieves pest status the same way. Any population density of medically important or structural insects classifies them as pests, as there is no acceptable or tolerable level permissible. In other words, the economic injury level and economic threshold are one and the same, and both equal zero. We have discussed the human perspective with regard to injury or loss of life: no level is acceptable to us. That view explains the attitude toward medically important species, but what about structural pests? Obviously wood-boring insects do not appear to warrant the same consideration as those that can transmit disease—true, from the standpoint that structural pests do not affect human health. They also differ in that a true economic value can be placed on the damage caused to a home or some other structure. For example, termites cause more than $12 billion in annual monetary losses worldwide. However, as with insect disease vectors, discovering any wood-boring

insect in a home or building confers instant pest status. The EIL of pests such as termites and carpenter ants is also equal to zero, which means that preventive or immediate action is needed. Putting this in perspective, a home is typically the single largest purchase and asset of any family, so protecting that investment from wood-boring insects is absolutely essential for the economic security of the family, as well as from a safety standpoint, since the insects' feeding action will eventually compromise the structural integrity of the infested wood.

Several insect species can be viewed as potentially serious pests at any population level, including blood-feeders like fleas, bed bugs, mosquitoes, and flies; nonbiting disease vectors like some cockroaches and flies; and venomous insects like bees, wasps, hornets, and ants. The Fly Spot on page 103 focuses on the venomous red imported fire ant, *Solenopsis invicta* (Hymenoptera: Formicidae), a household pest that represents perhaps the worst-case

Figure 4.12. Adult red imported fire ants (*Solenopsis invicta*). Photo by USDA-ARS available at http://bit.ly/1JrUbyp.

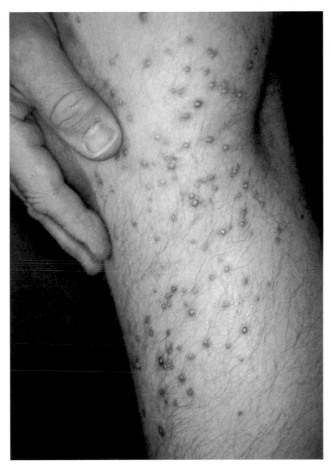

Figure 4.13. Bite marks and pustules resulting from fire ant attack. Photo by USDA available at http://bit.ly/1iyEEHi.

scenario for urban entomology: they aggressively seek out humans, can cause death, and almost nothing effectively controls their populations (figures 4.12, 4.13, 4.14).

They just "bug" me: Annoying insects

It is now time for a bonding moment. I can admit that some insects are flat-out annoying. Especially, for me, biting flies, specifically horse and deer flies (Diptera: Tabanidae) (figure 4.15) that constantly bite at the legs and ankles, then the arms and back, and eventually try for the back of the neck. All I want to do is relax on the beach, but no, *my body is being used as an open food cart servicing the multitude of green-eyed flying monsters terrorizing the eastern shores— they need to die, die, die!!!* Sorry about that, I touched off a few emotional memories. Let's recover our decorum.

Some insects, such as biting flies, fall into the category of annoying pests. They do not transmit disease,

really have little or no interest in our food, and cause no direct economic losses. What they do—and they do it well—is irritate us. Now, on the surface this may seem completely in-line with our discussions of household pests, which need merely exist. We appear to be again defining pests arbitrarily, by aesthetic injury levels. But it really is more than that. An individual or group is distressed or their comfort is disrupted as a result of the insect's activity. So simply ignoring or tolerating the insects is not an option; some type of action is warranted, which usually involves leaving the immediate area. Insects achieve the rank of annoying pests by falling into one of three categories:

1. They bite and/or sting us but really cause no injury or disease. Examples include deer and horse flies attacking at the beach, non-disease-causing mosquitoes,

Figure 4.14. Adult fire ant that has been decapitated by the fly parasitoid *Pseudacteon* sp. Photo by Sanford Porter available at http://bit.ly/1VbjtwS.

or most common species of ants that defend themselves when the colony is attacked (are you comfortable with ants in the pants?).

2. They won't leave us alone. These beasts do not bite, sting, or do anything other than buzz incessantly around our heads, or around a table lamp at night, bombarding the lampshade repeatedly while we are engrossed in a book or movie. A small swarm of gnats can achieve this status with anyone trying to walk or jog along a wooded trail on a hot summer day—you can play connect the dots with the bodies of these tiny insects that stick to your face and neck.

3. Their bizarre lives cause pest status. Some insects emerge in synchrony by the millions, leading to annoying end results. The seventeen-year cicada is one such example, in which the adults themselves can be destructive by their feeding action, but the mass insect choir belting out shrill-sounding songs

is also highly irritating to some individuals (https://www.youtube.com/watch?v=0s3qeZ41VJo). As is the case when they all die, seemingly at once, leaving dead bodies deposited on roadways, sidewalks, cars, and yards to be cleaned up by us. This habit is even more remarkable with insects like mayflies (Ephemeroptera) that die in close synchrony by the millions. The resulting death piles sometimes require shovels and even snowplows to clear the dead (figure 4.16)!

The examples given above share this in common: they rarely, almost never, cause direct economic losses. This appears to make the case that subjective determination is the deciding factor in calling annoying insects pests. A reasonable supposition, yet there is more to it than that. Take the example of biting flies at the beach. Direct monetary losses are not generally incurred, and the tolerance level for acceptable biting is based on the individual, yet constant attack by these flies will eventually affect anyone in the same location. What is disrupted is the time at the beach, which is meant to be relaxing and fun. It means that time spent on the beach is reduced, which does have an economic effect in terms of influencing whether you visit the same beach for your next vacation. This in turn has a trickle-down effect on a local economy that relies on tourism and vacationers. Obviously the beach community will invest in some type of control efforts for these flies (although the reality is that very little can be done). So pest status is achieved by a combination of features aligned with subjectivity and financial considerations.

For the other two categories of annoying insects, the pest situation is a bit different than for those that bite and sting, in that they clearly represent **ephemeral** scenarios, and control measures aimed at reducing the insect populations are not really appropriate remedies. Rather, you simply ride the storm out (https://www.youtube.com/watch?v=GVFgEBq0EKM; had to do it!). Any moth that gets in your house at night will likely die smashing into the windows by the next day anyway, and if you walk fast enough, the gnats usually move on in a few seconds to minutes. Insect mass emergence and dying is seasonal and short term. Mayfly adults live for approximately twenty-four hours. There can be minimal

The red imported fire ant, *Solenopsis invicta*, also known as RIFA, is one of the nastiest insect pests in the United States (**Figure 4.12**). Since its accidental introduction sometime in the early part of the twentieth century, the ants have expanded their range from the southeastern region of the United States into western and northern states, terrorizing anything in their path. The ants are classified as pests of agriculture and households, and are medically important. Worker ants defoliate and damage a wide range of vegetation, including crops. Mounds frequently are built in fields, creating numerous obstacles and difficulties for farmers, particularly during harvest. When attacking humans or other animals, the fire ants do not transmit disease, but instead offer a painful bite and sting that frequently leaves raised pustules (**Figure 4.13**). The effect is magnified hundreds to thousands of times because RIFA attack en masse, relying on chemical signals in the form of pheromones to trigger a colony to attack in unison once a single forager has stung. Group attack can lead to death. Perhaps even more disturbing is that in Texas,

Florida, and several other southern states, the ants have been reported to have become so bold as to move indoors. No dwelling is off limits, and they have invaded homes, businesses, schools, hospitals, and nursing homes. In the latter situation, forager ants have even cut through intravenous (IV) lines of patients to access the glucose solutions! Over $5 billion is spent annually for medical treatment and control measures in areas infested with fire ants.

Complementing the ferocity of their attacks is the altruistic behavior that RIFA display. Workers not only fight to the death for their own queen but also for unrelated queens. At times, unrelated colonies will work together to colonize new areas, allowing rapid expansion in a very short period of time. Now, this does not always work out in peaceful terms, as queens and workers will sometimes suddenly turn on each other, which is perhaps one of the more effective control methods. When threatened with the possibility of flooding, workers will form rafts with their bodies (linking together their mandibles), place the queen(s), eggs, and larvae in the center, and safely float downstream to dry land.

Fire ants are resourceful, altruistic, and venomous, characteristics that make them seemingly unstoppable. Thus far in the United States, this has been the unfortunate conclusion. Traditional control

methods for ants have had limited effect on RIFA. In infected areas, pest control companies make no promises (and thus avoid liability) for eradicating these ants, because insecticides barely make a dent in the populations. As a consequence, researchers have examined RIFA in their native habitats in South America, where several natural enemies keep the ants in check. What they have found is that protozoa, fungi, and parasitic flies all attack the fire ants and help maintain the populations at levels nonthreatening to humans. At least two species of the parasitic flies have been released into the United States as part of biological control programs targeting RIFA. Female flies lay a single egg on an adult ant, the egg hatches, and the larva immediately burrows into the body of the host to begin feeding. Eventually the fly larva reaches the head of the ant, and the increased weight causes the head to fall off. Fortunately, even fire ants cannot survive decapitation. In some locations, release of another invasive species, the crazy or rasberry ant, *Nylanderia fulva*, is reducing RIFA through direct competition. The new invasive species evokes some of the same problems as fire ants, but the hope is that if they are managed through controlled releases, crazy ants can be effective control agents of RIFA (**Figure 4.14**).

Figure 4.15. Face of an adult deer fly. Photo by Sam Droege available at http://bit.ly/1iOpYoi.

Figure 4.16. Mass emergence of mayfly adults and the subsequent synchronized death leaves a mess for this neighborhood to clean up. Photo by Joey Hulett/NOAA available at http://bit.ly/1WlcMWa.

costs associated with removing their bodies from streets, sidewalks, or anywhere else they accumulate. But again, this is a very short-term and highly localized problem. For the remainder of the year, mayflies (as juveniles) are important contributors to the functioning of freshwater ecosystems. They serve as an example of how pest status can change over time, or is achieved under only specific conditions. This is true of all type of insect pests, not just those that are classified as annoying.

Insects, disease, and human civilizations: Medically important pests

Undoubtedly you are fully aware that several species of insects are vectors of disease. Some of those insects were discussed in earlier chapters because they served as examples and reminders of the powerful influence that the class Insecta has had on shaping aspects of human civilizations. These beasts clearly are classified as medically important insects, but they are not the only type of pests that are grouped under this umbrella. Others can directly infest the tissues of humans, pets, or livestock, while several species garner the spotlight because they produce lethal compounds or induce immunological responses in the receivers, which, again, are usually us. Generally, insects classified as medically important can do one of the following:

1. Harbor parasites or pathogens that can be transmitted to humans, pets or other domesticated animals. Transmission is most often during blood feeding (the insects are therefore hematophagous), but can also be mechanical, i.e., physical transference.
2. Function as parasites in the absence of disease transmission. These insects are hematophagous, but generally not regarded as seriously debilitating.
3. Directly infest the tissues of humans or domesticated animals.
4. Sting, bite, or produce toxic secretions (table 4.2)

By now you should understand that defining insects as medically important is neither arbitrary nor based on anticipated economic losses. We have already examined this topic and know that human life is not a commodity that can be assessed in terms of monetary value. Try to place a value on one life. What is it worth? Is one person worth more than another? These are not questions that science or any discipline that I am aware of has the tools to address. With that said, the World Health Organization has been tasked with trying to make those determinations, mostly to put the loss of life in context for those

Table 4.2. Classification of most medically important insects. Some insects are blood-feeders and transmit disease-causing agents directly to man or other animals, others blood feed only or directly infest host tissues, and yet some species are medically important because they directly attack and/or produce substances that are potentially toxic or lethal to us.

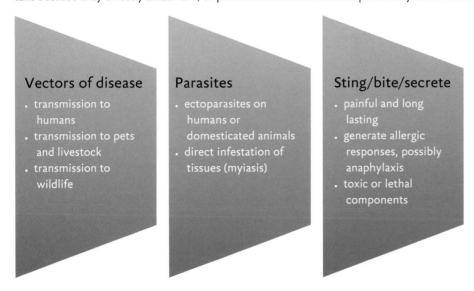

Vectors of disease
- transmission to humans
- transmission to pets and livestock
- transmission to wildlife

Parasites
- ectoparasites on humans or domesticated animals
- direct infestation of tissues (myiasis)

Sting/bite/secrete
- painful and long lasting
- generate allergic responses, possibly anaphylaxis
- toxic or lethal components

grappling with how to distribute the resources available to combat insect-borne disease. What they concluded is that the annual economic value associated with the effect of insect diseases on human existence is in the hundreds of billions of dollars. The estimate takes into account loss of an individual's contributions (including income) to the family, to the gross national product of the nation, control costs of the insect, and infrastructure costs (vaccine development, research, treatment, etc.). To understand how daunting the task is, remember that over 65% of the world's population resides in regions at risk for insect-borne diseases. Just three diseases (dengue fever, Chagas disease, and malaria) expose three billion people each year! It is also important to understand that the estimates do not take into account any of the other types of insects recognized as medically important, only those that transmit disease. What can be concluded is that far more than 65% of the world is living in exposure to some type of medically important insect, which, in turn, means that hundreds of billions of dollars is a gross underestimation of the financial effect these insects have on humans, assuming that we really could come close to formulating an accurate value.

The concept of insect-borne disease is one that most people are familiar with to some degree. Diseases not directly associated with insects also garner media and public attention, such as the 2014 outbreak of Ebola* in several West African countries, which illustrated how quickly disease can spread, how underprepared most nations are for highly communicable or insect-borne diseases, and how much still needs to be learned about their biology and epidemiology before we can hope to lessen their effects on human populations. The effects of non-disease-causing medically important insects are usually not as severe. For example, those that blood-feed but that generally do not vector diseases have much more modest influences on human populations, such as creating temporary discomfort, mild medical conditions, and small financial losses, usually in control efforts. Fleas, hair, body lice, and some biting flies and mosquitoes are examples of such hematophagous insects.

A few insects directly invade human tissues. The term "myiasis" is used for this type of insect invasion; it is usually restricted to flies in two families, Oestridae (bot

---

*Ebola virus disease or Ebola hemorrhagic fever is caused by the Ebola virus, which appears to move between animal hosts, often monkeys and fruit bats. The disease induces symptoms rapidly, causing nausea, diarrhea, and decreased liver and kidney function. Eventually, bleeding issues occur, and death is the typical outcome.

Figure 4.18. The beautiful and delicate monarch butterfly is actually one of the deadliest insects found in North America, owing to the sequestering of cardiac glycosides in its body. The orange and black colors are typical aposematic coloration, warning would-be predators of the potentially lethal meal. Photo by Clinton and Charles Robinson available at http://bit.ly/1Py0569.

Figure 4.17. Larva of the screwworm, *Cochliomyia hominivorax* (Diptera: Calliphoridae). The fly causes facultative myiasis in humans and other animals by infesting dead tissue and then moving to living. Photos by the Mexican-American Commission for the Eradication of the Screwworm (http://bit.ly/1Kxv1lJ) and John Kucharski (http://bit.ly/1PxZd1u), respectively.

flies) and Calliphoridae (blow flies and bottle flies) (figure 4.17). Fly infestation of tissues can induce a range of effects on the host, from relatively mild damage, if treatment occurs promptly and the afflicted individual (or pet) is otherwise healthy, to severe pathological consequences, including death. The pathogenicity of myiasis depends on the fly species involved, the host targeted (humans and many other vertebrate animals can serve as hosts), the relative health of the host, and the conditions in which the host resides. The latter is especially important considering that many cases of myiasis are associated with individuals who are living in poverty in tropical regions or are incapacitated in some fashion and rely on a caregiver. (We will examine this last aspect in cases of

neglect and abuse when exploring forensic entomology in chapter 15.) Interestingly, flies that are true parasites (Oestridae) do less damage to human hosts than those that opportunistically switch from feeding on dead tissues (like an open wound or gangrenous tissue) to feeding on living tissues (some Calliphoridae).

Other medically important insects often overlooked are those that produce toxic or potentially deadly compounds, which they generally bite, sting, or secrete. We tend to avoid most of them, because they are brightly adorned with warning colors, better known as **aposematic coloration**, in which patterns of orange, yellow, red, and black foretell an unfortunate encounter if disturbed (figure 4.18). Most animals heed the warning. Sometimes contact is unavoidable or accidental, and other times the nasty little beasts offer no warning at all. When this occurs, biting, stinging, or release of secretions can occur. Typically two avenues of injury can occur with such insects. In the more common case, an allergic reaction is involved; the insect venom or toxins are not considered lethal, but instead components in the fluids stimulate an acute systemic allergic reaction known as **anaphylaxis**. The victim is hypersensitive not to the venom but to insect proteins. The other avenue of injury is responsiveness to insect toxins and venoms. Some insects produce potent, lethal toxins, while others attack

Table 4.3. Examples of deadly insects toward man and other animals.

Insect (order)	Toxin release	Effect
Blister beetles (Coleoptera)	secretions	skin lesions, lethality
Paederus spp. (Coleoptera)	secretions	festering lesions, pain, lethality
Reduvius personatus (Hemiptera)	bite	pain, throbbing, burning sensations
Pogonomyrmex maricopa (Hymenoptera)	sting	intense pain, immune responses
Pepsis spp. (Hymenoptera)	sting	excruciating pain
wide range of wasps, bees, and hornets (Hymenoptera)	bite/sting	pain, repeated aggressive stinging
Lonomia spp. (Lepidoptera)	stinging hairs	pain, hemorrhage

as a group (e.g., social wasps and hornets), delivering a toxic payload much greater than that of a single attacking insect. Of the two scenarios, injury from anaphylaxis is far more common throughout the world than exposure to truly deadly insects. Insects that produce lethal toxins and venoms are in general rare, particularly in North America and Europe, and are most likely to be encountered in parts of Asia, Africa, Australia, the Middle East, and South America (table 4.3).

Bug bytes

Battle of the ants

https://www.youtube.com/watch?v=KTZf1qETlVE

Implications for modern societies

The preceding section may have convinced you to never go outside again, but there really is no need for that level of concern. Of all the pests discussed in this chapter, only those classified as annoying or medically important should generate much attention from you on a regular basis, but even then, remember that these critters represent a small number of species compared with all known insects, most of the annoying ones are ephemeral, most stinging and biting insects advertise with warning colors so they can be avoided, and there are treatments and remedies for almost all insect-borne diseases.

Advances in technology and control methodologies, vaccine development, and enhanced efforts in education about disease prevention, including management of insect populations, continue to be at the forefront of today's scientific and medical strategies for disease prevention. Such efforts have proven effective, as the global incidence of malaria has dropped by nearly 30% since 2003. In the United States, malaria was considered the highest risk factor for death until 1900, especially for those living in the southeastern region of the country. However, since the 1950s, malaria has essentially been eradicated from the nation. Similarly, focused control strategies based on the biology of the screwworm fly, *Cochliomyia hominivorax* (Diptera: Calliphoridae), led to the eradication of the myiasis-inducing fly from the United States, and continued efforts have driven it from Central America and most of South America. These examples serve as evidence that research efforts into understanding the biology of pests can lead to successful management and control strategies. So, as the outlook looks bleak at the moment in terms of Chagas disease, dengue fever, and attack by the red imported fire ant, there is also plenty of reason to be optimistic that the tide will turn.

Not to cast a cloud over the optimism, but there are also new challenges to be faced that favor the pest status of certain insect species. Specifically, the effects of climate change on current pests are not understood, nor whether pest status will change. What this means is it is entirely possible that some insects that have traditionally been considered pests only on occasion, or that rarely become severe pests, may in fact become more serious threats to the human condition. Likewise, species that were never pests before may find new opportunities as agricultural or medically significant insects. Of course any changes in one species may be countered by corresponding changes in other species currently considered pests, relegating them to benign status. That would be the best-case scenario; the worst case is that climate changes lead to increases in the total number of pest species. Research has begun to explore such possibilities and try to anticipate the outcomes so that prevention rather than stopgap strategies can be developed.

The other area of concern is the changing status of control efforts, namely, the use of insecticides. Several measures (e.g., the Food Quality and Protection Act) have been put in place in the United States, Europe, and elsewhere to improve food quality and safety by reducing pesticide residues. In practical terms what this means is decreased use of chemical control for all the pest situations discussed earlier in this chapter. The goal is to reduce the negative effects of chemical residues on humans, other life forms in the natural environment, and ecosystem functioning. However, the tradeoff may be that, in some situations, pest numbers will rise, along with the negative outcomes associated with each pest. As with climate change, this creates the potential for changes in pest status. For instance, the return of the bed bug, *Cimex lectularius* (Hemiptera: Cimicidae), in several countries in which the insect was considered a pest of the past is believed to be linked to reduced use of pesticides in urban environments. Similar speculation has been used to account for cockroach population outbreaks in cities and other municipalities throughout the United States. Only time will tell whether these concerns will become realities.

CHAPTER REVIEW

✦ You're making it tough to be friends!

- Of the roughly one million described species of insects, the vast majority, nearly 990,000 species, fall into the categories of essential, entertaining, or no threat to man. It is the remaining few, less than 1% of the whole, that cast a shadow of negativity on the entire class.

- Because insects take advantage of us, we classify them as pests. More specifically, humans define insects as pests based on our own subjective measures. In other words, our view of what is or is not a pest is based completely on prioritizing the human condition above all else, even over environmental concerns and ecosystem functioning.

- Entomologists are more pragmatic in defining when and whether an insect is a pest. Be that as it may, of insect pests are classified based on what they target. For example, insects that attack food in the form of crops are agricultural pests, whereas those that target food postharvest are stored product pests. Livestock pests or medically important pests frequently are blood-feeders on domesticated cattle, sheep, swine, poultry, pets, and us. Insects that invade our homes or other dwellings are broadly termed urban pests, but can be further distinguished as household or annoyance pests if invading human habitation, and structural pests if feeding on and/or burrowing into wood-based products in building construction, as well as overlapping into stored product and medically important pests.

✦ What does it mean to be an insect pest?

- The reality is that comparatively few insect species are considered pests of the human condition in any context, and even fewer are classified as serious pests on a regular or consistent basis. Are any ever pests all the time? Subjectively speaking, yes, some species never deviate from being a problem for humans. For instance, medically important insects always represent a threat when living in close proximity to man.

- Even for entomologists, an air of subjectivity surrounds defining pests—but not nearly as much as in the stance "They exist, therefore they are pests." Broadly defined, insects are pests whenever they affect human welfare or aesthetics. Of course, both "welfare" and "aesthetics" lie in the eye of the beholder, so a more tangible measure is needed for establishing a baseline. What this means is that it is not economically practical or physically feasible to react every time a perceived insect pest turns up in a particular area. A baseline or threshold must be established for making decisions on when to get involved, say by treating with insecticide or another type of artificial control method.

- A threshold level is established for a particular commodity from which to make decisions on whether injury or damage will occur or has occurred. One such threshold is the economic injury level (EIL), which is defined as the lowest population density of a specific insect that will cause economic damage, or the amount of pest injury that justifies using some type of control strategy for suppressing the pest. This measure is a function of both the cost of controlling the pest and the value of the crop or commodity. With this mind, it is important to recognize that some commodities have so little overall value that it is not worth the cost of trying to save or rescue them from insect damage. Correspondingly, the value of some products is so great that any level of insect damage justifies the use of control measures. This is also true for commodities in which any injury is considered damage, as in the case of fruits and vegetables sold as produce.

- In some instances, pest status is completely defined by the individual, based on personal reasoning. Any insect in question maybe a pest to one person but not to another. Entomologists define these pest determinations as aesthetic injury levels (AIL). Pest status is entirely subjective and individual, independent of cost analyses or other typical considerations used for determining an EIL or ET. Aesthetic injury levels certainly should not be dismissed, because they are a major factor underlying why the pest control industry in the United States is a multibillion-dollar annual entity.

Why are they so good at being bad?

- Insects tend to be extraordinarily good at any endeavor they put their little minds to. Being a pest is no exception. The attributes that make insects exceptional at achieving pest status, by any convention, are basically the same as those that contribute to their enormous success overall.

- Insects that are pests share some important features: they are winged; they have acute olfaction and well-developed chemical communication; they are efficient eating machines, with extraordinary efficiency in digestion and conversion of nutrients to biomass; and they have body defenses capable of detoxification and resistance.

Where has all the food gone? Agricultural pests

- Insects that attack our food supply are thought of as agricultural pests. In reality, however, we generally distinguish between these insects based on when and what they attack. For example, the traditional category of agricultural pests includes those species that make their mark during the growth phase, that is, the preharvest phase, of crops. Insect injury and damage to crops can be direct or indirect, yielding a host of negative outcomes. Insects drawn to plant-based foods postharvest are categorized as stored product pests. Meat-based food, meaning from livestock and poultry, falls under the umbrella of livestock entomology, and hence those insects are livestock pests. Depending on their interaction with livestock, these same pests can also be medically important insects. It is also important to note that not all agricultural products are intended for human consumption, but the pests of those crops are still agricultural.

- More than 50% of all known insects are herbivores, and that each and every type of plant humans deliberately grow generally have some type of insect that feeds on it. The majority of the estimated 10,000 insect pest species that inhabit the globe are herbivorous and agriculturally significant, and in terms of sheer numbers, likely represent billions of individuals feeding on plants on any given day. Needless to say, the potential for extensive damage is great. That said, it is still difficult to fully calculate the effect of insect injury and damage to agriculture because so many variables are at play. Do not misunderstand, the values (for estimated damage) put forth by agricultural economists and entomologists are based on sound models and years of observations and experience. Nonetheless, biological systems of any kind are incredibly complex, difficult to study precisely, and therefore not fully understood.

- In the United States, close to 350 million acres of land are planted in crops each year. Of those, just shy of 100 million acres, or approximately 28%, is dedicated to the production of corn. Corn, or maize, is used in literally thousands of corn and maize-based products, making it one of the most important agricultural crops in the United States and worldwide. With so much acreage dedicated to raising corn, it is clearly one of the crops most targeted by insect pests. Perhaps the most sinister of the corn pests are beetles in the genus *Diabrotica* (Coleoptera: Chrysomelidae), more notoriously known as corn rootworms or cucumber beetles. More than 35% of all maize in the United States is attacked by corn rootworms each year, resulting in nearly $2 billion in economic losses, and about the same amount of money is spent in control efforts, amounting to a nearly $4 billion measurable economic effect.

- The story of steep economic losses associated with agriculture is repeated with numerous pests. Take the Asian citrus psyllid, *Diaphorina citri* (Hemiptera: Psyllidae): since its relatively recent introduction into the North America from Asia in 1998, it has caused nearly $10 billion in damage to orange and grapefruit crops in Florida and California, feeding on new shoots and sucking sap, as well as transmitting disease to citrus. The Khapra beetle, *Trogoderma granarium*, is one of the most serious invasive species of grain products and seeds worldwide. Beetle larvae feed on damaged and whole grains, resulting in grain weight loss of 5%–30% but sometimes approaching 70% during severe pest outbreaks. These grain pests are capable of causing complete destruction of stored grains in a short period, yielding monetary losses exceeding $5 billion annually. In the mid-Atlantic region of the United States, the accidentally introduced brown marmorated stink bug, *Halyomorpha halys* (Hemiptera: Pentatomidae) has rapidly established itself as one of the most destructive pests of orchard fruits, causing annual losses to apple production alone exceeding $40 million in the eastern United States, with the total economic effect estimated to exceed $200 million each year. The accidental introduction of the Asian long-horned beetle, *Anoplophora glabripennis* (Coleoptera: Cerambycidae) has caused an estimated $3.5 billion in damage each year. Nearly $4 billion in damage is quite significant, but pales in comparison to one of the most destructive tree pests in North America, the European gypsy moth, *Lymantria dispar* (Lepidoptera: Erebidae). The annual economic losses associated with the gypsy moth are approximately $14 billion, surpassing the monetary damage caused by nearly every other insect pest in North America!

- In reality, the number of severe pests in the United States is relatively minor, as only 150 to 200 insects species are crowned with this title. Most agricultural pests cause losses that can be calculated in millions rather than billions. In many regions of the world, the damage is more severe than would occur in the United States or Europe; many countries lack the necessary resources to prevent or manage insect outbreaks.

Let's live together: Household pests

- Entomologists examine insects associated with human habitation and the human environment under the moniker of urban entomology. Within this area, insects can be household pests, in association with our homes, buildings, and yards; structural pests, of building materials, principally wood-based products; or agricultural pests that secondarily invade human spaces. These pests are still considered primarily agricultural pests, usually livestock pests in the form of flies.

- Insects can also achieve pest status through measures that at times are entirely different than those discussed for agricultural pests, meaning that both the AIL and EIL factor in to the determinations. For example, for many insects that invade human habitation, the beast is a pest simply because it is present. Thus, the arbitrary measure of the aesthetic injury level is the determining factor for attaining pest status and the threshold for pest versus no pest is completely based on individual tolerances. Non-disease-causing insects that attempt to cohabit fall into this category and range from benign moths that fly in attracted to light escaping past a door ajar, to a disgusting cockroach scurrying across the kitchen floor. Pest status can also be defined by traditional mathematical modeling, albeit in modified form, as discussed in agricultural contexts. The idea of the ET and EIL in a field situation relies on the ability to sample the insect of interest so that the population density can be monitored over time. Such sampling is nearly impossible in a home or business setting.

- The situation changes radically when our attention turns to medically important insects or structural pests. On the surface, there would appear to be nothing in common among a bed bug, a frothing-at-the-mouth cockroach, and a termite, other than the obvious physical attributes that classify them as insects. In reality, each achieves pest status the same way. Any population density of medically important or structural insects classifies them as pests, as there is no acceptable or tolerable level permissible. In other words, the economic injury level and economic threshold are one and the same, and both equal zero.

- Several insect species can be viewed as potentially serious pests at any population level, including blood-feeders like fleas, bed bugs, mosquitoes, and flies; nonbiting disease vectors like some cockroaches and flies; and venomous insects like bees, wasps, hornets, and ants.

They just "bug" me: Annoying insects

- Some insects, such as biting flies, fall into the category of annoying pests. They do not transmit disease, really have little or no interest in our food, and cause no direct economic losses. What they do—and they do it well—is irritate us.

- Insects achieve the rank of annoying pests by falling into one of three categories: they bite and/or sting us but really cause no injury or disease (e.g., deer and horse flies attacking at the beach, non-disease-causing mosquitoes, or most common species of ants that defend themselves when the colony is attacked); they won't leave us alone; or their bizarre lives cause pest status.

- Annoying insect pests share in common that they rarely, almost never, cause direct economic losses. This appears to make the case that subjective determination is the deciding factor in calling annoying insects pests. A reasonable supposition, yet there is more to it than that. Take the example of biting flies at the beach. Direct monetary losses are not generally incurred, and the tolerance level for acceptable biting is based on the individual, yet constant attack by these flies will eventually affect anyone in the same location.

- For the other types of annoying insects, the pest situation is a bit different than for those that bite and sting in that they clearly represent ephemeral scenarios, and control measures aimed at reducing the insect populations are not really appropriate remedies. Rather, you simply ride the storm out.

Insects, disease, and human civilizations: Medically important pests

- Several species of insects are vectors of disease. Some of those insects have been discussed in earlier chapters because they served as examples and reminders of the powerful influence that the class Insecta has had on shaping aspects of human civilizations. These beasts clearly are classified as medically important insects, but they are not the only type of pests that are grouped under this umbrella. Others can directly infest the tissues of humans, pets, or livestock, while several species garner the spotlight because they produce lethal compounds or induce immunological responses in the receivers, which, again, are usually us.

- Generally, insects classified as medically important harbor parasites or pathogens that can be transmitted to humans, pets, or other domesticated animals; function as parasites in the absence of disease transmission; directly infest the tissues of human or domesticated animals; or sting, bite, or produce toxic secretions.

- By now you should understand that defining insects as medically important is neither arbitrary nor based on anticipated economic losses. To understand how daunting the task is, remember that over 65% of the world's population resides in regions at risk for insect-borne diseases. Just three diseases (dengue fever, Chagas disease, and malaria) expose three billion people each year! It is also important to understand that the estimates do not take into account any of the other types of insects recognized as medically important, only those that transmit disease.

- A few insects direct invade human tissues, causing myiasis. Myiasis is usually restricted to flies in two families, Oestridae and Calliphoridae. Fly infestation of tissues can induce a range of effects on the host, from relatively mild damage, if treatment occurs promptly and the afflicted individual is otherwise healthy, to severe pathological consequences, including death.

- Other medically important insects often overlooked are those that produce toxic or potentially deadly compounds, which they generally bite, sting, or secrete. We tend to avoid most of them because they are brightly adorned with warning colors, better known as aposematic coloration, in which patterns of orange, yellow, red, and black foretell an unfortunate encounter if disturbed.

Implications for modern societies

- Of all insect pests, only those classified as annoying or medically important should generate much attention from you on a regular basis, but even then, remember that these critters represent a small number of species compared with all known insects, most of the annoying ones are ephemeral, most stinging and biting insects advertise with warning colors so they can be avoided, and there are treatments and remedies for almost all insect-borne diseases.

- Advances in technology and control methodologies, vaccine development, and enhanced efforts in education about disease prevention, including management of insect populations, continue to be at the forefront of today's scientific and medical strategies for disease prevention.

- The effects of climate change on current pests are not understood nor whether pest status will change. What this means is it is entirely possible that some insects that have traditionally been considered pests only on occasion, or that rarely become severe pests, may in fact become more serious threats to the human condition. Likewise, species that were never pests before may find new opportunities as agricultural or medically significant insects. Of course any changes in one species may be countered by corresponding changes in other species currently considered pests, relegating them to benign status.

- The other area of concern is the changing status of control efforts, namely in the use of insecticides. Several measures have been put in place in the United States, Europe, and elsewhere to improve food quality and safety by reducing pesticide residues. In practical terms what this means is decreased use of chemical control for pest situations. The goal is to reduce the negative effects of chemical residues on humans, other life forms in the natural environment, and ecosystem functioning. However, the tradeoff may be that, in some situations, pest numbers will rise, along with the negative outcomes associated with each pest. As with climate change, this creates the potential for changes in pest status.

MUSHROOM FARMING (SELF-TEST)

Level 1: Knowledge/Comprehension

1. Define the following terms:
   (a) humancentric
   (b) entomophobia
   (c) anaphylaxis
   (d) economic injury level
   (e) aesthetic injury level
   (f) ephemeral

2. Explain the difference between an economic injury level and the economic (or action) threshold.

3. Identify some of the physical and physiological characteristics of insects that make them well suited for being pests.

4. Describe the commodities attacked by insects that define them as agricultural pests.

Level 2: Application/Analysis

1. What makes an insect a pest in a household or building situation?

2. Is it possible for an insect to be considered medically important without being hematophagous? If so, under what conditions?

3. Construct a graph that shows the relationship between the economic injury level and the action threshold. Show the hypothetical relationship between the two when corn is (a) used for cereal or oil, (b) sold as produce.

Level 3: Synthesis/Evaluation

1. Describe any anticipated effects of climate change, especially the concept of global warming, on the distribution of the red imported fire ant, *Solenopsis invicta,* in North America.

2. If given the task of designing new methods of control to manage the populations of disease-causing insects, what aspects of their biology make the most sense to target? Base your answer on the attributes that make them so successful at being pests.

REFERENCES

Alston, D. G. 2011. Pest management decision-making: The economic-injury level concept. Utah Pests Fact Sheet IPM-016-11. Utah State University Extension and Utah Plant Pest Diagnostic Laboratory.

Centers for Disease Control and Prevention. 2006. Locally acquired mosquito-transmitted malaria: A guide for investigations in the United States. *Morbidity and Mortality Weekly Report,* September 8, 2006/Vol. 55/RR-13. http://www.cdc.gov/mmwr/PDF/rr/rr5513.pdf. Accessed July 24, 2014.

Griffin, C. 2014. Climate change may impact gender of insects that help farmers. *Science World Report,* May 22. http://www.scienceworldreport.com/articles/14905 /20140522/climate-change-impact-gender-insects-help -farmers.htm. Accessed August 3, 2014.

Haeussler, G. J. 1952. Losses caused by insects. In: Insects: The Yearbook of Agriculture. United States Department of Agriculture, Washington, DC.

Kettle, D. S. 1995. Medical and Veterinary Entomology. Oxford University Press, Oxford, UK.

Kiplinger Agricultural Newsletter. July 2011. https://service .kiplinger.com/pubs/KE/KAL/. Accessed February 14, 2013.

Lesky, T. C., B. D. Short, B. R. Butler, and S. E. Wright. 2012. Impact of the invasive brown marmorated stink bug, *Halyomorpha halys* (Stal), in mid-Atlantic tree fruit orchards in the United States: Case studies of commercial management. *Psyche* article ID 535062. doi:10.1155/2012/535062.

Meyer, J. R. 2006. Insects as pests. http://www.cals.ncsu .edu/course/ent425/text18/pestintro.html. Accessed July 26, 2014.

Nelson, W. A., O. N. Bjørnstad, and T. Yamanaka. 2013. Recurrent insect outbreaks caused by temperature-driven changes in system stability. *Science* 341(6147): 796-799.

Otranto, D. 2001. The immunology of myiasis: Parasite survival and host defense strategies. *Trends in Parasitology* 17:176-182.

Pedigo, L. P., S. H. Hutchins, and L. G. Higley. 1986. Economic injury levels in theory and practice. *Annual Review of Entomology* 31:341-368.

Pedigo, L. P., and M. E. Rice. 2006. Entomology and Pest Management. 5th ed. Prentice Hall, Upper Saddle River, NJ.

Rivers, D. B., and G. A. Dahlem. 2014. The Science of Forensic Entomology. Wiley-Blackwell Publishers, West Sussex, UK.

Trogoderma granarium. Global Invasive Species Database. http://www.issg.org/database/species/ecology.asp?si=142 &fr=1&sts. Accessed July 29, 2014.

Waldbaur, G. P. 1968. The consumption and utilization of food by insects. *Advances in Insect Physiology* 5:229–288.

Whitfield, J. B., and A. H. Purcell III. 2013. Daly and Doyen's Introduction to Insect Biology and Diversity. Oxford University Press, New York, NY.

THE ENTOMOLOGIST BOOKSHELF (SUPPLEMENTAL READINGS)

Greenberg, B. 1971. Flies and Disease, Vol. 1: Ecology, Classification and Biotic Associations. Princeton University Press, Princeton, NJ.

Karasov, W. H., and C. Martínez del Rio. 2007. Physiological Ecology. Princeton University Press, Princeton, NJ.

Kusek, K. 2014. New potential way to control the spread of insect-borne disease. *Science Daily*, http://www.sciencedaily.com/releases/2014/07/140717141959.htm. Accessed July 23, 2014.

Nikiforuk, A. 2011. Empire of the Beetle: How Human Folly and a Tiny Bug are Killing North America's Great Forests. Greystone Books, Vancouver, BC.

Primack, R. B. 2014. Walden Warming: Climate Change Comes to Thoreau's Woods. University of Chicago Press, Chicago, IL.

Rees, D. 2008. Insects of Stored Products. SBS Publishers, Dartford, UK.

Spielman, A., and M. D'Antonio. 2001. Mosquito: A Natural History of Our Most Persistent and Deadly Foe. Hyperion, Cambridge, MA.

Zumpt, F. 1965. Myiasis in Man and Animals in the Old World. A Textbook for Physicians, Veterinarians and Zoologists. Butterworths, London, UK.

ADDITIONAL RESOURCES

Agricultural insect pests
http://pubs.ext.vt.edu/category/agricultural-insects-pests.html

Centers for Disease Control and Prevention
http://www.cdc.gov/

Center for Urban and Structural Entomology at Texas A & M University
http://urbanentomology.tamu.edu/

Common agricultural pests
http://ipm.ncsu.edu/AG271/agpests.html

Screwworms as agents of myiasis
http://www.fao.org/ag/aga/agap/frg/feedback/war/u4220b/u4220b07.htm

Stored product pests in the pantry
http://www.ca.uky.edu/entomology/entfacts/ef612.asp

United States Air Force medical entomology
http://www.afpmb.org/content/united-states-air-force-medical-entomology

Urban and structural pests
http://urbanentomology.tamu.edu/urban_pests.html

Walter Reed Institute of Research
http://wrair-www.army.mil/

5   "Dressed" for Success

The Insect Body Plan

Don't accept the chauvinistic tradition that labels our era the age of mammals. This is the age of arthropods. They outnumber us by any criterion—by species, by individuals, by prospects for evolutionary continuation.

Stephen Jay Gould (1989)*

Insects dominate the planet: a theme repeated not just here, but over and over throughout the book. A closer look reveals that insects' true dominion is held over terrestrial ecosystems. They cede control of aquatic habitats, particularly marine environs, to their close cousins, other arthropods like crabs, shrimp, plankton, and lobsters. This is not a matter of strategic alliances between blood lineages, fostered like on *Game of Thrones*;[†] rather, it is an outcome of successful evolutionary adaptions. What this means is that, although all arthropods are derived from a common marine ancestor over 500 million years ago during the Cambrian period, they diverged into animals that are well adapted for life in the sea or for tackling the hostile world of land. A freshwater existence came about as an afterthought. Well, not really, but it definitely is tertiary to marine and terrestrial species. Arthropods look similar to each other but not identical, they share many internal and external features, but they most definitely cannot trade places in terms of habitats. Here, we will explore the body design of arthropods as a whole and insects specifically, with the intent of learning the features that permit extraordinary success in varied habitats. The characteristics that define the phylum and the class Insecta will be examined, particularly in the contexts of tagmosis, form and function, and evolutionary relationships between the various members of the phylum. Chapter 5 will also delve into the attributes that facilitate insect dominance in terms of species richness and sheer abundance over other animal groups, including other arthropods, as well as the special features that promote a terrestrial existence.

---

*From his book *Wonderful Life: The Burgess Shale and the Nature of History*. Of course, in his 1996 book *Full House*, Gould proclaims that we are now living in the "age of bacteria." Sorry Stephen, I'm old school and sticking with the original thesis!

[†]*Game of Thrones* is a fantasy drama television series airing on HBO. The series is based on the fantasy novels *Song of Ice and Fire* by George R. R. Martin, which explore the complex interactions between several noble houses engaged in a civil war for the Iron Throne. Arthropods are no less ruthless, even toward their kin, yet broad territorial lines have been drawn between those that are marine and those land-based; rarely do the kingdoms overlap.

Jointed, boneless, and proud of it: An introduction to the phylum Arthropoda

In the entire world, what can rival the enormous success of the class Insecta? I would argue, nothing, but of course, I am partial. If, however, the question was rephrased, to ask, what is the most successful group of animals, the answer becomes dependent on the definition of "group." The hands-down winner is class Insecta but so, too, is the larger group they belong to, the phylum Arthropoda. Pandora's box of questions just exploded open! What is a phylum? What is an Arthropoda (actually, an arthropod)? And how can there be two winners?

RICKY BOBBY:* You can't have two number ones.
CAL NAUGHTON JR.: Yeah, 'cause that would be eleven.

Slow down, shake and bake. Let's tackle each question separately. What is a phylum? A phylum is an organism classification that represents a subdivision of a kingdom. It is determined based on taxonomic features that include anatomical (morphological) and molecular characteristics. In the past, this meant that organisms that looked very similar or shared a distinguishing feature(s) were grouped together. However, the advent of advanced technologies allowing for detailed nucleic acid analyses has changed our thoughts on some relationships between organisms. In practical terms, this means that sometimes organisms that seem related, in appearance, really are not that close evolutionarily, or vice versa (figure 5.1). We will examine aspects of classification in more detail in chapter 7, as we explore how to distinguish different insect groups (e.g., orders). Hierarchically, a phylum lies just be-

low a kingdom and above a class (figure 5.2). In this case, the phylum Arthropoda is one of thirty-five recognized phyla in the kingdom Animalia and contains five subphyla, which each can be further subdivided into classes, which in turn are composed of orders, which are divided into . . . You get the point. Biologists are excellent at classifying and subclassifying, further dividing subclassifications, creating new terms to explain the bases for the classification scheme, which in turn needs more terms for a comprehensive explanation of the new term(s), and so forth and so on. By the time anyone with a passing interest in a group of animals like arthropods or insects tries to digest it all, they have aged ten years, violently hurled, or done themselves harm! **Taxonomy** and **phylogeny** are not the topics with which to *introduce* you to insects' kin. So from here, and on through the remainder of chapter 5, just enough taxonomy and terminology will be used to make you dangerous for *Jeopardy!*† but not so much as in a traditional introductory course in entomology, typically heavy on classification and piled high with terminology.

You may very well question the previous statement later in the chapter as we examine body plans of insects, because terminology will seem to be oozing from every crease in this text. Some new terminology is needed to lay a basic foundation of insect anatomy and place topics related to form and function (e.g., all arthropods are amazingly efficient with using their body parts) and evolutionary adaptation in appropriate context. After all, it is perhaps mildly interesting to learn that insects are the only invertebrates that can fly, so of course they have wings. But it is incredibly cool to explore the differences in wing structure and flight patterns that allow them to overcome problems of turbulence, to serve as inspiration in the design of helicopters, and to reach flight speeds approaching 90 miles an hour for a body that weighs less than fifteen milligrams! The point is that investing in just a smidge of terminology can open an endless world of amazing biological adaptation. Hopefully, for your sake, I'm not lying!

---

*Conversation between Ricky Bobby, played by Will Ferrell, and Cal Naughton Jr. (John C. Reilly) in the 2006 racing movie spoof *Talladega Nights: The Ballad of Ricky Bobby*.

†Long-running game show on American television, in which contestants demonstrate breadth of knowledge by forming their responses in the form of a question when answers are revealed on the game board. Successful contestants generally are filled with limitless trivia absent from "normal" people and often termed "useless information" by the spousal unit.

Figure 5.1. The mara and rabbit share strong anatomical similarities, but the mara is actually more closely related to a kangaroo than a hare. Photo of mara (by Hans Hillewaert) available at http://bit.ly/1OVq6yb, and of the cottontail rabbit (Justin Wilde) at http://bit.ly/1OVqlcA.

Classification of Animals

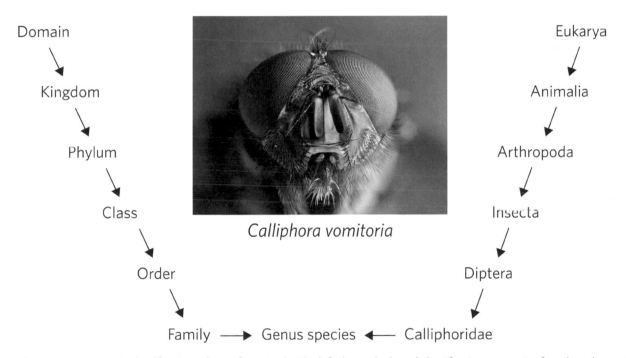

Calliphora vomitoria

Domain → Kingdom → Phylum → Class → Order → Family → Genus species ← Calliphoridae ← Diptera ← Insecta ← Arthropoda ← Animalia ← Eukarya

Figure 5.2. Taxonomic classification scheme for animals. The left shows the broad classification categories from broadest to increasingly specific, with the corresponding names on the right representing insect classification, using a blow fly as a specific example. Photo by JJ Harrison available at http://bit.ly/1iOumDN.

This now brings us to the question of what is an arthropod. Simple—an arthropod is a member of the phylum Arthropoda. Well that was helpful. Arthropods are defined by a set of unifying morphological characteristics (both internal and external) that all members share, which are therefore referred to as absolute characters, and several that most but not all possess, termed typical characters (this should recall our discussion of Johann Christian Fabricius in chapter 2) (figure 5.3). All arthropods are invertebrate animals with an **exoskeleton** composed largely

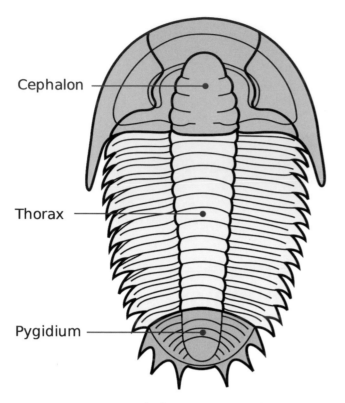

Cephalon

Thorax

Pygidium

Figure 5.3. Tagmosis, or body separation into functional regions of a trilobite. Image created by Sam Gon III available at http://bit.ly/1OwBAZC.

of the carbohydrate chitin, have segmented bodies organized into functional regions termed tagmata (the organization itself is called **tagmosis**) (figure 5.4), possess appendages that are jointed (which is literally what "arthropoda" means, "jointed foot") and positioned as opposing pairs, consistent with the **bilateral** or mirror symmetry of the body (figure 5.5). The body cavity is designed as a **hemocoel**, meaning the internal organs are bathed by the circulating fluid **hemolymph**, which is analogous to human blood. Other absolute characteristics of arthropods are listed in table 5.1. Understand that these features are present in at least one developmental stage for a given arthropod, or were at one time in their evolutionary history. Some traits are "lost" over evolutionary time, a process termed **regressive evolution** (as opposed to structures becoming vestigial,* as has occurred with some vertebrate animals).

In contrast to the absolute characters that unify all arthropods, typical characters are common among many members but certainly not all. For example, almost all arthropods have an open circulatory system, in which the circulating fluid (hemolymph) is not contained within vascular tissue (meaning blood vessels) and is pumped under low pressure via either a chambered heart (usually just two chambers) or a tubular heart. A tubular heart is designed as a linear array of pumping chambers, separated by weak valves that provide directional flow to the fluid. This example points to the nonabsoluteness of the trait, in that some arthropods have a closed circulatory system, but for most it is open; among the latter, the heart can be one of at least two types: chambered or tubular. Most adults in the phylum possess one or more pairs of antennae, yet some have none. Excretion typically, but not always, relies on tubules, and two types of "plumbing" are utilized, **coelomoducts** in some arthropods and **Malpighian tubules** in others. Eyes can vary from the simple **ocellus** (plural **ocelli**) that detects changes in light intensity to agglomerate eyes that recognize shapes and light and dark to the more sophisticated image-forming **compound eyes**, which in some insects have color vision. Table 5.1 lists typical and absolute characteristics of the phylum.

Now back to the tough question: How can you have two winners? It is definitely possible. In chapter 1, we made the case for insects being the most successful of all animals. You have reviewed the data and know the arguments. With over one million described species, the class Insecta is the clear number one. Right? Correct. But arthropods as a whole make up the most successful phylum. There over 1,170,000 described species of arthropods, which accounts for 80% of all animals on the planet. Impressive! Now take into account the diverse range of habitats that members of the phylum occupy. Insects dominate the terrestrial environment, inhabit freshwater to a much lesser extent, and have a minimal presence in marine environments. Where insects are weak, their cousins reign supreme: non-insect arthropods dominate the oceans and seas, several species exist in freshwater environs, and in the world of insects (e.g.,

---

*Vestigial structures are still present but appear to (at least as far as we can tell) no longer have functional purposes. In

humans, dentition in the form of molars is an example, as is the appendix, or, in my household, the brain.

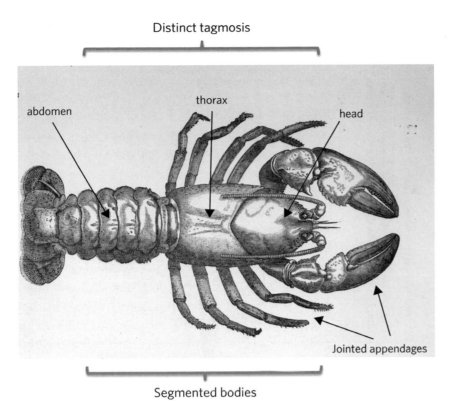

Figure 5.4. Sagittal (side) view of a crayfish with several basic traits (absolute characters) of an arthropod labeled. Illustration by Thomas Henry Huxley available at http://bit.ly/1VhoJhe.

on land) only a few non-insect arthropods are truly terrestrial. Collectively, arthropods rule everywhere! So, there clearly is another winner. Later in the chapter we will explore the characteristics of arthropods that favor dominion in water versus on land.

Arthropods are old!

Arthropods are old. Fossilized remains found in China indicate that members of this phylum date back to at least 541 to 539 million years ago, during the Cambrian period. This is a time when all animal life was restricted to the seas (i.e., marine) and the oceans were believed to have been warm, such as the conditions found today near the equator. An "explosion" of life forms came into being at that time, or so has been speculated based on the fossil record. Most likely the earliest arthropods and their ancestors are significantly older than the discovered fossils. Why? Arthropods lack bone, which is the material of vertebrate animals that fossilizes so well. Instead, the arthropod skeleton is their skin (hence the term, **exoskeleton**), and it is comparatively much softer

and composed of different materials than bone and generally not conducive to long-term preservation in the environment under natural conditions. There are of course always exceptions: the hardened shells fortified with calcium carbonate, as is commonly found in crabs and even the puparia (cocoon) of some flies, is seemingly indestructible and evident many years after death. Nonetheless, those examples were undoubtedly not typical of early arthropods. So no fossil record is left of the earliest ancestral forms to know just exactly when they first appeared. Generically we say they are "old," but more precisely the estimate is around 540 million years old.

Among the earliest fossils of arthropods were the familiar trilobites, which are similar in appearance to modern-day horseshoe crabs, and euthycarcinoids, critters that shared features of both insects and crustaceans (figure 5.6). Do not be concerned if you do not now recognize all these creatures, because we will examine the major groups of arthropods a little later in this section. The oldest crustacean fossils are estimated at about

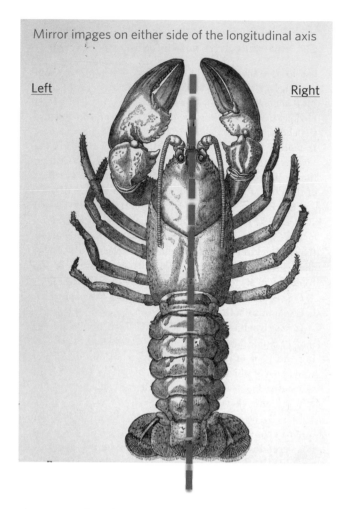

Mirror images on either side of the longitudinal axis

Left

Right

Figure 5.5. Bilateral symmetry is true of all arthropods. What this means is that on either side of the longitudinal axis of the body, the left and right sides are mirror images of each other. They taste the same on both sides too! Illustration by Thomas Henry Huxley available at http://bit.ly/1VhoJhe.

513 million years old, arachnids are a bit younger at 420 million years, and insect fossils date to 407 to 396 million years ago. Most experts agree that these early insect fossils are of winged species, which are believed to be the youngest of the insect lineages. Thus, insects are likely at least as old as the arachnids, if not more ancient. This would be consistent with the age at which the earliest arthropods are believed to have inhabited land, around 419 million years ago. Our take-home lessons from examining the fossil record are that (1) arthropods are among the oldest animals known, (2) the presumed origins of arthropods are associated with warm seas of the Cambrian period, (3) arthropods were the earliest land animals, and (4) this transition (from sea to land) occurred more recently than their existence in the oceans.

Insects and their kin

All arthropods are related and have roots to a common ancestor. They are therefore said to be monophyletic, as opposed to polyphyletic, which means having multiple ancestral lineages. This statement is almost universally accepted but not without some dissention. For example, some individuals have argued that arthropods must have evolved from multiple ancestors based on unique mechanisms associated with how the exoskeleton forms and hardens, differences in the compound eyes, and other features associated with appendage variation and construction. Multiple ideas or hypotheses are common with every aspect of biology. However, some are more testable than others, and in the case of evolutionary questions like the origin of arthropods, most of the debate is left to speculation. What that really means is that there is no way to experimentally test the questions. So such ideas can never be considered definitively right or wrong. Here we subscribe to the view that all arthropods originated from a common ancestor. This origin also influences how arthropods are classified. Generally, five subphyla are recognized in the phylum. This is a fairly recent departure from the three- or four-subphyla classification scheme (depending on what authors do with crustaceans) traditionally used, and also does not include velvet worms or onychophorans, which have been elevated to the status of a stand-alone phylum. (You may have missed the news if not invited to the promotion party!) In the current scheme, one subphylum, Trilobita, represents many of the ancestral characteristics and all members are now extinct; the remaining four are extant or living. What follows is a brief description of each subphylum and some examples of current and past members (figure 5.7).

1. Subphylum Trilobita (also known as Trilobitomorpha): Affectionately known as trilobites, all are extinct and only known to us today from fossils. They lived some 500 million to 600 million years

Table 5.1. Absolute and typical characters of the Phylum Arthropoda. Absolute characters are observed in all members of the phylum, whereas typical characters are found in many but not all.

Absolute

☐

- ☐ Bilateral symmetry
- ☐ Exoskeleton
- ☐ Externally segmented
- ☐ Jointed appendages
- ☐ No external cilia
- ☐ 1 pair of appendages per segment
- ☐ Distinct tagmosis
- ☐ Hemocoel
- ☐ Ventral nerve cord

Typical

☐

- ☐ 1 or more pairs of antennae
- ☐ Open circulatory system
- ☐ Excretion relying on tubules
- ☐ Ventilation using trachea, gills or "lungs"
- ☐ Reproduction is usually sexual
- ☐ Multiple forms of eyes

ago as bottom-dwelling inhabitants of the oceans, feeding as scavengers or ingesting mud to consume attached organic matter. Their bodies were arranged into three regions—a **cephalon** (head), thorax (middle section), and **pygidium** (posterior body region or shield)—that resembled more the modern-day chelicerates or crustaceans (horseshoe crab, specifically) than insects. Trilobites were also characterized as dorsoventrally flattened and covered by a protective covering or carapace. One of their most distinctive traits is that their appendages were biramous; that is, each terminated into two branches (figure 5.8). Trilobites were considered very successful animals, as they appeared to have lived for a long period of time, as evidenced by a rich fossil history, and diversified into possibly eight or nine orders (grouped in a single class). Extinction of the subphylum did not occur all at once, and is believed to have occurred in phases. It is generally accepted that a closely related ancestor of the

trilobites gave rise to all other arthropod subphyla. Modern-day horseshoe crabs strongly resemble trilobites but are actually members of the subphylum Chelicerata.

2. Subphylum Chelicerata: This is perhaps the scariest subphylum of all the arthropods, as it contains the spiders, scorpions, and mites. Oh my! Generically the group is referred to as the chelicerates, a reference to the characteristic of possessing tiny claws (or fangs, in the case of spiders), termed **chelicerae,** in the vicinity of the mouth (figure 5.9). As would be expected from examining the structures, chelicerae are used for grasping and ripping food. The bodies of chelicerates are divided into two regions: the **cephalothorax** (also called prosoma), representing a fusion of the head and thorax, and the abdomen (=opisthosoma). These animals lack antennae and possess agglomerate or compound eyes, and/or ocelli, traits that represent examples of typical but not absolute characters of the phylum. Four pairs of

Figure 5.6. Trilobites (left) and horseshoe crabs (right) share very similar morphology but do not belong to the same subphylum. All trilobites belong to the subphylum Trilobita and are only known from the fossil record, whereas horseshoe crabs are chelicerates that have changed very little since dinosaurs roamed. Photo of trilobite fossil (by Dwergenpaartje) available at http://bit.ly/1YDRvci, and of horseshow crab (Ricce) at http://bit.ly/1jePxyD.

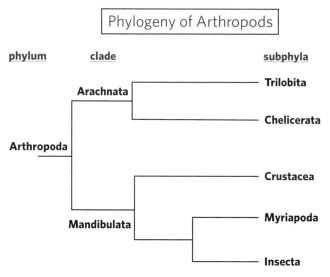

Figure 5.7. Evolutionary relationships between different groups within the phylum Arthropoda. Considerable debate still exists about many of the relationships, which affects the number of subphyla and precisely which groups insects are aligned with. Original phylogenetic diagram created by H. F. Paulus and available at http://bit.ly/1FiAwWG.

appendages used for locomotion (e.g., walking) are attached to the cephalothorax, which are not to be confused with the chelicerae or the **pedipalps,** which have the appearance of legs but are regarded as mouthparts (figure 5.10). All of the appendages are uniramous, consistent with all other arthropods except the extinct trilobites and members of the subphylum Crustacea. Chelicerates are thought to have evolved in marine environments but now only a few remain, as the vast majority of extant species are terrestrial and none are found in freshwater. The subphylum is divided into three classes.

a. <u>Class Merostomata</u>: horseshoe crabs.
b. <u>Class Arachnida</u>: scorpions, spiders, mites, ticks, and daddy-long-legs.
c. <u>Class Pycnogonida</u>: sea spiders (figure 5.11).

3. Subphylum Crustacea: This diverse subphylum is often controversial in terms of classification. It has some very recognizable members, including lobsters, shrimp, crabs, crayfish, and several not-so-common kin like ostracods and barnacles. Crustaceans are apparently similar to chelicerates in displaying two distinct tagmata or body regions (cephalothorax and abdomen), but share very little else in common in terms of appearance. In reality, crustaceans generally have three body regions (head, thorax, and abdomen), but the anterior two tagmata are frequently fused into a cephalothorax covered by a carapace. Each body region bears appendages, with the head possessing two pairs of antennae (as well as

Figure 5.8. Model of what a typical trilobite likely looked like running around the sea floors some 450 million years ago. Model from Trilobite tracks at the World Museum in Liverpool, England, image available at http://bit.ly/295q7hx.

Figure 5.9. A scorpion is a common member of the subphylum Chelicerata. The large pinchers represent one of the most distinctive traits of this group, a pair of chelicerae. Photo by אבי בן זקן available at http://bit.ly/1QB7PEQ.

compound eyes); the thorax is home to legs designed for either walking or feeding; and the abdomen contains the swimming legs (swimmerets or **pleopods**) and terminates with a telson. Most, but not all, appendages are biramous, others are uniramous, and some triramous (class Malacostraca). The majority of crustaceans are aquatic, with **nauplius* larvae**

---

*The nauplius stage is a juvenile form in which appendages on the head are used for locomotion in aquatic habitats. Larvae also possess a single unpaired eye that typically is absent as development progresses.

(figure 5.12); some are terrestrial; some are parasitic on vertebrate and invertebrate animals; and others are sessile, meaning they lack locomotor stages.

In most species, the exoskeleton is fortified with calcium carbonate that yields a very hard outer shell (you ever cracked open steamed blue crabs with a mallet?). However, the mechanisms involved in the process of forming a new skin and hardening after molting or shedding the exoskeleton is varied among crustaceans, contributing to disagreements about evolutionary relationships and classification schemes. Perhaps most important for us, this subphylum contains the most delectable of all the arthropods, with more than ten million tons produced by fisheries or caught for human consumption per year. This subphylum is typically subdivided into eight classes.[†]

    a. Class Cephalocarida: small, primitive shrimp.
    b. Class Cirripedia: sessile animals known as barnacles.
    c. Class Ostracoda: small crustaceans that have appearance of bivalve molluscs.
    d. Class Branchiopoda: very small, aquatic, and possess gills located on feet.
    e. Class Copepoda: small animals known as copepods.

Collectively are **zooplankton**

    f. Class Branchiura: ectoparasitic species.
    g. Class Mystacocarida: very small shrimp-like crustaceans.
    h. Class Malacostraca: largest group, containing shrimp, lobsters, crabs, pill bugs (wood lice) (figure 5.13).

4. Subphylum Myriapoda (also known as Atelocerata): Myriapods represent some very common animals

---

[†]Classification schemes for various animal and other organism groups can change as more research is conducted. As a consequence, a class (or any other division) may be renamed or combined with one or more others. It is important to note, however, that all researchers do not always acknowledge changes to classification schemes, resulting in different nomenclature in separate texts or articles.

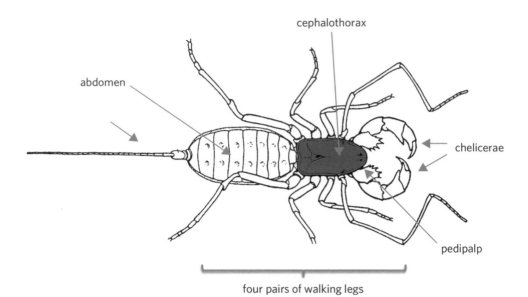

Figure 5.10. Basic characteristics of a chelicerate. Image by Deboccacino available at http://bit.ly/1NWOsbv.

Figure 5.11. Representative members of (from left to right) the class Pycnogonida, class Merostomata, and class Arachnida of the subphylum Chelicerata. Photos of the sea spider (Sylke Rorlach) available at http://bit.ly/1iyRHIX, horseshoe crab (Seth Cochran) at http://bit.ly/1MLurV6, and spider (Carlosar) at http://bit.ly/1G5h4rj.

such as centipedes and millipedes, some not so common like symphylans and pauropods, and one group that is extinct. They are characterized by having bodies with two obvious regions: a head with one pair of antennae and a "trunk" that contains many segments. The body segments are adorned with a large number of legs arranged in pairs, ranging from ten total legs to over seven hundred. Simple eyes (ocelli) are present on the head, as are mandibles. A tubular heart similar to the hexapods is used in an open circulatory system. All extant species are terrestrial and typically restricted to moist microhabitats because of their inability to close the spiracles they use for gas exchange. The subphylum comprises five classes.

a. Class Chilopoda: This class contains the centipedes, which are predatory and use venom dispensed through hollow fangs. Some species may reside in human habitation, but when they do so, they are feeding on insects.

b. Class Diplopoda: This class comprises millipedes, which feed on detritus and leaf litter. Trunk segments are fused together to give the appearance of two pairs of legs per segment, a feature that allows them to be easily distinguished from centipedes. And also they do not produce venom!

c. Class Symphyla: Symphylans are relatively rare; only about two hundred species exist worldwide and almost all reside in the upper layers of soil.

They resemble centipedes but are smaller, translucent, and have fewer legs.

d. Class Pauropoda: Pauropods are also not as common as centipedes and millipedes, though they reside in similar habitats to both and have the appearance of small millipedes.

e. Class Arthropleuridea: This class represents an extinct group that had at least one member that was herbivorous and achieved body sizes of ten feet in length (figure 5.14).

It should be noted that in many cases when the subphylum Atelocerata is recognized in place of Myriapoda, insects are classified as a class (or superclass) and are within this subphylum. The same is true with classification schemes in which Uniramia or Mandibulata are considered subphyla. Newer genetic evidence suggests that insects should be grouped in a separate subphylum with other six-legged arthropods, which is presented here as the subphylum Hexapoda. However, it is also important to emphasize that considerable debate is ongoing with regard to evolutionary relationships between and among the various groups of arthropods, and none can be viewed as definitively correct. Now, on to the last subphylum!

5. Subphylum Hexapoda: The insects and friends. 'Nuff said. Okay, you should already know several of the characteristics of this group from earlier chapters. Here is a brief recap. This group represents the

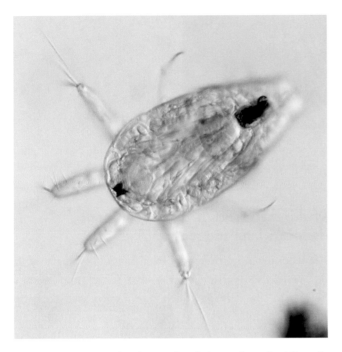

Figure 5.12. A nauplius larva of a shrimp. Photo by Micropix available at http://bit.ly/1Tnvhbz.

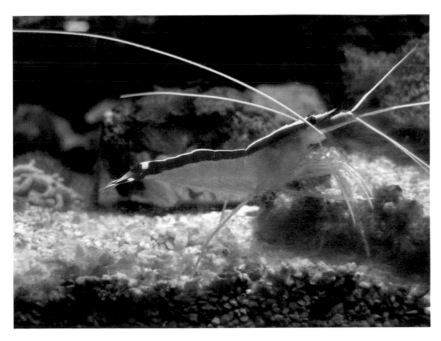

Figure 5.13. An example of a common crustacean, a shrimp residing in the benthic regions of an ocean floor. Photo by Paolo Neo available at http://bit.ly/1KxCtgO.

Figure 5.14. Examples of the two most common members of the subphylum Myriapoda, a centipede (left) and a millipede (right). Centipede photo by Mantanya available at http://bit.ly/1KC5FAd and millipede photo by ltshears available at http://bit.ly/1VbFPyf.

largest (by far) of all in the phylum, with members found in high abundance in terrestrial habitats, some in freshwater, and a few species in marine environments. The body displays distinct tagmosis, with a head, thorax, and abdomen. An adult insect typically possesses one pair of antennae and a pair of compound eyes on the head; six legs and one or two pairs of wings on the thorax; and the abdomen displays external genitalia and sensory structures. Primitive hexapods lack compound eyes, wings, and a few other structures that distinguish them from the insects. There is considerable variation among the insect groups in terms of morphology, size, the color of all appendages, and even body plans. We'll have a closer examination of insect characteristics later in this chapter, as well as in chapter 7 when we explore the major insect orders—and there are a lot! This subphylum is composed of two classes:

a. <u>Class Entognatha</u>: small, wingless animals that possess simple or no eyes and that display primitive antennae or none.

b. <u>Class Insecta</u>: as the name indicates, the insects—much more is revealed in the coming chapters (figure 5.15).

Beyond the text

Based on the characteristics of the five subphyla of the phylum Arthropoda, which one is likely the most closely related to the subphylum Hexapoda?

Phylogenetic surf 'n' turf: Arthropods dominate land and sea

Arthropods as a group have been incredibly successful. As we learned earlier, over 80% of all animal species belong to this phylum. Such dominance provokes the question: Why? What is it about arthropods that has allowed them to survive where others cannot? To do far more, in fact, than just exist, to dominate by a landslide anywhere they reside. Their extraordinary species proliferation is even more impressive considering that arthropod dominance occurs in two extremely different worlds: in water and on land. Considering that life evolved from marine ancestors means that all other groups of animals had the same evolutionary opportunities for success.* Yet arthropods evolved to dominance. In terms of their terrestrial presence, few other animal groups have followed suit by evolving to live on land. But in typical arthropod fashion, they came, they saw, they kicked evolutionary butt by adapting to the harsh conditions of living in air to reign supreme over all other animals. Okay, enough with the arthropod PR machine, and back to the biology. Among the features (absolute and typical) that define arthropods, discussed in the first section, and those that distinguish between the different groups, which ones afford them a skill set that puts them in the driver's seat of evolutionary adaption? The answer might be much simpler than you would expect.

---

*This is an incredible oversimplification of evolutionary processes, but the point is not diminished by such a dastardly deed.

Figure 5.15. Examples of non-insect hexapods: (A) collembolan (photo by Mvuijlst available at http://bit.ly/1G5ifHb), (B) dipluran (FlickreviewR at http://bit.ly/1gS07dg), and (C) proturan (Gregor ?nidar at http://bit.ly/1PyiKyV).

What is the secret of life? I have no idea. But, based on arthropods, the two key components are skin and legs. More precisely, their exoskeleton and jointed appendages stand out as the major features of arthropods that allow them to be so successful in any environment (figure 5.16). What the exoskeleton and appendages have in common is their composition and adaptability. Arthropod appendages are extensions of the exoskeleton, so there is an intimate relationship in terms of physical and chemical properties. The exoskeleton is mostly non-living cuticle, containing an array of lipids and proteins. Chitin is the dominant carbohydrate present and is cross-linked to proteins to produce a number of physical characteristics of the exoskeleton. For many species of arthropods, calcium carbonate is used to biomineralize the exoskeleton. In practical terms, this means the minerals fortify the skin and makes it stronger, as evident with the "shells" of crabs and lobsters. The only living component is the underlying **epidermis**, which, as discussed in the next chapter, is an amazing tissue in terms of all its duties in producing and maintaining the exoskeleton. Functionally, the exoskeleton is used for muscle attachment, protection, defense, locomotion, reproduction, and communication. Oh, and aesthetics (figure 5.17).

It is repairable and adaptable. The classic phrase from introductory biology is "form meeting function," and is more evident with the exoskeleton and associated structures of arthropods than any other animal adaptions. It is the exoskeleton that permits crustaceans to thrive in aquatic environments, preventing compression of internal organs from water pressure, functioning as a barrier to toxins, parasites, and predators, and permitting a range of locomotor activities, all at the same time. Muscles attach to the large internal surface area of exoskeleton, facilitating movement of a large number of appendages associated with feeding, locomotion, sensory perception, defense, and sex. Importantly, the composition can be modified so that the exoskeleton is adapted for life in marine, freshwater, and terrestrial habitats. The exoskeleton is a dynamic structure that has no rivals among other animal groups—one can only assume they have skin envy.

Complementing the extraordinary exoskeleton are jointed legs. Really this can be extended to include all arthropod appendages, because all share the feature of being jointed. Jointed legs essentially have points of articulation where each true segment of an appendage meets another (figure 5.18). This is impressive,

Figure 5.16. The arthropod exoskeleton and jointed appendages are the keys to the success of all members of the phylum. Photo by Enziarro/Joseph Stansburry Rosin available at http://bit.ly/29bLgZl.

Figure 5.17. Face of an adult male spider *Phidippus putnami*. Pretty! Photo (Opterser) available at http://bit.ly/1FjCuWO.

considering that articulated segments can function essentially like the ball-and-socket joints found in humans. For us, this allows a wide range of movement at the shoulder, elbow, hip, and knee. For arthropods, each point of articulation can therefore operate very similarly to a ball-and-socket joint, which permits an incredible range of locomotion of each appendage and remarkable control for starting and stopping, changing direction,

and using multiple appendages at the same time. Try to catch a scurrying cockroach (not some wimpy Madagascar hissing cockroach that seems to be sessile most of its life—I mean a real cockroach, like the regal American cockroach, *Periplaneta americana*; now that's a roach that can run!) as it is running across the floor or table, and you will witness locomotor control at its finest. Do not misunderstand the comparison; articulated appendage joints are not identical to ball-and-socket joints. They merely function similarly. Jointed legs also permit arthropods to be covered by hardened exoskeleton for protection, since the legs provide flexibility. The legs are remarkably adapted to meeting specific functions, permitting walking, running, jumping, digging, swimming, and hunting. If one gets damaged or yanked off during attack by a jealous rival or a hungry predator, no problem, another can form with the next molt.* The in-

*Molting is the process of shedding the old skin and synthesizing a new one during growth and metamorphosis. A new leg or other appendage can only be produced in insects that are still juvenile. In other words, once the adult stage has been reached, insects generally do not molt again, and thus cannot repair or replace exoskeleton and structures derived from it. Details of molting will be presented in chapter 6.

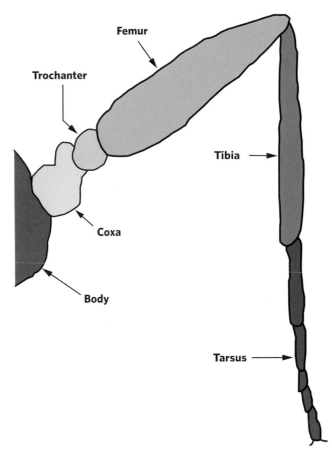

Figure 5.18. Basic insect leg (cursorial) that also reflects the typical arrangement in most arthropods. Individual leg segments come together at points of articulation. Image by Nwbeeson available at http://bit.ly/1MM0owu.

credible thing about the jointed legs is that they are not incredible at all, when placed in the context that all arthropod appendages share the same adaptable features.

Quick check

What features of the exoskeleton contribute to the success of arthropods?

Less is more: Characteristics of the Insecta

Insects are arthropods. Arthropods are the most successful phylum of all animals. Insects share all the absolute characters of other members of the phylum, yet outnumber all other species of arthropods by ten to one. How is that possible? They are all so closely related. The real question is, what features of the class

Insecta distinguish them from the rest. Or do they do more, or something unique, with the same tools? Before we can attempt to decipher what makes insects so successful over their kin, we need to first take a closer look at the characteristics that define them. Didn't we already do that in the last section? Only to a limited extent, and even then, the attributes discussed were for adult insects specifically. Here we will examine characters unique to adults and immatures (larvae) of insects that distinguish them from other members of the phylum. Because of the enormous size of the class, any trait discussed shows tremendous variation among the members. Remember, insects, like all arthropods, are extremely adaptable, and the form of any of their structures has evolved to meet a specific function(s). When we mention legs, all of them do not look or function the same. The same is true for antennae, wings, mouthparts, eyes, and even the number of segments that make up the body.

For the sake of comparison, the body plans of grasshoppers or cockroaches are typically used to present the general or basic designs and arrangement of insect structures. All deviations from the basic designs are considered evolutionary adaptations permitting specific groups to be successful at tasks that the general arrangement did not afford. What will also become evident as we explore insects in more depth are two other evolutionary trends: all insects have not been equally successful, and insect evolution typifies a reduction in number of structures. The former should already be evident to some extent from our discussions of insect diversity in chapter 1. Beetles, or members of the order Coleoptera, represent more than 50% of all insects. As we address the key components of insect success in the next section, we will attempt to shed light on why beetles have achieved dominance among the dominant. The trend of "reduction" is not unique to insects, as this is also associated with those animals considered to be at the higher or more complex end of the evolutionary tree. Perhaps an easier explanation is that with increased efficiency at any task, fewer structures are needed to complete the same function. A clear example comes in the order Diptera—home to flies, midges, and mosquitoes—which are the most efficient and fastest animals that can fly, and have become so by being the only insects with just two wings. All other insects that fly have two pairs of wings, and the fossil rec-

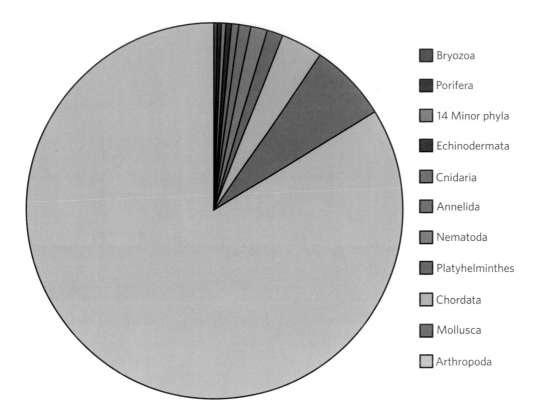

Figure 5.19. Estimation of animal life based on species numbers. One could argue that the phylum Arthropoda dominates all other animals. Image by Nick Beeson available at http://bit.ly/1KCGv4n.

ord suggests that ancestral species had far more* (figure 5.19).

To explore the insect body plan, we first start with the basic design or template that has served as the springboard for adaptation. The body of adults does not always resemble the juveniles. In fact, for many species, the offspring and parents look nothing alike. Consequently, one set of basic traits does not cover different developmental stages. A brief discussion of the body design of juveniles and adults is our starting point. From there, we'll explore a bit more detail regarding insect morphology by examining the body, tagma by tagma that is, separating each functional region, since unique struc-

tures are present to aid with common tasks. We move from head to thorax and, last, to the abdomen. It is important to note that the presentation that follows provides substantially less detail than many introductory entomology textbooks. The reasons for this deliberate decision are to avoid intellectual drowning (e.g., piling on of facts and terms) and to stay true to the proven success strategy of reduction in number. Huh? Several excellent texts already exist that provide extraordinary accounts of insect morphology; they should be consulted and used, instead of copied here.

The basics

All insects do not look the same, and even different developmental stages of the same species can display unique morphology. Therefore to understand insect biology and evolution (success), we need to have a foundation of what they are and how to recognize them. Five features represent the general body plan of an adult insect:

---

*Two additions arise from this statement. The first is that all insects do not have the ability to fly. In fact, it is probably close to a fifty-fifty split between those that do and those that don't. The second is that ancient insects had more than four wings. In fact, some interpretations of the fossil record argue as many as six or more pairs were present on some dragonfly species that approached *six feet* in length! Maybe a tiny house fly in your house is not that big a deal in the grand scheme of things?

1. *Tagmosis with three distinct body regions*. The body is separated into the functional regions of the head, thorax (three segments), and abdomen (6-11 segments). The cephalothorax, or trunk, of other arthropod groups does not occur with insects.

2. *The thorax bears the namesake (Hexapoda) three pairs of legs*. The thorax is divided into three body segments, with each segment possessing a pair of legs, positioned on either side of the body (bilateral symmetry). As we will learn later, tremendous variation in leg sizes and shapes occurs with different insects, often serving as distinguishing features of particular orders.

3. *The head contains one pair of antennae*. All insect adults have, or had at one time in their evolutionary history, a pair of antennae, usually positioned very close to a pair of compound eyes. As with legs, antennae have undergone extensive morphological modifications to perform specific functions.

4. *A single pair of compound eyes is present on the head*. Some groups have evolved (regressive) away from these image-forming eyes, but most adults own a pair. Compound eyes are usually accompanied by two or three simple eyes, otherwise known as ocelli.

5. *Wings are present on the thorax*. Insects are the only invertebrates that can fly, but not all insects fly. When they do, two pairs of wings are the norm, and a single pair occurs with one group of insects. As to be expected by now, all wings are not the same, and the variation results in differences in flight efficiency (figure 5.20).

Again, remember that these characteristics represent merely the basic arrangements for adult insects. Variations, adaptations, and exceptions abound with this group of animals. Some of the adaptations will be presented shortly.

Let's turn our attention to juvenile insects. Juvenile refers to the stage of insect development after egg hatch and before becoming either an adult or a pupa, the latter of which will eventually metamorphose to an adult. The anatomy of eggs and pupae will not be addressed here. Rather, we will characterize juvenile stages, also broadly termed **larvae** but often more narrowly defined as **nymphs**,

Figure 5.20. Two pairs of wings on an adult dragonfly. Photo by William Cho available at http://bit.ly/1R4ddkt.

naiads, or larvae. Specific definitions are provided in chapter 6 during discussions of insect growth. Six basic characteristics distinguish juvenile insects:

1. *For most, three pairs of legs are present, but a true thorax may not be*. The differences are associated with the type of growth demonstrated by an insect. The net effect is that some insects have legs similar to the adults, some possess unique legs, and several have no legs at all during this stage of development.

2. *Many are soft bodied, lacking tagmosis*. Some juveniles have an exoskeleton that is hard, but many more are soft bodied, with variation in body shape, such as that of a fly maggot (vermiform), caterpillar, grub, or wormlike (figure 5.21).

3. *Wings are absent*. No immature insects have wings. Some do form them as buds on the outside of their bodies, which gradually become more and more developed with each subsequent molt. Others do not form wings until transitioning from a pupa to an adult (figure 5.22).

4. *Compound eyes are not typical*. A pair of compound eyes is present with some insect groups (the same ones that form wing buds), but simple eyes or none

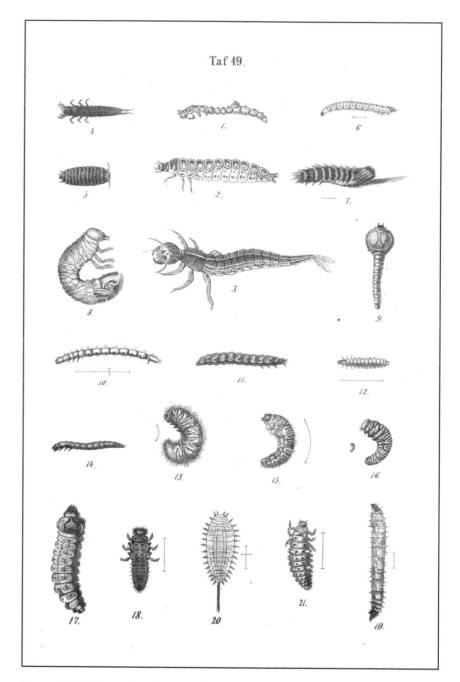

Figure 5.21. Different larval types of beetles. The diversity of juvenile body plans for just one group of insects is amazing! Images created by C. G. Calwer (1876) available at http://bit.ly/1FjF0wl.

at all are very common, especially among the juvenile forms with soft bodies.

5. *Antennae are not typical.* As with compound eyes, a pair of antennae may occur with juveniles that possess compound eyes. However, they generally do not occur with immatures that are soft bodied, especially when living in habitats not conducive to dangly appendages such as in manure, on a liquefying corpse, or when burrowing through soil.

6. *Frequently located in entirely different locations from adults.* This is true for some insects but not all. In many cases the juveniles and adults do not feed on

Figure 5.22. Juvenile milkweed bugs develop wings externally as buds (noted by arrow). After the final molt, the wings are fully formed in adults (right). Photo by Beatriz Moisset available at http://bit.ly/1KCH6TI.

the same food, many adult insects do not feed at all, and several species are aquatic in larval stages and terrestrial as adults, thus accounting for the apparent "lack of parental supervision" (the latter is a separate matter altogether, discussed in chapter 9).

Head and shoulders, knees and toes

Before we can really dive into the tagmosis of the insect, a few more general ideas of the exoskeleton need to be addressed. As already mentioned, insect appendages are derived or are extensions of the exoskeleton. This is true regardless of which region of the body is being examined. So whether an antenna, leg, wing, or sensory structure hanging from the abdomen, each is composed of the same materials as the skin, is jointed, and can move. The exoskeleton itself is composed of a series of external plates called **sclerites**. These structures warrant a chapter to themselves (depending on the interest of the reader), but we won't do that here. For our needs, it is important to understand that sclerites are external, vary in every imaginable morphological way (size, shape, color, and hardness), and are separated by membranous areas. The separation permits flexibility to the exoskeleton. Individual sclerites form from external grooves (also called **sutures**) in the exoskeleton that represent points of invaginations of the skin. In other words, wherever a suture is evident, immediately underneath or inside that point is inwardly pushed exoskeleton. Such

inward points are termed **apodemes** and represent sites of muscle attachment.

With exoskeletal housekeeping now complete, we can begin our examination of the functional regions of the insect body. Three distinct regions or tagmata are present in the class Insecta, a feature that distinguishes insects from all other arthropods. The key to this statement is "distinct." There is no better place to start than the head, and then we'll proceed toward the anus. Just a typical day!

HEAD

The head of an insect actually represents the fusion of six body segments with the primitive anterior region of the ancestral form (figure 5.23). Those six segments possessed one pair of legs each, and are evident in modern insects as appendages of the mouth and antennae. As a consequence, the head appendages are moved by vestigial leg muscles, which allows for much greater strength and control than typical vertebrate facial muscles. Internally, a series of apodemes fuse together to form an internal skeleton called the **tentorium**, the site of muscle attachment for the muscles that control the mouthparts and antennae. Externally, the head is frequently oval in shape and termed the **cranium**. Housed within the cranium is the brain, or **frontal ganglion**, a structure that works in conjunction with the **subesophageal ganglion** to coordinate all physiological systems.

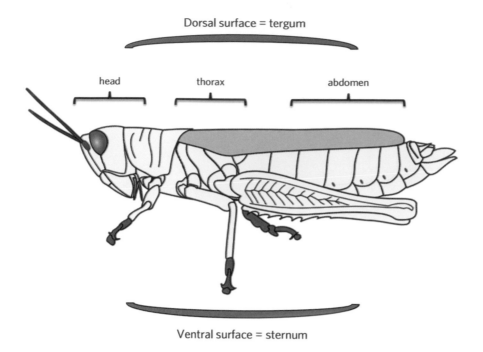

Dorsal surface = tergum

head   thorax   abdomen

Ventral surface = sternum

Figure 5.23. Basic external anatomy of an insect. Image by Giancarlo Dessi available at http://bit.ly/1VdhVSP.

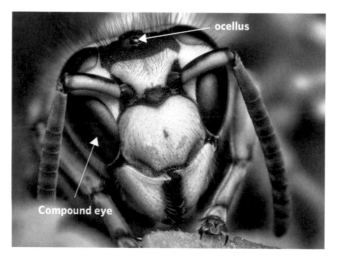

ocellus

Compound eye

Figure 5.24. Compound and simple eyes (ocelli) of a yellow-jacket. Photo by Opo Terser available at http://bit.ly/1JtoULi.

The head itself is associated with feeding and sensory functions: the accessory head appendages perceive the senses of taste (mouthparts), smell (antennae), sight (compound and simple eyes), and touch (antennae). Two types of eyes are common with most insects, one pair of large compound eyes that are image-forming, and two or three ocelli, which perceive changes in light intensity (figure 5.24). One pair of antennae is normally positioned between or just above the compound eyes. Long-tapering, filiform antennae represent the generic style found with grasshoppers and cockroaches. However, a wide range of adaptive shapes occur with antennae to permit a keen sense of smell, mechanical perception, defense, mate-finding, and courtship, and can even be used to saw through small tree limbs (figure 5.25).

The basic mouthparts of insects are **mandibles**, a reflection of the common ancestor shared with the myriapods and crustaceans. On the inside of each mandible are processes or dentition, which function similarly to human teeth. The morphology of the dentition reflects the typical diet of the insect: sharp edges for tearing and shredding and broad flat regions for grinding vegetation. On either side of the mandibles lie the accessory jaws, known as **maxillae**. These structures typically have sensory appendages or **palps** that aid in taste reception and food manipulation. A similar set of palps is attached to the lower or bottom lip, the **labium**. The upper lip is the **labrum**, which is attached to a hinged clypeus. Lips are underrated in animals: without them ingested food can easily roll back out of the mouth (figure 5.26). This arrangement of mouthparts works for insects that do not rely on specialized diets. As insects show divergence in almost every aspect of their biology, so too with their diets, and thus mouthparts. The mandibular mouthparts (also known as chewing or chewing/biting) have been modified to form piercing-sucking structures designed for sucking the fluids from plants and animals; siphoning structures that allow drinking of nectar from flowers and extra-floral nectaries; sponging structures that rely on adhesion and cohesion for ingesting liquid food; and chewing/lapping structures, so that mandibles are used for solid food and an elongate tongue (**hypoglossa**) laps liquid much like a dog (figure 5.27).

THORAX

The middle region of an insect is the thorax or thoracic region. It can be thought of as the region of muscle, because internally is almost entirely muscles, used for movement of wings and legs. Insect locomotion occurs principally via the appendages of the thorax. From a design standpoint, the thorax is divided into three segments: prothorax (anterior), mesothorax (middle), and metathorax (posterior). The name for any structure associated with any of the segments generally uses the prefix for the segment preceding the name of the structure. For example, a leg emanating from the prothorax is termed a prothoracic leg, a wing on the middle section is a mesothoracic wing (or forewing), and so forth. This region also has other terms for anatomical position not typical of the head. The top surface of the thorax is "dorsal" but is also termed the **tergum**. Correspondingly, the "belly" region is "ventral" but also called the **sternum**. The lateral side of the thorax is called the **pleural region**. Each side of the thorax is covered by sclerites, which are highly modified around legs and wings.

Several accessory structures are located on the thorax. Each segment possesses a pair of spiracles, positioned on opposite sides of the body. Spiracles are openings to the tracheal system, a network of air-filled tubes used in gas exchange. Tympanic membranes (**tympanum**) are present on some insects and are used in auditory communication, much like the beating of a drum. Appendages of

Figure 5.25. The morphology of several different types of insect antennae. Image by Bugboy54.40 available at http://bit.ly/1Vf0ewY.

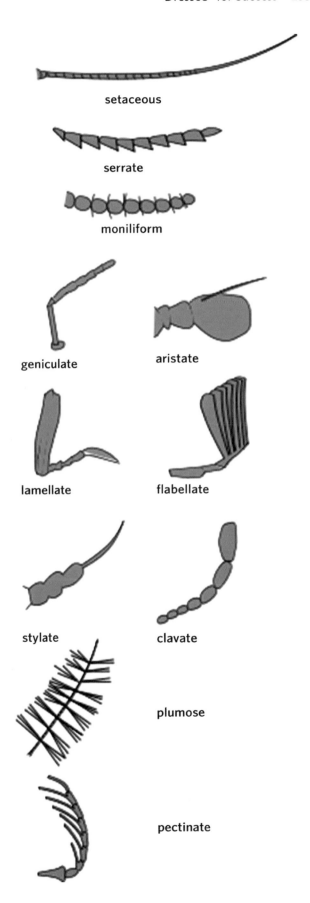

setaceous

serrate

moniliform

geniculate

aristate

lamellate

flabellate

stylate

clavate

plumose

pectinate

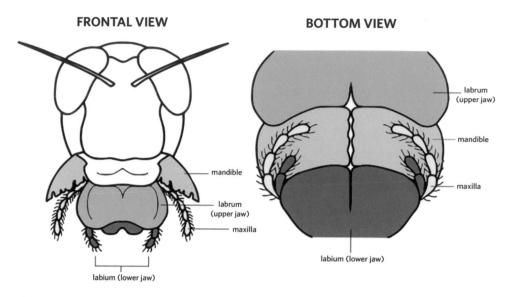

Figure 5.26. Figure 5.26. Basic mouthparts of a grasshopper. Image by Westeros91 available at http://bit.ly/1JtpN6A.

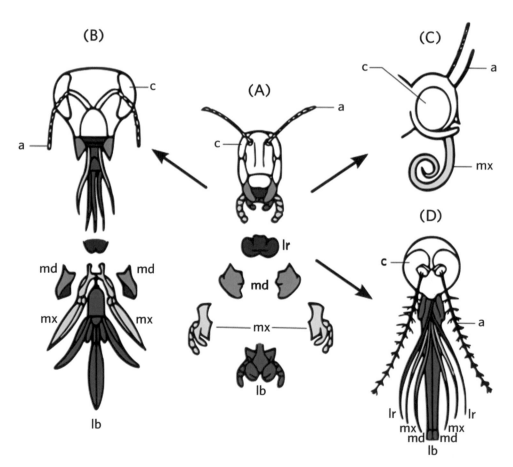

Figure 5.27. Types of insect mouthparts: (A) chewing/biting, (B) chewing/lapping, (C) siphoning, and (D) piercing/sucking (c = compound eye, md = mandible, mx = maxillae, lr = labrum, lb = labium). Image courtesy of Xavier Vazquez available in public domain at http://bit.ly/1KGKuhc.

Figure 5.28. Different wing types of insects (clockwise): **tegmina** are displayed by a grasshopper (Didier Descouens at http://bit.ly /1OX5dTs), **hemelytra** wing of a milkweed bug (Beatriz Misset at http://bit.ly/1NR1fut), **elytra** of an adult beetle (Mario Sarto at http://bit.ly/1QCP0kx), **scale** covered wings of a butterfly (chung-tung ye at http://bit.ly/1OX5eGY), **halteres** of a crane fly (Pinzo at http://bit.ly/1iAcMTe), and **membranous** wings of a dragonfly (Tim Bekaert at http://bit.ly/1iPUwG8).

locomotion occur here as well. Wings typically are present as two pairs, one pair each located on the meso- and metathorax, respectively. In the order Diptera, only one pair, mesothoracic wings, are present. The second pair has been reduced to a pair of elongate knobs, termed halteres, that function as gyroscopes. The basic insect wing is membranous in texture, clear, and displays a network of veins (derived from exoskeleton). The size, shape, and color of the wings are modified with different insect groups, as is the flight efficiency of different insects. Figure 5.28 shows examples of different types of insect wings.

Bug bytes

Fastest animals

https://www.youtube.com/watch?v=zcWxAfl0okE

Six pairs of legs occur on the thorax of an adult insect, with one pair per segment. The basic or generic leg is designed for walking or running (cursorial). The typical leg is long and slender and composed of five segments: the coxa (most proximal to the body), trochanter, femur, tibia and tarsi. By comparison, spiders have two more segments than insects, but more does not mean better, as chelicerates are far less mobile than their insect cousins. Modifications of individual leg segments can radically change function. For example, enlarged femurs are typical of legs specialized for jumping (saltatorial); enlarged tarsi facilitate digging (fossorial); hairs (**hydrofuge**) on the tibia and/or tarsi permit swimming (natatorial); and spines on the femur and tibia (raptorial) turn the insect into an efficient hunter (figure 5.29). In most cases, only one pair of legs is modified for one of these specialized functions; the others are cursorial.

ABDOMEN

The most posterior region of the insect body is the abdomen. It is a bag of surprises. Why? When viewed externally, the abdomen appears the least complex of all the tagmata. Few appendages are present except at the abdominal tip, and though largest by size, the region is mostly a series of repeating segments. Sclerites are evident on the tergal and sternal sides. Internally is where the

Figure 5.29. Comparison of three types of insect legs: (A) cursorial leg (Amada 44 at http://bit.ly/1MNY4Fo), (2) raptorial leg, (Siga at http://bit.ly/1YFIWxD), and fossorial leg (Siga at http://bit.ly/1FywLvZ).

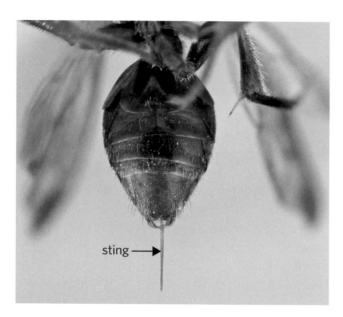

sting ⟶

Figure 5.30. Sting apparatus of bees (Hymenoptera: Apidae) that represents a modified ovipositor. Photo by Arunbc1987 available at http://bit.ly/1PCaaPE.

surprise comes: all major organ systems reside within this region. So any interest in physiological functioning of an insect should begin with the abdomen. This includes nearly all aspects of reproduction, which includes the external genitalia that occur on the last segments of the abdomen. In terms of arrangement, primitive insects possess eleven abdominal segments. More advanced insects, particularly those that are efficient fliers, contain between six and nine abdominal segments (again, the idea of reduction in number), with the last segments modified into plates around the anus.

Accessory structures are far fewer on the abdomen than on any other region of the body. A pair of spiracles exists on the first nine segments of the abdomen, and some species possess tympanic membranes on the sternum. At the abdominal tip of the female, the opening to the reproductive system (**gonopore**) is present, as is the **ovipositor**, a series of valves that form a tube for passing eggs out of the body into or onto a specific location. In some insects, the ovipositor has been modified into a sting apparatus, as with many social wasps, bees, and ants (figure 5.30). Males possess an **aedeagus** or **intromittent organ**, which is equivalent to the mammalian penis (figure 5.31). Sensory appendages are also present on both sexes, and most frequently include paired **cerci** and **styli**. Typically these structures function in mechanical and chemical perception but, in some instances, may aid in other functions, such as copulation. For example, male earwigs grasp a female by the abdomen with the cerci so that he can insert the adeagus into the gonopore (figure 5.32). As with other regions of the body, there is considerable variation in size, shape, and color of the abdomen.

Quick check
How can you tell the difference between an insect and a spider?

Figure 5.31. Female grasshopper displaying the ovipositor. Photo by Gilles San Martin available at http://bit.ly/1OyBeBR.

Why have insects been so successful?

In theory, you are now armed with some knowledge to tackle this question. The statement is not a knock against you. It reflects the difficulty in really being able to answer definitively the question of why insects have been so successful over other animals, including other arthropods. This is another of those evolutionary questions for which we cannot conduct experiments to test particular hypothesis. We are left again to speculation. In the preceding section we started by breaking down this question into two aspects:

a. What features of the members of class Insecta distinguish them from other arthropods?
b. Do they do more, or something unique, with the same tools?

The first question implies we should look for distinguishing anatomical characteristics, while the second suggests that efficiency is the secret to their success. A case can be made that both angles are part of the answer.

From a morphological perspective, the exoskeleton and jointed appendages are obviously major contributors to insect success. Skin and legs have led arthropods as a whole to the forefront of species dominance. Great, so we have narrowed the list of characters to those shared by just over one million species. The next step is identify-

ing whether any of the unique morphological features of insects give them a leg up on their kin. There appear to be three candidates:

1. *Wings*. Insects are the only invertebrate animals that can fly. Admittedly there are obstacles to flight, such as turbulence, drag, and initial muscle temperatures that must be achieved, but all are relatively minor compared to the benefits derived from this method of locomotion. In fact, there does not appear to be any evidence that insects were a particularly successful group, or at least more so than other arthropods, prior to the evolution of wings. Just the two advances of permitting insects to expand their range for seeking food and to escape would-be predators are reasons enough to argue what a powerful advantage they have gained over other animals. From an energetics standpoint, flight is so much more doable, via aerobic metabolism in air, than swimming in the oxygen-limiting environment of water, whether marine or freshwater.

2. *Exoskeleton made of chitin*. This may seem confusing, because all arthropods have exoskeleton made of chitin, as do some non-arthropod animals. True, but the statement emphasizes just how important the arthropod exoskeleton is overall and that any modifications in the composition, such as not using

Figure 5.32. Male earwig showing the sensory cerci at the abdomen tip. Photo by born1945 and available at http://bit.ly/1iQj2ad.

calcium carbonate as a component, can pay huge dividends in other ways. In this example, ammonia in relatively high concentrations is associated with the molting process of many arthropods using calcium carbonate biomineralization. Ammonia is toxic to nearly every cell type and inhibits many metabolic pathways at almost any concentration. For non-insect arthropods that rely on such mechanisms, the toxicity of ammonia is mostly averted by water dilution (i.e., they are aquatic species). Nonetheless, these animals must deal with an intentional threat during such a critical developmental process as molting, which undoubtedly results in physiological tradeoffs that insects do not have to contend with.

3. *Presence of an ovipositor.* The ovipositor is an egg-laying structure that permits eggs to be laid in specific and precise locations by the female. Though a few other animals (namely some fish) have similar structures, no other animal on the planet has perfected control of egg-laying like insects. Using the ovipositor, females have evolved to be able to

place their eggs buried in the ground; injected into a plant or animal; glued to any substrate of their choosing, including on the body of a mate; or dropped like bombs on the ground, over water, or on top of a host. The ovipositor is a structure that contributes to the incredible efficiency of reproduction observed with many insect species.

Complementing these distinguishing anatomical features is a remarkable efficiency at almost any physiological processes necessary to maintain homeostasis. Many animals are highly efficient at one particular physiological task, but very few have multiple standouts. Members of the class Insecta display efficiency in at least four areas that are benchmarks for all other animals. These should be familiar because they are also the attributes that allow them to be so good at being pests (chapter 4).

1. *Well-developed and acute olfaction.*
2. *Efficient and adaptable reproduction.*
3. *Body defenses capable of detoxification and resistance*
4. *Eating machines, with extraordinary efficiency in digesting and converting nutrients to biomass.*

We have already explored each of these efficiencies in the last chapter. An argument can also be made for the sophisticated chemical communication used by many insects, especially those that are socially organized, as well as two features that have not yet been mentioned: **complete metamorphosis** and a terrestrial existence. Neither fit nicely into the categories of morphology or efficiency that we have used to examine insect success. In complete metamorphosis, insects develop through four discrete stages: egg, larva, pupa, and adult. The larvae or immatures pass through several growth stages, characterized by molting, and almost never reside in the same location as the adults. Thus, intraspecific competition for food and other resources is greatly minimized. After the final larval stage, the next molt leads to the pupa. Pupae are unique to complete metamorphosis and represent a dynamic developmental stage in which nearly all larval structures are destroyed to be replaced by entirely new tissues designed for the adult. It also is a stage that can withstand extreme environmental conditions, permitting the insect to survive winter, all forms of precipitation, and seasonal changes. The pupa is

Table 5.2. Orders of Insects That Use Complete Metamorphosis

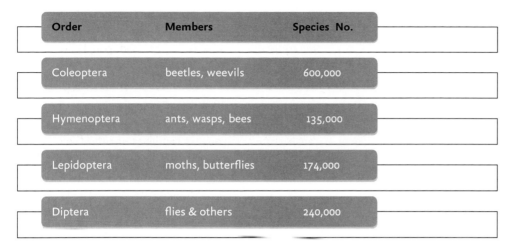

Order	Members	Species No.
Coleoptera	beetles, weevils	600,000
Hymenoptera	ants, wasps, bees	135,000
Lepidoptera	moths, butterflies	174,000
Diptera	flies & others	240,000

This is just a partial list of insects that use complete metamorphosis. However, these are the four most abundant and successful groups of insects.

therefore a developmental transition stage that provides protection from harsh climatic conditions. How important is complete metamorphosis to insect success? Very! The four most abundant insect groups (orders) in terms of numbers of species all rely on complete metamorphosis (table 5.2).

The final attribute is related to the evolution of insects as a whole. They evolved for life on land. No other animal group can make this claim. Sure, insect ancestors were marine animals. But insects are believed to have evolved, first, to be terrestrial creatures, and then secondarily some species have moved back to the sea or freshwater. The benefit of this is huge. First, no other animals were initially around to compete with insects for land resources. This allowed insects to establish dominance in terrestrial habitats and also to form intimate relationships with flowering plants (angiosperms), which already existed prior to the "insect invasion." This offers context for our earlier discussions of insects as pollinators and the fact that the majority of insect agricultural pests are herbivores. The second benefit is a matter of energetics. Insects evolved to be air breathers and to utilize aerobic metabolism under near-optimal conditions: for energy production, it is far easier to produce ATP, thermoregulate (which is only important to insects under specific conditions), and move in air than water. The details are not important to this discussion, but the net effect is. Insects can produce and utilize metabolic energy much more abundantly and efficiently than their aquatic cousins. The obvious extension is that insects can do any task requiring energy better than other arthropods, because they can produce more ATP more efficiently.

CHAPTER REVIEW

✴ **Jointed, boneless, and proud of it: An introduction to the phylum Arthropoda**

- A phylum is an organism classification that represents a subdivision of a kingdom. It is determined based on taxonomic features that include anatomical and molecular characteristics. In the past, this meant that organisms that looked very similar or shared a distinguishing feature(s) were grouped together. However the advent of advanced technologies allowing for detailed nucleic acid analyses has changed our thoughts on some relationships between organisms. In practical terms, this means that sometimes organisms that seem related, in appearance, really are not that close evolutionarily, or vice versa.

- A phylum lies just below a kingdom and above a class. In this case, the phylum Arthropoda is one of thirty-five recognized phyla in the kingdom Animalia, and contains five subphyla that each can be further subdivided into classes, which in

Why do insects have wings but their cousins don't?

An excellent question! Why do insects have wings but other arthropods do not relates back to our examination of insect success. Insects evolved for a terrestrial existence. Other arthropods are adapted for aquatic environments. It is true that a few species of non-insect arthropods dwell on land, but even then, all tend to be restricted to moist microhabitats. The point is that they have not truly separated themselves from an aquatic existence, whereas members of the class Insecta have not only become land-based, they also flourish in terrestrial habits, including taking to air as a means of locomotion. The evolution of flight is one of the major factors contributing to their success.

This is a nice segue to the next question, where did insect wings come from? A much tougher question to answer, because major gaps in the fossil record make it not that useful. For instance, fossils from about 390 million years ago indicate that insects then were wingless. Fast-forward sixty-five million years, and winged and wingless fossils occur together, but no evidence of intermediates has been found. So the evolution of wings remains an entomological mystery. Any good mystery leads to speculative theories, and the origin of insect wings is no different. A couple of prevailing theories exist to explain how wings came on the scene with insects. One is often referred to as the gill theory, which basically contends that insect wings originated from abdominal gills on aquatic insects. Lending support to this idea are modern-day crustaceans that have a pad-like arrangement on abdominal appendages; such pads are thought to be the precursors to wings. The second theory argues that wings evolved from side extensions of terrestrial insects. Such structures are theorized to have allowed gliding in tree canopies. Either theory contends that wings are extensions of exoskeleton, and that vestige leg muscles eventually became flight muscles. Without new fossil evidence, the origins of insect flight will remain theoretical. Regardless of whether the true story is ever revealed, modern insects do show how vitally important wings have been to the success of this group of animals.

turn are composed of orders, which are divided into . . . You get the point.

- All arthropods are invertebrate animals with an exoskeleton composed largely of the carbohydrate chitin, have segmented bodies organized into functional regions termed tagmata, and possess appendages that are jointed and positioned as opposing pairs, consistent with the bilateral or mirror symmetry of the body. The body cavity is designed as a hemocoel, meaning the internal organs are bathed by the circulating fluid hemolymph, which is analogous to human blood.

- Almost all arthropods have an open circulatory system in which the circulating fluid (hemolymph) is not contained within vascular tissue and is pumped under low pressure via either a chambered heart or a tubular heart. A tubular heart is designed as a linear array of pumping chambers, separated by weak valves that provide directional flow to the fluid. This example points to the nonabsoluteness of the trait, in that some arthropods have a closed circulatory system, but for most it is open; and among the latter, the heart can be one of at least two types: chambered or tubular. Most adults in the phylum possess one or more pairs of antennae, yet some have none. Excretion typically, but not always, relies on tubules, and two types of "plumbing" are utilized, coelomoducts in some arthropods and Malpighian tubules in others. Eyes can vary from the simple ocellus that detects changes in light intensity to agglomerate eyes that recognize shapes and light and dark to the more sophisticated image-forming compound eyes, which in some insects have color vision.

❧ Arthropods are old!

- Fossilized remains found in China indicate that members of this phylum date back to at least 541 to 539 million years ago, during the Cambrian period. This is a time when all life was restricted to the seas and the oceans were believed to have been warm, such as the conditions found today near the equator. An "explosion" of life forms came into being at that time or so has been speculated based on the fossil record. Most likely the earliest arthropods and their ancestors are significantly older than the discovered fossils. Why? Arthropods lack bone, which is the material of vertebrate animals that fossilizes so well.

- Among the earliest fossils of arthropods were the familiar trilobites, which are similar in appearance to modern-day horseshoe crabs, and euthycarcinoids, critters that shared features of both insects and crustaceans. The oldest crustacean fossils are estimated at about 513 million years old, arachnids are a bit younger at 420 million years, and insect fossils date to 407 to 396 million years ago. Most experts agree that these early insect fossils are of winged species, which are believed to be the youngest of the insect lineages. Thus, insects are likely at least as old as the arachnids, if not more ancient.

❧ Insects and their kin

- All arthropods are related and have roots to a common ancestor. They are therefore said to be monophyletic, as opposed to polyphyletic, which means having multiple ancestral lineages. This statement is almost universally accepted but not without some dissention.

- Generally, five subphyla are recognized in the phylum. This is a fairly recent departure from the three- or four-subphyla classification scheme traditionally used, and also does not include velvet worms or onychophorans, which have been elevated to the status of a stand-alone phylum. In the current scheme, one subphylum, Trilobita, represents many of the ancestral characteristics and all members are now extinct; the remaining four are extant or living.

- Subphylum Trilobita, affectionately known as trilobites, are all extinct and only known to us today from fossils. They lived some 500 million to 600 million years ago as bottom-dwelling inhabitants of the oceans, feeding as scavengers or ingesting mud to consume attached organic matter. Their bodies were arranged into three regions—a cephalon, thorax, and pygidium—that resembled more the modern-day chelicerates or crustaceans than insects. Trilobites were also characterized as dorsoventrally flattened and covered by a protective covering or carapace. One of their most distinctive traits is that their appendages were biramous; that is, each terminated into two branches.

- Subphylum Chelicerata is perhaps the scariest subphylum of all the arthropods, as it contains the spiders, scorpions, and mites. Generically the group is referred to as the chelicerates, a reference to the characteristic of possessing tiny claws, termed chelicerae, in the vicinity of the mouth. As would be expected from examining the structures, chelicerae are used for grasping and ripping food. The bodies of chelicerates are divided into two regions: the cephalothorax, representing a fusion of the head and thorax, and the abdomen. These animals lack antennae and possess agglomerate or compound eyes, and/or ocelli, traits that represent examples of typical but not absolute characters of the phylum. Four pairs of appendages used for locomotion are attached to the cephalothorax, which are not to be confused with the chelicerae or the pedipalps, which have the appearance of legs but are regarded as mouthparts. All the appendages are uniramous, consistent with all other arthropods except the extinct trilobites and members of the subphylum Crustacea.

- Subphylum Crustacea is a diverse subphylum and often controversial in terms of classification. It has some very recognizable members including lobsters, shrimp, crabs, crayfish, and several not-so-common kin like ostracods and barnacles.

Crustaceans are apparently similar to chelicerates in displaying two distinct tagmata or body regions, but share very little else in common in terms of appearance. In reality, crustaceans generally have three body regions, but the anterior two tagmata are frequently fused into a cephalothorax covered by a carapace. Each body region bears appendages, with the head possessing two pairs of antennae; the thorax is home to legs designed for either walking or feeding; and the abdomen contains the swimming legs and terminates with a telson. Most, but not all, appendages are biramous, others are uniramous, and some triramous (Class Malacostraca).

- Subphylum Myriapoda is composed of some very common animals like centipedes and millipedes, some not so common like symphylans and pauropods, and one group that is extinct. They are characterized by having bodies with two obvious regions: a head with one pair of antennae and a "trunk" that contains many segments. The body segments are adorned with a large number of legs arranged in pairs, ranging from ten total legs to over seven hundred. Simple eyes are present on the head, as are mandibles. A tubular heart similar to the hexapods is used in an open circulatory system. All extant species are terrestrial and typically restricted to moist microhabitats because of their inability to close the spiracles they use for gas exchange.

- Subphylum Hexapoda as a group represents the largest of all in the phylum, with members found in high abundance in terrestrial habitats, some in freshwater, and a few species in marine environments. The body displays distinct tagmosis with a head, thorax, and abdomen. An adult insect typically possesses one pair of antennae and a pair of compound eyes on the head; six legs and one or two pairs of wings on the thorax; and the abdomen displays external genitalia and sensory structures. Primitive hexapods lack compound eyes, wings, and a few other structures that distinguish them from the insects. There is considerable variation among the insect groups in terms of morphology, size, the color of all appendages, and even body plans.

✦ Phylogenetic surf 'n' turf: Arthropods dominate land and sea

- Arthropods as a group have been incredibly successful. As we learned earlier, over 80% of all animal species belong to this phylum. Such dominance provokes the question, Why? What is it about arthropods that have allowed them to survive where others cannot? To do far more, in fact, than just exist, to dominate by a landslide anywhere they reside. Their extraordinary species proliferation is even more impressive considering that arthropod dominance occurs in two extremely different worlds: in water and on land. Considering that life evolved from marine ancestors means that all other groups of animals had the same evolutionary opportunities for success. Yet arthropods evolved to dominance. In terms of their terrestrial presence, few other animal groups have followed suit by evolving to live on land.

- The exoskeleton is mostly nonliving cuticle, containing an array of lipids and proteins. Chitin is the dominant carbohydrate present and is cross-linked to proteins to produce a number of physical characteristics of the exoskeleton. For many species of arthropods, calcium carbonate is used to biomineralize the exoskeleton. In practical terms, this means the minerals fortify the skin and makes it stronger, as evident with the "shells" of crabs and lobsters. The only living component is the underlying epidermis, which is an amazing tissue in terms of all its duties in producing and maintaining the exoskeleton. Functionally, the exoskeleton is used for muscle attachment, protection, defense, locomotion, reproduction, and communication. It is repairable and adaptable. The classic phrase from introductory biology is "form meeting function," and is more evident with the exoskeleton and associated structures of arthropods than any other animal adaptions. It is the exoskeleton that permits crustaceans to thrive in aquatic environments, preventing compression of internal organs from water pressure, functioning as a barrier to toxins, parasites, and predators, and permitting a range of locomotor activities, all at the same time. Muscles attach to the large internal surface area of exoskeleton, facilitating movement of a large number of appendages associated with feeding, locomotion, sensory perception, defense, and sex.

- Complementing the extraordinary exoskeleton are jointed legs. Really this can be extended to include all arthropod appendages, because all share the feature of being jointed. Jointed legs essentially have points of articulation where each true segment of an appendage meets another. This is impressive, considering that articulated segments can function essentially like ball-and-socket joints found in humans. For arthropods, each point of articulation can therefore operate very similarly to a ball-and-socket joint, which permits an incredible range of locomotion of each appendage and remarkable control for starting and stopping, changing direction, and using multiple appendages at the same time. Jointed legs also permit arthropods to be covered by hardened exoskeleton for protection, since the legs provide flexibility. The legs are remarkably adaptable for meeting specific functions, permitting walking, running, jumping, digging, swimming, and hunting.

Less is more: Characteristics of the Insecta

- All insects do not look the same, and even different developmental stages of the same species can display unique morphology. Five features represent the general body plan of an adult insect: Tagmosis with three distinct body regions; the thorax bears the namesake (Hexapoda) three pairs of legs; the head contains one pair of antennae; a single pair of compound eyes is present on the head; and wings are present on the thorax. These characteristics represent merely the basic arrangements for adult insects. Variations, adaptations, and exceptions abound with this group of animals.
- Juvenile refers to the stage of insect development after egg hatch and before becoming either an adult or a pupa, the latter of which will eventually metamorphose to an adult. Six basic characteristics distinguish juvenile insects: for most, three pairs of legs are present, but a true thorax may not be; many are soft bodied, lacking tagmosis; wings are absent; compound eyes are not typical; antennae are not typical; and frequently juveniles are located in entirely different locations from adults.
- The exoskeleton is composed of a series of external plates called sclerites. Sclerites are external, vary in every morphological way imaginable, and are separated by membranous areas that permit flexibility to the exoskeleton. Individual sclerites form from external grooves (also called sutures) in the exoskeleton that represent points of invaginations of the skin.

Why have insects been so successful?

- From a morphological perspective, the exoskeleton and jointed appendages are obviously major contributors to insect success. Their skin and legs have led arthropods as a whole to the forefront of species dominance. The next step is identifying if any of the unique morphological features of insects give them a leg up on their kin. There appear to be three candidates: wings, exoskeleton made of chitin, and the presence of an ovipositor.
- Complementing these distinguishing anatomical features is a remarkable efficiency at almost any physiological processes necessary to maintain homeostasis. Many animals are highly efficient at one particular physiological task, but very few have multiple standouts. Members of the class Insecta display efficiency in at least four areas that are benchmarks for all other animals. These are well-developed and acute olfaction, efficient and adaptable reproduction, body defenses capable of detoxification and resistance, and they are eating machines, with extraordinary efficiency in terms of digesting and converting nutrients to biomass.
- An argument can also be made for the sophisticated chemical communication used by many insects, especially those that are socially organized, as well as two features that have not yet

been mentioned: complete metamorphosis and a terrestrial existence.

MUSHROOM FARMING (SELF-TEST)

Level 1: Knowledge/Comprehension

1. Define the following terms:

   (a) tagmosis   (d) absolute characters

   (b) exoskeleton   (e) typical characters

   (c) phylum   (f) complete metamorphosis

2. Describe the concept of tagmosis and explain how it differs throughout the phylum Arthropoda.

3. Identify the absolute characters that define the phylum Arthropoda.

4. What features of the class Insecta distinguish them from other arthropods?

5. Explain why arthropods have been so successful over other animal groups.

Level 2: Application/Analysis

1. Discuss the features of the exoskeleton of an arthropod that make it such a unique and dynamic structure from the standpoint of evolutionary success.

2. Compare and contrast a chelicerate, crustacean, myriapod, and insect using morphological characteristics.

3. Make a diagram of a basic insect leg located on the mesothorax and label the segments. Which segment is typically modified in a saltatorial leg? A fossorial leg?

Level 3: Synthesis/Evaluation

1. The success of insects in terms of species abundance has been unmatched by any other animal group, including other arthropods. Identify what you believe to be the single most important factor contributing to their evolutionary success and explain why you have taken this stance.

2. Divisions within the phylum Arthropoda have changed over the years from a three-, four-, and now five-subphyla classification scheme. Explain what controversies or possible conflicts in classification account for these different schemes. What has been the effect of molecular data on the classification of arthropods?

REFERENCES

Averof, M., and S. M. Cohen. 1997. Evolutionary origin of insect wings from ancestral gills. *Nature* 385:627-630.

Budd, G. E., Butterfield, N. J., and S. Jensen. 2001. Crustaceans and the "Cambrian Explosion." *Science* 294:2047.

Carter, J. S. 1997. Phylum Arthropoda. http://biology.clc.uc .edu/courses/bio106/arthrpod.htm. Accessed August 17, 2014.

Chapman, R. F. 1998. The Insects: Structure and Function. 4th ed. Cambridge University Press, Cambridge, UK.

Cotton, T. J., and S. J. Braddy. 2004. The phylogeny of arachnomorph arthropods and the origins of the Chelicerata. *Transactions of the Royal Society of Edinburgh: Earth Sciences* 94(3): 169-193.

Friedemann, K., R. Spangenberg, K. Yoshizawa, and R. G. Beutel. 2014. Evolution of attachment structures in the highly diverse Acercaria (Hexapoda). *Cladistics* 30(2): 170-201. doi:10.1111/cla.12030.

Garcia-Bellido, D. C., and D. H. Collins. 2004. Moulting arthropod caught in the act. *Nature* 429:6987.

Gillot, C. 1995. Entomology. 2nd ed. Plenum Press, New York, NY.

Giribet, G., G. D. Edgecomb, and W. C. Wheeler. 2001. Arthropod phylogeny based on eight molecular loci and morphology. *Nature* 413(6852): 157-161.

Gullan, P. J. and P. S. Cranston. 2010. The Insects: An Outline of Entomology. 4th ed. Wiley-Blackwell, West Sussex, UK.

Kukalova-Peck, J. 1992. The Uniramia do not exist: The ground plan of the Pterygota as revealed by Permian Diaphanopterodea from Russia (Insecta, Paleodictyopteroidea). *Canadian Journal of Zoology* 70(2): 236-255.

Pisani, D., L. L. Poling, M. Lyons-Weiler, and S. B. Hedges. 2004. The colonization of land by animals: Molecular phylogeny and divergence times among arthropods. *BMC Biology* 2:1.

Snodgrass, R. E. 1993. Principles of Insect Morphology. Cornell University Press, Ithaca, NY.

Waggoner, B. 1996. Introduction to the Myriapoda. http:// www.ucmp.berkeley.edu/arthropoda/uniramia /myriapoda.html. Accessed August 20, 2014.

Whitfield, J. B., and A. H. Purcell III. 2013. Daly and Doyen's Introduction to Insect Biology and Diversity. Oxford University Press, New York, NY.

Willmer, P. 1990. Invertebrate Relationships: Patterns in Animal Evolution. Cambridge University Press, London, UK.

Yanoviak, S. P., M. Kaspari, and R. Dudley. 2009. Gliding hexapods and the origins of insect aerial behavior. *Biology Letters*. doi:10.1098/rsbl.2009.0029. Accessed August 25, 2014.

THE ENTOMOLOGIST BOOKSHELF (SUPPLEMENTAL READINGS)

Alessandro, M., G. Boxshall, and G. Fusco. 2013. Arthropod Biology and Evolution. Springer, Berlin, Germany.

Berenbaum, M. R. 2009. The Earwig's Tail: A Modern Bestiary of Multi-Legged Legends. Harvard University Press, Cambridge, MA.

Gibb, T., and C. Oseto. 2005. Arthropod Collection and Identification: Laboratory and Field Techniques. Academic Press, San Diego, CA.

Grimaldi, D., and M. S. Engel. 2005. Evolution of Insects. Cambridge University Press, London, UK.

Johnson, N. F., and C. A. Triplehorn. 2004. Borror and DeLong's Introduction to the Study of Insects. Cengage Learning, New York, NY.

Ma, X., X. Hou, G. D. Edgecomb, and N. J. Strausfield. 2012. Complex brains and optic lobes in an early Cambrian arthropod. *Nature* 490(7419): 258.

Marden, J. H., and M. G. Kramer. 1994. Surface-skimming stoneflies: A possible intermediate stage in insect flight evolution. *Science* 266(5184):427–430.

ADDITIONAL RESOURCES

Arachnophobia
http://www.princeton.edu/~achaney/tmve/wiki100k/docs/Arachnophobia.html

Arthropod success story
http://evolution.berkeley.edu/evolibrary/article//arthropods_intro_05

Blue crab
http://www.chesapeakebay.net/fieldguide/critter/blue_crab

Chitin in the exoskeletons of Arthropoda: From ancient design to novel materials science
http://www.mpie.de/index.php?id=2957

Cyborg insects
http://www.popsci.com/category/tags/cyborg-insects

The horseshoe crab
http://horseshoecrab.org/

Insect wings
http://www.amentsoc.org/insects/fact-files/wings.html

North American spiders
http://www.insectidentification.org/spiders.asp

The worlds of David Darling: Arthropods
http://www.daviddarling.info/encyclopedia/A/arthropod.html

6   The Insect's New Clothes

Growing by Shedding

With scorpions, the bigger the better. But if you get stung by a small one, kid, don't keep it to yourself. Okay?

Indiana Jones
Indiana Jones and the Kingdom of the Crystal Skull (2008)*

Animals with an exoskeleton face a developmental quandary. The sleek, armored skin looks good with just about any type of accessory, gives the body its shape, and provides excellent protection from almost anything trying to get in or out of the body. The problem is the exoskeleton fits like a glove—a tight one—thereby preventing increases in size. Oh the joy many humans would have if wearing tight clothes prevented weight gain! For insects and other arthropods, the characteristics of the exoskeleton that confer protection—sclerite-ridden, thick, and hard—confirms the suspicion that growth is also limited by heavy armor. In the grand scheme of things, the exoskeleton could have been the undoing of arthropods. Instead, the ever-adaptive jointed ones have evolved an elaborately complex yet efficient solution: off with the old and on with the new. When the old skin becomes limiting for growth (i.e., too small), arthropods simply produce a new one and cast off the old one. The processes, known as molting and ecdysis, allow insects to be eating machines within the sclerite-clad exoskeleton, producing new, larger skin as needed, as well as able to repair damage and change physical appearance, all while progressing through development. In this chapter, we explore the processes associated with growth and development, highlighting the importance of the exoskeleton. To understand the significance of molting and the radical transformations that result, we dissect the skin to reveal the various living and nonliving layers, what each does, and what happens to each during molting and subsequent synthesis of the new skin. We'll discuss the factors that influence molting, including the hormonal control and regulation of the type of new skin produced. A brief discussion outlines the types of metamorphosis used by different insect groups, with special attention on holometabolous development, "the metamorphosis of choice" among the most successful insects.

---

*Indiana (Indy) Jones, played by Harrison Ford, is the star of four adventure films directed by Steven Spielberg, in which the teacher, archaeologist, and sometime hero chases after ancient treasures, lost children, and alien civilizations. The quote by Indy is to his son, who seems concerned that large scorpions are falling all over his body.

Key Concepts

+ Crunchy on the outside, gooey on the inside
+ Eat, grow, shed: The development plan
+ The insect's wardrobe: Exoskeleton
+ What to do when your pants don't fit: Molt
+ To metamorphose, or not to metamorphose, that is the question

Crunchy on the outside, gooey on the inside

Indiana Jones's wisdom extends beyond just archaeology. His entomological insight reveals two key features of scorpions: bigger is better in terms of venom toxicity, from our perspective (there tends to be an inverse relationship between body size and the strength of toxic venom products toward humans), and scorpions can get larger, suggesting a type of growth that allows their skin to enlarge. The latter aspect is our focus in this chapter. The centerpiece to insect growth is the exoskeleton. As we discussed in chapter 5, this amazing structure is one of the key factors in the success of the phylum Arthropoda and class Insecta. The skin affords protection from mechanical injury, is a barrier to keep things out or in, gives the body shape, serves as the site of muscle attachment, and functions in the greatest battle for all terrestrial animals: to prevent desiccation. The characteristics of the skin that permit such benefits also limit growth in arthropods. Actually, just stating that growth is limited does not really capture the full situation. Growth occurs but is infinitely more complex because of the exoskeleton.

For many arthropods, particularly those that fortify the exoskeleton with calcium carbonate, the body is hard and relatively inflexible. Growth implies expansion in size, getting bigger, which appears impossible when your body armor is rigid and covered with hard plates. You may say, "Now wait just a minute. I know for a fact that all insects are not hard. Many are soft bodied, particularly juveniles. And what about soft shell crabs: *soft* is in the name?" My response is yes, yes, and yes! Indeed, many insects, other arthropods, and other animals (with an exoskeleton) do not possess rigid skin, but growth is still hampered by the outer covering's limited capacity to expand to accommodate a larger body or mass. For animals that possess an exoskeleton composed primarily of the polysaccharide chitin, growth is achieved by shed-

Figure 6.1. An adult cicada emerging from nymphal exoskeleton. The old skin or exuvia is cast off during the process of ecdysis. Photo by Brian1442 available at http://bit.ly/1FyAGsI.

ding the old skin and replacing it with a new one that is larger. This growth process occurs in two phases, termed **molting** and **ecdysis**. Details of both phases are examined in later sections of this chapter. For now, suffice to say that molting is where digestion of a portion of the old skin occurs and the new and improved skin of the next stage of development is produced. Ecdysis is the physical removal of any remnants of the old exoskeleton. Growth actually occurs only during the period of time in which the old skin has been removed and the new skin is expanded to full size by increases in "blood" pressure (actually hemocoelic pressure) (figure 6.1).

Animals that grow by shedding—in other words, need to molt to grow—are grouped together as the clade Ecdysozoa. A clade is a group containing an ancestor and all of its descendants. It is not necessarily a term of true taxonomic position, like a phylum, class, or order, but it does represent a group evolutionarily connected by a unifying characteristic(s). In this instance, Ecdysozoa include the phyla Arthropoda, Nematoda (nematodes), Onychophora (velvet worms), Tardigrada (waterbears or moss piglets), Cephalorhyncha, and Nematomorpha (horsehair worms) (figure 6.2). All are invertebrate animals with an exoskeleton that limits growth, and thus the major linkage among each of these groups is that development progression requires molting to occur. This should be where you ask the question "If so many other animal groups (phyla) have an exoskeleton made of the

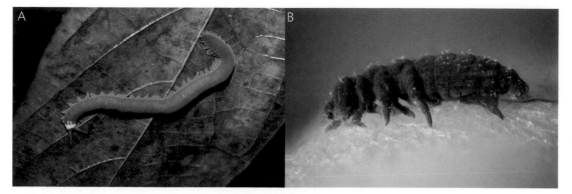

Figure 6.2. Examples of non-insect ecdysozoans that must molt to grow: (A) a velvet worm (phylum Onycophora) and (B) water bear (phylum Tardigrada). Photos by Geoff Gallice (http://bit.ly/24ZK3KE) and Darron Birgenheier (http://bit.ly/1VdPap3).

same major component as that of arthropods, then why are they not as successful, or at least not mentioned in the same conversation, as the most successful animals?" An excellent observation on your part! Evolutionary adaption is highly complex and depends on many interconnected factors that go far beyond the scope of our discussion of insect growth and molting. An oversimplified answer is that other ecdysozoans lack all the other exoskeleton features identified in chapter 5 as being associated with arthropods, including the diverse range of jointed appendages. Those same adaptions that promote success of the phylum Arthropoda add to the complexity of growth for these animals, particular with regard to maintaining functionality during the process of molting.

In the coming sections, we examine insect growth, from events that take place within the egg—embryonic development—to adult, including physiological and morphological changes that define adulthood in insects. Molting and ecdysis are explored from the perspective of repurposing the old skin and synthesizing a new exoskeleton based on cues derived from inside (endogenous) and outside (exogenous) the body of an insect.

Eat, grow, shed: The development plan

Insect development is a bit more complex than "just" molting. Yes, as mentioned, molting is a complicated process that requires significant coordination of physiological systems in an insect or any arthropod to ensure successful completion. However, molting is not the only component to insect development. In fact, the development plan used by most insects begins with the egg, a

stage in which molting does not occur, and continues through adulthood. The path that any organism takes to complete development from egg to adult is **ontogeny.** It is literally the developmental history of any organism. In the case of insects, the development path or history occurs in two broad phases:

1. Development from egg to adult
2. Maturity to adulthood

As we shall soon learn, the first comprises a series of cell divisions and molts to achieve developmental progression, while the second refers to developmental events that occur once an insect has molted to the adult stage to attain sexual maturity. Generally, most species require a few days after the molt to become sexually mature and achieve adulthood status.

Development from egg to adult

As would be expected, the period from egg to adult reflects the most significant developmental events that occur in any insect's ontogeny. It does not necessarily, however, represent the longest span of time spent in a developmental period for a given insect, in that some species may reach the adult stage in a relatively short period (seven to ten days) but live much longer (weeks) in adulthood. Conversely, others spend an inordinate (comparatively or in actuality) amount of time completing development, only to have very short adult lives. A mayfly of the order Ephemeroptera is one extreme, in which juvenile development lasts several weeks, but the adults survive just long enough to have sex than die (approxi-

Figure 6.3. Mass emergence of the seventeen-year periodic cicada. Cast skins can be seen mixed with newly emerged adults. Photo by Marg0marg available at http://bit.ly/1LD2AVA.

mately twenty-four hours). Of course periodic cicadas (order Hemiptera*) may be the most unbalanced, spending as much as seventeen years underground as immatures, yet a paltry two weeks as adults (figure 6.3). During this period of development, insects undergo massive changes in body size and physical appearance, to eventually display the characteristics typical of adult members of the class Insecta.

Most insects begin life in the egg stage, and once the egg hatches (termed **egg eclosion** or emergence), the newly minted juvenile essentially becomes an eating machine. As the immature feeds, it eventually reaches a point in which body mass outpaces the exoskeleton. Consequently, the old skin must be replaced with a new, larger one, and once installed, eating resumes. The process continues, with some major and minor twists and turns, until finally the adult stage is reached. This overview of

---

*The noticeable increase in references to taxonomic classification, namely order names, is deliberate, with two intentions: to encourage you to begin making associations between insects groups and their evolutionary adaptations, and as preparation for the next chapter, which introduces the major insect groups and how you can distinguish between them.

development actually reveals that the path to adult involves two key periods of life: development before and development after egg hatch. These periods reflect major development phases, referred to as **embryonic development** and **postembryonic development**. For simplicity, each developmental phase can be distinguished from the other based on whether or not molting occurs: molting is restricted to postembryonic development. We briefly discuss here some of the major events that occur during each phase of development.

EMBRYONIC DEVELOPMENT

For most insects, once fertilization occurs (if at all), the egg is deposited outside of the mother's body. Remember from our discussion in chapter 5 the importance of the ovipositor, the unique egg-laying structure that allows an adult female to dictate where her eggs will be placed during the process of **oviposition**, or the laying of eggs. Insect eggs are oviposited in just about any imaginable location, including in and on the bodies of parents. The significance of egg placement is that this location is generally where embryonic development begins. At some point after oviposition, a trigger or activation step initiates embryonic development. The egg cell responds by mitotically dividing the zygote nucleus. What is the zygote nucleus? It is the result of fertilization and forms when an egg cell essentially engulfs a sperm cell, ultimately fusing nuclei from both gametes in the egg cell. The finished product is a single cell with twice the chromosome content as the original gametes, and hence the nucleus is referred to as the zygote nucleus. The mitotic division of the zygote nucleus yields daughter nuclei contained within the same cell, which by definition results in the formation of a **syncytium**. Eventually additional cells form within the syncytium to produce a superficial layer to the periphery of the egg. This layer will become a one-cell-thick layer known as the **blastoderm**. In terms of embryo development, blastoderm formation signifies relatively rapid changes inside the embryo as this layer gives rise to all cells of the larval body, and each of the cells in the blastoderm can undergo mitosis to produce new daughter cells. Consequently, with each round of cell divisions, the embryo gradually morphs to look more and more like a young or neonate juvenile. Yolk within the egg provides the nutrients necessary to fuel mitosis and any other developmental processes occurring during embryonic development.

The length of time required to complete embryonic development depends on the insect species. Some species can pass through this phase in just a few days, while others may need weeks to months to even years—but to say "need" this long is not totally correct. It is true that the time spent within the egg stage may be very long, but the length is often part of a reproductive strategy tied to seasonal conditions, predator/parasite avoidance, and/or as a means to decrease **conspecific** or **allospecific** competition. For example, if the length of the egg stage is varied among members of the same clutch of eggs, eclosion is thus staggered so that all siblings are not directly competing with each other for the same resources. Without such influences, embryonic development can be completed in a much shorter period of time. Regardless of the reasons for lengthening this phase of development, the implications are that the mother can clearly influence the development of her progeny even after the eggs have passed from her body.

Embryonic development officially comes to an end with the hatching of the egg. In insects, egg hatch is termed eclosion or egg emergence, and results in a juvenile insect being "dumped" on the scene, ready to begin postembryonic development by feeding 24/7. Generically, immature insects are often referred to as larvae, but it is actually more correct to refer to them based on the type of development they display as a species. The details of specific types of development and metamorphosis are presented later in the chapter, but for those species that utilize complete metamorphosis, the juveniles are truly larvae. Those that rely on incomplete metamorphosis are called **nymphs** as immatures, unless they reside in aquatic habitats, in which case the young are known as **naiads**. So at egg eclosion, either a larva, nymph or naiad emerges, depending on development and habitat. Regardless of what you call them, they all begin the next phase of development as eating machines, and so molting will indeed become very necessary.

POSTEMBRYONIC DEVELOPMENT

Postembryonic development begins with egg eclosion and continues until an adult insect forms. Molting is the process that must occur for an insect to move from one stage of development to another. We have already discussed that shedding of the exoskeleton is necessary to permit increases in size; viewed another way, molting facilitates transition from one developmental stage to another. Let's use a blow fly as an example (figure 6.4). They are lovely beasts, best known for feeding and developing on dead animal remains, including humans. A **gravid** female, or one carrying eggs, oviposits on a corpse, the eggs eclose, and now very young larvae are ready to start feeding on tissues of the dead animal. Each new larva has entered the first **stadium** of larval development, which is also referred to as the first **instar**. An instar is simply a term for a developmental stage, but in practice is usually applied only to immature stages. So the blow fly larva that emerges from the egg is a first instar larva: "first" because it is the initial development stage of postembryonic development. This stage comes to an end when the old exoskeleton is shed via ecdysis, and the new, larger one underneath is revealed. The next stage is called the second instar larva, the one following is the third, and so on (figure 6.5).

Molting coupled with ecdysis signals the completion of one stage of growth and the beginning of the next.

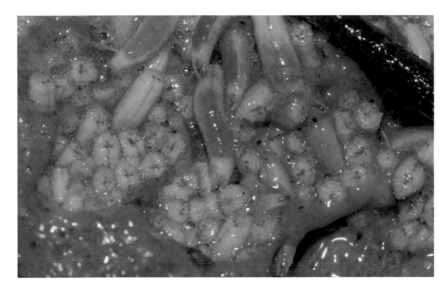

Figure 6.4. Blow fly maggots feeding on animal tissues. Photo by Susan Ellis, Bugwood.org.

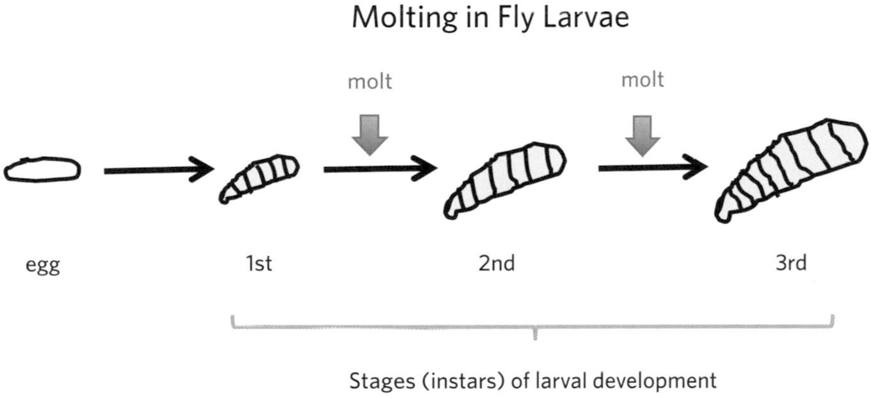

Molting in Fly Larvae

egg → 1st → molt → 2nd → molt → 3rd

Stages (instars) of larval development

Figure 6.5. Insect growth requires molting from one developmental stage to the next.

How many instars does each insect have? That depends on the species. Some mayflies, for example, molt forty times, while several fly species exhibit only four molts during the entire life span. Many insects have a fixed number of developmental stages, usually in terms of the immatures; these are said to display **determinate growth**. For these insects, the length of each larval or nymphal instar is influenced by temperature and food availability, but the total number of developmental stages remains constant. In contrast, some species display variable numbers of immature stages, termed **indeterminate growth**. For these insects, temperature, competition, and food availability all influence the total number of immature developmental stages.

Regardless of which type of growth an insect follows, molting will continue until the adult or **imago** stage is reached. If the insect in question has wings as an adult, or did so at one time in its evolutionary history (remember the concept of regressive evolution from chapter 5), then molting does not continue at this point. Development, in terms of expansion of size, ceases with the adult stage for these insects. However, if the species are more primitive and have never relied on flight, molting can continue even into adulthood.

Molting produces a not only larger skin, but also one that takes on the physical and chemical characteristics of the next stage of development. Each molt can produce changes in size, color, and morphology, as well as repair

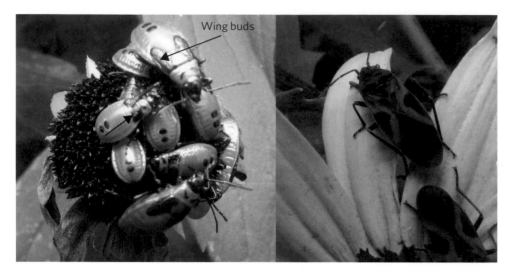

Figure 6.6. Development of the milkweed bug, *Lygaeus turcicus*, which undergoes gradual metamorphosis with each molt. The arrows point out the formation of wings as buds. With each molt, the wings become gradually more developed until reaching the adult stage. Photos by Beatriz Moisset available at http://bit.ly/1iQp3n7.

if any damage has occurred. Molting can even replace appendages that have been lost through fighting or attack. So for an insect like a milkweed bug (order Hemiptera), which are called nymphs as juveniles and form their wings from buds on the thoracic tergum, each molt of a nymphal stage produces the next nymphal instar with slightly more developed wings than the previous, until the final nymphal molt yields an adult with fully formed wings (figure 6.6). The point of this example is that not only is the next exoskeleton bigger, it also is anatomically different from the previous stage. The same is true when a pupal stage occurs in between the larval and adult stages of development: the morphology of the exoskeleton is unique and is produced with the last larval molt.

Details of the actual process of molting and the types of control mechanisms that allow such precisely timed physiological processes to occur are discussed a little later. Before we can fully appreciate how molting actually works, we first need to explore the composition of the exoskeleton and epidermis. In the next section, the functional layers of the exoskeleton are revealed, so that we can understand what changes during a molt, and what does not.

Quick check

Why is the term "instar" not relevant to embryonic development?

Maturity to adulthood

Postembryonic development "rounds the final turn and heads for home" with the synthesis of the adult exoskeleton and shedding of the last nymphal skin. This statement holds true for most insects that rely on incomplete metamorphosis. Development is not finalized, however, with this final ecdysial event (figure 6.7). A series of changes occur with the skin, including biochemical reactions that lead to hardening (**sclerotization**) and tanning (**melanization**) of the exoskeleton. These processes occur with each molting event, not just when an adult skin is formed. Why are they necessary? Because the newly formed exoskeleton is soft and pliable, a necessity since it is produced on top of the epidermis but under the old, rigid exoskeleton and does not yet possess the properties necessary for providing all forms of protection and identity needed for the insect to survive in a harsh terrestrial or aquatic environment. The adult insect is usually not sexually mature immediately following the adult molt. So, typically, a period of time is needed for the **gonads** to become fully functional. Most adult females are not receptive to a male's advances until adulthood is achieved. This is critical for males that can only mate once, and particularly so if the female is not capable of storing his sperm for fertilization of eggs.

Insects that rely on complete metamorphosis do not directly molt from the larval form to the adult. Rather, a

Phases of Insect Development

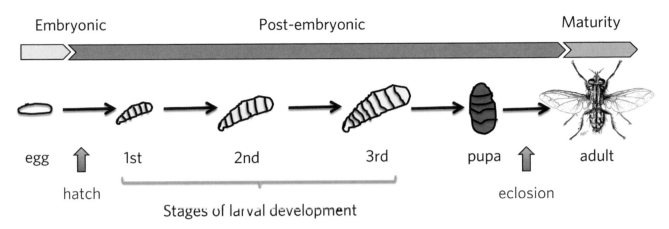

Figure 6.7. Phases of insect development from embryogenesis to sexual maturity. Illustration of the adult fly by R. E. Snodgrass available at http://bit.ly/1LSle7E.

transitional stage known as a pupa separates these stages. In this form of development, the last larval molt results in the formation of a pupal skin. As with seemingly every aspect of insect biology, there is more than one type, and each kind of pupa differs substantially in appearance from the others (figure 6.8). During pupal development, juvenile tissues are broken down and replaced by entirely new adult structures. This is in complete contrast to insects relying on incomplete metamorphosis, which gradually transform into adults using existing structures. Pupal development terminates when the adult insect emerges or ecloses from the pupal skin. Emergence can be a challenging and exhausting process, as often the pupa is buried in several inches of soil, within another organism, or even underwater. Successful extrication from the pupal environment is commonly followed by a period of grooming the body. The adult legs comb and smooth the adult structures, particularly the wings. Some species do not emerge from the pupal stage as full-size adults, so expansion of the adult body is required. This is typically achieved by gulping air to elevate hemocoelic pressure, thereby permitting full extension of the abdomen, wings, and other structures.

Bug bytes

Pupal eclosion

http://vimeo.com/42759483

The insect's wardrobe: Exoskeleton

The events described for postembryonic development centered on the outcomes of molting. The actual process of molting has not yet been discussed in detail. To do so requires some understanding of the structure and composition of the exoskeleton. We explore the exoskeleton by examining the layered arrangement of the skin, or integument, relating composition of each layer to functionality. The basic design of the exoskeleton relies on an outer **cuticle** and an inner **epidermis**. The cuticle is the "functional" exoskeleton and hence the key feature of arthropod and insect success. The epidermis lies beneath the cuticle and serves as the outermost cellular layer of an insect. Though the cuticle in the form of the exoskeleton receives all the credit for insect terrestrial dominance, the epidermis is perhaps the unsung hero, as it produces everything in the cuticle and drives molting.

Cuticle

The cuticle is the outer covering of an insect that defines its appearance and conveys numerous functional aspects to each individual. Cuticle also extends into the lining of the trachea used for gas exchange and into the foregut and hindgut of the digestive tract, as well as forms the wings and internal apodemes. Stop for just a moment to conceptualize what this last statement means during molting: cuticle in all of these locations is replaced

Exarate   Obtect   Coarctate

Figure 6.8. Types of insect pupae: (A) exarate, or "naked," pupae have the developing adult appendages free and not covered; (B) obtect pupae have the appendages covered and glued to the body; and (C) coarctate pupae are exarate pupae covered by the hardened skin (puparium) of the last-stage larvae. Photos by Gyorgy Csoka (http://bit.ly/1OIrk29), Evanherk (http://bit.ly/1Lbpd51), and Scott Bauer (http://bit.ly/1KUNcDy), respectively.

during each molt. The insect must literally rip its guts out to grow! No other animals experience such trauma as part of development. If molting is so disruptive, even potentially deadly considering the vulnerability it exposes an insect to, why or how can it still be the mechanism used for growth? The simplest explanation is that the benefits of molting must outweigh the costs. Certainly this is true from the perspective of the functional benefits of possessing an exoskeleton. We'll revisit this question once we finish examining the basic design of the skin. For now, we continue our focus on cuticular structure. The cuticle is a nonliving layer composed predominantly of chitin, which can account for up to 50% of its dry weight. Chitin is typically chemically bound to protein, and the chitinous portion (**procuticle**) of the exoskeleton is the first layer produced during the process of molting (figure 6.9). The remaining layers of the cuticle are composed primarily of various lipids and proteins (structural and enzymatic). Chemical composition matters, as the types of lipids and proteins present dictate the function(s) of each layer and also contribute to the phenotypic appearance of the insect. All layers of

the cuticle are synthesized by epidermis, which includes assembling all macromolecules needed for producing skin and responding to exogenous signals conveying information on the type, size, and composition of the new exoskeleton. In return, the cuticle functions as a protective barrier between the epidermis and the outside environment (figure 6.10).

The cuticle is a complex structure composed of ordered layers. At the onset of molting, the procuticle is synthesized first by the epidermis and contains the bulk of chitin found in the exoskeleton. In addition to being the thickest layer of the cuticle, it is also the hardest, although the degree of hardness varies tremendously among insects. Generally the procuticle is regarded as the structural arm of the exoskeleton, in that strength and rigidity are functionally derived from this layer. It is divided into two layers: the **endocuticle** and the **exocuticle**. The inner (meaning closer to the epidermis) endocuticle is thicker than the exoskeleton and is nearly completely digested enzymatically during a molt so that the released nutrients can be reabsorbed. What is digested? Mostly macromolecules in the form of proteins

Figure 6.9. The beautiful cuticle of the sweat bee, *Augochloropsis sumptuosa,* is the result of the ordered arrangement of cuticle layers and the chemical composition of each layer. Photo by USGS Bee Inventory and Monitoring Lab available at http://bit.ly/1LSmkAo.

and lipids. In contrast, the exocuticle is difficult to digest because of sclerotization. Proteins are cross-linked to chitin in this region, adding to the strength and hardness of the cuticle and also making protein nearly resistant to degradation by enzymes. Consequently, this layer is shed nearly intact during ecdysis.

Layered on top of the procuticle is the very thin **epicuticle**. This layer supports the contention that bigger is not better. How does that apply here? Well, the epicuticle is the thinnest layer of the integument, yet most of the functional properties of the exoskeleton are associated with this layer. For example, water permeability of the cuticle is regulated by the quantity and type of lipids (mainly hydrocarbons) in four layers (cement, superficial layer, outer epicuticle and inner epicuticle) of the epicuticle. The hydrophobic or hydrophilic characteristics of the exoskeleton are also influenced by the protein and lipid composition of these layers, as is the extent that the skin can expand or stretch. What is not found here, that is in the procuticle, is chitin. Therefore, "hardness" is not a feature of the epicuticle and is actually associated with the underlying layers below this outermost region.

An additional layer is sometimes present in the epicuticle, known as the wax layer. Or, it is more correct to say that there is disagreement among authors as to whether to consider the wax layer a true layer or to view it as a coating on the superficial layer. Either way, waxes and other lipids produced by glands in the epidermis are transported to the epicuticle via canals (filaments) that extend through the procuticle to openings in the outer epicuticle and superficial layers. Waxes contribute to dehydration prevention and hydrophobicity of the exoskeleton as a whole.

Epidermis

What does the epidermis do for the exoskeleton? Anything it wants! Okay no, not anything. But close. The epidermis is the only cellular layer of the exoskeleton. It is thus the last living outpost from the inside to the outside of an insect. Structurally, the epidermis is one-cell-layer thick, composed of epithelial cells that rest on an inner basal lamina or basement membrane. Cellular membranes (apical membranes) facing the cuticle often display projections or ridges that serve as the sites of secretion for chitin and outer epicuticle during molting. The intracellular environment of epidermal cells is distinctive in that it is packed with rough endoplasmic reticula and Golgi apparatus. Both represent cells that synthesize proteins for transport to extracellular locations. Sandwiched within the cellular layer are gland cells with either microvilli or ducts extending through the procuticle. Technically, all cells of the epidermis are glandular, in that they all release secretions associated with the construction of the new cuticle.

Section of Insect Integument
(Symbolic impression)

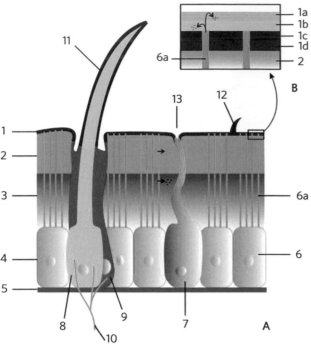

Figure 6.10. Cross section of the insect integument: (A) cuticle and epidermis in cross section, (B) layers of the outer epicuticle. Illustration by Xvazquez available at http://bit.ly/1MuIIYO.

A: Cuticle and epidermis;
B: Detail of the epicuticle.
1: Epicuticle
  1a: Cement
  1b: Wax layer
  1c: Outer epicuticle
  1d: Inner epicuticle
2+3: Procuticle
2: Exocuticle
3: Endocuticle
4: Epidermal epithelium
5: Basement membrane
6: Epidermal cell
6a: Pore Canal
7: Glandular cell
8: Tricogen cell
9: Tormogen cell
10: Nerve ending
11: Sensory hair (sensillum)
12: Seta
13: Glandular pore

An examination of cellular structure reveals that epidermal cells are heavily involved in secretion. Essentially any material used to produce any layer of the cuticle is either synthesized or transported to the epidermis. This is true for processes from cuticle synthesis to any processes associated with hardening, tanning, pigment deposition, waterproofing, repair, and defense. Several enzymes are distributed throughout the cuticle to aid in detoxification, body defense, and immune reactions. The epidermis plays a primary role not only in producing materials needed for molting, but also in receiving and responding to internal or endogenous signals so that the events of molting can proceed. This includes ramping up cellular activity to prepare for cuticle synthesis, producing enzymes to digest the old endocuticle, synthesizing a protective membrane to prevent autolysis of the epidermis, and interpreting hormonal signals so that the proper cuticle is produced with each molt. Molting essentially starts and stops with the epidermis, which is interesting from at least three perspectives:

1. A one-cell-layer thick tissue is responsible for all the complex events of molting. If any one does not proceed correctly, the insect will most likely die. Talk about pressure on the epidermis!
2. Molting removes the protective barrier of the cuticle, albeit for a short period of time, thereby exposing the epidermis to the harsh outside environment. Thus, the epidermis is deliberately exposing itself.
3. The epidermis itself does not physically change during growth despite orchestrating a massive external transformation of the insect. It most certainly increases in cell number, but the basic design is consistent throughout development.

As stated at the onset, the cuticle, in the form of the exoskeleton, is viewed as one of the major reasons that arthropods and insects have been so successful. However, it is readily apparent that without the amazingly efficient multitasking epidermis, world domination would not have occurred for the class Insecta.

What to do when your pants don't fit: Molt

When an adult human discovers that a pair of jeans no longer fits, several options exist for how to respond. One is to force-fit them over the likely culprit, meaning the expanded buttocks. This may continue for several minutes until exhaustion sets in. Another option is denial. The pants fit yesterday so they will fit today. This of course can lead to initiation of the first response until the ATP in your arms is spent, or it may be that the jeans are returned to the closet, with the intent of giving them a second chance to do the right thing at a later date, which of course is to slide properly over your hips. The final option, rarely used as the first response, is to admit that the pants no longer fit. This admission is in no way an acceptance of the view that unexpected growth has occurred. No, it is entirely possible that someone has broken into your room, quickly stitched the pants to a smaller size while you were asleep or away, and then returned them to the exact location undetected. Multiple working hypotheses must be considered, you say to yourself, so this explanation seems plausible.

Insects, on the other hand, do not deal with denial, and if they have mischievous friends, mischief most likely results in consumption, tight-fitting skin and all! When the exoskeleton becomes too small, thereby restricting growth, the insect solution is to molt. Though we have discussed the end result of molting, we have yet to examine the actual events of molting. This section explores the sequential steps of molting, including the role of the epidermis in each step, how the new cuticle forms, and what happens to the old skin. Hormonal signals are fundamental to orchestrating the events of molting to ensure that the processes occur with precise timing and the correct skin is produced.

The initiation of molting begins with recognition by the insect that a new skin is needed. The precise signals for molting are not known for all insects, but with some species, stretch receptors in the exoskeleton signal that maximum expansion has been achieved. Neurosecretory cells in the brain are activated to release neuropeptides, including the hormone **prothoracicotropic hormone** or **PTTH**, which is released into the hemolymph. The hormone will eventually bind to the prothoracic gland (the name indicates location for most insects), which in turn produces and releases another hormone, **ecdysone**. Ecdysone is a complex of multiple forms, termed **ecdysteroids,** which usually require an activation step. For our purposes, the name ecdysone suffices. This hormone is often referred to as molting hormone, because its binding to epidermal cells initiates molting. To do this, the hormone must circulate through the hemolymph to reach the epidermis; it then crosses the cellular membranes of epithelial cells to bind to DNA in the nucleus. Activation of the epidermal cells results,

beginning the ramping up of cellular activity necessary for molting to occur.

The first major step of molting is the separation of the epidermis from the cuticle. This process is termed **apolysis** and creates a space between the living layer and the cuticle. Muscles and trachea are still attached at this point, so the old cuticle is still a functional exoskeleton despite apolysis having occurred. Epidermal cells engage in two activities as the next steps to molting. The first is a series of mitotic divisions to increase the cell numbers of the epidermis: the epidermis will expand in size and surface area as a result of the cell divisions. The second activity is secretion of molting fluid into the newly created space. The fluid contains inactive forms of digestive enzymes that will eventually target chitin (chitinases) and proteins (proteases), but not until the new cuticle is synthesized to occupy the space between the epidermis and old cuticle. The new cuticle produced is initially undifferentiated, as it is not full sized, lacks pigment, and has not undergone sclerotization. It is flexible and pliable at this point, which is needed since it still resides beneath the old, hard skin. Enzymes within the molting fluid become active, targeting proteins, lipids, and chitin in the endocuticle. The released digestive products are absorbed by the epidermis and used to finish the synthesis of the new skin. Once the new cuticle is full length (but not expanded), the epidermis will reabsorb molting fluids and the old skin will be removed from the body. Before this can take place, however, muscles must detach from the old exoskeleton and connect to the apodemes of the new skin. The coordination must be precisely timed so that ecdysis is not initiated too soon. Otherwise, shedding of the old skin cannot be completed.

Ecdysis is initiated by the insect increasing hemocoelic pressure at weak points (called cleavage lines) in the skin. The exoskeleton splits at these points, permitting the insect to force its body through it to reveal the new, larger skin. The old exoskeleton is left behind, leaving an empty imprint of the old insect. If you look closely at the cast skin or **exuvia**, you will be able to see the cuticular linings of the trachea and possibly foregut and hindgut. At this point, the new exoskeleton is soft and lacks pigment and any special markings. The cuticle is also wrinkled and resembles an inflatable yard decoration that has collapsed (figure 6.11). By gulping in air,

Figure 6.11. A newly synthesized insect cuticle in many ways resembles a deflated lawn decoration. For the decoration and insect exoskeleton, intake of air results in expansion to the full size. Photos by Michael Rivera available at http://bit.ly/1syuwEd.

insects elevate their hemocoelic pressure, and the process of skin expansion begins. Tanning, pigment deposition, and hardening also occur during the following several minutes to hours. As you might expect, the period immediately following ecdysis is when the insect is most vulnerable, so molting frequently takes place in a protected locale. Many species groom or clean themselves following expansion of the skin, while others remain motionless for an extended period of time.

Growth is not complete with ecdysis. Most species continue to add protein and lipids to the cuticle for several days. Muscle attachment continues as well, in some cases to reestablish connections but also to form new muscle attachments. Juvenile forms will resume feeding immediately after completing a molt, with the process of feeding, enlarging, and molting continuing several more times. Molting ceases with winged species once the wings are fully formed. For non-winged insects, molting may continue throughout adulthood.

How is the type of skin determined after each molt? More specifically, how does the skin "know" when to produce another juvenile skin, or pupal, or adult? The entire process of molting—from initiation to ecdysis—is under the control of hormones. PTTH in combination with other neuropeptides initiates the events of molting; ecdysteroids activate epidermal cells; and the presence of **juvenile hormones** (as with ecdysteroids, multiple forms exist) in relation to ecdysteroids dictates what type of new skin will be produced. Generically speaking, when levels of juvenile hormones in the hemolymph are high in relation to levels of ecdysone, epidermal cells will synthesize a new juvenile skin. If juvenile hormone levels decline and ecdysone concentrations become higher, a pupal skin will be produced for species undergoing complete metamorphosis. Finally, if ecdysone is the only hormone present, the epidermis will be compelled to synthesize an adult cuticle. In reality, hormonal regulation is far more complex than depicted here, but the examples given do convey the principles of how skin fate is determined.

Quick check

Can an insect grow if it does not shed its skin?

To metamorphose, or not to metamorphose, that is the question

All insects do not develop the same way, which is in no way surprising. During growth, specifically when the

Table 6.1. Types of metamorphosis displayed by different insect orders. Information based on Gullan and Cranston (2010).

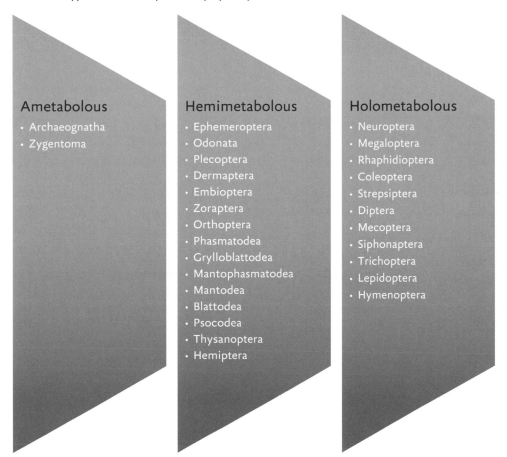

Ametabolous
- Archaeognatha
- Zygentoma

Hemimetabolous
- Ephemeroptera
- Odonata
- Plecoptera
- Dermaptera
- Embioptera
- Zoraptera
- Orthoptera
- Phasmatodea
- Grylloblattodea
- Mantophasmatodea
- Mantodea
- Blattodea
- Psocodea
- Thysanoptera
- Hemiptera

Holometabolous
- Neuroptera
- Megaloptera
- Rhaphidioptera
- Coleoptera
- Strepsiptera
- Diptera
- Mecoptera
- Siphonaptera
- Trichoptera
- Lepidoptera
- Hymenoptera

occasion to molt arrives, transition from one instar to the next can result in anything from virtually no changes to radical transformations. This is metamorphosis, or the process of change. With insects, multiple forms of metamorphosis are displayed, and the differences reflect how much uniqueness exists between juveniles and adults. For example, with some species, the juveniles (nymphs) and adults reside in the same locations, eat the same food, and basically look the same, with the exception of size differences and whether external genitalia are developed. Such insects are ametabolous (display **ametabolous metamorphosis**), literally meaning lacking change, and represent the most primitive species (table 6.1). Ametabolous insects also show the basic or primitive condition with respect to nearly all structures, including a reliance on cursorial legs for locomotion, simple eyes, and elongate abdomens (no reduction in numbers of segments).

Species that rely on a gradual transformation from juvenile (nymphs) to adult through each molt use **hemimetabolous (incomplete) metamorphosis** (figure 6.12). These juveniles often, but not always, occur in the same habitat as adults, but they do not appear as miniature imagoes. Instead, nymphs develop wings as external buds that increase in size by mitosis with each molt. Similarly, genitals of the adults develop gradually with each molt. Fully formed adults typically have functional wings (though some display regressive evolution in this regard), which permits departing from where the juveniles are located. Molting also ends with the formation of functional wings. Some species develop in aquatic habitats as juveniles (termed naiads) and emerge from water to molt to the adult stage. Thus, the young and the imagoes never occupy the same habitats. Hemimetabolous metamorphosis is the most commonly used by members of the class Insecta.

A new age of insect control: Growth regulators

Wouldn't it be a great if you could kill your favorite insect without running the risk of harming yourself? Well now you can! Sarcasm aside, it is possible to achieve insect control using chemical compounds not designed to kill. **Insect growth regulators** (IGR) take advantage of the physiology of growth and molting by mimicking naturally produced insect hormones or growth regulators: that is, a given IGR competes for the same receptor site(s) targeted by a specific hormone. Once an IGR is bound to a receptor, the targeted cell cannot carry out the function that the hormone would have instructed it to. How does this alter insect growth? Simple: IGRs compete with the hormones ecdysone and juvenile hormone. Both are intimately involved in the process of molting, so if these hormones are prevented from carrying out their roles during growth, the insect is seriously compromised.

Successful juvenile hormone mimics often prevent an insect from reaching the adult stage or induce premature molting in juvenile stages. If the timing of developmental events is not precise—for example, an ill-timed molt or production of an inappropriate cuticle—then the resulting instar will be deformed and/or unlikely to survive. IGRs that mimic the action of ecdysone commonly trigger precocious molts or disrupt ecdysis. When ecdysis occurs, the old skin remains attached, and depending on how the insect exits the exuvia, the head may remain covered, thereby preventing feeding. In some insects, ecdysone-like IGRs cause mortality during the pupal stages by triggering the development of larval rather than adult tissues.

Another group of growth regulators wreak havoc by inhibiting the synthesis of chitin. Chitin is a component of the cuticle, specifically, the procuticle. If the amount of chitin incorporated into the newly synthesized cuticle is reduced, the obvious effect is a reduction in the strength of the skin. For many insects, disruption of chitin synthesis is more devastating in that the new cuticle cannot be assembled properly, in which case death is the typical outcome. This can also result when eggs are exposed to IGRs that target chitin synthesis because, though molting does not take place during embryonic development, the first cuticle is produced there.

There are drawbacks to using IGRs for insect control. The most obvious is a lack of specificity. Any animal that possesses a chitinous exoskeleton must molt to grow. Thus any ecdysozoan is susceptible to the effects of IGRs, not just a pest insect. Another consideration is economics. Most IGRs are more expensive to produce, and sometimes to distribute, than traditional chemical insecticides. As a result, the cost of pest management is higher, changing the economic threshold decision when using an IGR versus a cheaper insecticide. Compounding the economics of IGRs is that many do not cause a rapid demise of the targeted insect. This is especially true if the IGR causes sterility in adults, but the crop damage is caused by feeding activity by immatures and/or adults—relief may not be instantaneous. Nonetheless, IGRs are a major weapon in the arsenal to combat pest insects because they are effective at disruptive growth, display almost no harmful effects on mammals, and have a modest negative effect on ecosystem functioning.

The most complex and radical transformations occur with **holometabolous (complete) metamorphosis**. In this form of development, juveniles (larvae) and adults look nothing alike and rarely reside in the same locations or feed on the same types of food. Larvae grow via molting, passing through a series of either fixed or variable numbers of instars before entering a transitional stage, termed a pupa (figure 6.13). The pupal stage is typically a nonmobile, nonfeeding stage characterized by the destruction and replacement of larval tissues with entirely new adult structures. Very few tissues remain from a larva; usually portions of the nervous system, including the brain, re-

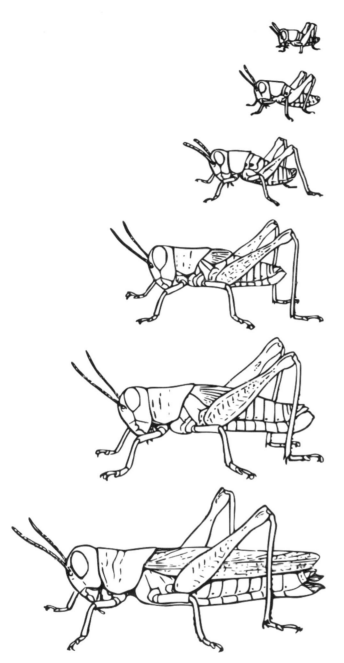

Figure 6.12. Example of an insect (grasshopper) that undergoes hemimetabolous or incomplete metamorphosis. Illustration by S. E. Snodgrass available at http://bit.ly/1FncKsF.

main intact. However, even these tissues will eventually undergo metamorphosis during pupal and adult development. Adult structures are derived from **imaginal discs**. Imaginal discs are essentially clusters of stem cells that form during larval development but that remain undifferentiated until pupal development. Hormonal signals transform these once-dormant tissues into wings, legs, antennae, and nearly every other structure in the adult.

Gulf Fritillary Life Cycle

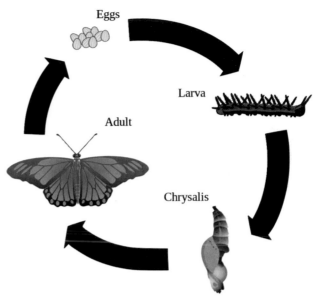

Figure 6.13. Example of an insect (butterfly) that undergoes holometabolous (complete) metamorphosis. Illustration by Bugboy52.40 available at http://bit.ly/1KVhINE.

Once the pupa is fully formed, pupal development climaxes with adult emergence: the insect sheds the pupal skin via ecdysis to emerge as an adult. As discussed earlier, a series of changes must occur to expand the body and achieve adulthood. The resulting molt reveals an adult that shows no morphological similarity to the larva. In fact, if the development events were not witnessed, it would be very difficult to believe that the larva and adult were related. The transformation is radical, abrupt, and complete. This form of metamorphosis is used by the insect groups (orders) that are the most successful in terms of species numbers and abundance.

CHAPTER REVIEW

✦ **Crunchy on the outside, gooey on the inside**
- The centerpiece to insect growth is the exoskeleton. It affords protection from mechanical injury, is a barrier to keep things out or in, gives the body shape, serves as the site of muscle attachment, and functions in the greatest battle for all terrestrial animals: to prevent desiccation. The characteristics of the skin that permit such benefits also limit growth in arthropods. Actually, just stating that growth is limited does not really

capture the full situation. Growth occurs but is infinitely more complex because of the exoskeleton.

- For animals that possess an exoskeleton composed primarily of the polysaccharide chitin, growth is achieved by shedding the old skin and replacing it with a new one that is larger. This growth process occurs in two phases, termed molting and ecdysis. Molting occurs when a portion of the old skin is digested and the new and improved skin of the next stage of development is produced. Ecdysis is the physical removal of any remnants of the old exoskeleton. Growth actually occurs only during the period of time in which the old skin has been removed and the new skin is expanded to full size by increases in "blood" pressure.

- Animals that grow by shedding—in other words, need to molt to grow—are grouped together as the clade Ecdysozoa. A clade is a group containing an ancestor and all of its descendants. In this instance, Ecdysozoa include the phyla Arthropoda, Nematoda, Onychophora, Tardigrada, Cephalorhyncha, and Nematomorpha. All are invertebrate animals with an exoskeleton that limits growth, and thus the major linkage among each of these groups is that development progression requires molting to occur.

Eat, grow, shed: The development plan

- Insect development is a bit more complex than "just" molting. In fact, the development plan used by most insects begins with the egg, a stage in which molting does not occur, and continues through adulthood. The path that any organism takes to complete development from egg to adult is ontogeny. It is literally the developmental history of any organism. In the case of insects, the development path or history occurs in two broad phases: Development from egg to adult and maturity to adulthood.

- Most insects begin life in the egg stage, and once the egg hatches, the newly minted juvenile essentially becomes an eating machine. As the immature feeds, it eventually reaches a point in which body mass outpaces the exoskeleton. Consequently, the old skin must be replaced with a new, larger one, and once installed, eating

resumes. The process continues, with some major and minor twists and turns, until finally the adult stage is reached. This overview of development actually reveals that the path to adult involves two key periods of life: development before and development after egg hatch. These periods reflect major development phases, referred to as embryonic development and postembryonic development.

- For most insects, once fertilization occurs, the egg is deposited outside of the mother's body. The significance of egg placement is that this location is generally where embryonic development begins. At some point after oviposition, a trigger or activation step initiates embryonic development. The egg cell responds by mitotically dividing the zygote nucleus. The mitotic division of the zygote nucleus yields daughter nuclei contained within the same cell, which by definition results in the formation of a syncytium. Eventually additional cells form within the syncytium to produce a superficial layer to the periphery of the egg. This layer will become a one-cell-thick layer known as the blastoderm. With each round of cell divisions, the embryo gradually morphs to look more and more like a young or neonate juvenile. Embryonic development officially comes to an end with the hatching of the egg.

- Postembryonic development begins with egg eclosion and continues until an adult insect forms. Molting is the process that must occur for an insect to move from one stage of development to another. We have already discussed that shedding of the exoskeleton is necessary to permit increases in size; viewed another way, molting facilitates transition from one developmental stage to another.

- Each new larva has entered the first stadium of larval development, which is also referred to as the first instar. An instar is simply a term for a developmental stage, but in practice is usually applied only to immature stages. So the blow fly larva that emerges from the egg is a first instar larva: "first" because it is the initial development stage of postembryonic development. This stage comes to an end when the old exoskeleton is shed via ecdysis, and the new, larger one underneath is revealed. The next stage is called the second instar larva, the one following is the third, and so on. Molting coupled with ecdysis signals the completion of one stage of growth and the beginning of the next.

- Many insects have a fixed number of developmental stages, usually in terms of the immatures; these are said to display determinate growth. For these insects, the length of each larval or nymphal instar is influenced by temperature and food availability, but the total number of developmental stages remains constant. In contrast, some species display variable numbers of immature stages, termed indeterminate growth. For these insects, temperature, competition, and food availability all influence the total number of immature developmental stages.

- Development is not finalized with the final ecdysial event to adult. A series of changes occur with the skin, including biochemical reactions that lead to hardening (sclerotization) and tanning (melanization) of the exoskeleton. These processes occur with each molting event, not just when an adult skin is formed. Why are they necessary? Because the newly formed exoskeleton is soft and pliable, a necessity since it is produced on top of the epidermis but under the old, rigid exoskeleton and does not possess yet the properties necessary for providing all forms of protection and identity needed for the insect to survive in a harsh, terrestrial or aquatic environment. The adult insect is usually not sexually mature immediately following the adult molt. So, typically, a period of time is needed for the gonads to become fully functional.

The insect's wardrobe: Exoskeleton

- The basic design of the exoskeleton relies on an outer **cuticle** and an inner epidermis. The cuticle is the "functional" exoskeleton and hence the key feature of arthropod and insect success. The epidermis lies beneath the cuticle and serves as the outermost cellular layer of an insect. Though

the cuticle in the form of the exoskeleton receives all the credit for insect terrestrial dominance, the epidermis is perhaps the unsung hero as it produces everything in the cuticle and drives molting.

- The cuticle is the outer covering of an insect that defines its appearance and conveys numerous functional aspects to each individual. Cuticle also extends into the lining of the trachea used for gas exchange and into the foregut and hindgut of the digestive tract, as well as forms the wings and internal apodemes. Stop for just a moment to conceptualize what this last statement means during molting: cuticle in all of these locations is replaced during each molt. The insect must literally rip its guts out to grow!

- The cuticle is a complex structure composed of ordered layers. At the onset of molting, the procuticle is synthesized first by the epidermis and contains the bulk of chitin found in the exoskeleton. In addition to being the thickest layer of the cuticle, it is also the hardest, although the degree of hardness varies tremendously among insects. Generally the procuticle is regarded as the structural arm of the exoskeleton in that strength and rigidity are functionally derived from this layer. It is divided into two layers: the endocuticle and the exocuticle. The inner endocuticle is thicker than the exoskeleton, and is nearly completely digested enzymatically during a molt so that the released nutrients can be reabsorbed.

- Layered on top of the procuticle is the very thin epicuticle. The epicuticle is the thinnest layer of the integument, yet most of the functional properties of the exoskeleton are associated with this layer. For example, water permeability of the cuticle is regulated by the quantity and type of lipids in four layers of the epicuticle. The hydrophobic or hydrophilic characteristics of the exoskeleton are also influenced by the protein and lipid composition of these layers, as is the extent that the skin can expand or stretch.

- The epidermis is the only cellular layer of the exoskeleton. It is thus the last living outpost from the inside to the outside of an insect. Structurally, the epidermis is one-cell-layer thick, composed of epithelial cells that rest on an inner basal lamina or basement membrane. Cellular membranes facing the cuticle often display projections or ridges that serve as the sites of secretion for chitin and outer epicuticle during molting. The intracellular environment of epidermal cells is distinctive in that it is packed with rough endoplasmic reticula and Golgi apparatus. Both represent cells that synthesize proteins for transport to extracellular locations. Sandwiched within the cellular layer are gland cells with either microvilli or ducts extending through the procuticle.

- Essentially any material used to produce any layer of the cuticle is either synthesized or transported to the epidermis. This is true for processes from cuticle synthesis to any processes associated with hardening, tanning, pigment deposition, waterproofing, repair, and defense. Several enzymes are distributed throughout the cuticle to aid in detoxification, body defense, and immune reactions. The epidermis plays a primary role not only in producing materials needed for molting, but also in receiving and responding to internal or endogenous signals so that the events of molting can proceed. This includes ramping up cellular activity to prepare for cuticle synthesis, producing enzymes to digest the old endocuticle, synthesizing a protective membrane to prevent autolysis of the epidermis, and interpreting hormonal signals so that the proper cuticle is produced with each molt.

What to do when your pants don't fit: Molt

- The initiation of molting begins with recognition by the insect that a new skin is needed. Neurosecretory cells in the brain are activated to release neuropeptides, including the hormone prothoracicotropic hormone or PTTH, which is released into the hemolymph. The hormone will eventually bind to the prothoracic gland, which in turn produces and releases another hormone, ecdysone. This hormone is often referred to as molting hormone, because its binding to epidermal cells initiates molting. To do this, the hormone must circulate through the hemolymph to reach the epidermis; it then crosses the cellular membranes of epithelial cells to bind to DNA in the nucleus. Activation of the epidermal cells results, beginning the ramping up of cellular activity necessary for molting to occur.

- The first major step of molting is the separation of the epidermis from the cuticle. This process is termed apolysis and creates a space between the living layer and the cuticle. Epidermal cells engage in two activities as the next steps to molting. The first is a series of mitotic divisions to increase the cell numbers of the epidermis. The second activity is secretion of molting fluid into the newly created space. The fluid contains inactive forms of digestive enzymes that will eventually target chitin and proteins, but not until the new cuticle is synthesized to occupy the space between the epidermis and old cuticle. The new cuticle produced is initially undifferentiated, as it is not full sized, lacks pigment, and has not undergone sclerotization. Enzymes within the molting fluid become active, targeting proteins, lipids, and chitin in the endocuticle. The released digestive products are absorbed by the epidermis and used to finish the synthesis of the new skin. Once the new cuticle is full length, the epidermis will reabsorb molting fluids and the old skin will be removed from the body. Before this can take place, however, muscles must detach from the old exoskeleton and connect to the apodemes of the new skin. The coordination must be precisely timed so that ecdysis is not initiated too soon. Otherwise, shedding of the old skin cannot be completed.

- Ecdysis is initiated by the insect increasing hemocoelic pressure at weak points (called cleavage lines) in the skin. The exoskeleton splits at these points, permitting the insect to force its body through it, to reveal the new, larger skin. The old exoskeleton is left behind, leaving an empty imprint of the old insect. At this point, the new exoskeleton is soft and lacks pigment and any special markings. The cuticle is also wrinkled and resembles an inflatable yard decoration that has collapsed. By gulping in air, insects elevate their hemocoelic pressure, and the process of skin expansion begins. Tanning, pigment deposition, and hardening also occur during the following several minutes to hours.

- The entire process of molting—from initiation to ecdysis—is under the control of hormones. PTTH in combination with other neuropeptides initiates the events of molting; ecdysteroids activate epidermal cells; and the presence of juvenile hormones in relation to ecdysteroids dictates what type of new skin will be produced. Generically speaking, when levels of juvenile hormones in the hemolymph are high in relation to levels of ecdysone, epidermal cells will synthesize a new juvenile skin. If juvenile hormone levels decline and ecdysone concentrations become higher, a pupal skin will be produced for species undergoing complete metamorphosis. Finally, if ecdysone is the only hormone present, the epidermis will be compelled to synthesize an adult cuticle.

To metamorphose, or not to metamorphose, that is the question

- All insects do not develop the same way, which is in no way surprising. During growth, specifically when the occasion to molt arrives, transition from one instar to the next can result in anything from virtually no changes to radical transformations. With insects, multiple forms of metamorphosis are displayed, and the differences reflect how much uniqueness exists between juveniles and adults. For example, with some species, the juveniles and adults reside in the same locations, eat the same food, and basically look the same, with the exception of size differences and whether external genitalia are developed. Such insects are ametabolous, literally meaning lacking change, and represent the most primitive species.
- Species that rely on a gradual transformation from juvenile to adult through each molt use hemimetabolous or incomplete metamorphosis. These juveniles often, but not always, occur in the same habitat as adults, but they do not appear as miniature imagoes. Instead, nymphs develop wings as external buds that increase in size by mitosis with each molt. Similarly, genitals of the adults develop

gradually with each molt. Fully formed adults typically have functional wings, which permits departing from where the juveniles are located.

- The most complex and radical transformations occur with holometabolous or complete metamorphosis. In this form of development, juveniles and adults look nothing alike and rarely reside in the same locations or feed on the same types of food. Larval grow via molting, passing through a series of either fixed or variable numbers of instars before entering a transitional stage, termed a pupa. The pupal stage is typically a nonmobile, nonfeeding stage characterized by the destruction and replacement of larval tissues with entirely new adult structures. Adult structures are derived from imaginal discs. Imaginal discs are essentially clusters of stem cells that form during larval development but that remain undifferentiated until pupal development. Hormonal signals transform these once-dormant tissues into wings, legs, antennae, and nearly every other structure in the adult. Once the pupa is fully formed, pupal development climaxes with adult emergence: the insect sheds the pupal skin via ecdysis to emerge as an adult.

MUSHROOM FARMING (SELF-TEST)

Level 1: Knowledge/Comprehension

1. Define the following terms:
   (a) molting   (d) eclosion
   (b) apolysis   (e) ecdysis
   (c) instar   (f) metamorphosis

2. Explain the concept that insects grow by shedding.

3. What are the major developmental events that occur within the egg?

4. Describe the major events that occur during (a) ametabolous metamorphosis,(b) hemimetabolous metamorphosis, and (c) holometabolous metamorphosis.

5. What does ecdysone do that has vaulted it to the status of "molting hormone"?

6. Describe the functions of the cuticle.

Level 2: Application/Analysis

1. Describe the hormonal regulation of molting, specifically detailing how insect hormones dictate the outcome of each molt.

2. Make a diagram that depicts the layers of the epicuticle. Provide a brief explanation of the functions of the epicuticle.

3. Explain how holometabolous metamorphosis contributes to the success of those insect groups that utilize this form of development.

Level 3: Synthesis/Evaluation

1. Ecdysozoans are a clade in which all members possess an exoskeleton composed of chitin and all share the need to molt for growth to occur. If the exoskeleton of insects did not contain chitin, would molting still be a necessity for growth?

2. Speculate on the mode of action of an ideal insect growth regulator from the standpoint of inhibiting larval growth of a pest insect while demonstrating minimal to no impact on nontarget insects.

REFERENCES

Chapman, R. F. 1998. The Insects: Structure and Function. 4th ed. Cambridge University Press, Cambridge, UK.

Ewer, J. 2005. How the ecdysozoan changed its coat. *Plos Biology* 3(10): e349. doi:10.1371/journal.pbio.003049.

Gilbert, L. I. 2009. Insect Development: Morphogenesis, Molting and Metamorphosis. Academic Press, San Diego, CA.

Gullan, P. J., and P. S. Cranston. 2010. The Insects: An Outline of Entomology. 4th ed. Wiley-Blackwell, West Sussex, UK.

Heming, B.-S. 2003. Insect Development and Evolution. Cornell University Press, Ithaca, NY.

Lockey, K. H. 1988. Lipids of the insect cuticle: Origin, composition and function. *Comparative Biochemistry and Physiology B: Comparative Biochemistry* 89(4): 595–645.

Simpson, S., and J. Casas. 2010. Insect Integument and Colour, vol. 38, Advances in Insect Physiology. Academic Press, San Diego, CA.

Snodgrass, R. E. 1993. Principles of Insect Morphology. Cornell University Press, Ithaca, NY.

Truman, J. W., and L. M. Riddiford. 1999. The origins of insect metamorphosis. *Nature* 401: 447–452.

Vincent, J. F. V., and U. G. K. Wegst. 2004. Design and mechanical properties of insect cuticle. *Arthropod Structure and Development* 33(3): 187–199.

THE ENTOMOLOGIST BOOKSHELF (SUPPLEMENTAL READINGS)

Adler, J. H., and R. J. Grebenok. 1995. Biosynthesis and distribution of insect molting hormones in plants—A review. *Lipids* 30(3): 257–262.

Chown, S. L., and J. S. Terblanche. 2006. Physiological diversity in insects: Ecological and evolutionary contexts. *Advances in Insect Physiology* 33:50–152.

Horst, M. N., and J. A. Freeman. 1993. The Crustacean Integument: Morphology and Biochemistry. CRC Press, Boca Raton, FL.

O'Reilly, D. R., and L. K. Miller. 1989. A baculovirus blocks insect molting by producing ecdysteroid UDP-glucosyl transferase. *Science* 245(4922): 1110–1112.

Stokstad, E. 2014. Insect molting is "like having you lungs ripped out." *ScienceNews.* August 29, 2014. http://news.sciencemag.org/climate/2014/08/insect-molting- having-your-lungs-ripped-out.

Trigger for Insect Molting. 1983. New York Times Science Watch. January 18. http://www.nytimes.com/1983/01/18/science/science-watch-trigger-for-insect-molting.html.

ADDITIONAL RESOURCES

Beyond pesticides: Insect growth regulators
http://www.beyondpesticides.org/infoservices/pesticidefactsheets/toxic/insectgrowthregulators.php

Insect growth regulators
http://npic.orst.edu/ingred/ptype/igr.html.

Insect hormones
http://users.rcn.com/jkimball.ma.ultranet/BiologyPages/I/InsectHormones.html.

Molting. 2011. Insects Science
http://insectsscience.blogspot.com/2011/04/molting.html.

Stick insect molting
http://wn.com/stick_insect_molting.

The insect's process of molting
http://www.insectidentification.org/process-of-molting.asp.

7   Insect Facebook

The Basics of Insect Classification

If insects were the size of birds, or people the size of mice, "bug watchers" would be as prevalent as bird watchers, and entomologists would command the budget of the Defense Department.

Dr. Howard Ensign Evans
The Pleasures of Entomology:
Portraits of Insects and the People Who Study Them (1985)

For most any individual, if a butterfly flies by, you know what it is. There is immediate recognition that the insect in question is in fact a "butterfly." But how did you know that? What characters did you use for instant recognition that the critter was a butterfly? Something served as a QR code* for the brain that led to the conclusion that a butterfly flew past. Obviously the choices are related to an external view of the body, which means that either the anatomy or the behavior, or a combination of the two, provided the clues. Could it be that the slow flapping of wings alternating with gliding effortlessly through the air cannot be confused with any other beast? Probably not, unless you believe that the only creatures that can fly are butterflies. The reality is that many species of insects possess wings, as do birds and bats, so flight alone does not serve as the instant recognition factor. In fact, even if the possession of wings led you to think of an insect, that would narrow the list to a million, the approximate number of described species of insects known to inhabit the planet. So a bit more information is necessary for concluding butterfly. What about the beautiful display of colors arranged artistically into distinctive patterns on the wings of the butterfly? That would work for some species, but not others, and it also presupposes that butterflies are the only insects with ornate coloration. But they do not, in fact, have sole ownership of beauty in the world of insects.

What allows us to recognize insects from other animals, and one group or species of insect from another, is a combination of anatomical features shared by a particular taxonomic group of insects. Usually the group is also distinctive in other aspects of its biology, such as the niche it occupies in the environment, its habitat, its reproductive strategies and behaviors, or some other aspect of its life history. This chapter introduces the basic ideas and concepts associated with grouping insects into particular assemblages; provides a brief introduction to traditional identification techniques, as well as to newer methods like DNA barcoding; and offers an overview of the most

---

*QR codes are matrix barcodes that were originally developed for the Japanese automobile industry but have now spread to a multitude of uses. One of the most common is for mobile phones; simply scanning the QR code found in print or electronic form opens a website for an advertisement, company, university, or any other imaginable entity.

USA: Alaska, woods near Kenai National Wildlife Refuge headquarters building 60.4618°N 151.0806°W 02.Sep.2010. Matt Bowser. KNWR: Ento: **10036**

KNWRC1254

Figure 7.1. Example of a unique QR code used on a taxonomic label for an insect collected in Alaska. QR code and label created by Matt Bowser available at http://bit.ly/1VhcgWM.

common insect orders. A discussion of why insect identifications are important to agriculture, controlling the spread of disease, legal investigations, and national security will also be explored, to answer the dread question, "Why should I care what kind of insect it is?" (figure 7.1).

Key Concepts

* Why should I care how to identify an insect?
* Thank you, Linnaeus! Binomial classification
* Insect crib notes: Key features used to identify adult insects
* Insect diversity
* Classification by taste. Seriously?

Why should I care how to identify an insect?

It seems we have been here before. Back in chapter 1, we broached the subject of whether it is critical for you to understand insect classification in order to study insects or appreciate them. The conclusion: No! You do not need to dive into taxonomy, systematics, or phylogeny to appreciate what insects do or to realize that they are important in a multitude of ways. This may be considered heresy to my entomological brethren. But what it means to you is that the ability to recognize an insect beyond the order or family level will not impede your understanding of the importance of insects to the human condition, or the types of reproductive strategies they use, methods of communication they employ to attract mates or send an alarm, or how they process food after ingestion. Or most any other topic associated with the endlessly fascinating world of insects. In many ways, studying insects devoid of taxonomy allows you to sim-

ply enjoy their biology, like reading an engaging novel with no regard for the writer's choice of syntax or unique grammatical style. It is learning for the sake of learning. Does this approach present limitations to studying insects? To a degree yes, but mostly for topics that are beyond this introduction to insect biology. Studying insects in the absence of taxonomy is like learning biology without a background in chemistry and physics. It certainly can be done, and a great deal of understanding is possible, even lacking a full natural sciences foundation. For some topics, though, this lack will limit the depth of learning; for other topics, so much so that only superficial skimming is possible.

For example, if an insect in question is considered a pest of a particular crop, knowing its identity is absolutely critical. Classifying the insect to genus and species is necessary to determine whether the crop is fed upon by this species, and if so, what type(s) of control should be applied and when. As we will learn in chapter 14, understanding the biology of an insect is essential to effective pest management and making responsible decisions that reduce the threat of the pest without harming mankind or the environment. Similarly, an insect found at a port of entry into the United States must be quickly and accurately identified to assess whether the critter in question is a potential pest, poses no threat, or is even possibly a threat to national security. The recent outbreak of West Nile virus in the mid-Atlantic and northeastern regions of the United States highlights the need for being able to identify insects: various species of mosquitoes are the prime vectors of the disease (figure 7.2). Essential to diagnosing and preventing insect-borne diseases like Zika virus, malaria, and dengue fever is recognizing the culprit so that its biology can be manipulated. From a criminal investigation perspective, the identity of an insect feeding on a corpse can reveal important development data that in turn can be used to determine a portion of the **postmortem interval**,* which requires

---

*The postmortem interval generally refers to the time since death. However, forensic entomology typically can reveal information about only a portion of the interval, termed the minimum postmortem interval, which is the minimum time that a body must have been available for the insect to have colonized the corpse. Thus, there is a period of time from the moment of death to insect colonization that cannot be determined from the insect evidence.

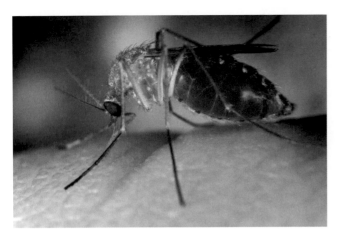

Figure 7.2. The main transmitter of the West Nile virus in the southeast is the mosquito species *Culex quinquefasciatus*. Species identification is essential to disease prevention programs. Photo by Jim Gathany available at http://bit.ly/1LdncFi.

being able to identify the **necrophagous** creatures to genus and species (figure 7.3). Is it not good enough to be able to recognize that the insect is commonly found on a corpse? No. Why? Because the postmortem interval depends on species-specific developmental data. After all, you might need to establish that you are an expert when testifying in court.

> *Defense attorney*: What type of insect evidence did you collect at the crime scene?
>
> *You, the "expert"*: A bug.
>
> *Defense attorney*: What type of "bug"?
>
> *You the "expert"*: A big one.
>
> *Defense attorney*: Is there anything else you can tell us about this "bug"?
>
> *You the "expert"*: Yes. It smelled really bad.
>
> *Jury*: Guilty!

Guilty, maybe—of stupidity!

Of more practical use, for the non-insect-expert, is that being able to recognize the types of insects in your home means the difference between *panic* (bed bugs), *action* (termites and carpenter ants), *disgust* (cockroaches), or *apathy* (why, that's just an insignificant little brown moth). The identity could also be important to you in determining whether the red and black "bug" that landed on your head is simply an insect such as a harmless ladybird beetle (commonly called lady bug) or an assassin bug (a true bug or member of the order Hemiptera) with a nasty dis-

position and a propensity to bite. The same can be said for the need to distinguish other kinds of biting and stinging insects, or any other insects with attributes you wish to avoid.

As a non-entomologist, do you really have to be able to taxonomically classify an insect to know how to avoid it? Absolutely not! And the goal here is not to convince you otherwise. This chapter presents some of the basics used to classify insects, so that we can better understand the relationships between different insect groups. Insect classification, taxonomy, and phylogeny are topics best appreciated by those enthralled by the beasts. Hopefully you will join those ranks, but the immediate goal is for you to appreciate why insect identifications are indeed important under the right circumstances.

Thank you, Linnaeus! Binomial classification

Any discussion of how to classify any organism, insect or otherwise, or why there is a need to do so, should begin with a tribute to Linnaeus. In much the same way that a crowd at a sporting event goes quiet as the Blue Angels* fly overhead, and then erupt into a loud roar once past, I do the same when speaking of the great Swedish naturalist Carl von Linné (better known as Carolus Linnaeus) (figure 7.4). Actually that's a lie; I do not pause in silence with each mention of his name, but I do contribute to the roar at a baseball game (more so when the vendor with the frozen lemonade walks past, but nonetheless the sentiment is genuine). Linnaeus was briefly introduced in chapter 2 and given credit for identifying thousands of insects and plants (4,400 insects and 7,700 plants) over his lifetime and, most importantly, for developing a classification system for naming botanical and zoological specimens. The system, known today as binomial classification, revolutionized organismal classification just by making it a simpler process. What a contrast to today, in which revolutionary advances in science all seem to be tied to either technology or molecular medicine.

---

*The Blue Angels formed in 1946 as the US Navy's flight demonstration squadron. The squadron of six pilots is known for performing death-defying acrobatic maneuvers that are awe inspiring for anyone. http://www.youtube.com/watch?v=HARMj_9L03E.

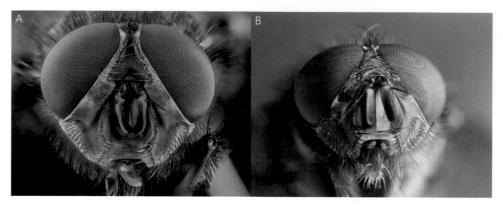

Figure 7.3. To the casual observer, these two common blow flies appear nearly identical, yet a close inspection of the facial and wing morphology reveals that (A) is *Calliphora vicina* and (B) is *C. vomitoria*. Photos by USGS Bee Inventory and Monitoring Lab (http://bit .ly/1Mw4Dte) and JJ Harrison (http://bit.ly/1iOumDN), respectively.

Figure 7.4. Portrait of Carl von Linńe, otherwise known as Carolus Linnaeus, by J. Chapman in 1812. Image available at http://bit.ly/1jkOzku.

Linnaeus's efforts centered on classification schemes in which organisms were grouped based on similar traits—mostly but not exclusively anatomical—within hierarchal levels. This approach differed from most naturalists of the day, who elected to place all terrestrial animals in one group and aquatic beasts in another. Similar logic was used for developing species names: almost any scientist could name an organism or alter the existing species name. In most cases, the name for a particular organism was as long as a sentence, and generally a different scientist in another location would not use the same name. Why? The process used for naming an organism prior to the binomial system was called **polynomial nomenclature**; this system used a generic name to label a particular species and a second name that was meant to describe the species. It was this latter name that created the unwieldiness of the entire system. To achieve the function of species description, multiple words and even phrases were used. The net effect was names too cumbersome to remember, and furthermore, no mechanism for others to share and use the same names. This would be akin to every smart phone on the planet today using a unique operating system, so that no one could communicate with anyone else even though everyone had a communication device.

How did Linnaeus solve this problem? He first developed a hierarchy in which all species were grouped into higher categories, termed genera, and the names used were all Latin, as a mechanism for standardization. Genera were in turn placed in orders, orders in classes, and classes in a kingdom* (table 7.1). Additional levels have since been added, but Linnaeus's basic design is still used in modern taxonomy. Classification into a particular level of the hierarchy was based predominantly on morphological characters. Linnaeus's next step was developing a simplified means to name a species. He chose to use one Latin name to identify the genus name. The name selected was based on a unifying characteristic of all species grouped together. He also used a second Latin name, which he called a trivial name, to identify a species. Here is where Linnaeus greatly simplified organism naming: he opted to use the trivial name as a tag or a label for a species rather than as a means to describe the species. The resulting naming system shortened the unwieldy names of the day to just two (binomial): the genus and species.

Widespread adoption of the binomial system occurred very rapidly, providing scientists from around the world the first real means to communicate easily about the same species. The original system developed by Linnaeus has since been refined several times, but the essence of the binomial system serves today as the mechanism for naming all organisms. The utility of the system is that

1. The binomial name of any species is shorter and far easier to remember than any developed through the polynomial system.
2. The system is used as the international standard for naming organisms.
3. The genus and species names developed today rely on specific information, such as morphological characters, about the organism that permits distinction from any other organism.
4. The binomial name avoids confusion that occurs frequently with common names.
5. Rules have been established for proper naming practices, which provide stability to the entire naming system.

It is important to note that the binomial system developed by Linnaeus, as well as its predecessors (polynomial

---

*In *Systema Naturae*, Linnaeus proposed that the top of the organismal hierarchy was a kingdom, and he proposed three kingdoms: one for plants, one for animals, and one for minerals. So his original classification system named both living and nonliving things.

Table 7.1.

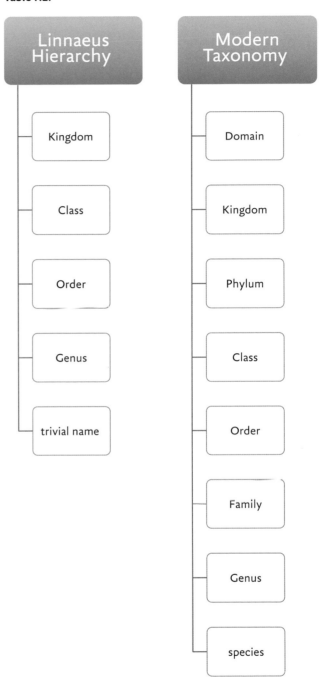

and others), depended almost exclusively on morphological characters for developing generic (genus) and trivial (species) names. Modern taxonomy is more holistic in the examination of organisms, with molecular and anatomical data being central to determining relationships. It thus becomes possible for species that appear nothing alike to be placed in the same genus. Consequently, naming a genus has evolved from the time of Linnaeus. In the next sections, we will put the binomial system to work to learn how to identify insects at different levels of the classification hierarchy.

Quick check

In which level of taxonomic classification, family or order, do members share the fewest number of characteristics?

Insect crib notes: Key features used to identify adult insects

As stated several times in this book already, insects have been incredibly successful. Perhaps the most astounding aspect of the insect success story lies in their species richness, with over one million species having been described and possibly three times that number yet to be discovered. Such success does come with practical problems, particularly for us. For example, how can we distinguish one insect species from another? And for that matter, how do *they* tell each other apart? Insect distinction relies on both sight and smell. Most insects rely on both for recognizing members of the same species (intraspecific) as well as those not related (interspecific).

We, as a species, depend predominantly on our acute sense of vision for recognizing nearly everything in our immediate environment. The appearance of an insect is therefore the starting point for recognition and identification. Aiding the process is that all insects, conveniently, do not look the same. Even different developmental stages of the same species can display unique morphology. Indeed, insect morphology has traditionally served as the chief means not only for recognizing individual insects, but also for naming them. At this point, you may be wondering, what is the relationship between the binomial system developed by Linnaeus and naming and identifying insects? Naming an insect species follows the rules associated with the binomial system. How the genus and species names were developed for a given insect is another matter altogether. Here, our major interest is on how to recognize and identify insects using existing names within the binomial system.

As mentioned, the morphology, or appearance, of an insect is generally the starting point for identifying an insect. However, it is not the only means. Insects can be identified using several unique features.

1. *Anatomy of the adult or juvenile stages.* External features are used to group and identify individual insects. This is the most commonly used feature for identifying and naming insects.
2. *The molecular profile.* This can include DNA fingerprinting of the genomic DNA, or more narrowly focusing on unique sequences, like short tandem repeats (STRs), single nucleotide polymorphisms (SNPs), and mitochondrial sequences, to identify uniqueness in a species at the molecular level.
3. *Bite or sting marks.* Not all insects bite or sting, but of those that do, some degree of classification can occur based on the marks left behind. This is not an approach typically used by an expert in insect taxonomy, but can be useful for anyone trying to find out "what bit me." This is especially important if the insect in question produces toxic or poisonous secretions, and/or if the person on the receiving end of the insect "gift" is prone to **anaphylaxis**.
4. *Signs of activity.* Sometimes an insect is not present but signs of past activity are evident. The signs may include webbing in food, tunnels in wood, skeletonized leaves, even remains of the digestive tract deposited on food (e.g., a corpse). Often, the evidence left behind by an insect is sufficient to recognize what type, if not species, was responsible for the material or damage found. Being able to identify an insect based on signs of activity is especially important for recognizing agricultural, stored product, and wood-infesting insects (figure 7.5).

In most instances, insect identification begins by examining morphological features. Adult insects are usually the focus of identification efforts because they are the most mobile stage, they carry out reproduction, and they often are the easiest of the developmental stages to characterize. These are basic external features used for adult identification:

Figure 7.5. Characteristic damage and silken webbing associated with feeding by the Indian mealmoth, *Plodia interpunctella*. Oftentimes the family of moths, if not the species, can be determined simply by the evidence of insect activity left behind. Photo by Clemson University-USDA Cooperative Extension Slide Series, Bugwood.org.

1. Tagmosis. The adult insect body plan is divided into functional body regions, or tagmata, which include the head, thorax, and abdomen. As discussed in chapter 5, significant variation exists among insects in terms of the shape, size, and color of each tagma, as well as the structures associated with each region (figure 7.6).

2. Head. The head contains several appendages associated with sensory perception and food ingestion. They vary in size and shape, which makes them useful diagnostic tools for distinguishing insects.

   a. Antennae. One pair on adults.
   b. Eyes. Two types: one pair of compound eyes, and two or three simple eyes or ocelli.
   c. Mouthparts. Mandibles designed for chewing are the generic type, but have been modified in several groups for sponging, sucking, biting, and lapping (figure 7.7).

3. Thorax. This region is designed for locomotion, as it contains legs and wings. Legs occur as three pairs: one pair per thoracic segment. Wings, in contrast, typically occur as a pair on the second and third thoracic segments. However, some species (order Diptera) have only one pair of wings and several do not have any wings at all as adults. Some species have secondarily lost such appendages (regressive evolution), which allows the absence of structures to also be diagnostic.

   a. Legs. One pair of legs occurs on each thoracic segment. The generic leg is designed for walking/running, but others exist for jumping (saltatorial), digging (fossorial), prey capture (raptorial), and swimming (natatorial).

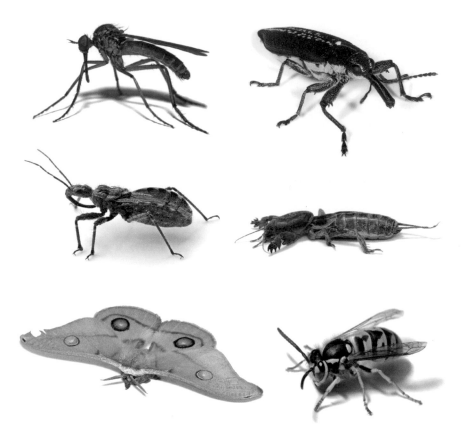

Figure 7.6. Adult insects display an array of sizes, shapes, and colors that can be useful in distinguishing between orders, families, and even species. Image by Bugboy52.40 available at http://bit.ly/1YHvFEC.

b. Wings. Two pairs of wings are typical of flying insects. Modifications of wing number and structure reveal the order of insects in question, and more specific wing characteristics can be used for more specific classification.

4. **Abdomen**. This region is known more for the absence of external structures in comparison to the other tagmata. The number of abdominal segments permits distinction between primitive and advanced species: a reduction in number of segments from the primitive condition of eleven to somewhere between six and nine among advanced species. External genitalia are complex and varied among insect groups. Consequently, they can be used to distinguish species. Significant abdominal structures used for identification include

a. Number of abdominal segments.
b. External genitals.

c. Sensory appendages such as cerci and styli (figure 7.8).

Generally, differences in external morphology are highlighted in identification guides or taxonomic keys so that each feature is compared using a systematic approach. The most common approach is using **dichotomous keys**, in which external features are presented in contrasting couplets. The feature in the first example in the couplet is examined, and if relevant to the specimen being examined, then instructions are given that indicate what feature to compare next. If the feature is not relevant, then the next example in the couplet is examined. If the specimen being examined is relatively common, one or the other example in the couplet should apply for that specific external feature. The process continues until the identity of the order, family, genus, or species is revealed in the last couplet reached (table 7.2).

A similar process is used to identify juvenile insects based on morphological characters. However, larval forms are frequently less distinctive between species than adults, and thus identification based on anatomical features alone can be very challenging. In practice, morphological identification of any stage of development of an insect beyond order, or possibly family, is not an easy task. To gain proficiency in successfully using dichoto-

Figure 7.7. Mandibular mouthparts are considered the ancestral or generic form in adult insects, from which all other types are derived. Photo by Thomas Quine is of a pair of adult stag beetles available at http://bit.ly/1TZsGVe.

mous keys, an individual needs to dedicate a great deal of time over several years to achieve the level of accuracy needed for species identifications. Even then, many morphological characters are highly variable between populations and individuals of the same species, requiring subjective determinations to be made with reference to certain features. The result is a lack of precision in species recognition when using morphological methods of identification. To overcome such limitations, molecular methods focused on DNA analyses have been employed. In many instances, molecular methods are more powerful tools for species identification than traditional morphological characterization, as evidenced by the ability to separate **cryptic species**—that is, those species considered to be anatomically indistinguishable.

Molecular techniques rely on the uniqueness of DNA; every species of organism possesses unique sequences of DNA that constitute its molecular fingerprint. These sequences remain the same throughout development, which means that an egg, larva, or adult insect can be identified to species based on DNA sequences. This consistency is the basis for the technique known as **DNA barcoding**. Short, unique sequences of DNA for a given

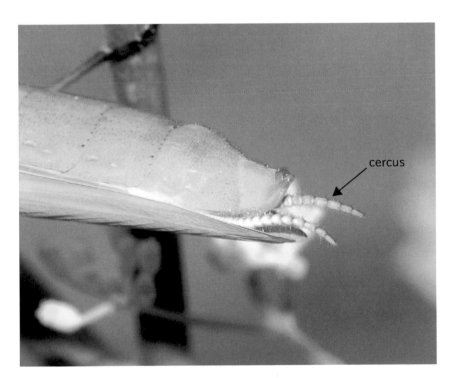

Figure 7.8. Abdominal sensory structures like cerci (singular = cercus) are often used to identify an insect species. Photo by Alves-gaspar available at http://bit.ly/1KKJfOf.

Table 7.2.

1a. Leaves usually without teeth or lobes: 2
    1b. Leaves usually with teeth or lobes: 5

2a. Leaves evergreen: 3
    2b. Leaves not evergreen: 4

3a. Mature plant a large tree—Southern live oak *Quercus virginiana*
    3b. Mature plant a small shrub—Dwarf live oak *Quercus minima*

4a. Leaf narrow, about 4-6 times as long as broad—Willow oak *Quercus phellos*
    4b. Leaf broad, about 2-3 times as long as broad—Shingle oak *Quercus imbricaria*

5a. Lobes or teeth bristle-tipped: 6
    5b. Lobes or teeth rounded or blunt-pointed, no bristles: 7

6a. Leaves mostly with 3 lobes—Blackjack oak *Quercus marilandica*
    6b. Leaves mostly with 7-9 lobes—Northern red oak *Quercus rubra*

An example of a dichotomous key used for some eastern oak tree species in the United States. The key uses leaf characteristics for contrasting species. The key is available at http://bit.ly/1OzPLx4.

insect species can be used to identify any developmental stage to the exact genus and species. Similarly, DNA from an insect's mitochondrial genome* contains species-specific variable sequences that can be useful for species identification. Mitochondrial gene sequences for cytochrome oxidase c (COI and COII) are commonly analyzed as part of molecular insect identification.

Identifications may also depend on analyzing unaligned DNA sequences from insects to compare with a reference database (like BLAST: basic local alignment search tool, http://blast.ncbi.nlm.nih.gov/Blast.cgi) of sequences from known insect species (figure 7.9). The technique is a **bioinformatics** approach to species identification, in which the total number of exact nucleotide base matches determines the probability of a species

match. Though at first glance it may sound complex, the approach is relatively straightforward and can be learned much more quickly than all the morphological features necessary for insect identification. Internal standards or controls are used with the molecular tools, something that cannot be done with identifications based on external anatomy, which allow increased precision in determining species. More recent techniques include the use of microRNA (miRNA) in species identification. These are noncoding RNAs that are highly conserved in a multitude of organisms. In insects, miRNAs are relatively easy to isolate through the use of RT-PCR, where the sequences are amplified and cloned for comparison. The technique has the potential to be much more sensitive than DNA analysis in identifying insect species, but has the drawback of being more expensive—definitely so compared to morphological techniques.

Insect diversity

The class Insecta is enormous! How else can you describe a group with a million members? Sure, Hair Club for Men† is popular, but is probably not relevant to this discussion. Enormity in terms of species numbers requires organizing related insects into groups that permit easy classification. Organizational specialists, or systematists, subdivide the class Insecta into orders, larger groupings of insects that share several morphological, molecular, and/or behavioral characteristics in common. Because orders were originally based almost entirely on morphological relationships, groupings have, or had, an element of subjectivity, and sometimes differences are in the eye of the beholder. In turn, this means not all experts have agreed on the order names.

Complicating the naming of orders is the requirement or rule‡ that an order represents **monophyly**. We won't delve into the precise definition here, but in brief, monophyly is grouping an ancestral species with all of its descendants. Again, opinion is not unanimous with regard to the relationships of different insect groups.

---

*Eukaryotic organisms possess both a nuclear genome (the DNA of the nucleus) and the mitochondrial genome (the DNA of the mitochondria).

†Hair Club for Men, now simply known as Hair Club, is a hair restoration and hair replacement company founded in 1976. Since insects do not have hair, the company does not have six-legged members.

‡All zoological species—in other words, animals—follow the International Code for Zoological Nomenclature (ICZN) for species names and other levels of animal hierarchy.

***ALU WARNING: HUMAN Alu-Sc subfamily consensus sequence

Sequence ID: gnl|alu|HSU14571 Length: 287 Number of Matches: 1

Range 1: 1 to 287 Graphics   ▼ Next Match   ▲ Previous Match

Score	Expect	Identities	Gaps	Strand
355 bits(192)	3e-99	259/291(89%)	5/291(1%)	Plus/Minus

```
Query  2731  ttttttttCTGAGACGGAGTCTCACTCTGTCGCCCAGGCTGGAATGCAGTGGCGTGATCTC  2790
             |||||||| |||||||||||||| |||||||| |||||||||| ||||||||| ||||||
Sbjct  287   TTTTTTT-TGAGACGGAGTCTCGCTCTGTCG-CCAGGCTGGAGTGCAGTGGCGCGATCTC  230

Query  2791  GGCTTTACTGCAGCCTCCGCTTCCTGGGTTCAAGCGATTCTCCTGCCTCAGCCTCCCGAG  2850
             ||| | |||||| |||||| ||| ||||||||||||||||||||||||||||||||||||
Sbjct  229   GGC-TCACTGCAACCTCCGCCTCCCGGGTTCAAGCGATTCTCCTGCCTCAGCCTCCCGAG  171

Query  2851  TAGCTGGGACTACAGGAGCGTG-TGCCACGCCCGGCTAATTTTTTGTATTTTTAGTAGAG  2909
             |||||||||||||||||||| |   |||||||| ||||| |||| |||||||||||||| |
Sbjct  170   TAGCTGGGACTACAGGCGCGCGCCACCACGCCCAGCTAA-TTTTTGTATTTTTAGTAGAG  112

Query  2910  ATGGGGTTTCACCATGTTACCCAGGATGGTCTCAATATCCTGAGTTCATGATCCACCCAC  2969
             | |||||||||||||||| |||||||||||||| || || ||| || ||||||| |||| |
Sbjct  111   ACGGGGTTTCACCATGTTGGCCAGGATGGTCTCGATCTCTTGACCTCGTGATCCGCCCCGC  52

Query  2970  CTTGGCCTCCCAAAGTGCTGGGATTACAGGCGTGAGCCACCACACCCAGCC  3020
             || |||||||||||||||||||||||||||||||||||||||| ||| |||
Sbjct  51    CTCGGCCTCCCAAAGTGCTGGGATTACAGGCGTGAGCCACCGCGCCCGGCC  1
```

Figure 7.9. An example of the type of data produced from a NCBI BLAST search. This is a BLAST-derived complementary sequence alignment between human c10orf76 mRNA and an ALU-Sc family consensus sequence. Image by the National Center for Biotechnology Information available at http://bit.ly/1JvDnGz.

Consequently, the exact number of orders in the class Insecta varies depending on which experts you subscribe to. The number of orders ranges from as few as twenty-six to as high as thirty-two. In reality, which classification scheme you follow is not as important as being consistent in using the same one; consistency helps avoid confusion.

Traditionally, insect orders have been broadly subdivided into two groups (Apterygota and Pterygota) based on wing evolution. Orders in which wings are primitively absent—that is, in which wings never evolved in adults—are classified as apterygotes. All other orders have evolved wings, although in some groups, the structures have been lost secondarily. Orders are also grouped based on the relationship of mouthparts (mandibles) to the head capsule: if mandibles have a single point of posterior **articulation**—meaning a joint, or moveable point of attachment—with the head, the orders are termed **monocondylic**; if two points, they are termed **dicondylic**. In practice, the types of classification lead to nearly identical divisions of the class Insecta. For example, a single insect order, Archaeognatha (bristletails), falls into the category of monocondylic. The same order is also classified as an apterygote, with only one additional member, the order Zygentoma (silverfish) (figure 7.10).

A more detailed examination of the characteristics that define apterygotes and pterygotes follows in the next two sections. The goal is to become familiar with the overarching features of each division/subclass. Instead of a detailed description of each order, the scientific and common names of the major orders of insects are presented, along with a website address that provides in-depth discussion of a given order as well as several references for further study. Figures 7.10, 7.11, and 7.12 also provide an overview of the insect orders following the twenty-eight-order classification.*

Apterygota

A lot can change in a short time. When I was an undergraduate learning about insects for the first time, the

---

*Insect orders in this chapter follow the system proposed by Gullan and Cranston (2014), in which they recognize twenty-eight distinct insect orders.

**What's in a name:
Scientific and common
names**

All described insects have a scientific name in the form of a binomial, which is the formal way of addressing a particular species. The scientific name functions as a precise label for a given species, thus avoiding confusion about the identity of the beast in question. Such formality is a must when discussing a particular species in scientific literature, technical bulletins, legal matters, or any other situation that requires the exact identity of the insect in question is known. An insect may also be identified by a nonscientific or common name, which usually reflects some distinctive trait of the insect, but not in the form of a Latin name. Common names make it easier for a nonscientist to discuss an insect. For instance, the common house fly, *Musca domestica*, is known throughout the United States by its common name, the house fly. Similarly, the necrophagous fly *Lucilia sericata* is often referred to as the green bottle fly.

The problem with using common names is that most have not been approved by any official organization, which would consider questions such as, is the common name actually descriptive or distinctive for that species, and, has the name been used for any other species. Consider *Lucilia sericata*, the green bottle fly: the adults are metallic green, but so

are several other species of fly in North America and across the globe. So the common name is descriptive but not distinctive. In addition, the same name is frequently used for other species of blow flies with similar body coloration. In other regions of the world, *Lucilia sericata* is sometimes called the common green bottle fly, the sheep blow fly, and the green blow fly. Which common name is correct? Technically, all of them! Again keep in mind that there are no internationally accepted rules that govern common names, so anyone can propose a name.

By contrast, scientific names must follow a strict set of rules governed by international organizations to ensure consistency and stability in the naming of a genus and species. This was not always the case, and arguably, even after Linnaeus developed the binomial system, many of the first genus and species names for organisms were nothing more than tributes to people (i.e., individual's names were used as the genus or species name). Thus, the names functioned as labels but offered nothing about the biology of the organisms. Today, genus and species names reflect information about either the morphology or behavior of an organism. The genus name is the first in the binomial and is always capitalized. The species name is lowercased. Both names are italicized in print and underlined when handwritten. In entomological journals, the last name of the author or the individual who originally described the species appears immediately after the species name.

For example, Linnaeus first described the common house fly, and thus is acknowledged in *Musca domestica* Linnaeus or L. Only Linnaeus and his famed student Johann Fabricius have their names abbreviated this way; they are considered so famous or important to taxonomy that they can be recognized by just their initials. The situation is a little different for the common green bottle fly, for which the name is written *Lucilia sericata* (Meigen) or *Lucilia sericata* (Meigen 1826). The author's name in parentheses indicates that a different genus name was used in the original description, and the organism has since been placed in a different genus. It can also mean that the genus name has been revised. Either way, the original genus name is no longer used for the species originally described, so the author's name is denoted parenthetically to reflect the change. Some scientific journals also expect the date of the initial description be included, hence the "1826" after Meigen's name.

In practical terms, does it really matter whether we call an insect by its common name or, more formally, by its scientific name? The answer is, yes it does. Which name to use really depends on the audience being addressed. Credibility in the courtroom for an expert witness depends on such precise aspects of proper use of scientific names, whereas a discussion with interested members of the broader community does not. In fact, sometimes, too much "science" can bog down the conversation and turn off those who might otherwise be quite interested.

Apterygota   Archaeognatha: (http://tolweb.org/Archaeognatha/8207)

Zygentoma: (http://bugguide.net/node/view/79)

Figure 7.10. The apterygote orders of insects. Photo of bristletail by Stemonitis available at http://bit.ly/1iBURvn, and of silverfish by Sebastian Stabinger available at http://bit.ly/1MwyRfz.

apterygotes were regarded as a subclass of the class Insecta, with five orders of insects making up the group. That number was then reduced to three orders, and today, many researchers believe that only two, the orders Archaeognatha and Zygentoma, should be regarded as extant representatives of primitively wingless insects (figure 7.10). What changed in a relatively short twenty-five-year window to lead to the reorganization of the Apterygota? Or, who is correct in terms of the classification scheme? The answers to both questions are intimately linked. Everyone is correct, in that exactness is not easily achieved when studying evolutionary relationships. In insect systematics and evolution, ideas and speculation are derived from careful study of the fossil record, morphological analyses of extant species, and molecular characterization of extant and, when possible, extinct species. The resulting ideas are developed in much the same way as those in archaeology. Some ideas cannot be experimentally tested. Speculation is therefore an important aspect of **insect phylogeny,** the study of evolutionary relationships among insects.

And, as would be expected, all investigators are not in agreement about the interpretation of the same available data. The end result, in terms of classifying insects, is differences in the placement of groups. Hence, a variable number of insect orders exist depending on which expert's classification scheme you follow (figure 7.13).

Many researchers agree with the notion that only two groups represent the primitively wingless insects, although the names used for the groups are not always the same. For example, the order name Thysanura is often recognized instead of Zygentoma. Others contend that the Thysanura should be viewed as more closely aligned with the Pterygota, because the mandibles are dicondylic. However, the counterargument is that thysanurans never evolved wings, so they display the major unifying characteristic of the apterygotes. The viewpoints all reflect logical arguments, which is why multiple opinions on classification schemes remain in effect and no single one is accepted as correct. It is also important to note that regardless of which order name is used—Thysanura or Zygentoma—the members are the same, and commonly

Pterygota	Ephemeroptera: (http://tolweb.org/tree?group=Ephemeroptera&contgroup=Pterygota)
Exopterygote Orders	Odonata: (http://tolweb.org/Odonata/8266)
	Plecoptera: (http://tolweb.org/Plecoptera/8245)
	Dermaptera: (http://tolweb.org/Dermaptera/8254)
	Embioptera: (http://bugguide.net/node/view/16969)
	Zoraptera: (http://www.tolweb.org/Zoraptera)
	Orthoptera: (http://tolweb.org/Orthoptera/8250)
	Phasmatodea: (http://www.phasmatodea.org/)
	Grylloblattodea: (http://tolweb.org/Grylloblattidae/8256)
	Mantophasmatodea: (http://tolweb.org/Mantophasmatodea/8251)
	Mantodea: (http://tolweb.org/Mantodea/8213)
	Blattodea: (http://www.bugguide.net/node/view/342386)
	Psocodea: (http://tolweb.org/Psocodea/8235)
	Thysanoptera: (http://tolweb.org/Thysanoptera/8238)
	Hemiptera: (http://tolweb.org/Hemiptera/8239)

Figure 7.11. Extant orders of exopterygotes.

known as silverfish. Other features of this subclass include one pair of compound eyes, ocelli, intrinsic musculature in only the first two antennal segments, simple chewing mouthparts that are partially retractable, **indirect insemination**, and ametabolous development in which juveniles and adults are virtually indistinguishable except for size. As discussed on the next section, apterygotes are considered simpler than or primitive in almost every way to members of the Pterygota.

Bug bytes

Insect wings improve surgery

https://www.youtube.com/watch?v=MVySM6ODRdo

Pterygota

The pterygotes represent all other insect orders not represented by the apterygotes. A simple unifying characteristic is that all members either possess wings as adults

or are secondarily **apterous**. Secondarily apterous species or groups display regressive evolution with respect to wings and therefore, in modern examples, are not winged as adults. More generally, pterygotes display more complex and advanced features in terms of morphology, development, behaviors, and nearly every other aspect of their life histories than wingless insects. This is evident in body designs adapted for flight, such as larger meso- and metathoracic regions, reduced number of abdominal segments, and gas exchange that favors sustained flight and rapid cooldown once the insect comes to rest (figure 7.14). Complex development and metamorphosis, which we discussed in chapter 6 as holometabolous development, is exclusive to pterygotes, as is **direct insemination**. The point is that with the evolution of wings, insects became exceedingly more complex in almost every way imaginable. The culmination of the complexity can be seen in the insects that display holometabolous development: members of these orders are

Pterygota	Neuroptera: (http://tolweb.org/Neuroptera/8220)
Endopterygote Orders	Megaloptera: (http://tolweb.org/Megaloptera/8218)
	Rhaphidioptera: (http://tolweb.org/tree?group=Raphidioptera&contgroup=Endopterygota)
	Coleoptera: (http://tolweb.org/Coleoptera/8221)
	Strepsiptera: (http://tolweb.org/Strepsiptera/8222)
	Diptera: (http://tolweb.org/Diptera/8226)
	Mecoptera: (http://tolweb.org/Mecoptera/8227)
	Siphonaptera: (http://tolweb.org/Siphonaptera/8228)
	Trichoptera: (http://tolweb.org/Trichoptera/8230)
	Lepidoptera: (http://tolweb.org/Lepidoptera/8231)
	Hymenoptera: (http://tolweb.org/Hymenoptera/8232)

Figure 7.12. Extant orders of endopterygotes.

Figure 7.13. Fossilized remains of a fly preserved in fossilized tree sap (resin), otherwise known as amber. Most insect fossils are not this easily discernable. Photo by James St. John available at http://bit.ly/1V9mrAL.

generally the most abundant in terms of species richness, are distributed in many **biogeoclimatic zones**, and are the most efficient of the insect flyers (figure 7.15).

The complexity of insect wings often leads to several subdivisions within the Pterygota. In one sense this is needed, since twenty-six insect orders are "lumped" in this one division or subclass. The subdivisions are based on how the wings develop or how they are folded when at rest. For example, if the wings develop externally and gradually during juvenile development, the order is termed an **exopterygote**. This division encompasses fifteen of the winged orders and all members are hemimetabolous.

By contrast, **endopterygotes** form wings internally from **imaginal discs** during the pupal development of complete metamorphosis. Eleven orders are classified in this division, with the Coleoptera, Lepidoptera, and Hymenoptera the most successful, and hence recognizable, of all the insects. Groupings based on wing folding include the **Palaeoptera**, which includes adult insects that cannot fold their wings against the body at rest, and **Neoptera**, in which they can. This classification is unrelated to the type of development or metamorphosis displayed by the insects, and thus results in a mix of orders at various levels of evolutionary relatedness being grouped together.

Figure 7.14. An insect adapted for efficient flight: only one pair of wings, reduced numbers of abdominal segments, and enlarged thoracic segments for locomotion. Photo by Alvesgaspar available at http://bit.ly/1O1ekDc.

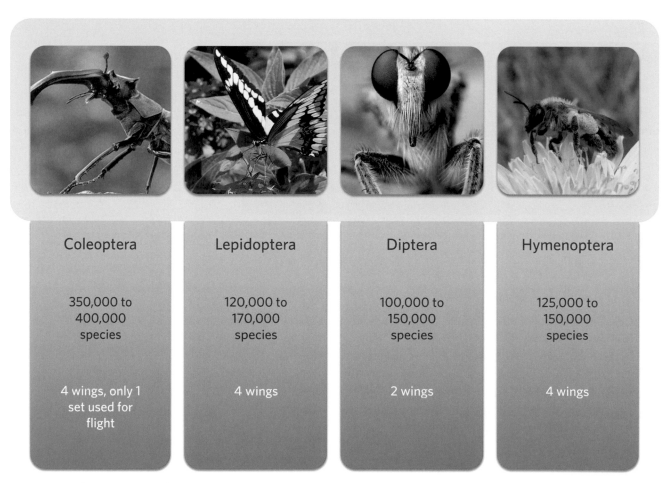

Coleoptera	Lepidoptera	Diptera	Hymenoptera
350,000 to 400,000 species	120,000 to 170,000 species	100,000 to 150,000 species	125,000 to 150,000 species
4 wings, only 1 set used for flight	4 wings	2 wings	4 wings

Figure 7.15. The most successful orders of insects. Photo of stage beetle (Pavel/krasensky) available at http://bit.ly/29bOeNC; photo of swallowtail butterfly (Umbris) available at http://bit.ly/295tFjJ; photo of male robber fly (Thomas Shahan) available at http://bit.ly/297GCyl; and photo of pollen covered bee (OilBac) available at http://bit.ly/299NzgS.

Figure 7.16. Deep-fried crickets served in Cambodia. Photo by Thomas Schoch available at http://bit.ly/1KFvMGq.

For readers interested in a more in-depth discussion of insect evolutionary relationships, particularly in the context of insect orders, the excellent texts by Gullan and Cranston (2014) and Johnson and Triplehorn (2004) should be consulted. Detailed natural history and systematic features of insect orders can also be found online at numerous sites, but two that stand out are http://www.bugguide.net and http://www.tolweb.org/Insecta.

Quick check

How does the presence of wings influence insect classification?

Classification by taste. Seriously?

As discussed early in the chapter, insects can be identified using several unique features, including morphology, molecular makeup, and behavior. Am I now insinuating that classification can be made based on taste? No! Well, maybe. Actually no one in his or her right mind would make such a suggestion. Of course, if you have read any preceding chapters in this book, you already have an opinion as to the status of my mind! That said, **entomoph-**

agy, or the practice of consuming insects (which we first explored in chapter 3) is usually restricted to just certain groups of insects. Others are avoided, either because they taste nasty (my wife is politely suggesting as I write that ALL insects fall into this classification!), are potentially harmful (bite, sting, contain noxious or toxic compounds), or are so repulsive to us for specific reasons that no amount of coaxing will every lead to us consuming them. Cockroaches almost always rate number one in this last category, and right up there with them are insects that feed on dung or corpses. The point is that an insect classification scheme based on taste can easily be developed:

1. Tasty
2. Nasty
3. Harmful
4. Repulsive

Not to be confused with Snow White's lesser-known favorite dwarfs.* Now, how does this intricate scheme

---

Snow White is a German fairy tale told by the Brothers Grimm in 1812, but really made famous by Walt Disney in his 1937 animated film *Snow White and the Seven Dwarfs*.

Figure 7.17. The highly toxic larvae of the monarch butterfly caterpillar, *Danaus plexippus*. Photo by Pseudopanax available at http://bit.ly/1LTYMv1.

compare with those we discussed earlier, based on morphological or molecular relationships? There is really no pattern at all. For example, some of the more commonly consumed insects are crickets belonging to the order Orthoptera, a group classified as an exopterygote. The house cricket, *Acheta domestica* (figure 7.16), can be dipped in chocolate, added to a stir-fry, or just sautéed in butter with garlic, and still taste delicious, with no more nasty side effect than a wing sticking to the back of your throat during swallowing. Yet other members of the same order sequester noxious components obtained from their herbivorous diet and should not be consumed.

Similar relationships exist within the endopterygote orders Lepidoptera and Hymenoptera, in which some members are eaten, such as caterpillars and honey ants, and others are avoided because they bite, sting, or contain harmful compounds. Larval stages of monarch butterflies (*Danaus plexippus*) serve as a clear example of the harmful classification: the caterpillars sequester cardiac glycosides from the milkweed plants they feed on, making consuming them likely your last meal (figure 7.17). We have already discussed the relationship between exopterygotes and endopterygotes, so clearly wing formation is not an indicator of tastiness among insects. That said, a lack of wings generally means a snubbing, for consumption purposes, in that apterygotes do not seem to be considered a food option by humans at all.*

Some members of the order Coleoptera, affectionately known as beetles and weevils, present a somewhat unique problem. Some species taste delicious when steamed over an open fire in palm leaves, others serve the same ingredient utility (e.g., yellow mealworm, *Tenebrio molitor*) that crickets do, but others pose the problem of

---

*In fairness to primitive wingless species, their small size and concealed habitats probably have more to do with these insects not being selected as food by humans than their lack of wings. It was far more convenient in the text, however, to blame their wing status.

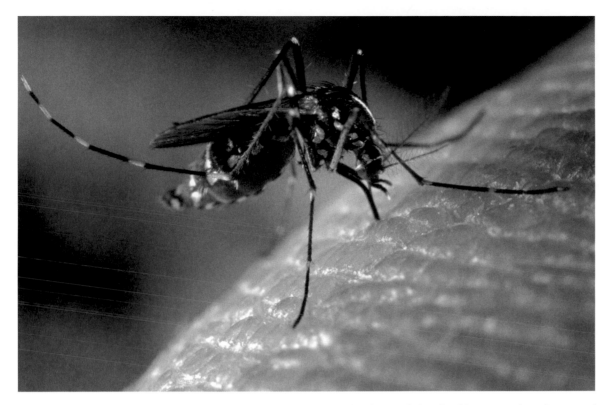

Figure 7.18. Insects engorged with human or any animal blood tend to be avoided as food by man and can be viewed as "repulsive." Photo by James Gathany available at http://bit.ly/1MQiYDL.

possessing relatively high concentrations of anabolic steroids. Yes, the same class of compounds that are referred to as performance enhancing drugs (PEDs) and that have caused such an uproar in such professional sports as cycling, Major League baseball and the National Football League. Overconsumption of these beetles can lead to aggressive, even violent behavior. Technically, this is an example of the harmful classification, even though the insect itself is not really causing the harm (figure 7.18).

About the only unifying trend in food-worthiness among all the insect orders is that if the insect blood-feeds, humans do not have an interest in consuming it. Several orders from both the exopterygotes and endopterygotes contain blood-feeding representatives, and man consumes not a one of them, at least not deliberately.

CHAPTER REVIEW

❋ **Why should I care how to identify an insect?**

- You do not need to dive into taxonomy, systematics, or phylogeny to appreciate what insects do or to realize that they are important in a multitude of ways. What this means to you is that the ability to recognize an insect, or not, beyond the order or family level will not affect your understanding of the importance of insects to the human condition, or the types of reproductive strategies they use, methods of communication they employ to attract mates or send an alarm, or how they process food after ingestion. Or most any other topic associated with the endlessly fascinating world of insects. In many ways, studying insects devoid of taxonomy allows you to simply enjoy their biology, like reading an engaging novel with no regard for the writer's choice of syntax or unique grammatical style.

- Classifying the insect to genus and species is necessary to determine whether the crop is fed upon by this species, and if so, what type(s) of control should be applied and when. For example, understanding the biology of an insect is essential to effective pest management and making responsible decisions that reduce the threat of the pest without harming mankind or the environment. Similarly, an insect found at a port of entry into the United States must be quickly and accurately identified to assess whether the critter in question is a potential pest, poses no threat, or is even possibly a threat to national security. Essential to diagnosing and preventing insect-borne diseases like West Nile virus, malaria, and dengue fever is recognizing the culprit so that its biology can be manipulated. From a criminal investigation perspective, the identity of an insect feeding on a corpse can reveal important development data that in turn can be used to determine a portion of the postmortem interval, which requires being able to identify the necrophagous creatures to genus and species.

- Of more practical use, for the non-insect-expert, is that being able to recognize the types of insects in your home means the difference between panic, action, disgust, or apathy. The identity could also be important to you in determining whether the red and black "bug" that landed on your head is simply an insect such as a harmless ladybird beetle (commonly called lady bug) or an assassin bug (a true bug or member of the order Hemiptera) with a nasty disposition and a propensity to bite. The same can be said for the need to distinguish other kinds of biting and stinging insects, or any other insects with attributes you wish to avoid.

Thank you, Linnaeus! Binomial classification

- Any discussion of how to classify any organism, insect or otherwise, or why there is a need to do so, should begin with a tribute to the great Swedish naturalist Carl von Linné (better known as Carolus Linnaeus). Linnaeus identified thousands of insects and plants over his lifetime and, most importantly, developed a classification system for naming botanical and zoological specimens. The system, known today as binomial classification, revolutionized organismal classification just by making it a simpler process. Linnaeus's efforts centered on classification schemes, in which organisms were grouped based

on similar traits—mostly but not exclusively anatomical—within hierarchal levels.

- The process used for naming an organism prior to the binomial system was called polynomial nomenclature; this system used a generic name to label a particular species and a second name that was meant to be descriptive of the species. It was this latter name that created the unwieldiness of the entire system. To achieve the function of species description, multiple words and even phrases were used. The net effect was names too cumbersome to remember, and furthermore, no mechanism for others to share and use the same names.

- How did Linnaeus solve this problem? He first developed a hierarchy in which all species were grouped into higher categories, termed genera, and the names used were all Latin, as a mechanism for standardization. Genera were in turn placed in orders, orders in classes, and classes in a kingdom. Additional levels have since been added, but Linnaeus's basic design is still used in modern taxonomy. Classification into a particular level of the hierarchy was based predominantly on morphological characters. Linnaeus's next step was developing a simplified means to name a species. He chose to use one Latin name to identify the genus name. The name selected was based on a unifying characteristic of all species grouped together. He also used a second Latin name, which he called a trivial name, to identify a species. The resulting naming system shortened the unwieldy names of the day to just two (binomial): the genus and species.

- It is important to note that the binomial system developed by Linnaeus, as well as its predecessors (polynomial and others), depended almost exclusively on morphological characters for developing generic (genus) and trivial (species) names. Modern taxonomy is more holistic in the examination of organisms, with molecular and anatomical data being central to determining relationships. It thus becomes possible for species that appear nothing alike to be placed in the same genus.

* **Insect crib notes: Key features used to identify adult insects**

  ■ Perhaps the most astounding aspect of the insect success story lies in their species richness, with over one million species having been described and possibly three times that number yet to be discovered. Such success does come with practical problems, particularly for us. For example, how can we distinguish one insect species from another? And for that matter, how do *they* tell each other apart? The appearance of an insect is the starting point for recognition and identification. Aiding the process is that all insects, conveniently, do not look the same. Even different developmental stages of the same species can display unique morphology. Indeed, insect morphology has traditionally served as the chief means not only for recognizing individual insects, but also for naming them.

  ■ The morphology, or appearance, of an insect is generally the starting point for identifying an insect. However, it is not the only means. Insects can be identified using several unique features, including the anatomy of the adult or juvenile stages, the molecular profile of the species, bite or sting marks, and signs of activity.

  ■ Adult insects are usually the focus of identification efforts because they are the most mobile stage, they carry out reproduction, and they often are the easiest of the developmental stages to characterize.

  ■ Generally, differences in external morphology are highlighted in identification guides or taxonomic keys so that each feature is compared using a systematic approach. The most common approach is using dichotomous keys, in which external features are presented in contrasting couplets.

  ■ To overcome limitations associated with morphological identifications of insects, molecular methods focused on DNA analyses have been employed. In many instances, molecular methods are more powerful tools for species identification than traditional morphological characterization, as evidenced by the ability to separate cryptic species. Molecular techniques rely on the uniqueness of DNA; every species of organism possesses unique sequences of DNA that constitute its molecular fingerprint. These sequences remain the same throughout development, which means that an egg, larva, or adult insect can be identified to species based on DNA sequences. This consistency is the basis for the technique known as DNA barcoding. Similarly, DNA from an insect's mitochondrial genome contains species-specific variable sequences that can be useful for species identification.

* **Insect diversity**

  ■ Organizational specialists, or systematists, subdivide the class Insecta into orders, larger groupings of insects that share several morphological, molecular, and/or behavioral characteristics in common. Because orders were originally based almost entirely on morphological relationships, groupings have, or had an element of subjectivity. In other words, not all experts have agreed on the order names. Complicating the naming of orders is the requirement or rule that an order represents a monophyletic group. Monophyly is grouping an ancestral species with all of its descendants. Again, opinion is not unanimous with regard to the relationships of different insect groups. Consequently, the exact number of orders in the class Insecta varies depending on which experts you subscribe to.

  ■ Traditionally insect orders have been broadly subdivided into two groups (Apterygota and Pterygota) based on wing evolution. Orders in which wings are primitively absent—that is, in which wings never evolved in adults—are classified as apterygotes. All other orders have evolved wings, although in some groups, the structures have been lost secondarily. Orders are also grouped based on the relationship of mouthparts (mandibles) to the head capsule: if mandibles have a single point of posterior articulation—meaning a joint, or moveable point of attachment—with the head, the orders are termed monocondylic; if two points, they are termed dicondylic. In practice, the types of classification lead to nearly identical divisions of the class Insecta.

- Apterygotes are primitive wingless insects that also can be characterized by other features of their subclass, including one pair of compound eyes, ocelli, intrinsic musculature in only the first two antennal segments, simple chewing mouthparts that are partially retractable, indirect insemination, and ametabolous development in which juveniles and adults are virtually indistinguishable except for size. Apterygotes are considered simpler than or primitive in almost every way to members of the Pterygota.

- The pterygotes represent all other insect orders not represented by the apterygotes. A simple unifying characteristic is that all members either possess wings as adults or are secondarily apterous. More generally, pterygotes display more complex and advanced features in terms of morphology, development, behaviors, and nearly every other aspect of their life histories than wingless insects. This is evident in body designs adapted for flight, such as larger meso- and metathoracic regions, reduced number of abdominal segments, and gas exchange that favors sustained flight and rapid cooldown once the insect comes to rest. Complex development and metamorphosis, which we discussed in chapter 6 as holometabolous development, is exclusive to pterygotes, as is direct insemination. The point is that with the evolution of wings, insects became exceedingly more complex in almost every way imaginable. The culmination of the complexity can be seen in the insects that display holometabolous development: members of these orders are generally the most abundant in terms of species richness, are distributed in many biogeoclimatic zones, and are the most efficient of the insect flyers

- The complexity of insect wings often leads to several subdivisions within the Pterygota. In one sense this is needed, since twenty-six insect orders are "lumped" in this one division or subclass. The subdivisions themselves are based on how the wings develop or how they are folded when at rest.

- ✦ **Classification by taste. Seriously?**
  - An insect classification scheme based on taste can easily be developed and might include the categories 1) tasty, 2) nasty, 3) harmful, and 4) repulsive.
  - How does this intricate scheme compare to those we discussed earlier, based on morphological or molecular relationships? There is really no pattern at all. For example, some of the more commonly consumed insects are crickets belonging to the order Orthoptera, a group classified as an exopterygote. The house cricket, *Acheta domestica*, can be dipped in chocolate, added to a stir-fry, or just sautéed in butter with garlic, and still taste delicious, with no more nasty side effects than a wing sticking to the back of your throat during swallowing. Yet other members of the same order sequester noxious components obtained from their herbivorous diet and should not be consumed. Similar relationships exist within the endopterygote orders Lepidoptera and Hymenoptera, in which some members are eaten, such as caterpillars and honey ants, and others are avoided because they bite, sting, or contain harmful compounds.
  - About the only unifying trend in food-worthiness among all the insect orders is that if the insect blood-feeds, humans do not have an interest in consuming them. Several orders from both the exopterygotes and endopterygotes contain blood-feeding representatives, and man consumes not a one of them, at least not deliberately.

MUSHROOM FARMING (SELF-TEST)

Level 1: Knowledge/Comprehension

1. Define the following terms:
   (a) phylogeny   (d) apterygote
   (b) monophyly   (e) DNA barcoding
   (c) endopterygote   (f) monocondylic

2. Explain the difference between an exopterygote, endopterygote, and apterygote insect.

3. Describe how mouthpart articulation is used to classify insects.

4. What features of an adult insect are typically used to distinguish different groups or species?

Level 2: Application/Analysis

1. Explain how wing evolution can be used to classify insect species.

2. Discuss how Linnaeus's binomial system of nomenclature has improved insect classification from the polynomial system used prior to Linnaeus.

3. Describe how the methods used in insect systematics are similar to those employed in archaeology.

Level 3: Synthesis/Evaluation

1. Describe how techniques like DNA barcoding or mitochondrial sequencing can be combined with traditional morphological classification to classify cryptic species.

2. Can hypothesis testing be done in the fields of insect systematics or insect phylogeny? Explain your answer.

REFERENCES

Blunt, W. T. 2002. Linnaeus: The Complete Naturalist. Princeton University Press, Princeton, NJ.

Caterino, M. S., S. Cho, and F. A. H. Sperling. 2000. The current state of insect molecular systematics: A thriving Tower of Babel. *Annual Review of Entomology* 45:1–54.

Evans, H. E. 1985. The Pleasures of Entomology: Portraits of Insects and the People Who Study Them. Smithsonian Institution Press, Washington, DC.

Gullan, P. J., and P. S. Cranston. 2014. The Insects: An Outline of Entomology. 5th ed. Wiley-Blackwell, West Sussex, UK.

Johnson, N. F., and C. A. Triplehorn. 2004. Borror and DeLong's Introduction to the Study of Insects. 7th ed. Cengage Learning, New York, NY.

Knapp, S. What's in a name: A history of taxonomy. Natural History Museum of London. http://www.nhm.ac.uk /nature-online/science-of-natural-history/taxonomy -systematics/history-taxonomy/session1/index.html. Accessed November 8, 2014.

Polaszek, A. 2009. Systema Naturae 250: The Linnaean Ark. CRC Press, Boca Raton, FL.

Rivers, D. B., and G. A. Dahlem. 2014. The Science of Forensic Entomology. Wiley-Blackwell, West Sussex, UK.

Swafford, A. L. Carolus Linnaeus: Classification, Taxonomy and Contributions to Biology. Education Portal. http:// education-portal.com/academy/lesson/carolus-linnaeus -classification-taxonomy-contributions-to-biology .html#lesson.

Virgilio, M., T. Backeljau, B. Nevado, and M. De Meyer. 2010. Comparative performances DNA barcoding across insect orders. *BMC Informatics* 11:206. doi:10.1186/1471- 2105-11-206.

Wheeler, W. C., M. Whiting, Q. D. Wheeler, and J. M. Carpenter. 2001. The phylogeny of the extant hexapod orders. *Cladistics* 17:113–169.

Wilson, J. J. 2012. DNA barcodes for insects. *Methods in Molecular Biology* 858:17–46. doi:10.1007/978-1-61779-591-6_3.

THE ENTOMOLOGIST BOOKSHELF (SUPPLEMENTAL READINGS)

Cranshaw, W., and R. Redak. 2013. Bugs Rule!: An Introduction to the World of Insects. Princeton University Press, Princeton, NJ.

Kjer, K. M., F. L. Carle, J. Litman, and J. Ware. 2006. A molecular phylogeny of Hexapoda. *Arthropod Systematics and Phylogeny* 64:35–44.

Kristensen, N. P. 1981. Phylogeny of insect orders. *Annual Review of Entomology* 26:135–158.

Trautwein, M. D., B. M. Weigmann, R. Beutel, K. M. Kjer, and D. K. Yeates. 2012. Advances in insect phylogeny at the dawn of the postgenomic era. *Annual Review of Entomology* 57:449–468.

Whitfield, J. B., and K. M. Kjer. 2008. Ancient rapid radiations of insects: Challenges for phylogenetic analysis. *Annual Review of Entomology* 53:449–472.

ADDITIONAL RESOURCES

Centre for Insect Systematics
http://insectsystematicukm.blogspot.com/

DNA barcoding of terrestrial insects
http://ibol.org/dna-barcodes-for-neons-terrestrial-insects/

Insect evolution
http://www.fossilmuseum.net/Evolution/evolution-segues /insect_evolution.htm

Insect groups
http://www.amentsoc.org/insects/fact-files/orders/

Insect identification
http://marketstock.hubpages.com/hub/insect-
    identification-2.

Insect orders
http://www.insectidentification.org/orders_insect.asp

Insect Systematics & Evolution
http://www.brill.com/insect-systematics-evolution

International Code for Zoological Nomenclature
http://www.nhm.ac.uk/hosted-sites/iczn/code/index.jsp

O. Orkin Insect Zoo, Smithsonian Natural History Museum
http://www.mnh.si.edu/education/exhibitions/insectzoo
    .html

Overview of insect orders
http://www.bugguide.net/node/view/222292

Tree of Life Project: Insecta
http://tolweb.org/Insecta/8205

8 Insects Are Phat but Not Fat

Diet, Nutrition, and Food Assimilation

Throughout the animal kingdom the selection of diet has become most highly developed among mobile species. It has attained finest discrimination among parasitic forms and, of free-living species, among insects which feed upon plants.

Dr. V. G. Dethier (1954)*

Insects are incredible eating machines. From the moment of egg hatch, juveniles ravenously consume food, appearing zombie-like in their incessant eating. But their feeding is not really mindless, as insects obey neural commands, feeding to meet nutritional needs and stopping when the need has been met. Many will not eat if the proper diet is not available. To the casual observer, it may seem like insects eat anything and everything, all the time. In reality, they can be quite selective in their dining habits, finicky even. Taste receptors are located in multiple locations on the body—feet, antennae, abdomen, and mouthparts—but all are external to the mouth. That's right—located exposed to the outside environment. The advantages are obvious, affording the ability to avoid toxic foods and eat only what is needed. In other words, insects do not get fat. What self-control! Here we dive into the world of insect nutrition, food processing, and assimilation of digested foodstuffs, making comparisons to other animals, including humans, to showcase the incredible efficiency displayed by insects in terms of food acquisition, dietary intake, digestive efficiency, and ultimate conversion of food into body mass.

Key Concepts

- What's on the menu? Nutrient requirements of insects
- Tools of the trade: Structures used for food collection
- Why insects don't get fat but people do
- Eating "crap" makes sense! Food processing depends on what is eaten
- It is only efficient if you can use it: Food assimilation

What's on the menu? Nutrient requirements of insects

What do insects eat? Of course the knee-jerk response is "Everything!" Or, "It would be easier to discuss what they don't eat." Are these fair and accurate assessments? From a global perspective, discussing insects as a whole, there are few naturally occurring

---

*From "Evolution of feeding preferences in phytophagous insects," *Evolution*: 8(1): 33-54.

organic-based creatures, living or dead, terrestrial or aquatic, that are not consumed by something with six legs. For one million species to not just survive but also thrive, the dietary menu must be quite large. So yes, insects appear to eat everything. However, individual species of insects have relatively selective diets. To argue that they are finicky eaters is perhaps not accurate, but it is true that for many species, dietary intake is limited. This can be attributed to the morphology of the mouthparts, taste receptor selectivity, and design of the gut tube itself. The net effect is that specific insects generally feed on specific types of food.

For example, the majority of insects "restrict" their diets to vegetative or plant matter and are classified as **phytophagous**. Those that consume animals are termed **carnivorous**. Several species may be grouped as true parasites (of either plants or animals) or the as the more specialized parasitoids. Some insects are fluid-feeders and specialize on liquid diets derived from another animal or a plant. And still others feed on decaying organic matter originating from any type of organism. These insects may be termed detritivorous or **saprophagous**, the latter referred to as **necrophagous** when feeding on dead animals exclusively. The point is that different insect species tend to feed on specific types or kinds of foods, and as we will soon see, the design of the digestive system reflects the food type.

Regardless of how each species is classified in terms of feeding or food types, insects share several unifying features of nutrition and digestion, all linked to their incredible efficiency.

1. Finding food (sensing and foraging)
2. Manipulating and consuming food (ingestion with specialized mouthparts)
3. Breaking down complex food into useable forms (chemical and mechanical digestion)
4. Absorbing nutrients and converting them to body mass (food assimilation)

What makes insects so good at these processes? So much of the efficiency of the class Insecta relates to the simple idea of form meeting function. Morphologically speaking, the design of the insect digestive system—from mouth to anus—is adapted for eating, processing, and using specific types of food. This literally means that di-

gestion is tailored to food of particular sizes, shapes, and chemical composition. Technically, so is ours. So why, then, are insects considered better or more efficient than humans? Easy—they eat what they are supposed to. In other words, insects respond to neural signals that signify specific nutritional needs, they seek out such foods, taste them before ingestion, and then consume them if the appropriate sensory feedback is received (figure 8.1). This means that consumption occurs only, at least in theory, if the nutritional need will be met by the discovered food. Once the need is met, the hunger drive is turned off, and in turn, the insect no longer forages. There are exceptions to the feedback loop, such as honey bees foraging for the entire colony, but for individual insects, the digestive system is driven by neural centers of hunger and **satiety**. Gluttony, or eating to excess, is absent from the world of insects. For clear contrast, take the celebration of the Thanksgiving holiday in North America. In the battle for planetary supremacy between mankind and insect, the clear winner in the category of self-control goes to creatures with six legs. We will further explore the topic of neuroregulation of nutrient intake later in the chapter, but first we turn our attention to the nutrients needed by insects.

As a starting point, the basic question that can be asked is, do insects have different nutritional requirements than most other animals, including us? The logic behind this question stems from at least two observations. The first is that, because insects outnumber all other animals by such a wide margin, it would seem reasonable to assume that perhaps they are able to eat foods that others cannot, thereby acquiring unique nutrients. The second observation is one we just discussed: insects do not overeat. Consequently, we can rule out the notion that insects simply out-consume, pound-per-pound, other animals. That's not to say they don't eat a lot, because indeed they do. But again, overeating or gluttony just simply does not happen with insects. So the idea that insects need different or unique nutrients is a logical premise. But in fact, insects require the same basic nutrients that we do, though not always in the same amount (usually less) or necessarily used in the same ways that other animals use them. Some nutriment needs are unique among members of the class Insecta, but not enough so to account for their extreme evolutionary

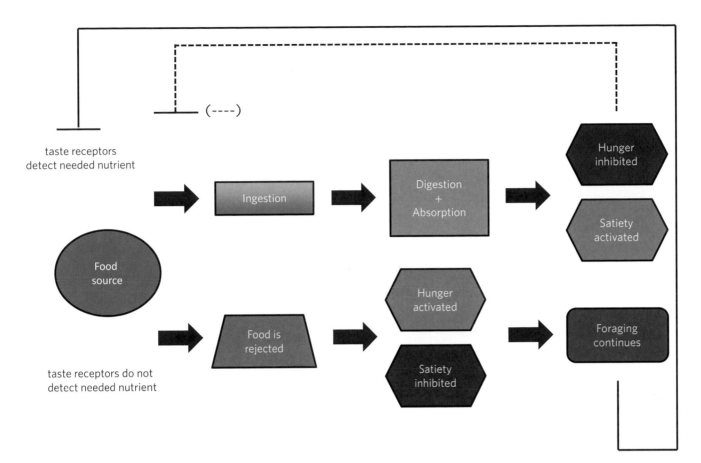

Figure 8.1. Negative feedback loop associated with insect feeding.

success story. What will become clear as we explore insect-food interactions is that insects are incredibly efficient at all phases of digestion. Insects eat for need, show extraordinary efficiency in breaking down food into useable forms, and then make excellent use of what they have consumed. However, it all begins with eating the right foods in the first place. Sound familiar?

Nutritional needs

Insects as a whole are chemically quite similar to each other in terms of the composition of their tissues and bodily fluids (intra- and extracellular). Not surprisingly, then, the nutrients needed to maintain homeostasis are also very similar among different insect species. Although some species do indeed display very unique diets, as a whole, the basic nutritional requirements of insects inhabit common ground. As with most animals, certain nutrients must be obtained in the diet of an insect, and those are termed **essential nutrients**. Since the

insect hunger drive is activated in response to nutritional needs, foraging and subsequent ingestion generally leads them to foods that contain the essential nutrients. Many of the needed nutrients can be synthesized from precursors found in ingested foodstuffs or derived from breakdown products. Such nutrients are called **nonessential nutrients**. Here is where insects stray from the path of many other animals: what is essential or nonessential does differ substantially from the human condition. The basic nutrient classes, however, are the same, regardless of whether a backbone is present or not, or how many legs an animal possesses.

Nutrient classes
1. *Amino acids*
2. *Carbohydrates*
3. *Lipids*
4. *Vitamins*
5. *Inorganic compounds*

Now, before we dive into the nutrients needed by an insect, let's address one question that may be in the back of your mind: How do you know what insects actually need nutritionally? It is not as if you can get them to complete a survey indicating likes, dislikes, and needs versus wants. Three common approaches for assessing dietary needs of insects and other organisms are (1) the *deletion method,* in which a specific component of a chemically defined diet is removed and the effects of that on growth and development are measured; (2) *substitution method,* in which an analog of a specific nutrient is used to replace the nutrient in question; and (3) *radiolabeled tracking,* which relies on tagging precursor molecules with radioactive labels so that biosynthesis using the precursors can be measured. All three methods essentially require being able to raise the insect in question in the laboratory on an artificial diet. Not only is this very expensive, but it is also labor intensive. Consequently, only a fraction of all insects have been tested by any of these methods. The discussion of insect nutrition that follows is based on a limited number of species, so undoubtedly many exceptions exist.

Amino acids

Amino acids are derived from protein digestion so that they can be used in turn to synthesize new proteins. Insects, like other animals, need to produce proteins for structural purposes, to serve as enzymes, for functional roles in membranes, and to circulate in hemolymph as transport proteins. Specific amino acids have roles independent of protein synthesis, including as energy sources (e.g., proline for flight), cryo- or stress protectants (alanine and proline), during sclerotization (tyrosine), as eye pigments (tryptophan), or to function as neurotransmitters (glutamate or γ-aminobutyric acid) (table 8.1). Glutamate takes on a special role in helping to synthesize other amino acids through a series of transamination reactions. Generally most amino acids are obtained from dietary protein, and the ability of an insect to meet an amino acid need is dependent on the digestibility of these proteins. Insects also use several nonessential amino acids, and though they are not required (i.e., essential), they can facilitate more rapid rates of growth and development. Nonessential amino acids are commonly synthesized from essential amino acids in the fat body. The downside

to producing amino acids in this fashion is that the biochemical reactions are typically energy demanding and can generate byproducts that require detoxification (which requires energy) or immediate excretion from the body. The amount of each type of amino acid needed by insects varies greatly between species and developmental stages.

Carbohydrates

Most insects do not have an absolute carbohydrate requirement in their diet yet do use simple sugars or monosaccharides as the primary cellular energy source. The above statement might seem to conflict with itself, but actually it does not, for at least two reasons. First, a particular macromolecule consumed for energetic needs is often not considered an essential nutrient. Classifying a macromolecule as a nutrient versus energy source comes down to deciding whether the consumed macromolecule simply sustains life (an energy source) or promotes growth and development (a nutrient). Second, carbohydrates are commonly synthesized from amino acids or fats found in food and therefore fall into the category of nonessential nutrients. This does not diminish the importance of ingested or synthesized carbohydrates: several monosaccharides function as primary cellular fuel; some have structural roles, particularly in the synthesis of chitin; and many can serve as the structural foundation to produce amino acids and fats. As with proteins, the digestibility of complex carbohydrates dictates the utility of these macromolecules in meeting nutritional needs. Enzymatic hydrolysis of polysaccharides and disaccharides generally liberates monosaccharides that are either absorbed or converted to other macromolecules. Insects do not directly digest large polymers produced by plants, such as starch and cellulose, since they lack the necessary digestive enzymes. In these instances, microorganisms residing in the digestive tract of phytophagous insects function as **symbionts** by synthesizing the appropriate enzymes and thus benefiting both the insect host and the microorganisms.

Social insects, particularly colonizing bees, represent unique exceptions to the dietary rules of carbohydrates in that they do have an absolute need for these macromolecules. Colonies demonstrate a division of labor based on reproductive status (i.e., queen versus workers)

Table 8.1. Comparison of essential amino acids in three orders of insects. Note that while insects require the same basic essential amino acids as most other animals, all insects do not have the same requirements. Data from Chapman (1998).

Essential Amino Acids		
Coleoptera	Blattodea	Hemiptera
Arginine	Arginine	
Histidine	Histidine	Histidine
Isoleucine	Isoleucine	Isoleucine
Leucine	Leucine	
Lysine	Lysine	Lysine
Methionine		Methionine
Phenylalanine		
Threonine	Threonine	
Tryptophan	Tryptophan	
Valine	Valine	

and also an age-based division of labor, all determined by the diet of young larvae. Larvae fed nutrient-rich, carbohydrate-laden royal jelly become queens, whereas larvae fed a less rich diet become workers. Newly emerged worker bees are "assigned" a variety of tasks inside the hive including tending eggs and larvae and honeycomb construction and food handling, while older workers forage for nectar and pollen. The point is that dietary carbohydrate, and not genetic difference, has a profound influence on caste determination and behavioral patterns in some social insects.

Lipids (fats)

Insects differ substantially in their dietary requirements for lipids. To a degree, the nutrient needs align with taxonomic relationships and feeding behavior. All insect orders studied to date display an absolute need for sterols (steroids), which has led to a broad generalization for all insects. Cholesterol is the most important dietary sterol needed by insects. Most species can synthesize other sterols from a cholesterol backbone, including the major molting hormones, **ecdysteroids** (discussed in chapter 6). Carnivorous, omnivorous, and **hematophagous** (blood-feeding) species generally can obtain cholesterol in their diets, whereas phytophagous insects cannot and must convert plant-specific sterols into a useable form.

For several insect groups, polyunsaturated fatty acids are essential dietary requirements. The nutritional requirement is generally met by consumption of linoleic or linolenic acids (figure 8.2). In the case of Diptera, the vast majority of species examined have no growth dependency on polyunsaturated fatty acids, yet all mosquitoes (family Culicidae) do. For some other orders (e.g., Lepidoptera and Hymenoptera), the fatty acid need is stage-specific: completion of adult development depends on linolenic acid, but a similar requirement is absent during larval stages. Some cockroach species (order Blattodea) require dietary polyunsaturated fatty acids for reproductive success, specifically needing the lipids to produce normal **ootheca**, the protective coating surrounding developing embryos (figure 8.3).

Nearly all other types of fatty acids and phospholipids needed by insects for constructing membranes or other structures can be synthesized from precursors in the diet. Therefore these lipid molecules are nonessential nutrients. The major exceptions are certain fat-soluble vitamins, discussed in the next section. It is also important to note that some insects demonstrate an entire lifestyle change—into that of a parasite—in which dietary lipids are obtained exclusively from the host and usually in a form that allows the parasitic insect to immediately assimilate the fats directly into its own biomass. In fact the lipid profile of the parasitic insect shows an incredibly high homology—greater than 90% in some cases—with the host.

Vitamins

Vitamins are a chemical hodgepodge, in that vitamin molecules share little in common structurally other than being organic substances. Most are needed only in small quantities, and each is typically considered essential for growth and development. The general view is that several vitamins must be obtained in dietary sources because insects cannot synthesize them. But for some insects, symbiotic microorganisms living within the digestive tract synthesize the required vitamins. Technically for these latter insects, the vitamins are essential but not obtained directly from the food.

Vitamins are classified based on their solubility properties: fat-soluble or water-soluble.

Fat-soluble vitamins. Carotenoids are the primary fat-soluble vitamin required by all insects. Perhaps most

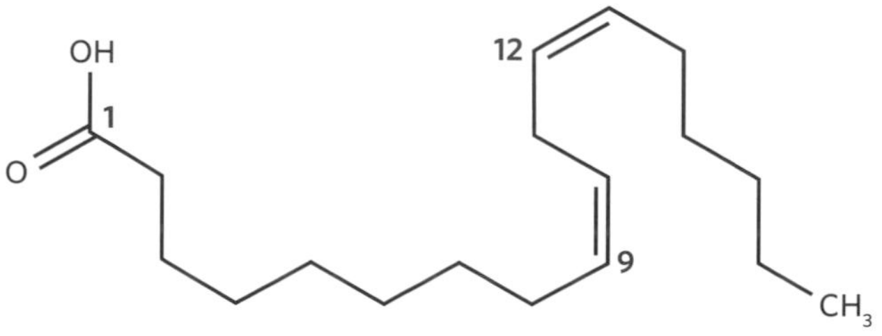

Figure 8.2. Chemical structure of linoleic acid. Image by Ju available at http://bit.ly/1FqfR2Q.

![photograph]

Figure 8.3. The ootheca of a hissing cockroach. Photo by Whitney Cranshaw available at http://bit.ly/1iBZqpx.

common among these is β-carotene or pro-vitamin A, which is used by insects in the formation of eye pigments and also in pigment deposition of the integument. Only microorganisms and plants synthesize these compounds. The non-carotenoid vitamin E appears to be essential to the reproductive success of some, but not all, insects and enhances the fecundity of several species of insects.

Water-soluble vitamins. At least seven water-soluble vitamins are essential for all insects. These include biotin, thiamine, riboflavin, folic acid, pyridoxine, nicotinic acid, and pantothenic acid. In essence, the class Insecta requires all members of the vitamin B complex. Water-soluble vitamins function as cofactors for numerous enzymes and also have structural roles.

Inorganic compounds

In the older literature, "salts" refers to nutrients critical for certain insect species, in particular aquatic insects. Practically, salts refer to several inorganic compounds

and elements, in particular sodium, potassium, calcium, chloride, phosphate, and magnesium. The precise amounts needed by different insects vary greatly among species and also between different developmental stages. Iron, zinc, and manganese are considered essential for all insects and must be obtained from dietary or microorganism sources. Several other inorganic compounds are known to dramatically accelerate the rate of growth of many freshwater species. This is especially evident in the larval stages of several mosquito species, in which the size and shape of anal papillae are modulated by availability of chloride, potassium, and sodium.

Beyond the text

Why is cholesterol not directly available in the diet of phytophagous insects?

Tools of the trade: Structures used for food collection

The primary tools for collecting and ingesting food are the mouthparts. As we discussed in chapter 5, insect mouthparts are quite varied. In fact, the structure of the mouthparts directly reflects dietary tendencies, particularly in terms of the physical form of the food. Mouthpart type, and hence diet, is also correlated with insect groups or specific orders. For example, liquid diets derived from plants are the norm for moths and butterflies (order Lepidoptera), and consequently adults of this order rely on siphoning mouthparts to partake of scrumptious carbohydrate-rich offerings from plants. In contrast, the larval forms of moths and butterflies are typically voracious plant feeders, utilizing mandibles with sharp and flat surfaces to chew and grind vegetative materials. Similarly, nearly all feeding stages of beetles (order Coleoptera) depend on mandibles for chewing, whether on animal or plant diets, a combination of the two, or dedicated to dining as saprophagous feeders. Beetles using mandibular mouthparts consume a wide range of foods. In fact, the majority of insect species and groups use mandibles for food acquisition and manipulation. Is there any special reason(s) for this trend? Well, actually, yes, there is! For one, mandibular mouthparts are quite adaptable, offering utility in obtaining, processing, and even tasting many types of food differing in physical compo-

Figure 8.4. Photo of a house cricket head with the mandibles on display. Photo by USGS Native Bee Inventory and Monitoring Laboratory available at http://bit.ly/1OYJGtx.

sition and structure. More important, though, mandibles are the ancestral mouthpart type of the class Insecta, the original mouthpart type that all others are derived from. So whether an insect crunches on a leaf, laps up nectar from a flower, or sucks blood from the butt of a host, the structure used is modified from the chewing mandible type (figure 8.4).

The way mandibles evolved—really, the entire head, for that matter—accounts for much of the remarkable efficiency that insects display in food acquisition and manipulation. The short story goes something like this: Insects evolved from a wormlike ancestor that was segmented and lacked a true head but did have a simple **acron** with an opening anteriorly located for food to enter and possessed several ventral appendages that aided in locomotion. The first six trunk segments are believed to have fused with the acron to form a complex head resembling that of modern-day insects, particularly that of a cockroach (order Blattodea) or grasshopper (order Orthoptera). Important in this evolutionary transition is that the pair of ventral appendages located on each trunk segment became head structures, including mouthparts

(mandibles and maxillae) and antennae, with each powered by the vestiges of the ancestral ventral appendage's muscles, meaning leg muscles. Unlike humans or other animals, when an insect eats, the mouthparts command the force or strength of legs, not of jaw! This may provide a new perspective as to why the bite of a tiny ant hurts so much when applied to the tender flesh of a human foot or leg.

Adding to the versatility of the mandibular mouthpart is the complex yet efficient design of the accompanying structures. A pair of large, relatively flat accessory jaws (**maxillae**) is present that allows the insect to manipulate food, along with leg-like **palps** that occur in pairs on maxillae as well as on the lower lip, called the **labium** (figure 8.5). The palps are also covered by thousands of taste receptors, as are the mandibles and maxillae, permitting insects the ability to "taste it before they eat it." The presence of lips (**labrum,** or front lip, and labium, or rear lip) is a remarkable evolutionary advancement in itself, as these structures allow insects to close their mouths once food is ingested (figure 8.6). By doing so, ingested food does not roll back out of the mouth, and a

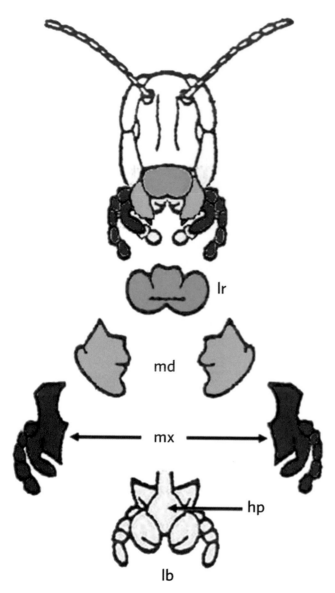

Figure 8.5. Cartoon of a grasshopper's mouthparts (lr = labrum, md = mandibles, mx = maxillae, hp = hypoglossa, and lb = labium). Image by Heds available at http://bit.ly/1KX5OTI.

Mandibles are good, in fact very good, in functions associated with collecting and processing several types of foods used by insects. But for some species, mandibles are not the right tools for the job. Mandibles are well suited for solid foods, whether derived from plants or animals. However, for those species that feed exclusively on liquid or semiliquid diets, the mandibular design needed to be adapted into something more functional. The evolved design for liquid feeding occurs in modern insects in three basic forms: as siphoning (order Lepidoptera), piercing-sucking (Hemiptera, some Diptera), and sponging mouthparts (most Diptera). Why so many mouthpart adaptions? Simple—liquid diets are not all the same. Such diets differ in source, composition, and viscosity, necessitating structures that maximize acquisition. Generally some sort of muscular pump is also required to promote intake, and as we will see later in this chapter, gut morphology for the liquid diet is also distinct from that used for processing solid diets. The take-home message is that all insects do not eat the same foods, and their mouthparts reflect this diversity in diet.

Bug bytes
Butterflies feeding
https://www.youtube.com/watch?v=MYWPWTme_YI

Why insects don't get fat but people do

Have you ever seen a fat insect? I mean fat! In other words, one that is overweight or obese. If being totally honest, the answer should be an unequivocal no. Insects in nature do not overeat, and hence do not get fat. This is not to say that it is impossible for them to become overweight. It is theoretically possible under unnatural conditions. For example, in the laboratory it has been demonstrated that if neural regulatory mechanisms that control food intake are inhibited or severed, **hyperphagia** or elevated appetite is induced, which may lead to a state of overeating. So a fat insect could occur, but only if the ventral nerve cord or recurrent nerve* is ablated or cut in front of or behind the brain. Either situation

unique internal environment can be maintained that facilitates the initiation of chemical digestion. Keep in mind that most animals do not have lips! Within this internal space is a tongue or **glossa** (or **hypoglossa**) that does function somewhat similarly to that of humans: it is covered with thousands of taste receptors and can also be used to lap up liquids. Mandibles and the associated mouthpart structures work very well under most situations, which is why so many insect species still use them.

---

*The recurrent nerve links the frontal ganglion (FG) of the stomatogastric nervous system with the hypocerebral ganglion, which lies posterior to the FG.

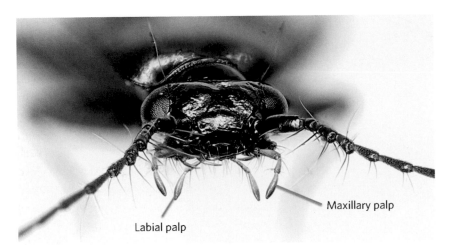

Figure 8.6. Maxillary and labial palps on display on this adult beetle. Photo by Siga available at http://bit.ly/1R92Mwc.

is likely encountered in nature only if a predator consumes an insect's head, a condition that can't commonly be corrected nor one in which obesity would any longer be of much concern.

Domestication can have this effect as well—that is, the overeating, not the beheading. Certain species of insects, as with pet dogs and cats, can become conditioned to eating whenever and whatever is supplied by the human caregiver. They can eat foods that they never would in the wild and likely have a surplus of food available 24/7. Several years ago in my General Entomology class, one of my students raised an adult female Madagascar hissing cockroach, *Gromphadorhina portentosa*, exclusively on water and Ritz Bits peanut butter crackers. After twelve weeks, that girl was fat (the cockroach)! Her abdomen was an almost perfectly round cylinder, her legs barely touched the ground since the abdomen was nearly as wide as the legs were long, and she made no attempt to run during the annual Madagascar Madness races that highlight the last day of class. She was overweight and Jim (her caregiver) was depressed (Dr. Phil* was not a television icon back then, so there was no appropriate counseling available for the unhappy couple).

Under natural conditions, however, insects do not overeat. How do they pull off what humans have never

Dr. Phil is an American television show starring Dr. Phillip McGraw, who provides psychological counseling to guests. To date, he has yet to provide counseling to any distressed insect, and it remains to be determined if this is simply a blatant disregard for patients with more than two legs.

mastered? Do they possess unique mechanisms of self-control that other animals lack? Most research examining insect nutrition, specifically those aspects associated with hunger and satiety, suggests that the mechanisms employed by insects are quite similar to vertebrate animals. Broadly speaking, diet is regulated throughout the insect digestive system and can be observed from the time of food ingestion, through food movement from one location to another (gut motility), to the release of digestive enzymes (in terms of both type and amount), and to the processes of absorption and assimilation. However, when discussing overeating, we tend to direct our attention to food intake. After all, the secret is eating right in the first place. So how do insects achieve this goal?

For one, unlike us, insects do not eat for pleasure. Rather, insects eat for need; insects eat when a nutritional need arises. How does an insect perceive this? Theoretically the same way we do, or should: via activation of the hunger drive, leading to the sensation of hunger. In humans, neural centers located in the hypothalamus serve to create the hunger drive after processing sensory information sent from throughout the body. An opposing neural center, the satiety center, is also located there, to essentially turn off the hunger drive. You have personal experience in knowing that these signals can be ignored. How many times have you been at a restaurant and ordered dessert even though you are completely stuffed? I know, that chocolate-caramel-covered cheesecake was calling you, so what choice did you have? You probably should even be commended because you ate it

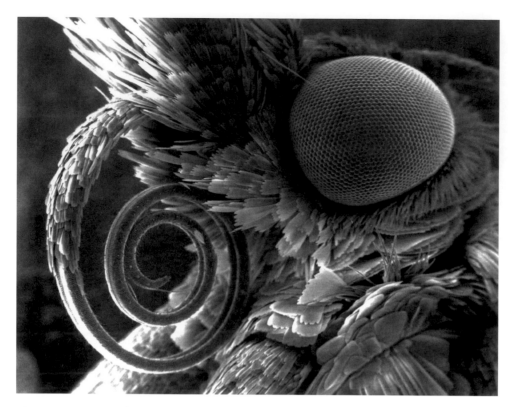

Figure 8.7. Scanning electron micrograph of the head of a pyralid moth (Lepidoptera). The coiled siphon mouthparts are covered with taste receptors, or chemoreceptors. Image by Svdmolen available at http://bit.ly/1G9IRYE.

to prevent someone else from doing so, who, no doubt would have faced an embarrassing pants explosion right in front of everyone. Thank you for your good deeds! Insects do not face such dilemmas.

In the world of insects, their sleek, trim figures are maintained through the same type of hunger and satiety controls. Only, they obey the messages! Basically, hunger and satiety are regulated through a series of interconnected mechanisms:

1. Taste receptors
2. Chemical modulation
3. Neural mechanisms

Taste receptors. As we have discussed earlier, taste receptors are located in multiple locations on the insect body, including on the various mouthpart structures, on tarsi, on antennae, and on the abdomen. In insects, it is more correct to refer to such receptors as **contact chemoreceptors,** in that foods can be tasted via sensing chemical signals in liquid or dry form, and the same receptors may detect chemicals independent of feeding activities.

However, for simplicity, we will still rely on the terminology of taste receptors, since our focus here is on nutrition and regulation of food intake. The basic feature that stands out is that hundreds to thousands of taste receptors are located outside the mouth, so any insect can taste its food before ingestion (figure 8.7). This permits detection of food composition so that an assessment of whether the nutritional need will be met can be made, and also whether the food item is potentially toxic. Sensory feedback from any of the taste receptors can inhibit the hunger center located in the frontal ganglion, activate satiety neurons, and/or result in hunger center neurons remaining active (figure 8.8).

Chemical modulation. Chemical regulation of hunger or satiety can result from both endogenous and exogenous chemical signals. Several biogenic amines, neuropeptides, and hormones have been identified from an array of insects that are capable of stimulating or inhibiting the hunger drive or satiety (figure 8.9). Biogenic amines, including octopamine, target octopaminergic receptors to modulate the insect hunger drive, particularly with

Figure 8.8. Light micrograph of an antenna of the vinegar fly, *Drosophila melanogaster*. The antenna is covered with a forest of chemoreceptors used for smelling (olfaction) food. Photo by AlphaPsi available at http://bit.ly/22gdre8.

regard to meeting protein needs. In contrast, compounds like sulfakinins and corticotropin-releasing factor-related diuretic hormone regulate the satiety neurons. In the case of sulfakinins and related kinin-like compounds, the signaling molecules are released from insulin-producing cells in the frontal ganglion and then bind specifically to G-protein coupled receptors in brain neurons to induce satiety. Neuropeptide F and short neuropeptide F both appear to stimulate appetite in some insects. Several other neuropeptides and hormones indirectly influence appetite and satiety by regulating muscle movement along the entire length of the gut tube, modulating digestive enzyme release and controlling the rate of crop emptying.

Neural mechanisms. Neural regulation of hunger and satiety relies on stretch receptors positioned critically along the digestive tract. For example, crop emptying stimulates activation of hunger in several insect species, generally to meet a carbohydrate need. Stretch receptors along the midgut influence feeding behavior of many he-

matophagous species; once a threshold "stretch" is reached, satiety neurons are activated and the hunger drive suppressed. Several studies suggest that a neuromodulator, termed cholecystokinin-like peptide, is released once the threshold is achieved, which in turn binds to neurons in the stomatogastric nervous system to trigger the onset of satiety. For most insects studied, stretch receptors positioned along the foregut modulate long-term feeding requirements, while those associated with the abdomen provide short-term regulation (figure 8.10). Nerves from the central nervous system and stomatogastric nervous system also innervate the gut tube and mouthparts. Consequently, sensory feedback from taste receptors, associated muscles, and those monitoring the internal gut environment contribute to modulating appetite and fullness.

Just how effective are these dietary control measures? Several studies have shown that if insects are not offered the right food, they will not eat. Starvation and death will result, rather than insects eating nonnutritious food. Some rely on complex foraging behaviors before even making an attempt to taste a potential food source. For any insect species, obeying the neural and chemical signals from the hunger and satiety neurons results in food intake that meets specific nutritional needs. Under normal or natural conditions, insects cannot ignore these signals, and consequently, they do not overeat or get fat. This is almost unimaginable for me, the proud owner of two bichon frise* dogs that proudly eat animal excrement out of the yard despite being well fed, with one showing all the morphological characteristics of a cracker-fed cockroach or a taxidermied sheep (figure 8.11).

Eating "crap" makes sense! Food processing depends on what is eaten

For an insect to take advantage of its ability to eat the right foods, it must have the capacity to release the valuable nutrients trapped inside ingested foods, through the process of digestion. Digestion relies on physical manipulation or breakdown of foodstuffs, termed **mechanical digestion** or **mastication**, and on **chemical digestion**, a

*Bichon frise is a breed of lap dogs reputed to be quite intelligent. In my experience, however, they rank just below freshwater sponges and slightly above river rocks.

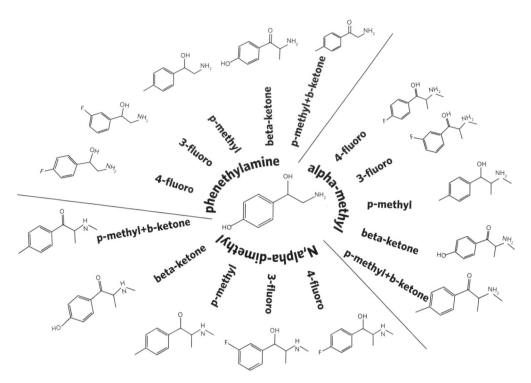

Figure 8.9. Octopamine and several of its analogues that are capable of modulating the insect hunger center. Image by Ezekiel.golan available at http://bit.ly/1YHDZ7z.

process that relies on enzymatic digestion and the chemical environment (i.e., pH and ions) in which the food is contained. Believe it or not, these are hallmarks of sophisticated digestive processes, and also typify animals that possess a gut tube. The insect gut tube is open at two ends (mouth and anus), permitting one-way flow of food. It is also designed for specialized functions to be carried out in different locations, thereby permitting multiple digestive events to occur simultaneously. While these may sound like unique adaptations, there are not. Compartmentalization of digestive functions is the norm with vertebrate animals and is also observed in the higher invertebrate phyla. With that said, however, some aspects of chemical and mechanical digestion are truly insect-owned, which will become apparent as we delve further into the topic of digestive functions.

The process of digestion is a mechanism that also ensures insects eat the right foods. This can be seen through the design of the gut tube, which has evolved to process foods of specific chemical and physical composition. The same can be said for the types and amounts of digestive enzymes produced by a given species; they match the food that should be eaten. An insect may be consuming

food that does not make sense to you, but it probably does to the insect; otherwise, it cannot extract nutrients from what is eaten. This means that those fly maggots that you may have seen dining on cow manure actually know what they are doing; their gut tubes were designed to, literally, eat crap. We explore these ideas in more detail by examining topics focused on the design of the insect digestive tract and the processes by which chemical digestion occurs primarily in the midgut, fluid flow in the ecto- and endoperitrophic spaces facilitates chemical digestion, and microbial symbionts aid digestive processes.

Insect digestive tract

As mentioned, the insect digestive tract is designed as a tube. The tube is portioned into three sections, each evolved for a specialized function: the **foregut** or **stomodeum**, the **midgut** or **mesenteron**, and the **hindgut** or **proctodeum**. Both the foregut and hindgut are derived from ectoderm and also lined with cuticle. Yes, the same cuticle discussed in chapter 6 as a major component of the insect exoskeleton. Cuticle, regardless of location, is shed during molting. So the lining of two-thirds of the gut tube is ripped out with each molt and replaced with

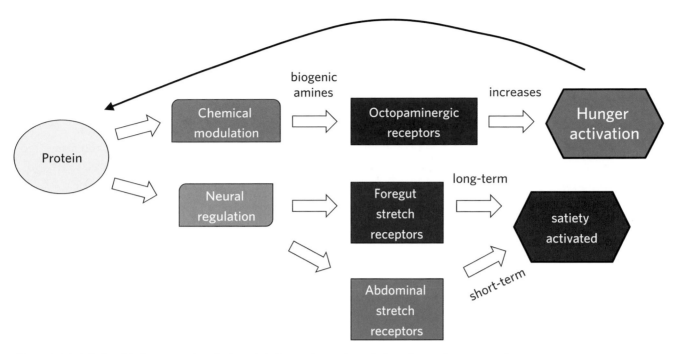

Figure 8.10. Relationship between chemical modulation and neural regulation of protein consumption.

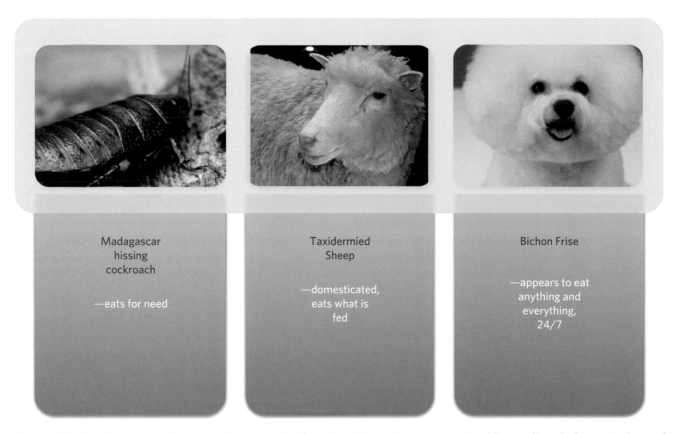

Figure 8.11. Feeding relationship among three animals of questionable intelligence (questionable actually only for two). Photo of cockroach by Alambes (http://bit.ly/1LdYk0h), Dolly the sheep by Toni Barros (http://bit.ly/1FA9nOO), and Bichon Frise by Akaporn Bhothisuwan (http://bit.ly/1OYLFhe).

The classic tale *The Very Hungry Caterpillar* by Eric Carle has probably been read to nearly every elementary-school-age child in the United States. The quaint yarn follows the **ontogeny** of an un-named caterpillar from egg hatch to adult emergence. Most of the story focuses on the voracious appetite of the young caterpillar. We are told that on Sunday, the neonate caterpillar hatches from the egg with the rising of the sun, a very plausible scenario considering that for many insects, egg hatch as well as adult emergence from the pupal stage are circadian events timed with daybreak. From this point on, however, the biology of *The Very Hungry Caterpillar*, specifically the description of feeding behavior, raises all kinds of red flags. For instance, the mother apparently did not lay her eggs (I am presuming mom laid a clutch and not just a single egg, yet there is no mention of siblings; maybe they're dead) on a host plant suitable for her young to eat, as the poor juvenile does not find food until Monday. Few newly hatched insects could survive for even a few hours without food, let alone a day, and the requirement for a journey from the site of egg hatch to an apple tree would surely be a death sentence in almost all cases. Mr. Carle depicts Monday's feast as a mature red apple, suggesting a late summer to early fall season. This implies that perhaps the young caterpillar is a codling moth, *Cydia pomonella* (Lepidoptera: Tortricidae). However, this implication is quickly dropped when we learn that, on day two (Tuesday), our young six-legged adventurer partakes of two fully mature pears, then three plums (Wednesday), followed by strawberries (Thursday), and on Friday consumes five oranges. Even if we were to believe that the hunger center was driving this young caterpillar to eat such a diverse range of fruits, it is highly unlikely that the gut tube design or digestive enzymes produced by one species of moth could accommodate this unique five-day smorgasbord, let alone the fact that these plants do not occur in the same geographical regions nor is the fruit of each ripe during the same growing seasons.

Let's say for the sake of argument, however, that this is possible—which it isn't—because as scientists, open-mindedness is a must. Our young caterpillar would have consumed a diet capable of support-ing the development of nearly one hundred individuals, assuming that each reaches an average weight of 150-200 milligrams prior to becom-ing a pupa. What does this mean? It should be obvious that Mr. Carle has implied that the young caterpillar practices gluttony, which we know insects simply do not do under normal conditions. And, since poor parenting is in fact the norm in the world of insects, we know it isn't that. Our trust of Mr. Carle is fleeting, and all compassion for the hungry caterpillar is completely lost, as Saturday's gustatory barbarity is finally unveiled. The young caterpil-lar is purported to consume a piece of chocolate cake, one ice cream cone, a pickle (presumed dill), a slice of cheese and a slice of salami, one lollipop (apparently just lying unwrapped next to the salami), a piece of cherry pie, one sausage (a carnivorous caterpillar to boot!), one cupcake, and one slice of water-melon. Utter dietary chaos.

Saturday's menu has too many issues to address them all, so let's go straight for the entomological jugular: to consume this gluttonous meal the hungry caterpillar would have had to ignore its taste recep-tors, the signals from its hunger and satiety centers, and the neural feedback from stretch receptors. None of these scenarios would make any sense. Insects eat for need, not for pleasure. There is no empirical evidence that any insects have a need for a piece of cake or an ice cream cone, or anything else on Saturday's menu, with the possible exception of watermelon. Of course, no explanation is provided for how this caterpillar came upon just a slice of each food item; no doubt it simply cut off the slices itself, despite lacking opposable thumbs (insert sarcastic face here). The truth of *The Very Hungry Caterpillar* is that the book is filled with impossibili-ties. Undoubtedly, countless numbers of children have been fraught with uncertainties about the dining preferences of caterpillars throughout their lives, and have not had the good fortune to seek counseling from a professional entomologist to alleviate the intellectual suffering. For future generations, we can only hope that children ignore the inaccurate portrayals of feeding caterpillars, and consult entomological texts at a very early age!

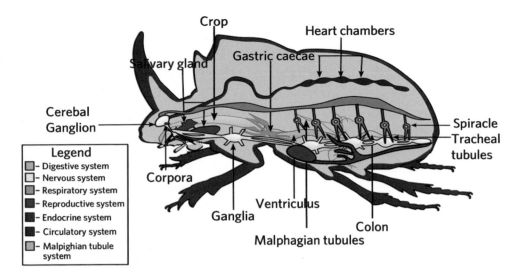

Figure 8.12. Internal anatomy of an adult insect digestive system. Image by Bugboy52.40 available at http://bit.ly/1KFzzU1.

a new one. That just sounds painful! The midgut is produced from endodermal tissue and does not possess a cuticular lining. Instead, epithelial cells synthesize a protective membranous tube (**peritrophic membrane**) that fills the inside of the midgut. The functions and associated structures are discussed for each gut region separately (figure 8.12).

Foregut. The first region of the gut tube, the foregut or stomodeum, is composed of the mouth, pharynx, esophagus, and crop. The region terminates posteriorly with a muscular valve, the **proventriculus**, which often possesses a series of cuticular teeth-like structures called **dendricles**. Contraction of the proventricular muscles can then drive the dendricles into the food emptying from the crop, resulting in partial mechanical digestion of solid food and cells. The mandibles and maxillae actually begin the process of mastication. Salivary glands (usually paired) reside at the anterior end of the foregut and release saliva that functions to lubricate mouthparts, and moisten food so that ingested materials form a bolus, and contains a few digestive enzymes that initiate chemical digestion. Enzymatic digestion targets chiefly carbohydrates, but some protein digestion occurs as well. The salivary enzymes work primarily in the crop, as ingested food is retained in this region until emptied into the midgut

for the bulk of chemical digestion. Some insect species regurgitate portions of the food, principally water, as a means to concentrate the food material to maximize enzymatic digestion.

Midgut. The basic design of the midgut, or mesenteron, is a thin tube lined by a single layer of epithelial cells. Extensive microvilli extend from the surface of the cells toward the lumen of the midgut, serving as sites for digestive enzyme release and also providing enormous surface area for absorption of released nutrients. The area between the surfaces of the epithelial cells to the tips of the microvilli is characterized by a pH gradient, which facilitates compartmentalized chemical digestion under differing conditions contained within microenvironments in the midgut. The anterior end of the midgut displays finger-like projections called **gastric cecae**, which functionally produce digestive enzymes and peritrophic membrane and represent maximized surface area for nutrient absorption. The entire length of the midgut is lined with a noncellular peritrophic membrane. This structure functions as a frictionless surface for food movement and protects the delicate epithelial cells from potentially abrasive food molecules. Posteriorly, the mesenteron terminates at a muscular pyloric valve.

Hindgut. A cuticular lining protects the hindgut, or proctodeum, but does not prevent absorption of water, ions, amino acids, and other solutes. The hindgut tube is functionally separated into the anterior **ileum**, followed by the **rectum**, which terminates as the **anus**, an opening to the outside of the body. A series of long, thin tubelike projections called **Malpighian tubules** extend from the anterior portion of the ileum, floating freely in the hemolymph or becoming embedded in tissues throughout the hemocoel. The hindgut and associated structures take on numerous roles, ranging from digestion and absorption of nutrients to ion and water balance to removal of undigested food materials (defecation) and metabolic wastes (excretion) to serving as host to a wide range of microorganisms that aid in chemical digestion and synthesis of valuable nutrients.

As with so many aspects of the insect condition, there are exceptions to every rule and, in this case, to the basic design of the gut tube. For example, some species, particularly those that feed on a liquid diet, lack a distinct crop and may instead possess blind end pouches, or **diverticuli**, as extensions from the esophagus. Liquids like nectar or blood, even plant resins, can be stored in the pouches until food is moved by muscular contractions (peristalsis) into the midgut. Other species may have either Malpighian tubules (**cryptonephridial arrangement** of Lepidoptera) or portions of the hindgut (**filter chamber** of Hemiptera) in intimate association with the midgut as a means to quickly and efficiently remove water from a liquid meal so that the solutes are concentrated before enzyme secretion. The hindgut is modified in some insects to facilitate storage of undigested food material (paunch of Isoptera) and/or to promote fermentation (fermentation chamber of Coleoptera). More subtle anatomical adaptations include the absence of salivary glands in some fluid-feeders, elaborate pumps in the pharynx of species that "suck," and elongate microvilli in the rectum of insects that absorb water through the anus. Now, that last statement deserves a moment of pause. . . . Adaptation to maximize digestion is the norm in insects, and what has been described here is just the basic design that all other designs have evolved from.

Chemical digestion occurs primarily in the midgut

As with the design of the gut tube, the type of digestive enzymes found in any given insect reflects the chemical composition of the primary food source. Most insects must digest the major nutrient macromolecules, meaning protein, carbohydrates, and lipids. Therefore, the same basic types of digestive enzymes are present in all insects. What differs is the exact form of the foodstuffs ingested. Insects with a solid diet of either plant or animal material will ingest predominantly proteins of different sizes, polysaccharides, and lipids in the form of glycerides, phospholipids, and glycolipids. By contrast, fluid-feeders benefit from tapping into the circulating fluids of their hosts, meaning that the dirty work of digestion has already occurred, so much less investment in chemical digestion is needed. When digestion does occur, it happens primarily in the midgut, regardless of the feeding habits of the insects in question. Epithelial cells lining the midgut synthesize the bulk of the digestive enzymes, although some are produced by the salivary glands and travel with the food bolus to the midgut. In most instances, the enzymes are produced in response to food arrival into the mesenteron. The value of this approach to individual insects is that enzymes do not accumulate over time and are synthesized in response to the chemical composition of the food. Released digestive enzymes must cross the peritrophic membrane to reach the food bolus. Continued digestion occurs outside of the peritrophic membrane, in the ectoperitrophic space, including along the surface of the epithelial cells.

What types of digestive enzymes do insects produce? For the most part, the same types that we make. There are subtle differences, such as molecular sizes, precise target sites on substrates, and of course enzyme names, but the basic enzyme types are very similar to ours. For instance, proteases are synthesized to digest proteins, and they tend to fall into two categories, endopeptidases and exopeptidases. They can of course be further characterized, but knowing the names of all the enzymes an insect uses really does not help much to understand what they eat and how they use food to meet their nutritional needs.

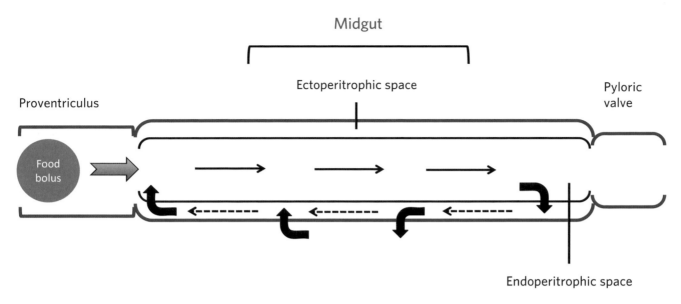

Midgut

Ectoperitrophic space

Proventriculus

Pyloric valve

Food bolus

Endoperitrophic space

Figure 8.13. Gut fluid current in the endo- and ectoperitrophic spaces of the midgut. Illustration based on Terra (1988).

With that said, it is worth mentioning that a few insects, namely some species of Diptera, possess cathepsin D-like proteinase, which is functionally similar to vertebrate pepsin but structurally nearly identical to cathepsin D. Pepsin-like enzymes have been found only in animals with an incredibly acidic stomach (midgut) to facilitate consumption of foods laden with microorganisms. In turn, carbohydrases are used to digest polysaccharides to liberate monosaccharides, the nutrient form that is absorbed. Several types of carbohydrases are produced by insects; the exact type depends on the source of the carbohydrate, that is, whether plant or animal. In some cases the primary polysaccharide is cellulose; insects do not directly synthesize cellulase, the enzyme needed for digestion of this large polymer and rely instead on endosymbiotic microorganisms to produce the necessary enzyme. Far less is known about the lipid-digesting enzymes in insects, largely because the dietary need for fat is miniscule in comparison to protein and carbohydrate. Lipases or esterases are the broad class of enzymes that target lipids. Most dietary lipid is in the form of neutral lipids, principally triacylglycerol. Midgut lipases release monoacylglycerol and fatty acids through the hydrolysis of triacylglycerol. However, in phytophagous insects with a highly alkaline gut pH, the digestion products are more apt to be free glycerol and fatty acids.

Fluid flow in the ecto- and endoperitrophic spaces facilitates chemical digestion

The presence of the peritrophic membrane in theory poses a potential problem to several aspects of digestion and absorption. The problem can be summed up succinctly as diffusion. Diffusion is the mechanism that accounts for digestive enzymes making contact with solutes—that is, the food particles to be digested. It is the same mechanism used for liberated nutrients to cross the peritrophic membrane into the ectoperitrophic space and then migrate to the epithelial cell surfaces for absorption. In an environment in which the fluids are not stirred or in motion, the concentrations of enzymes and substrates are the most important factors influencing the rate of diffusion, followed next by temperature.

Since insects are **poikilothermic**, the midgut temperature reflects ambient air or water conditions, which are ordinarily not predictable and therefore not dependable for increasing diffusional flux. Whatever is an insect to do, then, to help itself overcome the barrier of the peritrophic membrane? Amazingly, the solution is to stir the gut fluids. Any agitation, no matter how slight, of an unstirred fluid (or air) can greatly increase the rate of diffusion. Though there are likely several mechanisms employed by insects, one characterized for some Diptera and Orthoptera involves midgut cells in the posterior portion of the midgut releasing water into the ectoperitrophic space. The fluid flows anteriorly, to be absorbed by gastric cecae and then returned to the hemolymph. A unidirectional flow could be deleterious, in that food and nutrients would accumulate anteriorly, yielding limited absorption and an issue with removal of undigested and nonabsorbed food particles. Never fear, there is a solution. Cells located adjacent to the gastric cecae are also able to release water into the endoperitrophic space, creating a countercurrent to the one flowing in the anterior direction (figure 8.13). The effect is that the fluids are stirred on both sides of the peritrophic membrane, promoting enhanced rates of diffusion resulting in greater opportunities for enzymes to interact with food substrates, and, in turn, increased chance for liberated nutrients to contact epithelial surfaces for absorption to occur.

Exchanges also occur between the endo- and ectoperitrophic spaces, creating solute concentration shifts across the peritrophic membrane in both directions. Water will follow solute concentration shifts, inducing further movement of gut fluids. The solution, countercurrent flow in the midgut, is reminiscent of the same type of mechanism in the human kidney (loop of Henle). There, as in the insect midgut, water and solutes are absorbed along the entire length of exposed epithelial cells. Absorption of solutes further facilitates uptake of water, augmenting, though quite subtly, the fluid flow created by water secretion by midgut cells. It is quite likely that in some insects, the absorption of water and ions by Malpighian tubules accounts for all or some of the posterior movement of midgut fluid.

Quick check

Where does initiation of carbohydrate digestion begin in an insect's digestive tract?

Microbial symbionts aid digestive processes

Though it is sometimes difficult to admit (okay, maybe only for me), insects cannot always do it alone. "It," in this case, being digestion. Aid commonly comes in the form of microorganisms that take up residence within the gut tube of a host insect and take on the role of either digesting foodstuff that the insect lacks the necessary enzymes to digest or producing a nutrient needed by the host. In turn, the insect provides habitation and nutrients needed by the microorganism. The relationship is endosymbiotic, as both members of the association benefit. Generally, insects with simple straight gut tubes contain very few microorganisms, largely because their diet is not complex and therefore fairly straightforward to digest and then absorb. Insects with a complex, highly convoluted gut tube are more apt to harbor a wide range of microorganisms, and often rely on endosymbiotic organisms to fulfill certain aspects of digestion and nutrition (figure 8.14).

The types of endosymbionts residing in insects can be broadly classified as those that live in the digestive tract in or along the lumen, known as extracellular endosymbionts, and those that physically reside within cells of the digestive tract, termed intracellular endosymbionts. Regardless of classification, most of these organisms exist in the hindgut, most likely to avoid the milieu of digestive enzymes present in the midgut and to benefit from predigested food materials moving into their environment. Extracellular symbionts in the form of bacteria and protozoa are known to release digestive enzymes that break down plant polysaccharides like cellulose, pectins, and gums, which the insect host is not capable of doing itself. Some also fix atmospheric nitrogen, which the insect, in turn, can directly assimilate into its own tissues. Termites actually harbor protozoa that host bacteria at the cell surface that cooperatively ingest and digest cellulose for the benefit of all three organisms. The roles of intracellular endosymbionts are quite varied, with some functioning to convert stored uric acid into amino acids or to recycle nitrogen present in excretory products, others transforming nonessential amino acids into essential, and still others synthesizing vitamins that the insect host cannot acquire in its diet. In most cases, the endosymbionts are transferred from one generation to the next, often in specialized cells called **mycetocytes**.

It is only efficient if you can use it: Food assimilation

Insects demonstrate extraordinary regulation over what foodstuffs are consumed and then exceptional efficiency in digesting what has been ingested. Of course, none of these accolades matter if the highly sought after nutrients

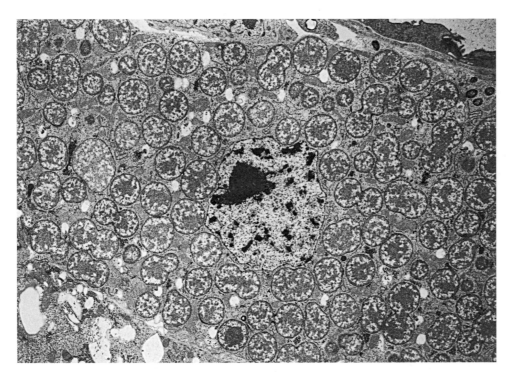

Figure 8.14. The endosymbiotic bacterium *Buchnera aphidicola* resides in aphid hosts within specialized cells (bacteriocytes) like the one shown here. Photo by J. White and N. Moran available at http://bit.ly/1FBK9PY.

never reach their targets. That is to say, highly efficient digestion is pointless if the nutrients are not absorbed with the same type of efficiency. Likewise, the absorbed molecules only become nutriment if they reach the tissues in need. In other words, the nutrients liberated by digestion cross the epithelial cells to enter the hemolymph, so that they eventually reach target tissues to become assimilated. The absorbed nutrients are used (assimilated) in cellular functions such as metabolism and the assembly of molecules or structures, or as fuels. Efficiency in food assimilation has two components: nutrient absorption by midgut cells and uptake and usage (assimilation) by cells and tissues throughout the body.

Absorption

Food is digested in the midgut to free or liberate nutrients that can be absorbed predominantly by midgut cells. This is a testament to the versatility of epithelial cells of the midgut, as many of these same cells produce digestive enzymes and function in creating the countercurrent fluid flow. Two main mechanisms drive absorption in insects: passive and active transport. Passive transport is diffusion, and as we discussed earlier, the mechanism for diffusion is concentration differences. Generally, absorption by passive transport relies on the solute in question moving down a concentration gradient: that is, from a region of higher concentration (the midgut lumen) to a region of lower concentration (intracellular environment). Obviously, under normal conditions, the cellular membrane serves as barrier to solutes moving freely back and forth, so a channel, gate, or transport receptor is required to allow a nutrient entry into the cell via diffusion. When operating properly, nutrients absorbed by passive transport move for free, meaning cellular energy is not required (figure 8.15).

Movement of solutes against a concentration or electrochemical gradient requires energy or active mechanisms for transport. It appears that ATP is the energy currency used by most insects to drive pumps (vacuolar type ATPase) located in apical membranes (those facing the lumen) of midgut epithelial cells to transport protons into the lumen in exchange for potassium and possibly calcium. Macromolecules in the gut must bind to membrane or transport proteins to be "pulled" into the cell against a concentration gradient. Likewise, ions moving

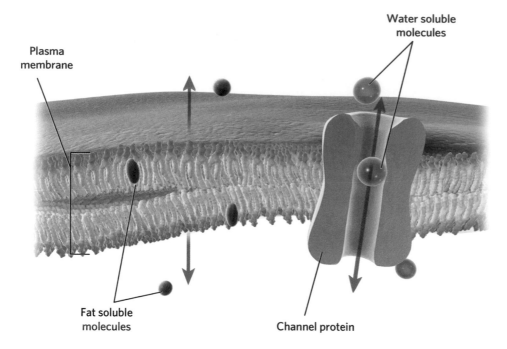

**Diffusion Across the
Plasma Membrane**

Figure 8.15. Passive transport, or diffusion, is the primary mechanism used to absorb nutrients across the midgut epithelial cells. Image by BruceBlaus available at http://bit.ly/1YIPuf1.

against an electrochemical gradient are pumped into the intracellular environment at the expense of ATP.

Most critical nutrients released in the gut are absorbed via passive transport. For instance, amino acids are usually in higher concentration within the gut lumen than in the intracellular environment of epithelial cells. Membrane proteins or **symports** facilitate the movement by first binding to specific amino acids, and then undergoing a conformational change that transports the amino acid into the cell. Carbohydrates also move by diffusion, and this transport is greatly enhanced by rapid conversion of glucose into trehalose within the fat body. The net effect is that a concentration gradient is maintained from gut to hemolymph, since the concentration of free glucose in the hemocoel is homeostatically low. Lipids are somewhat different than the other macromolecules in that they generally cross into epithelial cells by passive diffusion, often get modified within the cells, and then require active transport to be exchanged into the hemolymph. The exceptions are sterols, which move freely

across the cells into the hemolymph to fat body. Inorganic molecules are more varied, as potassium and water move into midgut cells via passive transport, whereas chloride, sodium, and calcium move against electrochemical gradients by active transport. All these inorganic substances, with the exception of water, are also absorbed across the ileum and rectum of the hindgut, and the mechanisms alternate between active and passive transport depending on the concentration and electrochemical gradients encountered. Water movement is always passive and occurs throughout the hindgut, including along the entire length of the Malpighian tubules.

Assimilation

Absorption is not the final step for macromolecules, for to truly be a nutrient, the substance must be used by a cell or tissue in need. Absorbed molecules circulate via hemolymph and move across target cells by active and passive mechanisms. Utilization of the nutrients occurs through metabolic reactions and energy use, or in the synthesis of other molecules, membranes, or other cellular components. This final destination is assimilation, which can be measured by the following equation:

$$\text{Assimilation efficiency} = \frac{\text{Rate of digestion} \times \text{Food retention time}}{\text{Concentration of food} \times \text{Midgut volume}}$$

The true or important efficiency of the digestive system is measured by whether the food ingested is incorporated into cellular uses. After all, it really means little overall that an insect eats the right foods if the nutrients never reach the audience (the cells) in need. With that in mind, just how good are insects in assimilating food? The answer depends on what kinds of feeders they are—that is, what types of foods are primarily consumed (e.g., phytophagous, carnivorous, etc.). Overall, most insects demonstrate assimilation efficiencies at or just below 50%, placing them much higher than most vertebrate animals but nowhere near the highest efficiencies known to be achieved. Well, just who in the animal world *is* most efficient? The answer may surprise you: insects. Now things have really gotten confusing! Certain insect species have adapted lifestyles that permit them to display extraordinary food assimilation capacities. These insects are parasitic on other invertebrates as well as vertebrate animals. Assimilation is maximized because the parasites obtain concentrated, predigested food materials from the host, which requires very short processing time in the insect's gut before it can be absorbed and assimilated. Consequently, food assimilation efficiencies can exceed 90% under these conditions.

Do parasitic insects have any rivals in the animal world for the title of kings of assimilation? Actually, yes they do. Other parasitic invertebrates display similar prowess in food assimilation. The other insect contenders are fly maggots that feed necrophagously on carrion or dead animals. Larvae from the families Calliphoridae and Sarcophagidae form large feeding aggregations or maggot masses as they consume an animal carcass. Flies in the maggot masses generate heat that can elevate the internal temperatures of the aggregations to more than 30° to 50°C above ambient conditions. Remember that temperature is an important factor influencing the rate of diffusion, and in this case, it can accelerate enzy-

matic digestion as well as the rate of absorption. As long as the maggot temperatures are not deleterious to fly development and survival, the rate of food assimilation increases with maggot mass temperatures. As a result, assimilation efficiencies typically exceed 75% and for some species, approach 90%. These values far surpass other saprophagous insects, highlighting the importance of self-produced temperatures in enhancing digestive efficiency of insects.

CHAPTER REVIEW

✳ **What's on the menu? Nutrient requirements of insects**

- What do insects eat? Insects generally feed on specific types of food. The majority of insects "restrict" their diets to vegetative or plant matter and are classified as phytophagous. Those that consume animals are termed carnivorous. Several species may be grouped as true parasites or as the more specialized parasitoids. Some insects are fluid-feeders and specialize on liquid diets derived from another animal or a plant. Still others feed on decaying organic matter originating from any type of organism and are called detritivores, or saprophagous.

- Regardless of how each species is classified in terms of feeding or food types, insects share several unifying features of nutrition and digestion. Among the most significant are that insects are among the most efficient of all animals with respect to finding food, manipulating and consuming food, breaking down complex food into useable forms, and absorbing nutrients and converting them to body mass. What makes insects so good at these processes? So much of the efficiency of the class Insecta relates to the simple idea of form meeting function. Morphologically speaking, the design of the insect digestive system—from mouth to anus—is adapted for eating, processing, and using specific types of food.

- Insects as a whole are chemically quite similar to each other in terms of the composition of their tissues and bodily fluids. Not surprisingly, then, the nutrients needed to maintain homeostasis are also very similar among different insect species. Although some species do indeed display very unique diets, as a whole, the basic nutritional requirements of insects inhabit common ground. As with most animals, certain nutrients must be obtained in the diet of an insect, and those are termed essential. Since the insect hunger drive is activated in response to nutritional needs, foraging and subsequent ingestion generally leads them to foods that contain the essential nutrients. Many of the needed nutrients can be synthesized from precursors found in ingested foodstuffs or derived from breakdown products. Such nutrients are called nonessential.

■ The basic nutrients needed by insects are the same as those needed by other animals, including us, and fall into the categories of amino acids, carbohydrates, lipids, vitamins, and inorganic compounds.

Tools of the trade: Structures used for food collection

■ The primary tools for collecting and ingesting food are the mouthparts. As we discussed in chapter 5, insect mouthparts are quite varied. In fact, the structure of the mouthparts directly reflects dietary tendencies, particularly in terms of the physical form of the food. Mouthpart type, and hence diet, is also correlated with insect groups or specific orders. The majority of insect species and groups use mandibles for food acquisition and manipulation. Is there any special reason(s) for this trend? Well, actually, yes, there is! For one, mandibular mouthparts are quite adaptable, offering utility in obtaining, processing, and even tasting many types of food differing in physical composition and structure. More important, though, mandibles are the ancestral mouthpart type of the class Insecta, the original mouthpart type that all others are derived from.

■ The way mandibles evolved—really, the entire head, for that matter—accounts for much of the remarkable efficiency that insects display in food acquisition and manipulation. Insects evolved from a wormlike ancestor that was segmented

and lacked a true head but did have a simple acron with an opening anteriorly located for food to enter and possessed several ventral appendages that aided in locomotion. The first six trunk segments are believed to have fused with the acron to form a complex head. Important in this evolutionary transition is that the pair of ventral appendages located on each trunk segment became head structures, including mouthparts and antennae, with each powered by the vestiges of the ancestral ventral appendage's muscles, meaning leg muscles.

■ Adding to the versatility of the mandibular mouthpart is the complex yet efficient design of the accompanying structures. A pair of large, relatively flat accessory jaws is present that allows the insect to manipulate food, along with leg-like palps that occur in pairs on maxillae as well as on the lower lip, the labrum. The presence of lips (labrum, or front lip, and labium, or rear lip) is a remarkable evolutionary advancement in itself, as these structures allow insects to close their mouths once food is ingested. By doing so, ingested food does not roll back out of the mouth, and a unique internal environment can be maintained that facilitates the initiation of chemical digestion. Keep in mind that most animals do not have lips! Within this internal space is a tongue or glossa (or hypoglossa) that does function somewhat similarly to that of humans: it is covered with thousands of taste receptors and can also be used to lap up liquids.

■ Mandibles are well suited for solid foods, whether derived from plants or animals. However, for those species that feed exclusively on liquid or semiliquid diets, the mandibular design needed to be adapted into something more functional. The evolved design for liquid feeding occurs in modern insects in three basic forms: as siphoning, piercing-sucking, and sponging mouthparts. Why so many mouthpart adaptions? Simple—liquid diets are not all the same. Generally some sort of muscular pump is also required to promote intake. The take-home message is that all insects do not eat the same foods, and their mouthparts reflect this diversity in diet.

❋ Why insects don't get fat but people do

- Under natural conditions insects do not overeat. How do they pull off what humans have never mastered? Do they possess unique mechanisms of self-control that other animals lack? Most research examining insect nutrition, specifically those aspects associated with hunger and satiety, suggests that the mechanisms employed by insects are quite similar to vertebrate animals. Broadly speaking, diet is regulated throughout the insect digestive system and can be observed from the time of food ingestion, through food movement from one location to another, to the release of digestive enzymes, and to the processes of absorption and assimilation.

- In the world of insects, their sleek, trim figures are maintained through hunger and satiety controls. Basically hunger and satiety are regulated through a series of interconnected mechanisms, which include feedback from taste receptors, chemical modulation, and neural mechanisms.

- Just how effective are these dietary control measures? Several studies have shown that if insects are not offered the right food, they will not eat. Starvation and death will result, rather than insects eating nonnutritious food. Some rely on complex foraging behaviors before even making an attempt to taste a potential food source. For any insect species, obeying the neural and chemical signals from the hunger and satiety neurons results in food intake that meets specific nutritional needs. Under normal or natural conditions, insects cannot ignore these signals and, consequently, they do not overeat or get fat.

❋ Eating "crap" makes sense! Food processing depends on what is eaten

- For an insect to take advantage of its ability to eat the right foods, it must have the capacity to release the valuable nutrients trapped inside ingested foods, through the process of digestion. Digestion relies on physical manipulation or breakdown of foodstuffs, termed mechanical digestion or mastication, and on chemical digestion, a process that relies on enzymatic digestion and the chemical environment (i.e., pH and ions) in which the food is contained. The insect gut tube is open at two ends (mouth and anus), permitting one-way flow of food. It is also designed for specialized functions to be carried out in different locations, thereby permitting multiple digestive events to occur simultaneously.

- The tube is portioned into three sections, each evolved for a specialized function: the foregut or stomodeum, the midgut or mesenteron, and the hindgut or proctodeum. Both the foregut and hindgut are derived from ectoderm and also lined with cuticle. Yes, the same cuticle discussed in chapter 6 as a major component of the insect exoskeleton. Cuticle, regardless of location, is shed during molting. So the lining of two-thirds of the gut tube is ripped out with each molt and replaced with a new one. The midgut is produced from endodermal tissue and does not possess a cuticular lining. Instead, epithelial cells synthesize a protective membranous tube (peritrophic membrane) that fills the inside of the midgut.

- The type of digestive enzymes found in any given insect reflects the chemical composition of the primary food source. Most insects must digest the major nutrient macromolecules, meaning protein, carbohydrates, and lipids. Therefore the same basic types of digestive enzymes are present in all insects. What differs is the exact form of the foodstuffs ingested. Insects with a solid diet of either plant or animal material will ingest predominantly proteins of different sizes, polysaccharides, and lipids in the form of glycerides, phospholipids, and glycolipids. By contrast, fluid-feeders benefit from tapping into the circulating fluids of their hosts, meaning that the dirty work of digestion has already occurred, so much less investment in chemical digestion is needed. When digestion does occur, it happens primarily in the midgut, regardless of the feeding habits of the insects in question. Epithelial cells lining the midgut synthesize the bulk of the digestive enzymes, although some are produced by the salivary glands and travel with the food bolus to the midgut.

- What types of digestive enzymes do insects produce? For the most part, the same types that we make. There are subtle differences, such as molecular sizes, precise target sites on substrates, and of course enzyme names, but the basic enzyme types are very similar to those produced by humans.

- The presence of the peritrophic membrane in theory poses a potential problem to several aspects of digestion and absorption. The problem can be summed up succinctly as diffusion. Diffusion is the mechanism that accounts for digestive enzymes making contact with solutes—that is, the food particles to be digested. It is the same mechanism used for liberated nutrients to cross the peritrophic membrane into the ectoperitrophic space and then migrate to the epithelial cell surfaces for absorption. Whatever is an insect to do, then, to help itself overcome the barrier of the peritrophic membrane? Amazingly, the solution is to stir the gut fluids.

- Insects sometimes need help with digesting certain types of foods or in acquiring specific nutrients. Aid commonly comes in the form of microorganisms that take up residence within the gut tube of a host insect and take on the role of either digesting foodstuff that the insect lacks the necessary enzymes to digest or producing a nutrient needed by the host. In turn, the insect provides habitation and nutrients needed by the microorganism. The relationship is endosymbiotic, as both members of the association benefit. Generally, insects with simple straight gut tubes contain very few microorganisms, largely because their diet is not complex and therefore fairly straightforward to digest and then absorb. Insects with a complex, highly convoluted gut tube are more apt to harbor a wide range of microorganisms, and often rely on endosymbiotic organisms to fulfill certain aspects of digestion and nutrition.

It is only efficient if you can use it: Food assimilation

- Insects demonstrate extraordinary regulation over what foodstuffs are consumed and then

exceptional efficiency in digesting what has been ingested. Of course, none of these accolades matter if the highly sought after nutrients never reach their targets. The nutrients liberated by digestion cross the epithelial cells to enter the hemolymph, so that they eventually reach target tissues to become assimilated. The absorbed nutrients are used (assimilated) in cellular functions such as metabolism and the assembly of molecules or structures, or as fuels. Efficiency in food assimilation has two components: nutrient absorption by midgut cells and uptake and usage (assimilation) by cells and tissues throughout the body.

- Two main mechanisms drive absorption in insects: passive and active transport. Passive transport is diffusion, and as we discussed earlier, the mechanism for diffusion is concentration differences. Generally, absorption by passive transport relies on the solute in question moving down a concentration gradient: that is, a from a region of higher concentration (the midgut lumen) to a region of lower concentration (intracellular environment). When operating properly, nutrients absorbed by passive transport move for free, meaning cellular energy is not required.

- Movement of solutes against a concentration or electrochemical gradient requires energy or active mechanisms for transport. It appears that ATP is the energy currency used by most insects to drive pumps (vacuolar type ATPase) located in apical membranes (those facing the lumen) of midgut epithelial cells.

- Most critical nutrients released in the gut are absorbed via passive transport. For instance, amino acids are usually in higher concentration within the gut lumen than in the intracellular environment of epithelial cells.

- Absorption is not the final step for macromolecules, for to truly be a nutrient, the substance must be used by a cell or tissue in need. Absorbed molecules circulate via hemolymph and move across target cells by active and passive mechanisms. Utilization of the nutrients occurs through metabolic reactions, as energy use, or in the synthesis of other molecules, membranes, or other cellular components.

- The true or important efficiency of the digestive system is measured by whether the food ingested is incorporated into cellular uses. After all, it really means little overall that an insect eats the right foods if the nutrients never reach the audience (the cells) in need. With that in mind, just how good are insects in assimilating food? The answer depends on what kinds of feeders they are—that is, what types of foods are primarily consumed. Overall, most insects demonstrate assimilation efficiencies at or just below 50%, placing them much higher than most vertebrate animals, but nowhere near the highest efficiencies known to be achieved.

MUSHROOM FARMING (SELF-TEST)

Level 1: Knowledge/Comprehension

1. Define the following terms:
   (a) assimilation
   (b) maxillae
   (c) countercurrent flow
   (d) essential nutrient
   (e) hyperphagia
   (f) satiety

2. Describe the basic design of the insect digestive tract.

3. What is the function of the peritrophic membrane?

4. Explain how midgut fluids flow during digestion and absorption.

5. What structures are involved in tasting food?

6. Describe the mechanisms used for absorption of (a) amino acids, (b) fatty acids, and (c) monosaccharides.

Level 2: Application/Analysis

1. Explain how insects regulate food intake to avoid overeating.

2. Make a diagram of an insect head, say, that of a cockroach, including the basic mouthparts.

3. Diagram the path that a protein follows through the gut tube, identifying the regions in which chemical digestion occurs.

4. Explain the advantages a gut tube provides an insect in terms of digestion.

Level 3: Synthesis/Evaluation

1. Speculate on how an insect can increase food assimilation efficiency under natural conditions.

2. Speculate on the state of the hunger and satiety centers in a caterpillar that has initiated apolysis.

3. If insects only eat what is good for then, meaning they obey the hunger and satiety centers, then explain why certain species, like the Indianmeal moth *Plodia interpunctella,* can be found infesting candy bars.

REFERENCES

Canavoso, L. E., Z. E. Jouni, K. J. Karnas, J. E. Pennington, and M. A. Wells. 2001. Fat metabolism in insects. *Annual Review of Nutrition* 21:23–46.

Carle, E. 1969. The Very Hungry Caterpillar. Scholastic, Inc., New York, NY.

Chapman, R. F. 1998. The Insects: Structure and Function. 4th ed. Cambridge University Press, Cambridge, UK.

Dethier, V. G. 1954. Evolution of feeding preferences in phytophagous insects. *Evolution*: 8(1): 33–54.

Downer, K. E., A. T. Haselton, R. J. Nachman, and J. G. Stoffolano Jr. 2007. Insect satiety: Sulfakinin localization and the effect of drosulfakinin on protein and carbohydrate ingestion in the blow fly, *Phormia regina* (Diptera: Calliphoridae). *Journal of Insect Physiology* 53:106–112.

Fraenkel, G., and M. Blewett. 1943. The vitamin B-complex requirements of several insects. *Biochemical Journal* 37:686–692.

Karasov, W. H., and C. Martínez del Rio. 2007. Physiological Ecology. Princeton University Press, Princeton, NJ.

Klowden, M. J. 2013. Physiological Systems in Insects. 3rd ed. Academic Press, San Diego, CA.

Lehane, M., and P. Billingsley. 1996. Biology of the Insect Midgut. Academic Press, San Diego, CA.

Nation, J. L. 2008. Insect Physiology and Biochemistry. 2nd ed. CRC Press, Boca Raton, FL.

Rahmathulla, V. K., and H. M. Suresh. 2012. Seasonal variation in food consumption, assimilation, and conversion efficiency of Indian bivoltine hybrid silkworm, *Bombyx mori. Journal of Insect Science* 12:82. Available online at insectscience.org/12.82.

Reynolds, S. E., and S. F. Nottingham. 1985. Effect of temperature on growth and efficiency of food utilization in fifth instar caterpillar of tobacco hornworm, *Manduca sexta*. *Journal of Insect Physiology* 31:129–134.

Scriber, J. M., and F. Slansky. 1981. The nutritional ecology of immature insects. *Annual Review of Entomology* 26:183–211.

Simpson, S. J., F. J. Clissold, M. Lihoreau, F. Ponton, S. M. Wilder, and D. Raubenheimer. 2015. Recent advances in the integrative nutrition of arthropods. *Annual Review of Entomology* 60:16.1–16.19.

Slansky, F., Jr. 1982. Insect nutrition: An adaptationist's perspective. *Florida Entomologist* 65:45–71.

Spit, J., L. Badisco, H. Verlinden, P. Van Wielendaele, S. Zels, S. Dillen, and J. Vanden Broeck. 2012. Peptidergic control of food intake and digestion in insects. *Canadian Journal of Zoology* 90:489–506.

Stoffolano, J. G., Jr., M. A. Lim, and K. E. Downer. 2007. Clonidine, octopaminergic receptor agonist, reduces protein feeding in the blow fly, *Phormia regina* (Meigen). *Journal of Insect Physiology* 53:1293–1299.

Waldbauer, G. P., and S. Friedman. 1991. Self-selection of optimal diets in insects. *Annual Review of Entomology* 36:43–63.

Wei, Z., G. Baggerman, R. J. Nachman, G. Goldsworthy, P. Verhaert, A. De Loof, and L. Schoofs. 2000. Sulfakinins reduce food intake in the desert locust, *Schistocerca gregaria*. *Journal of Insect Physiology* 46:1259–1265.

Widmaier, E. P. 1999. Why Geese Don't Get Obese (and We Do): How Evolution's Strategies for Survival Affect Our Everyday Lives. W. H. Freeman Company, New York, NY.

Winston, M. L. 1987. The Biology of the Honeybee. Harvard University Press, Cambridge, MA.

THE ENTOMOLOGIST BOOKSHELF (SUPPLEMENTAL READINGS)

Cohen, A. C. 2004. Insect Diets: Science and Technology. CRC Press, Boca Raton, FL. Dethier, V. G. 1976. The Hungry Fly: A Physiological Study of the Behavior Associated with Feeding. Harvard University Press, Cambridge, MA.

Gaio, A. de O., D. S. Gusmão, A. V. Santos, M. Berbert-Molina, P. F. P. Pimenta, and F. J. A. Lemos. 2011. Contribution of midgut bacteria to the blood digestion and egg production in *Aedes aegypti* (Diptera: Culicidae) (L.).

Parasites & Vectors 4:105. http://www.parasitesandvectors.com/content/4/1/105.

Kunieda, T., T. Fujiyuki, R. Kucharski, S. Foret, S. A. Ament, A. L. Toth, K. Ohashi, et al. 2006. Carbohydrate metabolism genes and pathways in insects: Insights from the honey bee genome. *Insect Molecular Biology* 15:563–576.

Nässel, D. R., and M. J. Williams. 2014. Cholecystokinin-like peptide (DSK) in *Drosophila*, not only for satiety signaling. *Frontiers in Endocrinology* 5:219. doi:10.3389/fendo.2014.00219.

Panizzi, A. R., and J. R. Parra. 2012. Insect Bioecology and Nutrition for Integrated Pest Management (Contemporary Topics in Entomology). CRC Press, Boca Raton, FL.

Sapolsky, R. M. 1998. Why Zebras Don't Get Ulcers. W.H. Freeman Company, New York, NY.

Terra, W. R. 1990. Evolution of digestive systems of insects. *Annual Review of Entomology* 35:181–200.

Waldbauer, G. 1996. Insects through the Seasons. Harvard University Press, Cambridge, MA.

Wigglesworth, V. B. 1982. The Principles of Insect Physiology. 7th ed. Springer, London, UK.

ADDITIONAL RESOURCES

Alien Empire
http://www.pbs.org/wnet/nature/alien-empire-introduction/3409/

Insect adaptation to plant defenses
http://www.reeis.usda.gov/web/crisprojectpages/0186708-insect-adaptation-to-plant-defense-decryption-of-gut-protease-deployment-and-development-of-synthetic-resistance.html

Insect digestive and excretory systems
http://www.cals.ncsu.edu/course/ent425/tutorial/digest.html

International Centre for Insect Physiology and Ecology
http://icipe.org/

Spider digestion and food storage
http://www.atshq.org/articles/Digestion.pdf

Termite digestion and biofuels
http://www.purdue.edu/newsroom/releases/2013/Q1/understanding-termite-digestion-could-help-biofuels,-insect-control.html

9

Sex in the City and Everywhere Else

Insect Reproductive Strategies

Methods of fertilization used by insects and their relatives, raw material for a *Kama Sutra* for arthropods, are bizarre from the human point of view, but they do make good evolutionary sense when they are examined more closely and in context.

Dr. Gilbert Waldbauer
Insects through the Seasons (1996)

The sex lives of insects would certainly blow the mind of Freud. Or at the very least, cause him to blush. Consider that insects display several forms of sexual reproduction; females commonly have multiple sexual partners, in some cases their sons; after sex, a female may store the sperm of her lover(s) for life, choosing when or if to use his seed; some impatient males strike an awaiting partner with their penis and then drive it into the body cavity to ejaculate, all while the unsuspecting female is in copula with another male; and still others mate with juveniles that have been hormonally induced to sexually mature. Jealous males seal the vagina of their mate to ensure "loyalty," while others induce ovulation after mating so that their sperm is used immediately, and some males remain in copula for hours after mating and even fly with their true love to a site for egg-laying, removing the penis only when oviposition is complete. Talk about trust issues! In some species, males have evolved a secondary penis, a backup if you will, to perform the duties that the original won't; a few species forgo the need for a penis altogether by depositing stalked droplets of semen on the ground for females to find for themselves; and still others display role reversals in which the male cares for the young while the female leads a promiscuous lifestyle. Here we focus on the many reproductive strategies employed by insect species, examining the evolutionary trade-offs associated with each as well as exploring the basics of sex education pertaining to insects—the nuts and bolts of sexual reproduction in the class Insecta. The process of sex will be explored from calling to courtship to copulation (or other mechanisms) to the mechanisms of egg dispersal.

Key Concepts

- Sex or no sex: Methods of reproduction used by insects
- Evolving sex: Adaptive trade-offs of sexual reproductive strategies
- Insects are sexually dimorphic inside and out
- CCC: Calling, courting, and copulation
- The kings (queens) of fertilization
- Get me out of here: Methods of egg dispersal
- And now for something completely different: Novelty behaviors

Sex or no sex: Methods of reproduction used by insects

Insects are amazingly prolific. This is evident from the multitudes that swarm in your face when taking a walk outside on a hot, humid summer day, or as you attempt to relax in the great outdoors only to have one, two, perhaps a million mosquitoes serenading your ears with a low-level buzzing sound. We examined the sheer abundance of insects in chapter 1, revealing that over one million species have been described, with a combined existence exceeding one quintillion individuals. What we did not discuss at that time was, where did all those insects come from? In other words, how do insects proliferate that leads to such extreme population densities? Do they follow the example of worms that produce thousands of eggs per body segment, and then simply release them wherever they choose? No, they do not. Those worms merely generate clones of themselves, which do not fare well as a whole in the potentially cruel world of nature. Nor do they show the species richness or diversity in individuals as do these dominant arthropods, the insects.

Insect reproduction can be summed succinctly with three points: (1) insects are like humans in that they have sex, (2) insects do not use contraception (obviously), and (3) insects never practice the missionary position. That's right, insects are sexual beings and, it seems, exist to accomplish two things in life: eat and have sex. (A lifestyle perhaps recognizable in at least some portions of another familiar species.) Sex relies on oxygen (as related to meiosis), which is so much more abundant in a terrestrial environment than in an aquatic one. To take full advantage of evolving toward a land existence, therefore, insects "needed" to maximize sexual reproduction to make it all worthwhile. Considering that members of this class of animals account for over 75% of all known species, it is safe to conclude they have maximized.

Just exactly how do insects have sex? Or perhaps the better question is, what do they do differently during sex or through a mechanism that is unique from other animals? There is something unique, even bizarre and a little freaky, when it comes to sex with insects (for clarification, insects' sex with each other). But they also have a great deal in common with most other sexually reproduc-

ing animals. In the following section, we examine mechanisms of reproduction—and not just sex—used by insects. This sets the stage for exploring complex behaviors and physiological mechanisms associated with finding mates and winning them over, fertilizing eggs and then getting them out of the body, and the insect version of parental care. Our discussion concludes with an examination of a whole host of bizarre or novel sexual behaviors, including incest, sperm competition, and even, perhaps, the basis for the movie *The Matrix*.*

Insect love, aka sex

Insects are sexually dimorphic. That already sounds like something kinky out of the *Kama Sutra*,† but it simply means that separate sexes or genders exist in the world of insects, and they can be distinguished based on anatomical features. More formally, insects are gonochoristic, meaning that separate sexes occur (figure 9.1). This is no different than most any other animal, including us, that engages in sexual reproduction. What does differ widely among insects is the structure used for sex; the genitals of insects show tremendous variation, although that is more true for males than for females.

The male **intromittent organ,** or **aedeagus**, akin to the human penis, exists in many shapes and sizes, sometimes in different locations but usually on the ninth abdominal segment; for a few, more than one is present (figure 9.2). There is so much variation in the anatomy of the male "package" that species identification can be made from the appearance of the aedeagus. Why so much variation? The general idea is that insects display species-specific male genitalia to ensure that interspecific mating or fertilization does not occur. This is a lock-and-key hypothesis to insect reproduction, and it does seem to fit many, but not all, species. Females are much less complex in terms of external genitalia, but make up for this simplicity with complex internal reproductive morphological features. More of those details are provided a little later in the chapter. For now, suffice to say that both

---

*The movie trilogy starring Keanu Reeves as Thomas Anderson (Neo), a computer hacker living in a dystopian future in which perceived reality is actually a computer simulation.

†The *Kama Sutra* is an ancient Indian Hindu text written sometime between 400 BC and 200 BC, detailing human sexual behavior in great detail.

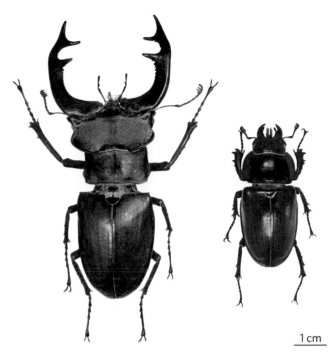

Figure 9.2. The aedeagus, or intromittent organ, of a male fly (*Pseudolyciella pallidiventris*) is folded flat against the abdominal sternum when not in use. Photo by Sarefo available at http://bit .ly/1LfeGpl.

Figure 9.1. Insects are sexually dimorphic, as evidenced by the male (left) and female (right) stag beetle, *Lucanus cervus*. Photo by Didier Descouens available at http://bit.ly/1h0EMy7.

Figure 9.3. Ovipositor of a female grasshopper, in use inserting eggs in the ground. Photo by Adam Majewski available at http:// bit.ly/1LVHIVo.

males and females produce appropriate gametes and possess the necessary plumbing to provide a union (fertilization) of the gametes either inside or outside the body (figure 9.3).

Inside or internal fertilization is a requirement for living on land. Otherwise, the sperm and eggs would never survive the desiccating conditions of unsaturated air. Much of what occurs during the process of sex with insects facilitates bringing the two sexes together. Mates need to size each other up to determine whether suitors are worthy of such a valuable commodity as genetic investment. Either gender can attempt to attract a mate, but ordinarily the male has a role in winning over the female. With most species, the adult female makes the decision whether to accept the sexual advances of a suitor and ultimately permit copulation. A male transferring spermatozoa to his partner consummates a sexual relationship. This is done most commonly via **direct insemination,** in which the male inserts the intromittent organ into the female's reproductive system (figure 9.4). There is more than one way to accomplish this task, so for now, we will merely address the fact that the missionary position is

not used. In more primitive species, direct contact does not occur. Rather, the male packages his sperm into silken **spermatophores** that can be left on the ground, or droplets of semen (sometimes on stalks) are deposited so that females can find them later (figure 9.5). This dispassionate approach to sex is termed **indirect insemination.** While the latter does not seem appealing from a human perspective, two key ideas must be remembered; insects do not have sex for pleasure, and indirect insemination does fulfill the need to protect sperm from desiccation.

Figure 9.4. Male and female anthomyiid flies in copula via direct insemination. Photo by Andre Karwath available at http://bit.ly /1LI3C2G.

Figure 9.5. Spermatophore in the form of stalked semen are used by some insects and closely related arthropods, like this springtail (class Collembola). Photo by Peter Bockman available at http://bit.ly/1YIRqUJ.

Our discussion so far has focused on the basics of insect sex. Where there are differences from other animals, most seem relatively subtle, and nothing so far really leaps out as a clear advantage that insects have over anyone else. So the question, what do they do differently, is still unresolved. What really does set insects apart is the timing of sex, the ability of females to store sperm indefinitely, and the presence of an ovipositor. In addition, insects employ other behavioral and physiological adaptations that, though not unique, are employed with the same remarkable efficiency they apply to nearly every other task in their lives.

For example, reproduction is intimately tied to environmental cues. It has to be, as most insects live for a very short time (less than two weeks). Adult males and females must be synchronized in their reproductive readiness to ensure that mating is possible before they die, and in turn, production of offspring must be timed for when conditions favor survival or progeny. How do insects achieve such synchrony? They respond to **environmental tokens** or cues that trigger behavioral and physiological mechanisms to turn on or off. The cues include a wide range of abiotic changes in the environment, including temperature, photoperiod, and humidity, and are even linked to lunar cycles. The advantage of timing reproduction to the environment is that generally the same conditions that favor sex also are ideal for egg-laying.

In most cases this timing also corresponds to when food is available for larvae, so that eggs can hatch almost immediately after oviposition. Such synchrony is achieved with any type of feeding preference: egg deposition and hatch occur when prey are available for carnivorous species, plants are in the appropriate stage of growth for phytophagous species, and hosts are available and of desired age for parasitic insects.

Most females can store sperm once mated and then release it to regulate fertilization. In some species, a mother can actually control the sex of her offspring via release of sperm; eggs that are fertilized become females, and those that are not become males. Remarkable! We will examine fertilization and sperm storage in more detail later, but to put this in immediate context, other animals cannot regulate fertilization like insects can. Other animals also lack the ovipositor of insects, an abdominal appendage used to place eggs basically wherever the mother chooses. The ovipositor allows eggs to be positioned outside of the female's body into a location that favors the progeny in some way, such as near food, concealed from potential predators and parasites, or protected from adverse weather conditions. This ingenious adaption permits most insects to forgo nest-building and parental care (insect parents are not up late waiting for their juvenile offspring to return home from a night of carousing around the neighborhood with god-knows-what type of arthropod). Sex the insect way works wonderfully well, but it is not foolproof. There are disadvantages to sex, and not just minor points, either. No, in some cases, sex can be deadly. But those are topics for the next section.

Beyond the text

Aquatic insects do not face desiccation yet do not rely on external spermatophore release. Why not?

Alternatives to sex

Sex is not always the method of choice for reproduction. It does dominate with all insect groups, regardless of taxonomic position, but some circumstances necessitate alternatives to sex. As humans, we cannot ever imagine a scenario in which sex is not preferred. Of course we also struggle with the idea that sex is not predominantly a tool of pleasure. Modern human society also sees sex intended for replicating as occurring later in life. Insects do not have a "later in life" stage. Time is of the essence, and maximum reproductive output is the goal. However, some insects face severe obstacles to having sex, such as living in an environment in which finding a mate is very challenging. This can be true in extreme climatic conditions, such as the Arctic, where exposure to the elements generally must be avoided. Mate-finding in such conditions can be costly, like, in terms of survival. Subterranean microhabitats also represent unique challenges to sex, particularly because chemical signals such as **pheromones** cannot effectively be dispersed, making attracting and finding the opposite sex nearly impossible. Artificial habitats like silos filled with grain are similarly impervious to pheromonal signaling (figure 9.6).

At other times, mate-finding is not the limitation for sex; instead it is a matter of speed, or, rather, time. Sometimes, performing sex and all its complicated behaviors—finding a mate, courting, and copulating—before oviposition is too slow to permit rapid colonization of an area, like a field planted in a particular crop (**monoculture**). The initial colonizer can sometimes populate the location by relying on asexual reproduction, a method in which genetic material is inherited from one parent only and that produces clones of the parent. In the case of insects, this parent is almost always the mother, via **aga-**

Figure 9.6. Steel silos storing sunflower seeds do not permit pheromonal signaling by insects that feed on the seeds. Photo by Leaflet available at http://bit.ly/1KNHz6r.

mogenesis, which is reproduction that occurs without male gametes being involved. The resulting offspring are essentially exact copies or clones of the parent (the mother), although in some forms of **parthenogenesis**, the progeny possess half as much genetic information as the mother. In those instances, the offspring are said to be **haploid** and the mother is **diploid** with respect to chromosome content.

The nonsexual route to reproduction allows an insect to elevate the population density rapidly as well as efficiently, since only one individual is required for the process. Some species revert back to sex once a critical population density is reached and/or environmental conditions favor the production of genetically diverse offspring. But asexual reproduction has its downsides, major issues that make sex always the attractive first choice: lack of genetic recombination, potential for genetic erosion (i.e., accumulation of deleterious genes), and **homozygosity** in all individuals. We will address these problems as we explore the different asexual methods of reproduction employed by insects.

1. Parthenogenesis
2. Polyembryony
3. Hermaphroditism
4. Regeneration

PARTHENOGENESIS

Parthenogenesis is the most common form of asexual reproduction among all animal populations. In this type of agamogenesis, an unfertilized egg develops into a new individual. If the egg is unfertilized, it would seem reasonable to conclude that the resulting embryo, and then insect, that develops, is haploid. Right? The logic is indeed sound, but remember, we are dealing with insects, so exceptions and surprises are the norm. In most instances, yes, the new individual is a haploid clone of the mother. This form of parthenogenesis is termed **arrhenotoky** and is used by members of the order Hymenoptera to produce sons. By contrast, daughters are produced sexually and are diploid, since fertilization represents the union of the egg nucleus from the mother with the sperm nucleus from the father.

Many species practice **outbreeding**, meaning they mate with partners that are not genetically related to avoid the potentially deleterious effects of inbreeding. However, incest is quite common with arrhenotokous Hymenoptera, with the females mating with sons or brothers. What benefit can come from something considered socially unacceptable in human society? Daughters produced sexually can share up to 75% of their genes in common with the mother, which is the underlying basis for socialization, a division of labor based on reproductive status, in ant and bee colonies (figure 9.7).

A second form of parthenogenesis, **thelytoky**, occurs with some parasitic wasps (Hymenoptera), aphids (Hemiptera), phasmids (Phasmatodea) and a few fly species (Diptera). Offspring are clones of the parent—again, the mother—but they are always diploid, at least in the case of Hymenoptera. Generally thelytoky occurs by one of two mechanisms: premeiotic doubling (**apomictic** meaning without meiosis), which generates daughters identical to their mother, or gamete duplication (**automictic** or postmeiotic) that leads to homozygous individuals, which also are daughters but may demonstrate some genetic diversity from the parent. Interestingly, some symbiotic bacteria (e.g., *Wolbachia*) residing within the wasps induce gamete duplication; treating the wasp with antibiotics like tetracycline eliminates the bacteria and terminates the thelytokous condition. Then what happens? The wasps rely on sexual reproduction to produce daughters and arrhenotokous parthenogenesis to generate sons.

A third type of parthenogenesis produces both sons and daughters asexually. The condition, **deuterotoky**, occurs with some Hymenoptera and also twisted-wing parasites in the order Strepsiptera. If parthenogenesis is not strange enough with the three examples already given, then consider **pseudogamy** or **gynogenesis**. Gynogenesis is a form of parthenogenesis in which sexual reproduction via direct insemination appears to have occurred. Both the male and female possess the necessary anatomical features for sexual reproduction, and sperm does physically make contact with an egg cell. The difference is that the sperm cell merely stimulates the egg to initiate mitosis so that embryonic development of the unfertilized egg occurs. No genetic material from the male is transferred.

POLYEMBRYONY

Polyembryony is a unique mode of reproduction in which two or more individuals can be produced from a single egg, with the resulting individuals being genetically identical. Four families of parasitic wasps (Hymenoptera) are known to rely on polyembryony, but they vary substantially in the total number of clonal individuals produced. In the most extreme example, with the family Encyrtidae, a single fertilized egg can yield one thousand to two thousand individuals. Here's how it works: after the first mitotic cleavage, a portion of the primary cell mass divides into a mass of developing embryos, termed the polymorula. Amazingly, the embryos will become both male and female adult wasps, despite the fact that they originated from the same egg! During larval development, the juvenile wasps feed on tissues of a caterpillar host.

From a parasitism standpoint, there is nothing unusual about this arrangement, with one exception: the wasp larvae are not alone. A second portion of the primary cell mass gives rise to serpentine-like precocious larvae that swim throughout the host's hemolymph on patrol. What are they looking for? The precocious larvae function as soldiers, seeking out nonrelated, or **allospecific,** competitors to kill. Interestingly, they are also tasked with chasing down male larvae to eliminate. That is correct; the soldiers attempt to trim back the number of males so that more nutrients are available for females. For many species of Hymenoptera, the goal of reproduction is to maximize production of females, generating only enough males to mate with their sisters or mother, or sometimes to outbreed. In turn, sons try to avoid detection by hiding in host tissues or moving frequently from one location to the next. Some escape detection, but

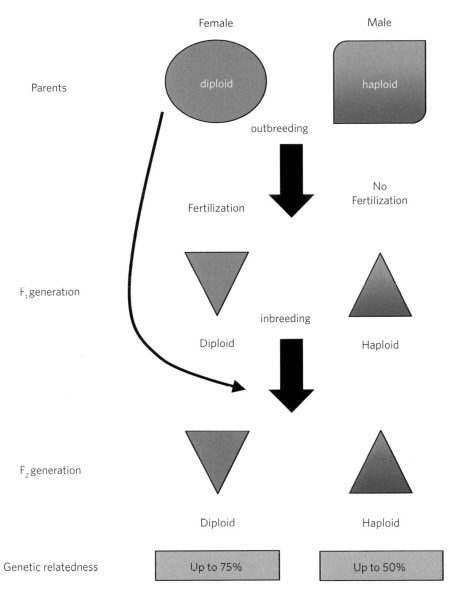

Figure 9.7. Genetic relatedness of progeny produced via arrhenotokous parthenogenesis. By the second generation, sons that mate with their sisters or mothers may yield daughters that share up 75% of their genetic composition in common with the mother. Sons are haploid clones of their mother and all their genes are shared with her.

many are killed. Again, keep in mind that these two forms of larvae are derived from the same egg! What allows this type of differentiation from a single fertilized egg? The exact mechanism(s) has not been fully deciphered, but what is known is that a third mass of cells forms from the initial mitotic division, and the mass functions as essentially a temporary endocrine organ, producing hormones that allow one mass to differentiate into embryos destined to become reproductive wasps and the other to become precocious larvae. The third

mass appears to also control the sex of the reproductive-destined wasps.

This crazy, magnificent form of reproduction begs the question, why do some insects reproduce by polyembryony? Several hypotheses have been posed, including that polyembryony serves as a means for females with limited egg-producing capacity to still maximize **fitness** or reproductive output. Another contends that polyembryony is a result of a conflict that exists between the mother and offspring, in that mom benefits more from larger clutch sizes than do her offspring. Under this scenario, the precocious larvae might be viewed as "conceived" by the children to get back at mom. Perhaps the most accepted hypothesis is that adult females are not particularly good at evaluating the amount of available resources (nutrients in the host) for the developing larvae. The precocious larvae essentially allow the offspring to make this assessment and then trim back the numbers. But observations with nonpolyembryonic species of parasitic wasps appear to confirm that most females are very good at evaluating the host condition before laying eggs, and rarely do they make poor decisions in terms of how many eggs can be supported by a host. Consequently, the third hypothesis is not consistent with what we know about other parasitic wasps, or insects as a whole, in terms of efficiency in using a resource.

HERMAPHRODITISM

Hermaphroditism represents a bit of a controversy with insects. Why? Some experts believe true hermaphroditism—that is, a single individual possessing both male and female reproductive structures—does not exist. In reality, a few species are indeed hermaphroditic. With that said, all seem to be simultaneous hermaphrodites, which have the ability to produce sperm and eggs at the same time, as opposed to sequential hermaphrodites, which start out life as one sex and then later become the other. Sequential hermaphrodites are not rare in the animal world, but no insects are reported to use this mode of reproduction.

Perhaps the best-studied example of hermaphroditism occurs with the cottony cushion scale, *Icerya purchasi* (Hemiptera). This insect relies on an **ovotestis** for gamete production; the structure can produce both sperm and eggs. Adults fertilize their own eggs, a process that permits them to use clonal reproduction to rapidly colonize an area. Self-fertilization produces offspring that are hermaphrodites. Even more bizarre, once the embryos form, sperm from the parent can invade the "daughter"* to fertilize the daughter's eggs. The result is production of progeny that are either hermaphroditic or males, and the original parent is the mother, father, grandmother,

---

*Females do not exist in this insect. However, hermaphrodite's offspring are termed daughters by convention.

and grandfather to their grandchildren. Nothing strange about that! (figure 9.8).

REGENERATION

Regeneration is not a topic usually discussed during animal reproduction. So why include it here? Well, it seems appropriate in this case since regeneration is a mechanism that leads to asexual production of tissues to replace those that were accidentally lost, taken by a predator, or purposefully dropped to avoid predation. Regeneration relies on mitosis, and in most instances it is an appendage that needs replacing. Typically, structures composed of cuticle can be regenerated only during a molt. Consequently, regeneration is restricted to larval stages, for as we learned in chapter 6, hemimetabolous and holometabolous insects do not molt as adults. Regeneration is believed to be restricted to a limited number of groups, including cockroaches (Blattodea), grasshoppers (Orthoptera), stick insects (Phasmatodea), true bugs (Hemiptera), and larval stages of many holometabolous insects.

Quick check

Which would seemingly be more beneficial to a female insect, parthenogenesis, hermaphroditism, or sexual reproduction?

Evolving sex: Adaptive trade-offs of sexual reproductive strategies

Sex is not free. No, for insects, evolutionary trade-offs have required adaptation to ensure that sex is not too costly, or even deadly. Too costly? Isn't sexual reproduction considered superior to asexual mechanisms? A very interesting question! The answer depends very much on how the question is framed. Traditionally, sex is viewed through the lens of long-term benefits. When it is, sex is deemed far superior to asexual reproduction for two main reasons: (1) sex leads to diploidy, the condition in which an organism has duplicate copies of its genetic material, and (2) sex leads to genetic diversity in a population.

The value of genetic diversity is obvious on many levels, but in particular, individuals of the same species are not clones and therefore can adapt to changing environmental conditions, so survival of the whole is much more

Figure 9.8. The hermaphroditic cottony cushion scale, *Icerya purchasi* (Hemiptera). Photo by Graham Wise available at http://bit.ly /1t76Wy0.

likely. Compare that to the clonal situation of asexual reproduction, in which, in theory, all individuals are identical: any rapid or unexpected change could be deleterious to the majority and may even lead to localized extinction. What about diploidy—how is this beneficial to insects? Diploidy means that an individual has two copies of each gene, though they might not be identical (an **allele** is a different form of the same gene), which permits an insect the ability to mask phenotypic expression of mutations or resist disease. If the insect is haploid, any mutation will be expressed, since a second copy of the gene is not present to potentially override or be dominant to the mutation.

In the short run, however, sex is expensive, and arguing that it is superior to all forms of asexual reproduction is not easy. For instance, sexual reproduction, with all its complicated behaviors—finding and courting a mate, then trying to win over the female to the idea of copulation—is a slow process. If competing for an ephemeral or **patchy** resource, insects practicing parthenogenesis will easily win the race of reproductive output against sexually producing competitors. Sex also requires complex reproductive systems in both the male and female; although that may also be true for some

forms of asexual reproduction, others, like hermaphroditism, can be quite simple. Sexual reproduction actually has several short-term problems that require insects to adapt to be successful in the long term.

1. Too slow for colonization
2. Requires complicated reproductive systems
3. Involves complex sexual behaviors
4. Needing to find partners limits distribution
5. Mortality risk

One issue of particular importance is the mortality risk associated with sex. Risk, in the form of transmission of disease to partners through physical contact during any aspect of courting and copulation, is most likely to happen during direct insemination. Similarly, parasite transference can occur during sex. It is also during the act of copulation that adult insects let their defensive guard down and thus become most susceptible to predation. A heck of way to die! It probably is safe to say that humans rarely face this form of risk during sex, but I'm still trying to get numbers from last year on the number of human copulation attempts in bear dens and on the open plains of the Serengeti. Even considering the long-term benefits that allow us to be smug about sex's superiority, there are kinks in the armor. After all, if one of the primary reproductive goals of an individual is to pass on its genetic blueprint to the next generation, then sex is not in the same league as any form of parthenogenesis that generates clones of the parent. And then there is the clonal mastery of polyembryony. The point is that sex does pose several challenges that are absent from asexual reproduction. If nothing else, the negatives of sex are why asexuality remains a viable option for several insects and other animals.

Just exactly how do insects overcome the obstacles to or with sex to justify its continued method-of-choice status? A lot of those details will be revealed in the section exploring sexual behaviors, but adaptations such as parental care, colony formation, modifications in sperm transfer, use of mood-changing chemicals (pheromones), and even the practice of midair copulation help reduce the risks, increase the speed of the entire process, and improve efficiency in the overall reproductive effort. In the long run, if sex did not work well for insects, they would have moved on to something else ages ago.

Insects are sexually dimorphic inside and out

Insects are gonochoristic, that is, sexes are distinguishable based on morphological features. Upon first inspection of an adult insect, this may be difficult to believe. Obvious external signs of gender are not readily apparent to the casual observer. Not unless you are an avid entomologist or practice **formicophilia**.* Nonetheless, insects are sexually dimorphic, inside and out. However, as with many features of insect biology, tremendous variation exists with regard to morphological form.

The variability has been attributed to reproductive isolation—that is, to prevent interspecies breeding. That only explains some of the differences. For example, with most species, females display an ovipositor that is used for placing eggs in locations specific for that species. So variability in ovipositor anatomy is a reflection of uniqueness in egg deposition, not a tool for intraspecies insemination. Likewise, the intromittent organ of the male

generally reflects species-specific adaption associated with reproductive isolation but also functions related to male-male competition. The degree of genital modification is so large that taxon groups, and even individual species, can be identified almost exclusively on the look of the genitalia. Still, despite this variability, sufficient commonality in reproductive structures occurs between groups that a generic plan can be used to describe the internal and external anatomy of male and female insects. For the sake of simplicity, we will examine the sexes separately, beginning with the anatomy of the male reproductive system, which is less complex overall.

Male reproductive structures

How can you tell the difference between male and female insects? Externally, both sexes may have the genitalia concealed by portions of the exoskeleton. So telling the sexes apart can be challenging. For male insects, the intromittent organ, (the aedeagus, or penis) can be displayed by an increase in hemolymph pressure, forcing circulating fluid to fill the penis, causing extension, the insect version of an erection.[†] The aedeagus is often lying flat under the abdominal sternum on the ninth segment, unfolding like on a hinge when hemolymph moves into the structure. The opening to the aedeagus is referred to as the **gonopore** (the same name for the opening to the female's reproductive system), and is the site of release of spermatozoa either in spermatophores or in liquid (semen) form. Usually the only other external structures used by males for sex-specific activities are **claspers** (other names also apply), which the suitor uses to grab or clasp the female to prevent her from leaving prior to or during copulation. A few species (e.g., Neuroptera) also rely on their mandibles to clasp or detain a female during copulatory behavior (figure 9.9).

Internally, the male reproductive system is little more than paired gonads (testes) that produce sperm, and ductwork that is the plumbing to move the sperm out of the male's body through the gonopore. Okay, the male's system is somewhat more complex than that, but the

---

*Formicophilia is a fetish in which sexual arousal is achieved from insects, usually from them walking on or biting a person, particularly when placed on the genitals.

[†]Which may get you thinking, what would Viagra or Cialis do to an insect? I honestly do not know, but I can assure you that I have no intention of asking my physician for a prescription so that I can experimentally arouse a male insect. You try it!

Figure 9.9. A male dobsonfly (Neuroptera) uses his saber-like mandibles as claspers during copulation. Photo by Geoff Gallice available at http://bit.ly/1FBW3cL.

functional overview is on the mark (figure 9.10). The testes are positioned on either side of the dorsal midline or longitudinal axis of the insect's body. In other words, if you cut the insect along the long axis, a gonad would lie in each half of the body. This is consistent with the bilateral symmetry displayed by higher animals, including all members of the class Insecta (a feature we first discussed in chapter 5). To each testis is connected a thin duct, the **vas deferens**, which transports newly produced spermatozoa from the gonads to the **seminal vesicle**. Seminal vesicles store sperm until ejaculation occurs by peristaltic contractions of muscles along the storage organ and length of the **ejaculatory duct**, which, in turn, extends through the intromittent organ to the gonopore. Insect sperm may be motile or nonmotile, depending on whether a flagellum is present or not. Typically, paired diverticuli pouches form along the vas deferens to become **accessory glands**. The glands have varied functions across the insect orders, including spermatophore formation and several roles in sperm competition behavior.

Female reproductive structures

Females are much more complicated than males. This is reflected more in the internal organs than the external genitalia. In fact, the primary structure involved with sex that is located externally, the **ovipositor**, does not show nearly as much variation in form as the aedeagus of the male. The major difference in ovipositor anatomy is linked with its origin: for primitive orders, the structure is derived from abdominal appendages; in all other groups, it originated from abdominal segments. From our discussion of mouthpart evolution in chapter 8, the level of muscular control associated with the former type of ovipositor should be apparent (figure 9.11). Eggs move out of the female's body through the shaft of the ovipositor. The shaft is formed from three pairs of valves. When the valves come together, a long thin tube is created that extends posteriorly from the abdomen or is held flat against the sternum. With several species, the valves can be moved separately, and when not used for egg laying, may give the appearance of sensory structures. Although not readily visible, the opening to the female reproductive tract is positioned at the base or proximal end of the ovipositor. In several species of Hymenoptera, the ovipositor is modified into a sting apparatus from which venom is injected into would-be attackers. Generally these insects are social species that have a division of labor dependent on reproductive status; the workers do not need the ovipositor for egg-laying functions. Interestingly, parasitic wasps use the ovipositor for both venom injection and egg deposition.

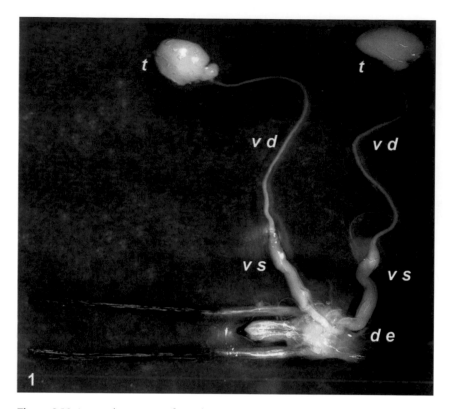

Figure 9.10. Internal anatomy of a male insect's reproductive system (t = testis, vd = vas deferens, vs = seminal vesicles, de = ejaculatory duct). Photo by Grozeva Snejana, Kuznetsova Valentina, Simov Nikolay, Langourov Mario, and Dalakchieva Svetla available at http://bit.ly/1Rad0ML.

The female insect's internal anatomy is far more complex than the male's, although the primary sex organs (gonads) are comparable. Females possess a pair of **ovaries** that are positioned along either side of the body. Ovaries produce eggs, referred to as oocytes, which are contained in individual sheaths or **ovarioles**. The number of ovarioles present in an ovary varies with species. Each ovariole contains several oocytes at different stages of maturity, which are positioned so that the most immature cells are distal to the oviducts, with the most mature located closest to the opening of the **lateral oviducts** (figure 9.12). The lateral oviducts are connected to the ovaries at one end and the **common oviduct** at the other. During ovulation, mature oocytes move through the lateral oviduct by peristalsis, into the common oviduct, and toward the genital chamber. The chamber is the receiving site for the male's aedeagus as well as spermatozoa. It is open to the outside via the **vulva** and connected to the common oviduct via the gonopore (note, for the female, the gonopore is the opening to the common ovi-

duct and does not open directly to the outside of the body as with the male). The genital chamber is also referred to as the **bursa copulatrix** in some species. In others, the vulva and genital chamber are much more narrow, and the term **vagina** is used in those species instead of genital chamber. The functions, however, do not differ between the two.

Several pairs of accessory glands are commonly associated with female insects. Among the most prominent are the **spermatheca** and **spermathecal gland**. For those species that are proud owners of such glands, sperm from the male is stored in the spermatheca after mating. In fact, some males can extend the penis through the genital chamber to the opening of the spermatheca for ejaculation. The spermathecal gland provides nourishment for sperm, so for many species, a female must only mate once. Other accessory glands produce egg coatings, adhesives, lubricants, and marking pheromones. Table 9.1 provides an overview of some of the functions of female accessory glands.

Figure 9.11. Examples of different types of ovipositors: (A) long ovipositor of a parasitic wasp (Hymenoptera)(Bruce Marlin, http://bit.ly/1FBWWlw), (B) female fruit fly (Diptera)(Ton Rulkens, http://bit.ly/1WrToqo), (C) Carolina mantis (Mantodea)(Kaldari, http://bit.ly/1VhyU6x), and (D) bush cricket (Orthoptera)(Pudding4brains, http://bit.ly/1YlY9xS).

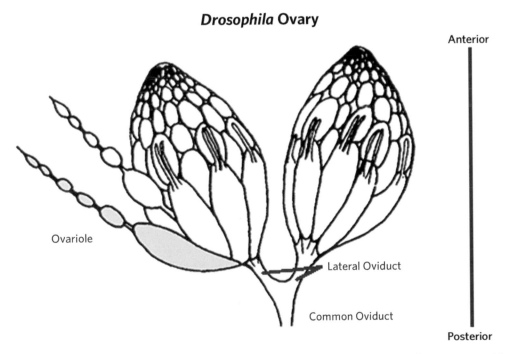

Figure 9.12. Ovaries of an adult vinegar fly from the genus *Drosophila* (Diptera). The common oviduct leads to the genital chamber, which is where the spermatheca is attached. Illustration by El Mayimbe available at http://bit.ly/1GaVtO5.

Table 9.1. Functions of insect reproductive accessory glands. Not all insects display the functions listed.

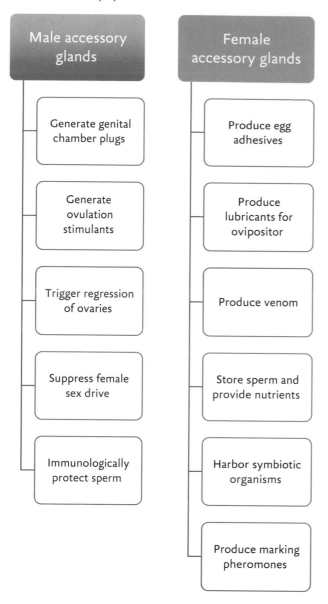

Male accessory glands	Female accessory glands
Generate genital chamber plugs	Produce egg adhesives
Generate ovulation stimulants	Produce lubricants for ovipositor
Trigger regression of ovaries	Produce venom
Suppress female sex drive	Store sperm and provide nutrients
Immunologically protect sperm	Harbor symbiotic organisms
	Produce marking pheromones

CCC: Calling, courting, and copulation

Oh the joys of courting and dating. I wonder if, on an insect's first date, the father calls the son in for chat, as mine did. "Son," pontificated my father, "remember, dippin' costs." In hindsight, there was nothing useful or comprehensible in the advice my father offered me. No, I'm pretty sure the advice from a sage paternal insect would be more to the point. Particularly since father and son will likely be dead soon, since the entire lifespan of an insect typically lasts about two weeks. For an insect,

sex cannot occur until the mates find each other and then unite. Since a male and female often do not develop in the same location, and are sometimes separated by more than a mile, there is an absolute need for the potential lovers to find each other. Insects rely on smell, visual cues, sound, and even taste (once in close range) to locate mates. Some species depend on multiple senses for mate-finding, others depend almost exclusively on chemical cues (pheromones) or auditory signaling. Several species display courting behaviors remarkably similar to humans, as in a male who cannot resist the harmonic singing of a beautiful young female desiring to bring children into the world, or male suitors trying to win the love of a female by bringing her gifts (not bling but food), or a couple that engages in dancing as sultry as a Latin salsa to check each other out before making a commitment.

Mate-finding and courting relies on multiple means to attract and impress a mate. The process is part of a complex set of assessments associated with sexual selection, in which both the male and female attempt to choose the best parent for their potential offspring. This also represents some of the slowest-moving aspects of sexual reproduction, as the behaviors may take hours, possibly more than a day, before the process is complete, meaning a decision for copulation or rejection. In this section, we examine some of the common mechanisms used for finding a mate (calling), nuptial gift-giving (courting), and sealing the deal (copulation).

Calling

To have sex, it is generally helpful for both members of the potential union to be in the same zip code. When they are not, or even if they are, one of the sexes, usually the male (surprise), advertises a desire to have sex. The process is termed calling, with one individual signaling to another an eagerness to mate as well as advertising a location to meet. Calling is really a means for soliciting sex. Bringing the sexes together when they are in close proximity to one another depends on visual cues more often than not. By contrast, when the male and female are separated by a substantial distance, chemical signals that can permeate through air or water are necessary, or making noise (auditory signaling) is needed to get the attention of a potential suitor. Visual cues can range from individual displays to a mass effect; for the latter,

Figure 9.13. The wing patterns of a butterfly like *Heliconius erato* (Lepidoptera) are perceived uniquely when detected by infrared, permitting species and sex identification and also enticing the opposite sex. Photo by Lilly M available at http://bit.ly/1KZ0BdQ.

swarming or aggregation of males are the two most common approaches.

Swarming occurs when several males of the same species take flight together, flying as a large group with the sole purpose of gaining the attention of interested females. The problem of course is for the individual male, trying to win the undivided affection of a female that approaches the swarm. You may have encountered one of these swarms (most likely gnats, Diptera) firsthand when out on a walk in a wooded area or along a lake or pond; the black mass of tiny insects flies into your face or seems to follow you. Good to know it's just a harmless group of hundreds of horny males that you just swallowed! Males of some species attempt to accomplish the same goal by forming large aggregations on an object, like a particular species of tree, even on a rock. The males know where to aggregate based on chemical signals given off by the first male arriving at the destination. The signals are made with aggregation pheromones (pheromones in general are used for intraspecific communication, so that members of the same species can chat). In this case, all the males that detect the odor follow its chemical trail back to the emitter. The idea of mass aggregation is that many are more likely to draw the attention of a female than just a lone individual. Again, the pitfall for members of the aggregation is that they will not all find a mate, but this approach does increase their odds of at least finding a female to attempt to court.

Sometimes visual attraction is just a matter of looking sharp. Color patterns or markings on the body and/or wings are how a male catches the eye of a female. The patterns that insects see may be quite different from our color vision, as patterns visualized in infrared are important for species recognition and mate identification for many species (figure 9.13). Visual calling can also be associated with movements, as in the way the male or female slowly and seductively flaps its wings, or in crazed, incessant Pitbull-style* dancing, usually performed by the male. Male dancing is quite common with flies (Diptera). Studies examining wing mutations in the vinegar fly, *Drosophila melanogaster*, have shown that when male wings are deformed (e.g., curly wing or apterous wing mutations) so that they do not beat with the same frequency or create vibrational sounds, females reject the suitors for normal dancing males. I can relate to being rejected for my dancing skills! The alpha male in a Madagascar hissing cockroach harem relies on antennal movements and

---

*Pitbull is the stage name for singer Armando Christian Perez from Miami, Florida. He is well known for constantly jumping and dancing during his concerts. https://www.youtube.com/watch?v=SmM0653YvXU.

Figure 9.14. Fireflies (Coleoptera) using bioluminescence to signal their desire to mate. Photo by T. Kiya available at http://bit.ly/1JxpyYf.

abdominal constrictions to make his intentions known to a female.

Perhaps the most familiar visual cue (to us) used for attracting mates is associated with **bioluminescence**. Fireflies, or lightning bugs (I offered my self-confessions on this topic back in chapter 1), are actually beetles that produce light via chemical reactions in the abdomen. During the warm summer months in North America, males belonging to the genus *Photinus* begin the process of courting females with the onset of dusk. Their familiar short flashes of light dot the landscape, as they fly in search of love (figure 9.14). An impressed female will direct a short light flash in the direction of the male's light, which, if detected, will cause him to orient in her direction. The two will flash light back and forth, first to evaluate each other, and then to guide the male to her. Of course the signaling draws the attention of other courters, so the initial male faces competition, and eventually the female will be forced to make a decision on who will be her mate, possibly more than one. Chemical cues may be used when the male is in close range to provide the final details about the female's precise location, for she is usually positioned on vegetation, watching the light show provided by all of the male suitors. For some un-

fortunate males, a rude and deadly surprise is waiting when they land next to their love. Instead of engaging in passionate lovemaking, the male soon realizes that he is the victim of insect phishing;* it was a male or female of another species that mimicked his species light flashes with the sole intention of luring him for dinner, with him serving as the main course!

For some insects, the way to true love is through song. Well, maybe not always singing per se, but at least by making noise that the courted simply cannot resist. Noise-making is almost always the job of the male. Members of the order Orthoptera, specifically crickets, grasshoppers, and katydids, "sing" using their wings. Each forewing has a hard edge called a **file** that is rubbed over a comb-like **scraper** on the other wing, a technique called **stridulation**. This rubbing action of the wings generates the chirping of crickets, or the sound "katy did, katy did" from well, katydids. Male crickets use the file and scraper to sing multiple songs to females, who can only listen, as

---

*Phishing is a practice of trying to gain information about or trust from an unsuspecting victim via electronic communication. In recent years, individuals have posed as the electronic boyfriend or girlfriend of their email partner, regardless of the sex of the other individual.

they lack the ability to sing back. Songs range from "this is my land (territory) and I will kick the butt of any other male dumb enough to get close" to one-on-one serenading of a targeted female, the latter being akin to the insect version of Barry White* (https://www.youtube.com/watch?v=ypyiAT1RelU). Male cicadas (Hemiptera) also sing to court females, using vibration of abdominal membranes (**tympanum**) to produce the characteristic shrill-pitched, war of the worlds† sound that girls cannot resist. Thousands of males will join in the chorus at once, a sound that can be heard from over a mile away. Many species rely on the same mating call to attract females. They avoid interspecies confusion by singing at specific times of day; thus, males can be lustful only at a designated time.

Male mosquitoes (Diptera) gain the attention of females via the vibrational noise of the wings beating. Once the two sexes get close enough, they taste each other to confirm that they are about to have sex with the appropriate species. What did they taste? Sex pheromones, a subject we will address in just a moment. Several species of flies rely on a combination of dancing (as we discussed earlier) and the vibrational frequency generated from wings flapping. Mechanical objects that operate at the same pitch and frequency can be mauled by horny females, ready to mate with outdoor HVAC equipment that really looks nothing like male insects (not all insects are that bright). Male Madagascar hissing cockroaches use hissing to convey multiple messages, just like male crickets. The hissing sound is created by air held in the tracheal system of the male being forced out of small openings or **spiracles** to the outside of the body. Hissing is used to relay irritation from being disturbed or threatened; to indicate territorial boundaries to other males, particularly when females are present; and also as part of courting a female. The latter presumably conveys a message of the male's masculinity and prowess, so the

hope is that she will be duly impressed. I know I am; any male that can win a female over by releasing gas deserves respect!

Bug bytes
Cicada love songs
https://www.youtube.com/watch?v=ubjoM75cYBY

Undoubtedly the most powerful tools insects use to convey a desire to have sex are chemical signals. The cues are sex pheromones. Both sexes can produce them, but more often than not, a female, through the release of her chemical signature, calls a male. The receiver in the relationship can readily be determined in most cases by an examination of the antennae: large "fluffy" or plumose antennae are a reflection of the need to detect miniscule concentrations of sex pheromone (figure 9.15). The typical course is for a sexually receptive female to secrete her scent from **exocrine glands** located on the abdomen into the air. Air currents then dispense the odorant, so that detection by a male occurs downwind. Sometimes the male picks up the scent from more than a mile away. Sex pheromones bind to olfactory receptors that transmit a sensory signal to the central nervous system that cannot be ignored. In other words, the male *will* be interested in the female and seek her out. To find her, he must use his antennae to trace the odor plume against the air current, flying back and forth to stay in a high concentration of his true love's scent. Once he gets close, the strength of the odor can become overwhelming and

Figure 9.15. Males of the cecropia moth (*Hyalophora cecropia*) display large plumose antennae, the type used to detect female sex pheromone from over a mile away. Photo by Brandyhouk available at http://bit.ly/1PI6N9O.

---

*No doubt this is another of those age divides across which you just said, "Who is Barry White." Sadly, you probably have immediate recognition if I answer, the baritone singer of the Burger King commercials. He was an American icon, a soulful singer with a bass baritone voice and the ability to generate seductive lyrics that made him the voice of love and romance.

†As in the beloved sound of aliens invading Earth in H. G. Wells's 1953 classic *The War of the Worlds*.

the precise location of the female usually must be revealed by some other means, such as through visual cues or sound.

Sex pheromones shared via air are termed volatile and confer several advantages in finding a mate: sex pheromones are highly concentrated and therefore can be released in small amounts; they are very stable in the environment, permitting the chemical signals to travel over long distances; and because they need to bind to olfactory or gustatory receptors, they provide species-specificity in terms of function. For species that live in very close proximity or that rely on other methods of calling, volatile sex pheromones are of no use. When chemical signaling is a component of mate-finding for such insects, contact sex pheromones are used. In these instances, secretion occurs from contact, so that immediate recognition of the signal occurs. Correspondingly, copulation occurs soon after. This is in contrast to volatile signals that attempt to draw attention to the female, and then the real courting begins once the male arrives. Several species of flies (Diptera) and cockroaches (Blattodea) make use of contact sex pheromones.

Courting: nuptial gifts

Some insects resort to gift-giving to win over a female. The gifts, termed nuptial gifts, are usually in the form of food. Use of nuptial gifts is not a form of calling, because the sexes are already in contact. Rather, this is good old-fashioned courting. A single male, or several, competes to be chosen by a receptive female. The competition is a matter of presenting the best gift. The winner gets to copulate; the loser must seek out another competition, or be devious by waiting for the "winner" to mate and then returning to couple with the same female. Who said insects had morals? In those instances, the sperm of the last male mated with is used first during fertilization. So who really is the winner? Regardless, nuptial gifts are usually food and, based on the quality of the food presented, allow the female to gauge certain qualities of the suitors, such as foraging ability. The insect maiden shows a great deal of patience before making her decision, as her choice influences the potential success of her offspring. Dragonfly (Odonata) males frequently capture flying prey as nuptial gifts. A particularly impressive gift is a fly (Diptera), since most display some of the most efficient and powerful wing movements of all insect groups. The courted female enjoys a feast provided by all male suitors. So being picky may also be a means to ensure satiety!

Hangingflies (Mecoptera) show a whole host of interesting behaviors associated with nuptial offerings and calling. The adult male releases sex pheromone to entice a female to find him. Upon her arrival, he offers a nuptial gift, but does not surrender the offering; rather he allows the female to taste it. If she is interested, the female permits copulation. She then dangles her body in a head-down position, while still in copula, her weight supported by the male's genitals and possibly a leg. Once in position, the male hands off the food gift so that his partner can consume it while copulation is completed. The duration of copulation correlates with the quality of the gift: give a good gift and the consummation is long; a poor offering means either a "quickie" or outright rejection.

Several other insect species rely on nuptial gifts of some sort, usually in the form of nutritious food, but not always. Some species of dance flies (Diptera) offer silk bubbles or balloons or even leaves of no nutritional benefit to the female, but she accepts them anyway and copulates. Apparently it's the thought that counts. In other species, the bubbles are extremely nutritious. Similarly, some species of beetles (Coleoptera) and crickets (Orthoptera) offer females spermatophores as gifts during copulation. In addition to permitting sperm transfer, the spermatophores contain a gelatinous mass of nutrients that, when consumed by a female, increases her **fecundity**. Thus, both the male and female benefit, since the male's sperm is used to fertilize the eggs. Fireflies in the genus *Photinus*—yes, those same beetles that rely on bioluminescence for finding mates—also convey messages of spermatophore size through the length of their flash. A long flash is indicative of large spermatophores and therefore more nutrients available for female consumption. Females preferentially seek out males with longer flashes.

For many insects, a final step in decision making is mechanical stimulation, in the form of taste. Tasting can simultaneously provide species recognition (is this the right species?) and sexual arousal (sex pheromone). What do they taste? Usually each other, by sampling the sex pheromones via gustatory receptors, consuming the male's sperm or spermatophores or tasting secretions from salivary glands or some other glandular secretion. Gustatory feedback may be the final information the female needs to determine whether this will be a successful date. For an insect, a successful date of course means that copulation occurs.

Copulation

Copulation is the required next step for species relying on direct insemination. The male mounts a female from behind and inserts his aedeagus into her genital chamber to complete the event. For some species, the male must then turn 180° so that his penis can move through the bursa copulatrix and release spermatozoa into the entrance of the spermatheca. A mating pair will commonly remain in copula, end-to-end, for several hours before terminating the date. By contrast, some fly species require less than five seconds to complete the deed. Dragonflies and damselflies display extreme acrobatic maneuvering to achieve copulation. The male first uses abdominal cerci as claspers to grab the neck of his lover. He then must transfer sperm from the penis, located on the ninth abdominal segment, to the secondary penis located just posterior to the thoracic sternum. Did you say two penises? Yep! One aedeagus releases sperm from the ejaculatory duct and the second transfers the gametes to the female. The female must bend her abdomen under the male so that he can insert his secondary penis, all while his cerci are still holding her by the neck. He does not release the female until after they fly to a freshwater location and she completes oviposition, using his sperm for fertilization (figure 9.16).

A queen honey bee, *Apis mellifera* (Hymenoptera), engages in copulation with one of her sons during flight. Talk about the need to concentrate! For Madagascar hissing cockroaches, *Gromphadorhina portentosa* (Blattodea), males must obey a pecking order for the chance to copulate. Failure to do so leads to fighting initiated by the alpha male, which frequently results in injury and possibly (though rarely) death for the defeated suitor. In winged species of cockroaches, the female controls when copulation will occur because her forewings must be moved to permit the male to mount and insert his intromittent organ. This is actually a common phenomenon observed with many species of insects.

Figure 9.16. A mating pair of damselflies (Odonata), the female grasped by the neck while she consumes a nuptial food gift from her beloved male companion. Photo by Aniruddha dhmorikar available at http://bit.ly/1O3ys7F.

Any discussion of insect copulation is not complete until we examine sex with mantids. You are probably familiar with the story, at least the parts that matter. Boy meets girl, girl flirts with boy, boy flirts back, they end up on a tree branch in the throes of passion, and then girl rips boy's head off. A tragic ending to what seemed to be a promising relationship. This story has circulated for decades, appearing in many general biology and entomology textbooks, even inspiring creative T-shirt and bumper sticker designs in a form of entomological sophomoric humor. As it turns out, the tale of copulation via decapitation is basically an urban legend, or mostly so. The story appears to have originated in an article in *Science** in the late 1800s. A single observation was made of mating Caribbean praying mantis, during which, while the male was mounted on his partner, she rotated her head nearly 180° to bite his head off. It was later surmised that the Caribbean male cannot release sperm when in copula until juvenile hormone secretion has stopped. When freed of the burden of a cranium, the male ejaculates. Subsequent studies of species from the United

---

*L. O. Howard's observations reported in *Science* 8 (1886) may have started the reputation-tarnishing of praying mantis females.

States reveal that they are quite boring; males ejaculate with the head still attached. Of course, once the act is complete, his date might eat him, head and all, but not until after he transfers his seed.

The kings (queens) of fertilization

Successful copulation for an insect ends with sperm being transferred from the male to the female. The process ordinarily involves the two principals, the male and female, making physical contact, as outlined in the previous section describing copulation, so that the male's aedeagus is inserted into the female's reproductive tract. When the male ejaculates, sperm is released into the female's body. This method of sperm transfer is termed direct insemination, obviously owing to the need for direct physical contact. Direct insemination is considered an evolutionary adaptation that permits sexual reproduction for terrestrial animals, specifically to facilitate a union of sperm and egg.

Ancestral forms of insects, as well as many modern-day marine invertebrates, relied on external fertilization, taking full advantage of their marine existence. For those species, gametes are simply deposited into the aqueous environment to be dispersed by water currents. Fertilization mostly occurs by chance, with sperm and egg colliding into one another. This method allows unrelated individuals to reproduce without the requirement of finding or liking each other. Importantly, external fertilization requires mass production of gametes since most will be consumed or destroyed in the external environment; only a few actually locate an opposing gamete. By contrast, direct fertilization methods are more precise in the union of sperm and eggs, so the overall production of gametes is several orders of magnitude less than that for external fertilization. In terms of habitat, living on land means that gametes cannot be released externally, since they will face almost immediate desiccation and lack a mode of locomotion appropriate for a terrestrial terrain.

Direct insemination is one adaptive strategy for fertilization on land. At least two others strategies for sperm transfer are employed by insects: **indirect insemination** (if there is "direct" then there has to be an "indirect") and **traumatic insemination**. Indirect insemination, as the name implies, does not require the sexes to meet, although they might. The method is considered the tran-

sitional form used by early land-based insects that had evolved from externally fertilizing ancestors. The male packages his sperm into a protective pouch or sac, called a spermatophore. Spermatophore production is the responsibility of a pair of accessory glands. Often, the structure is assembled as a gelatinous (and nutritious) mass into which sperm is secreted. The spermatophore has the appearance of an opaque to whitish balloon, with sperm tucked inside to avoid desiccation. Males can leave the mass on the ground or in some other location that a female will find, or he can hand it off to her directly as an initial nuptial gift. In reality, most insect groups package sperm in some sort of spermatophore, regardless of whether they depend on direct or indirect insemination. The distinction is whether the sexes interact to deliver the package. Females that must find sperm on their own (indirect insemination) utilize some sort of evaluation process, typically involving tactile and gustatory stimulation, to determine whether the male's gametes should be used for fertilization. If the spermatophore passes the test, mating is completed by the female inserting the sperm package into her vulva. Any remaining coating of the spermatophore is dissolved by enzymes, thereby releasing spermatozoa, which then migrate to the spermatheca.

Some arthropods have modified approaches to direct insemination that involve droplets of semen. The male deposits the drops on stalks so that that semen is held above the ground surface. At a later time, a gravid female (one possessing mature eggs ready for fertilization) collect the droplets in a pouch near the mouth. When she then ovulates, the mother will smear semen from the pouches over the eggs to fertilize them. Though this method of indirect fertilization has not been reported for any insect species, it has been for some closely related arthropod groups that occupy habitats just above and below the soil surface.

Perhaps the most unusual method of sperm transfer occurs in a few select families of bugs; this means, of course, from the order Hemiptera. In species with such appropriate common names as ambush bugs and pirate bugs, the male bypasses the norm of direct insemination by piercing the body wall of a female with his aedeagus to ejaculate (figure 9.17). Appropriately enough, the method is called traumatic insemination, as it surely is

Figure 9.17. A male bed bug (bottom) traumatically inseminating a female (top). Photo by Richard Ignell available at http://bit.ly/1MSBA62.

for the female! At least two questions come to mind: (1) Why does the male do this? and (2) How does the female tolerate the sexual attack? The "why" seems to be the result of attempting (successfully) to forgo the female-dominated process of sexual selection associated with normal mechanisms of direct insemination. These species are equipped with all the necessary parts and tools to engage in sex via insertion of the male penis into the female's vulva. In fact, a few observations have been made of a male traumatically inseminating a female who is actually in copula with another male via old fashioned mating techniques. Ejaculated sperm circulates through the female's hemolymph; some are absorbed by her spermatheca, some digested, and the released nutrients are used for egg production. How does she tolerate it? Remarkably well, although it does place her in a compromised immunological state. In most cases, the male's aedeagus is close to the size of a leg. So traumatic insemination creates a large wound in the body wall of the recipient and minimally deposits a host of microorganisms along with spermatozoa that the female must contend with. Insects have efficient body defenses, an aspect we will discuss in chapter 12, so the female is well equipped to fend off an infection under normal situations. However, there is nothing normal about traumatic insemination. The female will typically ovulate shortly after the male encounter. Consequently, her energetic resources

are divided between reproductive and immunological events, on top of regular homeostatic regulation. Despite these challenges, traumatically inseminated females typically survive the ordeal and have a lifespan as long as those that engage in more tame versions of direct insemination.

No matter how sperm is transferred to a female, the end goal is still the same. That aim, of course, is for fertilization to occur. For most species of insects, egg fertilization takes place within the common oviduct. As eggs move from ovarioles to the common oviduct, a process called **ovulation**, mature eggs are turned via muscular contractions so that the egg openings face the spermathecae. The openings are structured as tiny canals, or **micropyles**, which permit entry of sperm. Females display remarkable control over the events of fertilization, particularly in terms of sperm release from the storage gland. In some species, fewer than two dozen sperm are freed per egg, resulting in conservation of the male's gametes by the female, and also in sex regulation. Remember, from our earlier discussion, that in some insects fertilized eggs become females and unfertilized eggs develop into males. In those species, controlling sperm release allows the mother to control the sex of her offspring. For most insects, however, fertilization is required to produce both sons and daughters. For these species, the advantage of sperm storage by the female is enhanced efficiency of egg fertilization, achieving fertilization rates—well over 75%—unimaginable for primates.

Get me out of here: Methods of egg dispersal

Eggs are removed quickly from the female's body once fertilization has been attempted. Why emphasize "attempted" rather than "fertilization"? Simple—fertilization rates are rarely 100%, but all eggs entering the common oviduct will be moved to the outside of the body. The process of egg dispensing is termed oviposition; egg-laying also refers to the passage of eggs from the mother to the environment. The vast majority of insect species practice **oviparity**, and the mothers are oviparous. In practical terms, this means that individual eggs possess an outer covering or shell, called the **chorion**, and the eggs are deposited out of the body and placed in a specific location with the aid of an ovipositor. The egg is laid prior to significant embryonic development, which means that most aspects of embryonic growth and development, including hatch, occur independent of the mother. The implication is that eggs receive a finite amount of nutriment (in the form of yolk) from the parent. Egg placement often can be inferred from the shape of the ovipositor. Oviposition may result in egg deposition on an exposed surface such as the ground, a tree leaf, water surface, hair or fur of a vertebrate animal, or any of a multitude of other locations, artificial or natural. Many species insert their eggs into concealed locations, such as buried under ground, under tree bark, or in an animal or plant host. Secretions from the mother's accessory glands can function as adhesives, permitting attachment to any type of surface, or function as modulators (e.g., venom) of a live host so that the eggs are not destroyed (figure 9.18).

As we have learned with so many topics in entomology, insects do not rely on a single method or mechanism to accomplish any task. Oviposition is no different. Though oviparity is the norm for most species, other mechanisms are also employed, and all the alternative approaches depend on significant retention time of the eggs within the mother. Vivipary is a broad term that refers to the retention of eggs in the reproductive tract until egg hatch. More than one mechanism of dispersal falls under the viviparous umbrella, but the net result is basically the same: the appearance of live birth. Larvae rather than eggs leave the mother, although an ovipositor is not always required. No, in some cases, the children simply eat their way out of mom. Vivipary is characterized by significant embryonic development occurring while the eggs are maintained in the parent, as opposed to oviparity, in which most aspects of embryonic development takes place outside the mother's body. With some forms of vivipary, such as **viviparity**, the mother continues to provide nourishment to the developing embryo and possibly larvae beyond the initial contribution of yolk. With other forms, the mother generally does not provide additional nutriment to the embryos, displaying **ovoviviparity**.

Viviparity. Viviparity is relatively rare with insects. Viviparous insects do not possess a chorion around the egg and use specialized structures akin to a uterus, milk gland, or similar adaptations that allow the mother to provide nourishment to the embryo. A reduced number

Figure 9.18. Eggs of a white butterfly (Lepidoptera) that were deposited by oviparity using an ovipositor. Photo by Rolf Brecher available at http://bit.ly/1WCSGcZ.

of ovarioles typifies viviparous insects, resulting in fewer offspring being produced than with other forms of reproduction. In some cases, only one larva is produced at a time, such as occurs with the tsetse fly, *Glossina morsitans*, in which a single larva develops within the uterus of the mother, feeding on milk derived from an accessory gland; at the time of parturition, the mother gives "birth" to a larva nearly three-quarters the length of her own body.*

Ovoviviparity. Ovoviviparity involves the mother retaining the eggs with a chorion within the reproductive tract, usually in an ovisac or common oviduct, until egg hatch. Egg hatch occurs inside the mother, either immediately before deposition or at some earlier prescribed time that requires the mother to find a suitable substrate for deposition or else face being consumed by her offspring. Ovoviviparity is essentially oviparity with the exception that eggs hatch inside the mother. All nutriment for the embryo is derived from the egg, and no special anatomical structures are associated with nutrient transfer from mother to progeny following egg provisioning.

---

*Note to future husbands/dads: it is a tremendously poor decision to point out to a woman in labor, particularly the mother of your child, that her birthing experience really pales in comparison to that of a female tsetse fly, which must contend with an infant nearly as big as itself. Unfortunately, I have empirical data to support this supposition.

Bug bytes
A big baby: tsetse fly birth
https://www.youtube.com/watch?v=Pj4zkqrWf3E

And now for something completely different: Novelty behaviors*

Normal or typical sexual behavior of insects may seem to the casual observer to be filled with the unusual. Granted, several topics we have addressed so far are indeed novel from the human condition. But the reality is, you ain't seen nothing yet! Sex, and all the physiological and behavioral mechanisms that accompany it, generally does not support homeostatic systems in most animals. Consequently, many aspects have evolved independently, through wildly different paths that often manifest themselves as behavioral adaptations that appear simply bizarre, or at least extremely unique and thus interesting to discuss. With a million species in existence, there are many novel sexual behaviors to choose from. We will look at just a few but enough to hopefully whet your appetite to discover more on your own. Those briefly discussed here include sperm competition/sperm precedence, ovicidal behavior, widow makers, parental care, role reversal, larval competition, and slave makers.

The quest for mating with a female so that she can use the sperm of a male for fertilization is highly competitive. The competition is most severe for males belonging to species in which the female mates more than once and stores sperm prior to fertilization. Though not covered by ESPN,† the phenomenon is called **sperm competition**. The behaviors displayed by males to compete for sperm usage can be quite bizarre, ranging from a male knocking another male in copula free from a female so that he can mate to secreting juvenilizing hormones along with the spermatozoa in an attempt to suppress the sexual appetite of his lover to forming plugs that seal the vulva

---

*Monty Python's "And now for something completely different," https://www.youtube.com/watch?v=FGK8IC-bGnU.

†ESPN (Entertainment and Sports Programming Network) is a global television channel providing coverage of nearly every type of sporting event imaginable. That said, they currently do not cover sperm competition in insects, nor have they ever responded to my request to televise the cockroach racing in my General Entomology course.

and bursa copulatrix—eggs exit via a separate opening from the vulva, otherwise that strategy would be pointless. That said, this insect version of a chastity belt is only moderately effective at preventing secondary matings, as the next suitor can insert his mandibles into the vulva to pull out the plug and then insert his aedeagus. Dragonfly males can use the secondary penis as a spoon to scoop out the semen from any previous suitors. This may actually require insertion into the bursa copulatrix to ensure that all of his rival's seed has been removed.

Death is a common outcome of some sexual behaviors. **Ovicidal** behavior, for example, is at the expense of eggs or larvae of a competitor. In several species of parasitic wasps, if a female detects eggs or larvae from another species as she initiates her stereotypic host evaluation checklist, she uses her ovipositor to puncture the allospecifics repeatedly, with the intent of destroying any competition to her own offspring. Some species will also perform this ritual with conspecifics, particularly if the host is small and/or either species is solitary. In other cases, what the mother does not accomplish, the children do: in **larval competition**, the juveniles eliminate the competition. We have already discussed an extreme example, during polyembryony, in which precocious larvae seek out any allospecifics as well as male conspecifics to confer a death sentence. In other groups, an adult female delivers the deathblow to the unsuspecting male who will be the father of her progeny. This behavior is termed **widow making**, and comes in a few versions. One form is the head removal technique employed by at least one species of tropical mantids as a necessary step for male ejaculation (figure 9.19). The more common scenario, especially with spiders, is for the male to consummate his love and then be injected with paralytic venom that preserves his body. In turn, the female uses his sperm to fertilize her eggs during ovulation and then oviposits on the father so that, at egg hatch, the kids will eat his body as their first meal. Oh the joys of parenting!

With some insects, the traditional roles of male and female are reversed (role reversal). Sometimes this is manifested during the calling phase: the male releases sex pheromones so that interested females search for him. At other times, roles are reversed during copulation, when the female mounts the male and then curls her abdomen to find his penis. One of the more impressive

Bullying, incest, and flowers in the attic

Reproduction in a honey bee colony represents an entire curriculum for a social work major. Parental favoritism, incest, sibling rivalry, bullying, and even murder are all part of the reproductive cycle of honey bees. The trouble begins from the outset, when the young larvae are developing in the honeycomb. Workers tending the nursery show favoritism for some of the daughters, feeding them royal jelly throughout the duration of juvenile development. By contrast, their sisters are fed the bee equivalent to porridge, just nutritious enough to ensure that they grow to become the working class of the hive. Each will devote their lives to the colony, chemically bullied (through pheromones) to do the queen's bidding, to die for her if necessary. The queen is also their mother, though she never shows interest in any of them, and is the sole female permitted to reproduce. Daughters treated as royalty at an early age are destined to one day become queens, although this may be a short-lived crowning. Why? New queens are only produced when workers detect that the old queen's power is fading. This detection comes in the form of queen's substance, the pheromone emitted by the reigning monarch to control the workers. When the concentration of pheromone declines below a threshold level, workers in the hive recognize that the old queen is dying. Consequently, royal jelly is used to preferentially select several new queens. Once they emerge from their pupal cells, only one can assume the crown of the hive. The siblings are thrust into a struggle for power. One or more may leave the colony, taking a portion of the working class with them to establish new hives elsewhere. Those that remain fight; the victor will be queen, the loser(s) dies or is cast out. The remaining step to the throne is for the new queen to kill the old one. With that task complete, the new queen can now turn her attention to her primary role, reproduction.

Honey bees practice arrhenotokous parthenogenesis, with a son as the sexual partner. The queen oviposits unfertilized eggs that will be sons, waits for their emergence as adults, and then is courted by the **drones** (males) until she mates during flight with one of her own sons, who happens to be a haploid clone of his mother. The hive matriarch can then sexually produce daughters using her son's sperm. As a consequence, daughters share a very high percentage of their genes in common with their mother, almost to the point of calling them sexual clones. Daughters make up the working class and cannot reproduce. Instead, they work for their mother's reproductive success, which makes sense from the vantage point that they all share so many genes in common. In essence, when the queen reproduces, it is as if they have as well, since so many of their genes will be in the next generation. What about the sons? Those not chosen to mate usually drift from hive to hive, seeking opportunities to mate with another queen. When autumn approaches, they are all pushed out of the hive to live and die on their own. And the chosen one? He is allowed to remain in the hive, taking on menial tasks to justify his existence in the colony. But as the end of summer approaches, he too, is given his walking papers.

examples of role reversal occurs with parental care (figure 9.20). Parental care, or parents raising the young, is not common in the world of insects. Certainly species that form colonies (termites, ants, social wasps, bees) are extremely good at mass producing offspring, but the number of insect species that invest in their offspring after oviposition is very limited. One group of nonsocial insects that does is giant water bugs (Hemiptera: Belostomatidae). In some species, following copulation, the female quickly fertilizes her eggs (via stored sperm) and then oviposits on the back of her mate. The father takes on the role of caring for the eggs, protecting them from

Figure 9.19. Praying mantis couple in copulation. Neither seems interested in decapitation. Photo by Oliver Koemmerling available at http://bit.ly/1VhBlpT.

Figure 9.20. A male giant water bug carrying his future offspring as eggs on his back (tergum). Photo by noisecollusion available at http://bit.ly/1GaXXfo.

predation, and keeping them alive by breaking the water surface at regular intervals so that gas-exchange with air can occur. He keeps up this vigil until the neonate nymphs hatch from the eggs. Of course his parenting skills are less than perfect, for if food is limited, he may eat his progeny after protecting them for several weeks.

Ants (Hymenoptera) are a source of countless fascinating biological adventures, including during reproduction. For example, several species of ants have evolved to become highly aggressive, waging war on other colonies with the sole purpose of kidnapping the nursery. The warring ants have evolved saber-like mandibles that pierce the bodies of the brave but ill-equipped soldiers of the other colony. The larger more powerful species make quick work of their foes. They then invade the nursery of the conquered species, collecting eggs and developing larvae, and take them back to their own colony (figure 9.21). The warring species is a type of slave-making ant; the developing young will become slaves to them. Pheromones released in the colony control the behavior of the slave ants, ensuring that they complete all tasks needed to maintain the slave makers, including feeding them. The major trade-off with the saber anatomy of the mandibles is that the mouthparts are essentially useless for food collection and mastication.

An equally impressive behavior in ant colonies involves the ants being duped by a beetle. The beetle, *Atemeles pubicollis* (Coleoptera: Staphylinidae), relies on ants for reproduction by tricking them into believing that beetle larvae are in fact young ants. Considering that the beetle larvae may be as much as ten times the size of ant larvae, how is it possible the ants cannot tell the difference? The answer is pheromones. The beetles produce the identification pheromones of the ants. As far as ants are concerned, if you smell right, you're an

Figure 9.21. Soldiers of a slave-making ant (*Polyergus lucidus*) returning with eggs and larvae from another ant species. Photo by James Trager available at http://bit.ly/1PI8Vyj.

ant. Worker ants feed the beetle larvae in their own nursery, and the beetles beg aggressively for food, out-eating their ant stepsisters. Once larval development is complete, they pupate in the colony and, upon adult emergence, walk right out, unmolested by soldier ants, to seek out a mate. The beetles switch between two ant species as seasons change to ensure that at least one of the ant colonies is actively raising young, so that the beetles have free day care year-round.

CHAPTER REVIEW

✤ Sex or no sex: Methods of reproduction used by insects

- Insects are sexually dimorphic. This simply means that separate sexes or genders exist in the world of insects, and they can be distinguished based on anatomical features. More formally, insects are gonochoristic, meaning that separate sexes occur. This is no different than most any other animal, including us, that engages in sexual reproduction. What does differ widely among insects is the structure used for sex; the genitals of insects show tremendous variation, although that is more true for males than for females. The male intromittent organ, or aedeagus, akin to the human penis, exists in many shapes and sizes, sometimes in different locations but usually on the ninth abdominal segment; for a few, more than one is present. There is so much variation in the anatomy of the male "package" that species identification can be made from the appearance of the aedeagus.

- Much of what occurs during the process of sex with insects facilitates bringing the two sexes together. Mates need to size each other up to determine whether suitors are worthy of such a valuable commodity as genetic investment. Either gender can attempt to attract a mate, but ordinarily the male has a role in winning over the female. With most species, the adult female makes the decision whether to accept the sexual advances of a suitor and ultimately permit copulation. A male transferring spermatozoa to his partner consummates a sexual relationship.

- Adult males and females must be synchronized in their reproductive readiness to ensure that mating is possible before they die, and in turn, production of offspring must be timed for when conditions favor survival or progeny. How do insects achieve such synchrony? They respond to environmental tokens or cues that trigger behavioral and physiological mechanisms to turn on or off. The advantage of timing reproduction to the environment is that generally the same conditions that favor sex also are ideal for egg-laying. In most cases this timing also corresponds to when food is available for larvae, so that eggs can hatch almost immediately after oviposition. Such synchrony is achieved with any type of feeding preference: egg deposition and hatch occur when prey are available for carnivorous species, plants are in the appropriate stage of growth for phytophagous species, and hosts are available and of desired age for parasitic insects.

- Most females can store sperm once mated and then release it to regulate fertilization. In some species, a mother can actually control the sex of her offspring via release of sperm; eggs that are fertilized become females, and those that are not become males.

- Some insects face severe obstacles to having sex, such as living in an environment in which finding a mate is very challenging. At other times, mate-finding is not the limitation for sex; instead it is a matter of speed, or, rather, time. Sometimes, performing sex and all its complicated behaviors—finding a mate, courting, and copulating—before oviposition is too slow to permit rapid colonization of an area, like a field planted in a particular crop. Asexual reproduction is a mechanism of reproduction in which genetic material is inherited from one parent only. In the case of insects, this parent is almost always the mother, a process referred to as agamogenesis.

- The nonsexual route to reproduction allows an insect to elevate the population density rapidly as well as efficiently, since only one individual is required for the process. Some species revert back to sex once a critical population density is reached and/or environmental conditions favor the production of genetically diverse offspring. But asexual reproduction has its downsides, major issues that make sex always the attractive

first choice: lack of genetic recombination, potential for genetic erosion, and homozygosity in all individuals.

Evolving sex: Adaptive tradeoffs of sexual reproductive strategies

- Sex is not free. No, for insects, evolutionary trade-offs have required adaptation to ensure that sex is not too costly, or even deadly. Traditionally, sex is viewed through the lens of long-term benefits. When it is, sex is deemed far superior for two main reasons: (1) sex leads to diploidy, the condition in which an organism has duplicate copies of its genetic material, and (2) sex leads to genetic diversity in a population.

- In the short run sex is expensive, and arguing that it is superior to all forms of asexual reproduction is not easy. For instance, sexual reproduction with all its complicated behaviors—finding and courting a mate, then trying to win over the female to the idea of copulation—is a slow process. If competing for an ephemeral or patchy resource, insects practicing parthenogenesis will easily win the race of the reproductive output against sexually producing competitors. Sex also requires complex reproductive systems in both the male and female; although that may also be true for some forms of asexual reproduction, others, like hermaphroditism, can be quite simple. Sexual reproduction actually has several short-term problems that require insects to adapt to be successful in the long term.

- One issue of particular importance is the mortality risk associated with sex. Risk, in the form of transmission of disease to partners through physical contact during any aspect of courting and copulation, is most likely to happen during direct insemination. Similarly, parasite transference can occur during sex. It is also during the act of copulation that adult insects let their defensive guard down and thus become most susceptible to predation.

Insects are sexually dimorphic inside and out

- Insects are gonochoristic, that is, sexes are distinguishable based on morphological features. Obvious external signs of gender are not readily apparent to the casual observer. Nonetheless, insects are sexually dimorphic, inside and out. However, as with many features of insect biology, tremendous variation exists with regard to morphological form. The variability has been attributed to reproductive isolation—that is, to prevent interspecies breeding. Variability in ovipositor anatomy is a reflection of uniqueness in egg deposition, not a tool for intraspecies insemination. Likewise, the intromittent organ of the male generally depicts species-specific adaption associated with reproductive isolation but also functions related to male-male competition

- Externally, both sexes may have the genitalia concealed by portions of the exoskeleton. So telling the sexes apart can be challenging. For male insects, the intromittent organ can be displayed by an increase in hemolymph pressure, forcing circulating fluid to fill the penis, causing extension. The aedeagus is often lying flat under the abdominal sternum on the ninth segment, unfolding like on a hinge when hemolymph moves into the structure. The opening to the aedeagus is referred to as the gonopore (the same name for the opening to the female's reproductive system), and is the site of release of spermatozoa either in spermatophores or in liquid form. Usually the only other external structures used by males for sex-specific activities are claspers, which the suitor uses to grab or clasp the female to prevent her from leaving prior to or during copulation.

- Internally, the male reproductive system is little more than paired gonads that produce sperm, and ductwork that is the plumbing to move the sperm out of the male's body through the gonopore. Okay, the male's system is somewhat more complex than that, but the functional overview is on the mark.

- Females are much more complicated than males. This is reflected more in the internal organs than the external genitalia. In fact, the primary structure involved with sex that is located externally, the ovipositor, does not show nearly

as much variation in form as the aedeagus of the male. The major difference in ovipositor anatomy is linked with its origin: for primitive orders, the structure is derived from abdominal appendages; in all other groups, it originated from abdominal segments. Eggs move out of the female's body through the shaft of the ovipositor. The shaft is formed from three pairs of valves. When the valves come together, a long thin tube is created that extends posteriorly from the abdomen or is held flat against the sternum. In several species of Hymenoptera, the ovipositor is modified into a sting apparatus from which venom is injected into would-be attackers.

- The female insect's internal anatomy is far more complex than the male's, although the primary sex organs are comparable. Females possess a pair of ovaries that are positioned along either side of the body. Ovaries produce eggs, referred to as oocytes, which are contained in individual sheaths or ovarioles. Each ovariole contains several oocytes at different stages of maturity, which are positioned so that the most immature cells are distal to the oviducts, with the most mature located closest to the opening of the lateral oviducts. The lateral oviducts are connected to the ovaries at one end and the common oviduct at the other. During ovulation, mature oocytes move through the lateral oviduct by peristalsis, into the common oviduct, and toward the genital chamber. The chamber is the receiving site for the male's aedeagus as well as spermatozoa. It is open to the outside via the vulva and connected to the common oviduct via the gonopore. The genital chamber is also referred to as the bursa copulatrix in some species. In other others, the vulva and genital chamber are much more narrow, and the term vagina is used in those species instead of genital chamber.

CCC: Calling, courting, and copulation

- Sex cannot occur until the mates find each other and then unite. Since a male and female often do not develop in the same location, and are sometimes separated by more than a mile, there is an absolute need for the potential lovers to find each other. Insects rely on smell, visual cues, sound, and even taste to locate mates. Some species depend on multiple senses for mate-finding, others depend almost exclusively on chemical cues or auditory signaling. Several species display courting behaviors remarkably similar to humans, as in a male who cannot resist the harmonic singing of a beautiful young female desiring to bring children into the world, or male suitors trying to win the love of a female by bringing her gifts, or a couple that engages in dancing as sultry as a Latin salsa to check each other out before making a commitment.

- Calling is a process in which one individual signals to another an eagerness to mate, as well as advertising a location to meet. Calling is really a means for soliciting sex. Bringing the sexes together when they are in close proximity to one another depends on visual cues more often than not. By contrast, when the male and female are separated by a substantial distance, chemical signals that can permeate through the air or water are necessary, or making noise (auditory signaling) is needed to get the attention of a potential suitor. Visual cues can range from individual displays to a mass effect; for the latter, swarming or aggregation of males are the two most common approaches.

- Color patterns or markings on the body and/or wings are how a male catches the eye of a female. The patterns that the insects see may be quite different from our color vision, as patterns visualized in infrared are important for species recognition and mate identification for many species. Visual calling can also be associated with movements, as in the way the male or female slowly and seductively flaps its wings, or in crazed dancing, usually performed by the male, and also through the use of bioluminescence.

- For some insects, the way to true love is through song. Well, maybe not always singing per se, but at least by making noise that the courted simply cannot resist. Noise-making is almost always the job of the male. Members of the order Orthoptera "sing" using their wings. Each forewing has a hard edge called a file that is rubbed over a comb-like scraper on the other wing, a technique called stridulation. Male cicadas (Hemiptera) also sing to court females, using vibration of abdominal membranes to produce the characteristic shrill-pitched sound that girls cannot resist.

- Undoubtedly the most powerful tools insects use to convey a desire to have sex are chemical signals. The cues are sex pheromones. Both sexes can produce them, but more often than not, a female, through the release of her chemical signature, calls a male. The receiver in the relationship can readily be determined in most cases by an examination of the antennae: large "fluffy" or plumose antennae are a reflection of the need to detect miniscule concentrations of sex pheromone. The typical course is for a sexually receptive female to secrete her scent from exocrine glands located on the abdomen into the air. Air currents then dispense the odorant, so that detection by a male occurs downwind. Sometimes the male picks up the scent from more than a mile away.

- Some insects resort to gift-giving to win over a female. The gifts, termed nuptial gifts, are usually in the form of food. Use of nuptial gifts is not a form of calling, because the sexes are already in contact. Rather, this is good old-fashioned courting. A single male, or several, competes to be chosen by a receptive female. The competition is a matter of presenting the best gift.

- Copulation is the required next step for species relying on direct insemination. The male mounts a female from behind, and inserts his aedeagus into her genital chamber to complete the event. For some species, the male must then turn 180° so that his penis can move through the bursa copulatrix and release spermatozoa into the entrance of the spermatheca. A mating pair will commonly remain in copula, end-to-end, for several hours before terminating the date. By contrast, some fly species require less than five seconds to complete the deed. Dragonflies and damselflies display extreme acrobatic maneuvering to achieve copulation.

☀ The kings (queens) of fertilization

- Successful copulation for an insect ends with sperm being transferred from the male to the female. The process ordinarily involves the two principals, the male and female, making physical contact, as outlined in the previous section describing copulation, so that the male's aedeagus is inserted into the female's reproductive tract. When the male ejaculates, sperm is released into the female's body. This method of sperm transfer is termed direct insemination, obviously owing to the need for direct physical contact. Direct insemination is considered an evolutionary adaptation that permits sexual reproduction for terrestrial animals, specifically to facilitate a union of sperm and egg.

- Direct insemination is one adaptive strategy for fertilization on land. At least two others strategies for sperm transfer are employed by insects: indirect insemination and traumatic insemination. Indirect insemination, as the name implies, does not require the sexes to meet, although they might. The method is considered the transitional form used by early land-based insects that had evolved from externally fertilizing ancestors. The male packages his sperm into a protective pouch or sac, called a spermatophore. Spermatophore production is the responsibility of a pair of accessory glands.

- In traumatic insemination, the male bypasses the norm of direct insemination by piercing the body wall of a female with his aedeagus to ejaculate. Why does the male do this? The "why" seems to be the result of attempting (successfully) to forgo the female-dominated process of sexual selection associated with normal mechanisms of direct insemination.

- No matter how sperm is transferred to a female, the end goal is still the same. That aim, of course, is for fertilization to occur. For most species of insects, egg fertilization takes place within the common oviduct. As eggs move from ovarioles to the common oviduct, a process called ovulation, mature eggs are turned via muscular contractions so that the egg openings face the spermathecae. The openings are structured as tiny canals or micropyles, which permit entry of sperm. Females display remarkable control over the events of fertilization, particularly in terms of sperm release from the storage gland.

☀ Get me out of here: Methods of egg dispersal

- Eggs are removed quickly from the female's body once fertilization has been attempted. Why emphasize "attempted" rather "fertilization"? Simple—fertilization rates are rarely 100%, but all eggs entering the common oviduct will be moved to the outside of the body. The process of egg dispensing is termed oviposition; egg-laying also refers to the passage of eggs from the mother to the environment. The vast majority of insect species practice oviparity, and the mothers are oviparous.

- Though oviparity is the norm for most species, other mechanisms are also employed, and all the alternative approaches depend on significant retention time of the eggs within the mother. Vivipary is a broad term that refers to the retention of eggs in the reproductive tract until egg hatch. More than one mechanism of dispersal falls under the viviparous umbrella, but the net result is basically the same: the appearance of live birth. Larvae rather than eggs leave the mother, although an ovipositor is not always required.

- Vivipary is characterized by significant embryonic development occurring while the eggs are maintained in the parent, as opposed to oviparity, in which most aspects of embryonic development takes place outside the mother's body. With some forms of vivipary, such as viviparity, the mother continues to provide nourishment to the developing embryo and possibly larvae beyond the initial contribution of yolk. With other forms, the mother generally does not provide additional nutriment to the embryos, displaying ovoviviparity.

☀ And now for something completely different: Novelty behaviors

- Sex, and all the physiological and behavioral mechanisms that accompany it, generally does not support homeostatic systems in most animals. Consequently, many aspects have evolved independently, through wildly different paths that often manifest themselves as behavioral adaptations that appear simply bizarre, or at least extremely unique and thus interesting to discuss. With a million species in existence, there are many novel sexual behaviors to choose from.

■ The quest for mating with a female so that she can use the sperm of a male for fertilization is highly competitive. The competition is most severe for males belonging to species in which the female mates more than once and stores sperm prior to fertilization. The phenomenon is called sperm competition. The behaviors displayed by males to compete for sperm usage can be quite bizarre, ranging from a male knocking another male in copula free from a female so that he can mate to secreting juvenilizing hormones along with the spermatozoa in an attempt to suppress the sexual appetite of his lover to forming plugs that seal the vulva and bursa copulatrix.

■ Death is a common outcome of some sexual behaviors. Ovicidal behavior, for example, is at the expense of eggs or larvae of a competitor. In several species of parasitic wasps, if a female detects eggs or larvae from another species as she initiates her stereotypic host evaluation checklist, she uses her ovipositor to puncture the allospecifics repeatedly, with the intent of destroying any competition to her own offspring. In other cases, what the mother does not accomplish, the children do: in larval competition, the juveniles eliminate the competition. In other groups, an adult female delivers the deathblow to the unsuspecting male who will be the father of her progeny. This behavior is termed widow making, and comes in a few versions.

■ With some insects, the traditional roles of male and female are reversed. Sometimes this is manifested during the calling phase: the male releases sex pheromones so that interested females search for him. At the other times, roles are reversed during copulation, when the female mounts the male and then curls her abdomen to find his penis. One of the more impressive examples of role reversal occurs with parental care.

MUSHROOM FARMING (SELF-TEST)

Level 1: Knowledge/Comprehension

1. Define the following terms:
   - (a) direct insemination
   - (b) oviparity
   - (c) calling
   - (d) sex pheromone
   - (e) aedeagus
   - (f) bursa copulatrix

2. Describe the function of the spermatheca.

3. Explain how the mechanism of arrhenotokous parthenogenesis generates male and female embryos.

4. What is the purpose of nuptial gifts?

5. Describe how the following mechanisms of egg dispersal are used to move progeny out of the mother's body: (a) oviparity, (b) viviparity, and (c) ovoviviparity.

6. Compare and contrast insect use of sound versus chemicals for calling mates.

Level 2: Application/Analysis

1. Sex dominates with all groups of insects, but it does come with some disadvantages. Discuss several of the problems with sex and explain how insects have overcome the issues.

2. Male genitalia vary widely in terms of morphology, so much so that species identification can be made based on the external anatomy of the aedeagus. Describe some of the prevailing theories as to why so much variation exists with regard to the structure of the male genitals.

3. Explain how some insects, like honey bees, can control the sex of their offspring.

4. In several insect species, females can mate more than once and have the ability to store sperm. This creates intense competition between males to be the last to mate with a female before she fertilizes eggs. Discuss some of the behavioral and physiological adaptations males employ during the phenomenon of sperm competition.

Level 3: Synthesis/Evaluation

1. What are the evolutionary benefits conferred from the control of sex in a honey bee colony?

2. Asexual reproduction is generally considered an evolutionary dead end. Yet several forms of nonsexual reproduction continue to exist, even flourish, in some insect groups. Why?

3. External fertilization is quite common in many modern-day marine invertebrate species and was used by ancestors of insects. However, freshwater insects do not use this method. Speculate on why they do not.

REFERENCES

Chapman, R. E. 2013. The Insects: Structure and Function. 5th ed. (J. Simpson and A. E. Douglas, eds.). Cambridge University Press, Cambridge, UK.

Gardner, A., and L. Ross. 2011. The evolution of hermaphroditism by an infectious male-derived cell lineage: An inclusive-fitness analysis. *American Naturalist* 178:191-201.

Leather, S. R., and J. Hardie (eds.). 1995. Insect Reproduction. CRC Press, Boca Raton, FL.

Leopold, R. A. 1976. The role of male accessory glands in insect reproduction. *Annual Review of Entomology* 21:199-221.

Lundmark, M., and A. Sura. 2006. Asexuality alone does not explain the success of clonal forms in insects with geographical parthenogenesis. *Hereditas* 143:23-32.

Morrow, E. H., and G. Arnqvist. 2003. Costly traumatic insemination and a female counter-adaptation in bed bugs. *Proceedings* B 10.1098. doi: 10.1098/rspb.2003.2514.

Romoser, W. S., and J. G. Stoffolano Jr. 1998. The Science of Entomology. 4th ed. W.C. Brown/McGraw-Hill Publishing, Dubuque, IA.

Segoli, M., A. R. Harari, J. A. Rosenheim, A. Bouskila, and T. Keasar. 2010. The evolution of polyembryony in parasitoid wasps. *Journal of Evolutionary Biology* 23:1807-1819.

Simonet, G., J. Poels, I. Claeys, T. Van Loy, A. de Loof, and J. Vanden Broeck. 2004. Neuroendocrinological and molecular aspects of insect reproduction. *Journal of Neuroendocrinology* 16:649-659.

Smith, R. L. 1984. Sperm Competition and the Evolution of Animal Mating Systems. Academic Press, New York, NY.

Thornhill, R. 1976. Sexual selection and paternal investment in insects. *American Naturalist* 110:153-163.

Waldbauer, G. 1996. Insects through the Seasons. Harvard University Press, Cambridge, MA.

Wilson, E. O. 1963. The social biology of ants. *Annual Review of Entomology* 8:345-368.

THE ENTOMOLOGIST BOOKSHELF (SUPPLEMENTAL READINGS)

Hurd, H. 1998. Parasite manipulation of insect reproduction: Who benefits? *Parasitology* 116:S13-S21.

Klowden, M. J. 2013. Physiological Systems in Insects. 3rd ed. Academic Press, San Diego, CA.

McMonigle, O. 2013. Keeping the Praying Mantis: Mantodean Captive Biology, Reproduction, and Husbandry. Coachwhip Publications, Greenville, OH.

Rolff, J., and M. T. Siva-Jothy. 2002. Copulation corrupts immunity: A mechanism for a cost of mating in insects. *Proceedings of the National Academy of Sciences (USA)* 99:9916-9918.

Shuker, D., and L. Simmons. 2014. The Evolution of Insect Mating Systems. Oxford University Press, Cambridge, UK.

Simon, J.-C., F. Delmotte, C. Rispe, and T. Crease. 2003. Phylogenetic relationships between parthenogens and their sexual relatives: The possible routes to parthenogenesis in animals. *Biological Journal of the Linnean Society* 79:151-163.

Thornhill, R., and J. Alcock. 1983. The Evolution of Insect Mating Systems. Harvard University Press, Cambridge, MA.

ADDITIONAL RESOURCES

Calling in crickets
http://www.brisbaneinsects.com/brisbane_grasshoppers/MoleCricket.htm.

Hermaphrodite insects fertilise daughters with parasitic sperm
http://blogs.discovermagazine.com/notrocketscience/2011/07/15/hermaphrodite-insects-fertilise-daughters-with-parasitic-sperm/#.VK2UPicnI3g

Insect reproduction
http://cronodon.com/BioTech/Insect_Reproduction.html

Massachusetts cicadas calling
http://www.masscic.org/videos/calling/.

Mating and reproduction of dragonflies
http://www.brisbaneinsects.com/brisbane_insects/Mating.htm

Reproductive system
http://www.cals.ncsu.edu/course/ent425/tutorial/repro.html.

The tiny insect with massive sperm
http://www.newscientist.com/article/dn23585-zoologger-the-tiny-insect-with-the-massive-sperm.html#.VL2jWCeG7V0.

10   You Can Teach an Insect New Tricks

Learning and Memory in Six-Legged Beasts

Many people would likely consider "insect intelligence" a contradiction in terms, viewing insects—when they think of them as anything more than pests—as something like hardwired tiny robots, not adaptive, not intelligent, and certainly not conscious. However, research over the last few decades have shown that a number of well-studied insects are capable of performing amazing intellectual feats, from recognizing individuals to employing a symbolic language in a behavior known as a "bee waggle dance."

Jeremy Hance (2010)*

An insect has a brain? No, that's crazy talk. They have several! (*Gulp*) "Excuse me?" Yes indeed, insects possess a well-developed nervous system, brains included, that permits them to accomplish a number of sophisticated acts perhaps thought reserved only for humans and other primates. They can learn, store memories, communicate, and even make decisions. This really should come as no surprise, considering how immensely successful the class Insecta has been from an evolutionary perspective. Or maybe it is! Some people struggle with the notion that an insect has a brain at all. To say that they have more than one likely conjures up images of science fiction or alien creatures. To then suggest that insects can learn from experiences and then store the information as long- and short-term memories to make better and more informed decisions, well that is just frightening. It is just a matter of time before they figure out how kill us all! Perhaps. Insects are intelligent creatures. In league with humans? Probably no, at least not with all humans, but insects are definitely not mindless robots. Certainly a wide range of insect behaviors is innate, requiring no prior experience to initiate the activities. This appears to be especially critical for dealing with unpredictable variations in the environment.

For many insects, however, learning is essential to major activities of everyday life, including but not limited to feeding, predator avoidance, mate-finding, aggression, and interactions with kin and nonrelatives alike. Learning and memory recall are particularly important to behaviors dependent on time, location, or interactions. Social learning even exists, where a novel behavior can spread rapidly through a colony. The capacity for learning and the types of behaviors that can be expressed are dependent on the genetic composition of a species. Individuals within a group often display variability such that some siblings are considered "smarter" than others, but remedial classes in the world of insects are usually not necessary, because slow learners tend to die quickly! Here in chapter 10, we tackle the range of behaviors displayed by insects

---

*From an interview by the author with Dr. Lars Chittka, titled "Uncovering the intelligence of insects, an interview with Lars Chittka" appearing on Monabay.com (http://news.mongabay.com /2010/0629-hance_chittka.html).

that relate to learning, from complex behaviors and social organization to life history and fitness. An examination of nervous system design, particularly the ganglia (including the brain) responsible for modulating insect behavior, is included, as well as discussions of some of the factors that influence and shape learning in insects, including how these creatures perceive changes in the environment.

Key Concepts

- Darwin and insect behavior
- Ganglionic architecture: New age building designs or the key to insect neurological functions?
- The genetic basis for learning, memory, and innateness
- Born to do it: Innate behaviors
- Learning in insects: There is no remedial class!
- How insects can tell it's raining and other environmental fun facts

Darwin and insect behavior

Many animals beyond just primates are capable of learning in some capacity. Take dogs, for instance. No doubt you have heard of Pavlov's dogs. Ivan Pavlov (1849-1936) was a Russian physiologist who is known primarily for his work with **classical conditioning**, which led to his winning the Nobel Prize for Physiology or Medicine in 1904. The technique is used to train or condition an animal to do or not do a behavior as a result of receiving a reward (positive reinforcement) or punishment (negative reinforcement) whenever the desired outcome is displayed. Pavlov was particularly interested in gastric and salivary function in dogs. Through his well-designed classical conditioning experiments, he demonstrated that dogs could be taught to associate the ringing of a bell with the presentation of food. Once enough training had occurred, Pavlov's dogs would respond to bell ringing by salivating, anticipating a food reward even when none was provided.

It turns out that not every dog breed is necessarily equally suited for this form of conditioned learning, as my beloved bichon frises readily demonstrate: every time a bell rings in the house, they either bark or urinate. The same response occurs when someone knocks at the door, a car drives past, the wind blows, someone in the house

exhales, or during the overpowering sound of silence. Actually what this demonstrates is that Charles Darwin (Chuck to those who don't know him) was right (figure 10.1). Not with regard to bichon frise dogs (no, they likely would have stumped him in terms of certain features of his theory of evolution), but rather his ideas regarding behavior being subject to natural selection and patterns of inheritance. As we will see, many species of insects can be conditioned in a way similar to Pavlov's dogs—not to salivate when a bell tolls, because that's not physically possible, but in a manner appropriate for the type of reward or punishment training they endured. A case will be made (not directly) that no insect species demonstrate intelligence comparable to the bichon frises raised in my home. Thank god! For now, let's turn our attention to the ideas of Darwin, specifically to see the linkage between natural selection and insect behaviors.

Figure 10.1. Portrait of Charles Darwin from the late 1830s. The watercolor portrait originally appeared in *Origins* by Richard Leakey and Roger Lewin and is now available in public domain at http://bit.ly/1KNZLNf.

Darwin was fascinated by animal behavior; his interest was fostered during childhood and reached public notice in his *On the Origin of Species by Means of Natural Selection, Or, the Preservation of Favoured Races in the Struggle for Life*. It was in *Origin of Species* that Darwin first addressed one of the major challenges to his theories of evolution: whether instincts can be acquired and modified through natural selection. He believed they could be and were in fact as much influenced by natural selection as any anatomical feature of an animal. With that said, Darwin was careful to shy away from commenting on the human condition, at least in 1859 (his later writings did tackle the highly controversial ideas of human behavior and natural selection).

Today we know that the instincts referred to by Darwin are **innate behaviors**. Innate behaviors are those that an animal (or any organism) is genetically preprogrammed to do, so that no learning is required to perform the behaviors. One of the key links to entomology is that Darwin referred to several examples of insect innate behaviors, as well as complex interactions (sociality) in *Origin of Species*, helping to establish that insects do indeed display a range of behaviors previously thought to be reserved for higher animals. He went on to suggest that the difference between human behavior and intellect and that of other "higher" animals is a matter of degree, not kind. In other words, humans do not have sole ownership among animals of complex behaviors, the capacity to learn, and perhaps not even to emotions and intellect. His writings also cast light on patterns of inheritance and the evolution of behavioral traits, the latter implying that primitive forms of complex behaviors should be evident in lower life forms. While the contention that insects are a lower life form is debatable, the concept of evolved behavioral complexity has strength when considering how similar insects are to primates: insects display rudimentary to complex forms of innate behaviors, communication, parental care, and social organization, as well as waging war and making slaves of their own kind. Hide the identity of the species, and the examples given may be mistaken entirely for human behaviors. Not all insect behaviors take a backseat to those of primates, as is the case with insects using chemical signaling to communicate, which arguably is the most evolved and sophisticated of all animal groups. It

should be obvious that Darwin was right, that natural selection does apply to the behaviors of insects, humans, and any other organism.

Darwin's intimations regarding animal behavior also contributed to the establishment of entomology as a discipline. His observations and theories regarding insect behavior helped spawn new research, launching several insect species into the role of model organisms for behavioral investigations. The trend has continued through today, with several insect species (e.g., ants, bees, fruit flies) used to examine a number of questions related to behavioral evolutionary theory. We will examine several of those ideas in the coming sections, paying particular attention to common examples of innate and learned behaviors displayed by different insect groups. Before we launch further into the types of behaviors used by insects, the genetic basis of learning and memory, and how these creatures perceive their environment, we need to develop some understanding of the basic design of the insect nervous system and how it operates in relation to sensory perception, integration, and motor output. We begin this journey by examining the architecture of the nervous system.

Ganglionic architecture: New age building designs or the key to insect neurological functions?

Is the design of the insect nervous system the same as that of higher animals? The answer obviously cannot be a straightforward "yes," because we already know that more than one brain is present in an insect. But a flat-out "no" is not accurate either. We therefore rest solidly on "similar, with differences." No doubt some explanation is needed. Insects share in common with humans a central nervous system with an anteriorly located brain (in the head or cranium) serving as the principal region for integration of sensory feedback, and nerve cells, or **neurons,** as the basic functional unit of the entire nervous system. There is also some overlap in the types of chemical messengers (e.g., neurotransmitters) used by insects, humans, and other higher animals in relaying information throughout the body for sensing and responding to environmental (internal and external) changes and coordinating physiological systems. For the most part, that pretty much sums up the common ground. The remaining design features of the insect nervous system deviate substantially from our understanding of the human plan. As a consequence, we need to spend some time exploring the basic layout of the insect nervous system. To do so, we examine the structure of the "main" or "big" brain, as well as the other, or "mini." brains. The insect nervous system is also composed of a series of systems (central, peripheral, and visceral), so we also decipher the components of each and what they actually do for the insect.

Main brain

Before we can fully appreciate the structure of the insect brain, we need to explore the simplest functional units of the nervous system, neurons. This also includes support cells known as **glial cells**. Neurons have a familiar morphology: the main or central structure is the cell body, and extending from opposite or even multiple regions of the cells are **axons** and **dendrites** (figure 10.2). Dendrites function essentially as electrical antennae, receiving incoming electrical and/or chemical messages from sensory pathways. By contrast, axons function as the electrical conduits that transmit the message farther along the pathway, to other neurons, for eventual interpretation and integration, or to effector tissues that generate a response, such as stimulating muscles to contract or targeting specific organ systems to turn on or off, or elevate or decrease activity.

A wide range of responses can be generated depending on the neurons involved in signal transmission. Neuron design permits almost unimaginable speeds in terms of the rate of cell to cell communication that occurs (figure 10.3). From a functional perspective, insect neurons

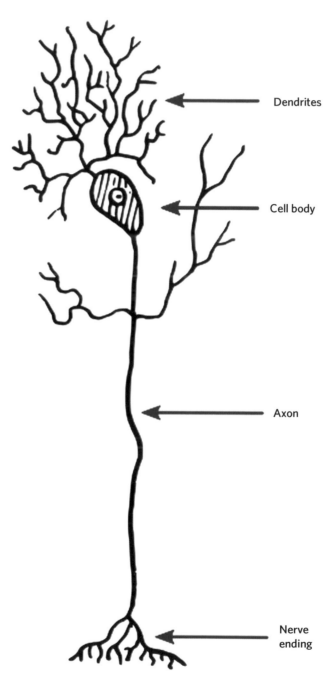

Figure 10.3. The basic anatomy of an insect neuron is nearly identical to that of a human neuron. Illustration by Pearson Scott Foresman in public domain at http://bit.ly/1LW1O1Z.

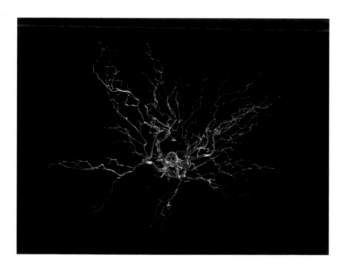

Figure 10.2. An isolated neuron displaying a tremendous number of projections or dendrites. Photo by Nicolas Rougier available at http://bit.ly/1MSGfVQ.

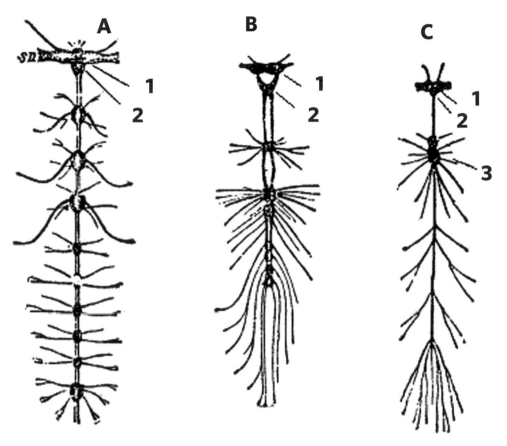

Figure 10.4. Comparison of the nervous systems of three insects: (A) termite (Isoptera), (B) beetle (Coleoptera), and (C) fly (Diptera). The major ganglionic centers are distributed in the head, thorax, and abdomen. Holometabolous insects display a reduction in number of ganglia. Illustrations available in public domain at http://bit.ly/1KZ7jjK, author unknown.

operate in a manner nearly identical to ours. Aiding the function of insect neurons are glial cells. They easily outnumber neurons in the central nervous system, forming folds around axons, including at **synaptic junctions** between neurons. Large clusters of neurons and glial cells, along with all of their processes, form tissues masses known as nerve centers, or **ganglia**. An individual ganglion is composed of thousands of cells cooperatively functioning together to receive, interpret, and respond to stimuli. Ganglia also specialize in the types of stimuli they process and are thus distributed in specific locations in the insect body. It is the assemblage of ganglia and the connections between them that form the "backbone" (how ironic!) of the insect nervous system (figure 10.4).

The brain best illustrates the significance of ganglia to insects. In fact, the insect brain is a ganglion. Correction—the brain represents the fusion of three, possibly four,

large ganglia. The structure is referred to as the **supraesophageal ganglion** (meaning, basically, above the esophagus), or dorsal ganglionic center. In simplest form, the brain is composed of the protocerebrum, deutocerebrum, and tritocerebrum—each region a separate ganglion that fused to form one large nerve center all functionally distinct from each other. Some experts also contend that a fourth ganglion, the **subesophageal ganglion** (below the esophagus), is a functional division of the brain. Why is this view not considered universal? Largely because the subesophageal ganglion is not fused to the others and is physically separated from the "main" brain, often located ventrally in the head capsule or in the anterior prothorax rather than the head. The protocerebrum forms distinct lobes (optic) that function in vision, while other cells in this region are associated with learning, memory, and time keeping (biological clocks).

The protocerebrum is a very complex region of the brain, and consequently several functional areas can be identified. For example, the pars intercerebralis is a mass of cells located on either side of the midline of the brain (body) that contains neurosecretory cells that regulate the activity of several organs throughout the body, including endocrine tissues like the corpora cardiaca and corpora allata. The optic lobes innervate the compound eyes and ocelli. Of particular importance are functional areas called **mushroom bodies** or corpora pedunculata, which are critical to olfactory learning and memory storage (both short-term and long-term). The size of the mushroom bodies is directly related to the complexity of behavior demonstrated by an insect group. Thus, the expectation is that the mushroom bodies are smallest in primitive insects and most well developed in social insects, particularly with the social Hymenoptera. The deutocerebrum is central to interpreting sensory signals emanating from the antennae, as this region of the brain possesses the antennal lobes and motor and mechanoreceptor centers of the antennae. Sensory feedback derived from the labial palps and body wall (e.g., stretch receptors) is also processed in the deutocerebrum. The tritocerebrum is the smallest of the fused ganglia that make up the brain. It occurs as a pair of small lobes positioned below (ventral) the deutocerebrum, and acts primarily as an integration center for signals received from the other regions of the brain. The tritocerebrum also serves as a relay center for sensory signals arriving from the visceral nervous system to the other brain lobes, and from the foregut and labrum to the **frontal** and hypocerebral ganglia.

The supraesophageal ganglion and subesophageal ganglion are physically in contact by two connections (esophageal connections) that permit electrical and chemical transmission between the two major ganglionic centers. In turn, the subesophageal ganglion functions to control feeding behaviors, including movements of all of the mouthparts. It also serves as a relay station for the neurons positioned anywhere else in the body, as it is physically connected to other nerve centers or mini brains via the **ventral nerve cord**. As the name implies, the ventral cord is located ventrally and represents a series of nerve connectives linking each ganglion with one positioned to the anterior and/or posterior. Two connections occur between each ganglion, thus forming a double ganglionic cord along the ventral side of the insect's body (figure 10.5).

Mini brains

So what is the story with the "other" brains? Smaller brains or ganglia occur outside the head. In other words, mini nerve centers are located in each segment of the thorax and abdomen in most insects. Each is a ganglion, although they do display size differences. Those in the thorax are composite ganglia, meaning that they represent multiple smaller ganglia that have fused together into a functional unit. Abdominal ganglia are not composite and therefore smaller. Each ganglion is connected to adjacent ganglia via the double ganglionic cord arrangement discussed earlier. One exception to this rule is the mini brains positioned in the terminal segment of the abdomen. These ganglia typically fuse to form the composite caudal ganglionic center, or just caudal ganglion, which is much larger than other abdominal ganglia, owing to the fact that it innervates the anus, genitals, and sensory structures such as cerci and styli, as well as muscles controlling the ovipositor.

The thoracic and abdominal ganglia also have nerve connections radiating into muscles and organs associated with the specific body segment in which they occur. As a consequence, a mini brain provides local control over effector tissues within a specific thoracic or abdominal segment. For example, the prothoracic ganglion regulates movement of legs located on the prothoracic segment; the mesothoracic ganglion controls the legs on that segment, and so on. To coordinate the movements of all six legs, the supraesophageal and subesophageal ganglion must be involved. Master control is required to ensure that leg movements result in locomotion and not just random, independent movement of each pair of legs.

Additional mini brains are found in the cranium. The frontal ganglion is positioned between the two major ganglionic centers in the head and is connected to the hypocerebral ganglion via the recurrent nerve. Both structures are components of the stomatogastric nervous system (a division of the visceral nervous system) and have roles in regulating appetite and satiety. From an evolutionary perspective, the presence of ganglia in each body segment is considered a primitive condition. More advanced species display a reduction in number of mini

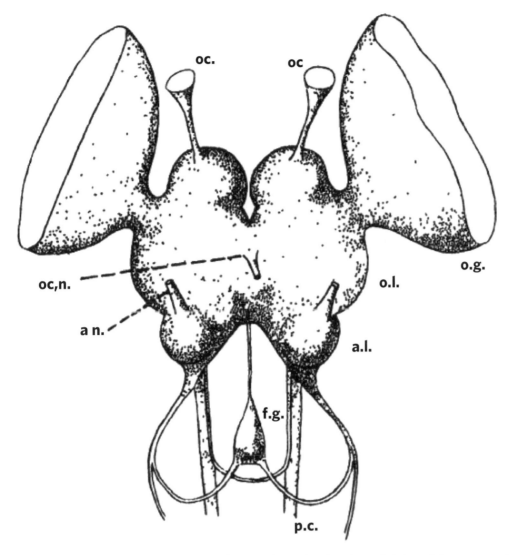

Figure 10.5. Image of an adult locust (Orthoptera) brain showing the optic lobes of the protocerebrum, the brain proper resulting from the fusion of the protocerebrum, deutocerebrum, and tritocerebrum. Diagram by A. D. Imms available in public domain at http://bit.ly/1FC3IaV.

brains, a result of ganglionic fusion into larger nerve centers that regulate effector functions in a particular tagma (i.e., head, thorax, or abdomen). The trend follows the different forms of development discussed in chapter 6 (table 10.1).

The systems

The nervous system of any insect is divided into functional systems, or divisions. Each system has defined roles in terms of physiological mechanisms or tissues they regulate and the sensory signals perceived and interpreted. The three divisions are the central, visceral,

and peripheral nervous systems. Collectively, the three recognized divisions of the insect nervous system provide overall coordination for all physiological systems in the body, permitting an insect to respond to stimuli derived from the internal or external environment within a time frame that maximizes the chances of not only surviving but thriving in terrestrial and aquatic habitats. The central nervous system is the major coordinating unit of the body and is represented by the ventral nerve cord and all the ganglia connected to it. This includes supraesophageal and subesophageal ganglia, and all the mini brains. The visceral nervous system functionally

Table 10.1.

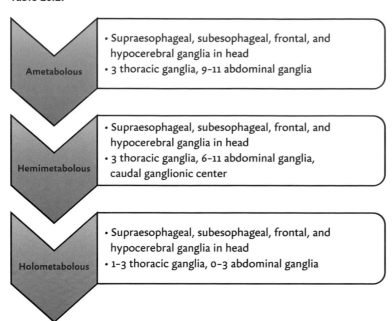

regulates most of the digestive tract, reproductive and most endocrine organs, and the tracheal system, including opening and closing of spiracles. Ganglia and nerves associated with these organ systems make up the visceral nervous system. The peripheral nervous system is predominantly the nerves associated with muscles and that connect to ganglia of the central nervous system, as well as the sensory neurons located in structures sensing the external environment, such as eyes, tympanic membranes, cerci, and styli.

The genetic basis for learning, memory, and innateness

The capacity to perform any behavior, simple or complex, has genetic underpinnings. Of course this implies that insect behaviors are inherited, which is not totally correct. Some behaviors are indeed inheritable, as Darwin predicted over 150 years ago. Those are manifested as innate actions of insects. However, for other types of behaviors, specifically those that are learned, what is learned cannot be passed on from generation to generation via genetic transfer. No, learning must occur anew with each generation. That said, the capacity to learn is based on the DNA composition of the species and indi-

vidual. Most behaviors displayed by insects are innate, meaning that they are genetically predisposed or programmed to do them. Insects are born with the capacity to display the behavior with the first encounter of the appropriate stimulus. This is absolutely essential when the behavior is critical for survival, such as responding to environmental change or an impending threat from a predator. Experience may modify the behavior if the same stimulus is present in the future, but the initial response is genetically hard-wired.

Learned behaviors also require a genetic blueprint for expression of the activity or response to stimuli. However, the behavior itself is learned as a result of experience, trial and error, or even through social learning, in which other members of the hive or colony essentially teach it. The genetic component includes the ability to learn and what can be learned. Learning in insects is subject to evolution via natural selection, but only in instances in which heritable variation in individuals leads to differences in fitness. In other words, a clear adaptive significance is associated with the variation. Variation does occur among individuals, so that the degree of learning or even the length of time needed to learn is not the same for all members of the same

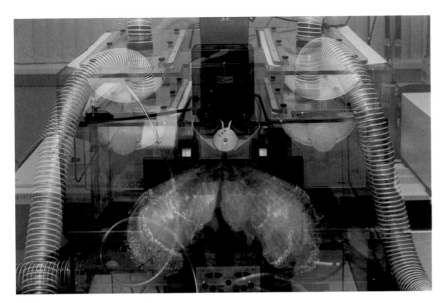

Figure 10.6. Confocal laser microscopy 3-D imaging of the adult brain of the vinegar fly, *Drosophila melanogaster* (Diptera). Image by АркадЗахаров available at http://bit.ly/1TjZEgM.

species. Similarly, insects can only learn certain behaviors in which they have the capacity to perform. For example, an adult insect obviously cannot be trained to drive a car. Not only does it lack the appropriate body size and structures, it also does not retain learned experiences as memories nearly as long as we do.

Short-term and long-term memories are features of insect learning, even to being stored in regions of the brain (e.g., mushroom bodies) that display commonality to the primate condition. However, long-term memories can decay within a matter of minutes in many species. This is not really a shortcoming of the insect condition; rather, it reflects the genetic programming that corresponds to life cycles that typically are very short (a few weeks). In fact, learning is considered beneficial only to species that live long enough to reap the benefits. Which insects are most likely to fall into this category? Those that form colonies and work cooperatively together to collect food, raise their young, and build structures that help resist adverse weather. These characteristics are features of social insects. Social insects are also genetically predisposed for memories created from experiences and learning, in the form of enlarged mushroom bodies. Specifically, social insects are born with bigger mushroom bodies, so the capacity to store short- and long-term

memories is genetically determined. It also appears that well-developed mushroom bodies contain multiple types of neurons, each displaying distinct gene expression to facilitate memory storage. In the case of the vinegar fly, *Drosophila melanogaster* (Diptera: Drosophilidae), three distinct types of neurons are present: α/β, α'/β', and γ. Which of these neurons are involved in each phase of learning and memory is still to be determined, as is how differential gene expression results in short- versus long-term memories (figure 10.6).

Quick check
Where are memories stored in insects?

Born to do it: Innate behaviors

Darwin called them instincts. We refer to them as innate behaviors: behaviors that animals can do in response to stimuli without being taught or having to learn. Are they the same, and does it matter? The answer is "yes" to both questions. Innate behaviors are genetically encoded to occur once the insect is exposed to the appropriate stimulus; no prior experience is required to perform the action or activity. "Instinctive behavior" was used by Darwin to describe the same types of behaviors, but experts in behavior (ethologists) avoid the term "instinctive" altogether. The word implies an acquired aptitude or talent, somewhat similar to a gut instinct or a hunch; neither has a biological basis and therefore they are not particularly useful in discussing insect behavior. Innate also distinguishes inherited from learned behaviors or complex behaviors. The distinction is a matter of genetics: innate behaviors are considered entirely inherited, learned behaviors are not, and complex behaviors represent an intertwining of the two.

Innate behaviors are considered the least complicated of the behaviors insects can demonstrate. While certainly many (e.g., reflexes) are indeed simplistic compared with complex behaviors that are based on experiences or observations, not all innate behaviors are simple or quickly completed actions. Some are certainly detailed responses to environmental or other stimuli. With that said, of course insect innate behaviors appear quite simplistic, even mindless, when held against the complexities of human behaviors. Whenever comparisons are made between what insects do and what other animals do (namely, humans), we tend to do two things: (1) we **anthropomorphize**,* and (2) we make observations and comparisons out of context, particularly with respect to the natural world. What I mean is that innate behaviors make perfect sense for insects. How so? First, consider that parental care is not practiced by the vast majority of insects. Once oviposition is complete, that brings to an end the investment by either parent in the offspring. So learning behaviors by observation or from parental instruction does not happen with insects. This necessitates that juveniles immediately have the capacity to initiate behaviors essential for survival—like feeding, responding to the environment, and recognizing predators—at the moment of egg hatch.

New innate behaviors are required as the insect matures: changing what it eats as it ages, seeking concealment to pupate, or initiating calling, courting, and copulation behaviors. Consequently, the need for innateness does not go away; rather it changes over time.

---

*Anthropomorphism is the practice of assigning human characteristics or purpose to other animals in attempt to explain what we observe. Insects do not have the same sense of purpose or belonging that humans do, so it is perfectly appropriate for them to not display many behaviors that humans are accustomed to.

Some experts contend that the small size of the brain is also a factor that limits the types of behaviors insects can display, including learning and complex behaviors. This argument views the insect brain as being a limited, finite structure. Consequently, keeping it simple (that is, innate), is all that can be expected. This supposition is not really supported by any empirical evidence. Though compact, the brain of an insect contains tens of thousands of neurons, more than sufficient to develop specialized regions dedicated to learning, including latent learning, and short- and long-term memories associated with sensory signals responding to visual, auditory, mechanical, and chemical stimulation. No, the size of the brain does not restrict the types of behaviors (nor does it in humans). To paraphrase the words of Charles Darwin, insect behaviors are not limited in the *kind* of behaviors that can be displayed, only to the *degree* to which they can be fully realized. There is also no evidence that any animal has a saturation point with regard to the brain; new experiences supplement rather than replace the older ones.

How can innate behaviors be distinguished from other types of insect behaviors? Obviously one answer is that they are inherited, and the others are not. True, but you cannot measure inheritability by simply observing the behavior. Innate behavior can be characterized as heritable, stereotypic, intrinsic, consummate, and possibly unchangeable. What this means is that the behaviors are genetically preprogrammed and passed from parents to offspring (heritable); performed the same way each time the same stimulus is encountered (stereotypic), which suggests that innate behaviors are inflexible regardless of experience (unchangeable), although there is evidence that some innate responses can be modified by experience; occur the same way whether the insects are raised in isolation or grouped together (intrinsic); and are expressed in their fullest form based on a particular stimulus with the first encounter (consummate). Does this help us recognize innate behaviors? Probably not. The characters allow you to examine various behaviors in context. The best way to recognize innate behaviors is to observe insects in their natural environment, watching for the features that typify the broad classes of inherited behaviors: **reflexes, kinesis, taxis,** and **fixed action patterns**.

Reflex. A reflex is not really any different for an insect than it is for us. Generally, a simple or subtle stimulus triggers an almost immediate reaction from an isolated body region or appendage, such as retraction of a leg when mechanoreceptors are brushed lightly. Just a few neurons are involved in the response and typically no interpretation/integration occurs in the central nervous system to elicit an effector response. This can be easily demonstrated in a headless cockroach that readily retracts a leg every time it is touched (of course the response no longer occurs once dead). Fly proboscis extension is another reflex triggered by chemosensory receptors on footpads making contact with food, provided the fly is not satiated (as perceived by the hunger center). The righting reflex of cockroaches and beetles is also quite common; if flipped upside down on the dorsum for any reason, a reflex is tripped that causes muscular contractions of the abdomen and/or thorax so that the individual turns itself right side up (figure 10.7). As with any type of innate behavior, an insect is born with the ability to use its reflexes with no learning required. Most are essential to survival, so they need to be able to do it early in the lifespan, usually once emerging from the egg.

Kinesis. Many innate behaviors involve locomotion, or at least a change in body position, in response to a stimulus. In the case of kinesis, there is no directionality in response to the stimulus, and thus the activity of a given individual appears random. Perhaps the most commonly recognized example occurs with pavement ants when their mounds are disturbed. Individual ants respond by running frantically in all directions, eventually seeking shelter from light anywhere it can be found. Similarly, many species of cockroaches, particularly those that reside in human abodes, move slowly and randomly within the area that they call home. If suddenly exposed to light, their activity increases but again with no definite directional purpose. Reaching a place of dark seclusion brings about a renewed sense of calm and slowed down activity.

Taxis. In contrast to kinesis, a taxis is an innate reaction to a stimulus in which the insect changes physical

Figure 10.7. Extension of fly proboscis occurs during feeding and also as part of a reflex when chemoreceptors on footpads detect food. Photo by Richard Bartz available at http://bit.ly/1VhEjdX.

position either toward or away from the stimulus. This response may or may not include actual locomotion. A familiar example is a moth that flies toward an outdoor light at night. It is attracted to the light, or more appropriately shows a positive phototaxis (figure 10.8). A negative response would be to fly or move away from the light stimulus. An anemotaxis is a response to air currents, which are used by insects (positive anemotaxis) to follow a sex pheromone plume when looking for love. Similar responses can occur with just about any type of stimulus imaginable (table 10.1).

Fixed action pattern. Some innate behaviors are more complex than a simple immediate response to a stimulus. Such is the case with fixed action patterns in which

Figure 10.8. This mass of insects attracted to an outdoor light is displaying a positive phototaxis. Photo by CGP Grey available at http://bit.ly/1RbkBuz.

the overall behavior occurs in a series of sequential steps, each with a threshold and/or requirement for an internal readiness before a response is evoked. Often, each subsequent step has a higher threshold than the previous one. Once the behavior has been initiated, it cannot be modified or aborted until completion. Fixed action patterns are commonly associated with various aspects of calling and courting, and also in prey capture of some insects like mantids (Mantodea). It is also reflected in migratory behavior of some locust species, in which an internal releaser is needed before the behavior can be expressed. Similarly, silk spinning behavior of some species of moths depends on hormonal titers of ecdysone and juvenile hormone before the behavior is initiated (figure 10.9).

Learning in insects: There is no remedial class!

When an organism changes its behavior as a result of experience, it is demonstrating learning. Modification of behavior through experience is a very simple definition of learning. Generally, the expectation is that such behavioral changes are adaptive, meaning that insects or other animals are making better choices, or "decisions," through experience. The idea that insects are capable of

decision-making is not universally accepted. In the view of some ethologists, insects are more akin to complex computers that respond to input (stimuli) in relatively predictable ways with every encounter of the same stimulus. To many others (including myself), experience demonstrates that insects not only learn but can also make informed choices (see Fly Spot 10.1 for one perspective). This is not to be confused with cognitive functioning or the complex higher-order behaviors associated with primates, which no insects are capable of doing. Nonetheless, learning does occur in the class Insecta and can be tangibly recorded as acquisition of neuronal representations, specifically in the form of stored memories associated with all sorts of environmental stimuli as well as relationships with kin and nonrelatives. Memory is of course how experiences are stored, and thus both short- and long-term memories are necessary components to learning (figure 10.10).

Two questions need to be addressed with regard to learning in insects. The first, how can you tell whether a behavior is learned versus innate? And the second, more evolutionary, why do insects need to learn? With regard to the latter, I can ask you the same question. Maybe your first response is "Because I want to get an 'A'

Table 10.2. Types of taxis behaviors displayed by insects. In many cases, the taxis can be demonstrated in both the adult and juvenile, in others, more than one developmental stage show the behavior.

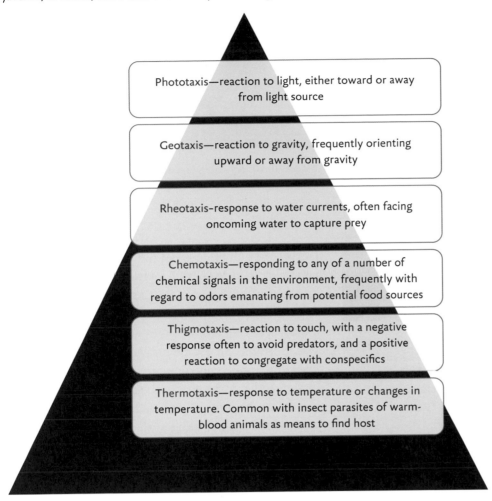

Phototaxis—reaction to light, either toward or away from light source

Geotaxis—reaction to gravity, frequently orienting upward or away from gravity

Rheotaxis-response to water currents, often facing oncoming water to capture prey

Chemotaxis—responding to any of a number of chemical signals in the environment, frequently with regard to odors emanating from potential food sources

Thigmotaxis—reaction to touch, with a negative response often to avoid predators, and a positive reaction to congregate with conspecifics

Thermotaxis—response to temperature or changes in temperature. Common with insect parasites of warm-blood animals as means to find host

in this class." A decent reason, but not one that changes your chances for survival or improves your fitness. On the other hand, once you learned to drive a car, you needed to learn how to navigate the streets of the town or city in which your home is located. As a youngster, memorizing your home phone number and address were two simple learning tasks that you could not possibly have been born with. Making decisions on the types of foods you like or dislike through trial and error, and learning which household products are poisonous, and not to play with matches* all had to be learned. Even recognizing the faces of family members had to be acquired so that memories would form, stored as neuronal repre-

sentations. These same types of needs occur for insects. Insects must learn where they live (e.g., nest location), where and when to find food, and which other organisms to interact with, and how. In this respect, learning is of primary importance for dealing with aspects of time, place, and individuals. Because each of these features can change over time, the response of the insect needs to be adaptable, which, for the most part, cannot occur with innate behaviors. So the capacity to learn is just as relevant to an insect as it is to humans.

Now that we have answered the "why" question, we'll move on to the "how": how can you distinguish learned behavior from innateness? Learned behaviors have certain characteristics that make them distinct from those that are innate. For example, learning is not inherited. Of course, as we have already discussed, the capacity to

---

*As it turns out, you may never have been told about matches if born in the last two decades! In short, don't play with them!

Figure 10.9. Web-spinning behavior of the wax moth, *Galleria mellonella* (Lepidoptera) is an example of a fixed action pattern requiring an internal releaser before this innate behavior can be initiated. Photo by the USDA/NIFA available at http://bit.ly/1KFxESL.

Figure 10.10. An adult female *Nasonia vitripennis* (Hymenoptera) using the ovipositor to drill a hole through the puparium of a potential fly host. Photo by M. E. Clark available at http://bit.ly/1jblHes.

learn *is* heritable. However, experiences of parents cannot be passed down to offspring genetically, not unless the behaviors represent variable traits in a population that are influenced by natural selection. Learned behaviors are extrinsic, in that they do not occur in insects that are isolated from others; they can be modified by experiences and trial and error, so learned behavior is adaptable; an insect may also become more proficient at a task with experience, so learning is progressive; and learned behaviors are not fixed, as with innate behaviors, meaning they can change over time—they are permutable, even when encountering the same stimulus multiple times. The types of behaviors displayed by insects that share these features range from relatively simple learning (**habituation** and **sensitization**), to conditioned responses (**classical conditioning/associative learning, and instrumental learning**), to rudimentary forms of complex behavior (**latent learning**). Each type of learning has a distinct purpose, so the capacity to demonstrate a particular form by an insect is critical to its survival;

lacking the ability does not indicate remedial learning ability but reflects no need for the behavior.

Habituation. This is a simple form of learned behavior in which insects stop responding to a stimulus that previously triggered a reaction. In most cases the response was to assume an escape or avoidance behavior when detecting a noise, vibration, visual cue, chemical signal, or some other stimulus. Some species of stick and leaf insects demonstrate habituation when they no longer respond to leaves rustling in the tree they are residing by mimicking the same movements. If no threat is present, it becomes energetically wasteful to continue to engage in the same behaviors, so they learn to stop.

Sensitization. This is also a simple behavior that is nearly a complete contrast to habituation. It occurs when a repeated stimulus evokes an increasingly progressive response. Often, the insect shows enhanced responsiveness to several related stimuli as a result of the repeated stimulus.

Classical conditioning or *associative learning.* Insects can be trained, or conditioned. This type of learning occurs when two stimuli become linked, causing either stimulus to evoke a particular response. Classical conditioning occurs when an insect is exposed to a stimulus, followed in close succession by a second stimulus, so that eventually the insect displays the same response to either. For example, larvae of *Drosophila melanogaster* (Diptera: Drosophilidae) can be conditioned to chemical odorants using negative reinforcement. The larvae are trained by exposing them to one odorant and, each time it is detected, an electrical current is applied that shocks the larvae. Similar exposure to a second odorant does not result in shock. After three or four training sessions, the larvae are placed in a container in which both odorants are released from opposite sides, and within a short period, the vast majority of larvae crawl toward the odor not associated with the negative reinforcer. Performing the same experiments with the same larvae, but switching the odorant linked to the electrical shock yields identical results, indicating that the larvae did not initially demonstrate habituation (i.e., stop responding to one of the odors). The learned behavior also lasts for several minutes, indicating that even fly larvae store experiences as memories. Associative learning appears to be quite common in a number of insects that rely on foraging for food acquisition, especially honeybees and bumblebees, some parasitic wasps, phytophagous insects, vinegar flies, and even some mantids.

Instrumental learning. This form of learning depends on an insect using memory from past experiences to modify or even improve performance, when subjected to an identical stimulus. In many ways instrumental learning is similar to conditioning in that, if an insect experiences a positive outcome, such as finding food or a mate, as a result of the stimulus, then it is more likely to do the behavior again. Likewise, if the outcome is not favorable, then the individual is expected to not demonstrate the behavior again, or to modify the response from the original.

Latent learning. Latent learning occurs when an insect demonstrates a learned behavior with no apparent reward or punishment. Generally the best examples are associated with social Hymenoptera that live in a nest or hive. Individual foragers learn landmarks or trail identifiers that allow them to remember the location of the colony. While this is indeed an example of learning, it is also difficult to say there are no consequences for not learning where home is; failure to find your way back to the nest most certainly will result in death! (figure 10.11).

Bug bytes

Bee waggle dance

https://www.youtube.com/watch?v=bFDGPgXtK-U

How insects can tell it's raining and other environmental fun facts

Living organisms can respond to stimuli in their environment. Pretty exciting stuff, don't ya think? Maybe not for you, because like most humans, we hide inside when environmental conditions are not to our liking. Not an option for insects. They must deal with the environment 24/7. As a consequence, it is imperative that they can perceive changes in their natural world, no matter how subtle, so that they can adapt if needed. If insects are able to anticipate impending change so that they can prepare the necessary behavioral, physiological, and/or morphological adaptations prior the onset of the environmental disturbance, even better. Insects are very good at "reading" the environment, being finely in tune with seasonal and even aseasonal changes. Their success as a group and as individual species is due in part to their extraordinary ability to sense what is occurring or will occur in the environment, which, in any habitat in any biogeographic location, is composed of a milieu of abiotic and biotic cues that must be interpreted so that appropriate responses can be made. For an insect residing in a temperate zone, the climatic conditions undergo seasonal change, and the signals are not static. Adding to that complexity, individual insects change over time, even if only by the molting associated with growth and development. Thus, the structures used by an insect to perceive its environment are modified simply because of aging. And these are only some of the factors that insects must contend with to live outdoors. Is it any wonder they always seem to be trying to get into our homes?

Considering the complexity of environmental cues that must be detected and interpreted by insects, how do they accomplish what meteorologists with sophisticated equipment often cannot—accurately predict the weather? Broadly speaking, they rely on the nervous system. The insect version of Doppler radar* relies on sensory

---

*Doppler radar is a device used by many meteorologists to inaccurately predict weather conditions. The instrument was apparently designed to be accurate, so it seems to be a matter of who is using it that creates the limitation. In practice, the device works by emitting microwave signals that bounce off a target. The reflection is recorded and compared to the original microwave to determine how the signal has been modified as a result of the target.

Thinking through oviposition

Decisions, decisions. So many decisions that insects must make. Take for example sexual reproduction. Female insects shoulder an enormous burden in the continuation of any given species. A given female is often tasked with finding a mate—not just any lustful male, but one that possess the traits that will give her future offspring the greatest chance of survival and reproducing themselves. During calling, courting, and copulation, it is the female that bears so much of the responsibility of making choices. Even after the sexual deed is complete, decisions by the mother-to-be must be made for when to ovulate, which eggs to fertilize, and where to place them during oviposition. A good mother will not stick her ovipositor just anywhere! No, she evaluates the situation before making a commitment to her final investment in child care.

Perhaps nowhere in the insect world is this better illustrated than with the parasitic wasp *Nasonia vitripennis* (Walker)(Hymenoptera: Pteromalidae). Females of this wasp use the puparial stages* of necrophagous and manure-breeding flies as hosts (**Figure 10.11**). After a period of searching for hosts (which admittedly seems not too difficult since they appear to track the gases emanating from manure and decomposing carrion), a successful foraging session results in discovery of one or more fly puparia. The female begins a series of stereotypic behaviors focused on assessing external host cues (size, shape, texture) that, if they meet a critical threshold, trigger the next level of commitment: drilling a hole through the puparium. Drilling is accomplished by driving the tip of the ovipositor into a weak portion of the puparial exoskeleton, typically at an intersegmental membrane. Lubricant trickles down the outer shaft of the ovipositor while the determined female thrusts and rocks the drill, relying on muscular contractions of the abdomen to

*Once a fly completes larval development, it initiates pupariation, the process of producing a new pupal skin, while separating the old larval skin (a process termed apolysis) from the body but not shedding it. The old larval skin hardens, forming a protective cover surrounding the pupa inside. An additional molt (producing the adult skin) occurs before adult emergence. Thus, multiple life stages are available within puparia for wasps to parasitize.

neurons that readily detect changes in day length (photoperiod), temperature, humidity, atmospheric pressure, and most any other abiotic feature associated with climatic conditions. This is even more impressive when we remember that the body of an insect is essentially isolated from the external environment via the chitinous exoskeleton.

For sensory neurons to function as environmental radar, these delicate cells, or portions of them, must poke through the exoskeleton to be exposed to the harsh world outside. The typical arrangement is for the cuticle to be modified into sensory organs so that environmental cues in the form of chemicals, light, sound (vibration), touch, moisture, or heat make contact directly or indirectly with a sensory neuron or just a portion of a dendrite, which itself is encased by cuticle. Insect senses are highly acute, so detection can occur with minute signals, hence, minimal exposure of neurons is required to trigger a response that will be sent directly to the central nervous system. The basic sense organ just described is a **cuticular sensillum** (plural = **sensilla**) and can be associated with almost any sensation (other than vision) and found in nearly every conceivable location on the body. Generally, sensilla are highly concentrated on mouthparts, antennae, footpads (**pulvilli**), cerci, styli, and even between body regions. Shape and location of sensory structures are modified for specific stimuli, so morphology of the sensory organs can be used to identify chemoreceptors (olfaction), photoreceptors, mechanoreceptors, and sound receptors. Some can even sense changes in temperature, moisture, and air currents.

Chemoreceptors can detect chemical signals that are airborne (olfaction) or that travel through aqueous solutions (taste). Though many of these receptors are posi-

provide the force. This action continues nonstop for a couple of hours; less time if the exoskeleton is relatively thin. The newly created hole is large enough for the ovipositor to be inserted, first just to skim over the surface of the fly inside, searching for eggs and larvae of potential competitors. If present, the female uses her ovipositor to pierce their delicate membranes in an attempt to commit ovicide or larvicide.

The mother's commitment to her progeny continues by plunging the ovipositor into the pupal tissues. Chemosensory receptors on the shaft are used to evaluate the host condition. What can the female detect? The exact chemical signals in the host are still a mystery, although specific amino acids and proteins circulating in the hemolymph do appear to influence the oviposition decisions of female wasps. What is clear is that this process permits females to distinguish between different fly species, host age, physiological condition (i.e., non-diapausing versus diapausing), and whether the fly has been previously parasitized or not. If the answer to the latter is yes, wasps also can distinguish how long ago the host was parasitized, whether wasp larvae are developing, and the gender (or proportion of females to males) of the competitors that are feeding on the fly. Based upon the sensory feedback, the mother will then inject proteinaceous venom into the fly, the exact quantity in response to the perceived conditions of the host, which functions to modify the fly's metabolism to produce nutrients for her offspring and also turns off host body defenses so that circulating hemocytes do not insult eggs or larvae. She then lays a clutch of eggs. Clutch size and gender of the offspring are dependent on the sensory information the mother has processed. How successful is "wasp mom" in making decisions? She's good. Real good! Females generally lay a clutch size that will complete development and emerge as adults; usually less than 10% of the total egg investment does not reach adulthood. Obviously other factors may come into play that the mother cannot predict, such as later-arriving parasites or aseasonal changes in the weather. However, when conditions are favorable, females of this wasp are amazingly successful at evaluating potential fly hosts and making decisions to maximize reproductive output. Even with insects, mom does know best!

tioned on the mouthparts and are associated with ingestion of food, chemoreceptors are also located on antennae, as well as most any other region of the body, for detecting pheromones and an array of other chemical cues. Thus, insects can "smell" or "taste" with multiple body parts. The basic arrangement of a chemoreceptor is that of a cuticular sensillum with one or more pores located along the shaft. Airborne chemicals become trapped in the pores, essentially concentrating the signals on or in the receptor. Dendrites from sensory neurons positioned in the epidermis extend through the cuticle and into the tip of the sensillum shaft. It is here that the chemical signal can bind to the neuron to elicit an excitation of the cell membrane—that is, provided that enough of the cue is present to exceed a critical threshold level that triggers the nerve cell to respond (figure 10.12). In reality, chemoreceptors exist in many morphological forms, including as hairs, pegs, plates, and pits. Despite what they look like or where located, the basic functional design is nearly identical to those described for cuticular sensilla.

Mechanoreceptors show a similar appearance to those used for olfaction and taste but differ in that the cuticular shaft (also called a hair or **seta**) does not have a dendrite running through it. Instead, one or more sensory neurons are located at the base of the hair within the epidermis. Any type of bending or touching of the seta applies pressure to the sensory neuron, which in turn triggers generation of an action potential, again provided that a threshold is surpassed. Mechanosensory organs are termed **trichoid sensilla** and are useful for perceiving tactile stimulation. Hair plate sensilla and **campaniform sensilla** are different forms of mechanoreceptors that can detect changes in body position

Figure 10.11. Hornets and other social insects rely on landmarks to learn where the nest or colony is located. Photo by Bernard Dupont available at http://bit.ly/27A20So.

(**proprioception**); they are positioned in different locations on the body, which constitutes different sensilla morphologies (figure 10.13).

All adult insects possess **Johnston's organs**, a type of proprioceptor that detects movement of antennal flagella. Changes in antennal position can be attributed to a multitude of stimuli, and consequently these proprioceptors may record such features as flight speed, body position during flight or traversing through or across water, or sound detection. Even sound detection relies on mechanoreceptors in the form of **chordotonal sensilla** (vibration) and **tympana**. Tympana are cuticular membranes that can detect vibrations. In some cases, the membranes work in conjunction with chordotonal sensilla by covering the receptors; vibrations of the membranes can in turn stimulate the sensilla. Often many chordotonal sensilla are present, each with different vibrational sensitivities, thus permitting the insect to detect a wider range of sounds than when only one type of sound receptor is present. Tympanic membranes also can create sound through alternating muscular contractions that change the tension of the membranes. Such sound producing

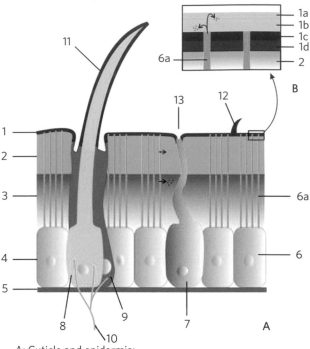

A: Cuticle and epidermis;
B: Detail of the epicuticle.
  1: Epicuticle
    1a: Cement
    1b: Wax layer
    1c: Outer epicuticle
    1d: Inner epicuticle
  2+3: Procuticle
  2: Exocuticle
  3: Endocuticle
  4: Epidermal epithelium
  5: Basement membrane
  6: Epidermal cell
  6a: Pore Canal
  7: Glandular cell
  8: Tricogen cell
  9: Tormogen cell
  10: Nerve ending
  11: Sensory hair (sensillum)
  12: Seta
  13: Glandular pore

Figure 10.12. Basic arrangement of a cuticular sensillum in relation to the layers of the exoskeleton. The sensory organ is derived from epithelial cells in the epidermis. Cartoon by Xvazquez available at http://bit.ly/1FD3FLZ.

structures are most frequently found on the abdominal sterna (Hemiptera) and on the legs (Orthoptera).

Thermoreception appears to depend on peg-in-pit **coeloconic sensilla**, located most commonly on legs or antennae. These receptors can detect air temperature

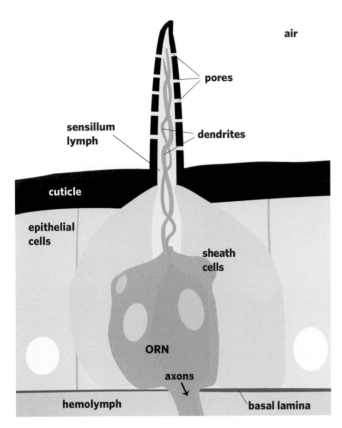

Figure 10.13. Schematic of a basiconic sensillum, a type of chemoreceptor commonly found on the antennae of flies. Cartoon by AlphaPsi available at http://bit.ly/1iWae2q.

changes, radiant heat, and also respond to environmental temperatures that deviate from a set point. There is no evidence that insects have similar receptors for evaluating body temperature, which is expected since all are poikilothermic. Hygroreception, or the detection of moisture in the environment, also occurs with some insects. The location of the hygroreceptors is believed to be on antennae, maxillary palps, and along the exoskeleton, but structural details of the sensory organs are poorly understood. There is some evidence that humidity can be perceived using the same receptors that detect temperature changes.

Vision and light detection depends on photoreceptors. The receptors are located in the compound eye, ocelli, stemmata, along the exoskeleton, and, in some species or specific developmental stages, light is directly perceived by neurons in the brain. An array of different photoreceptors permits detection of light (as opposed to darkness), objects, movement, and changes in light intensity, as well as photoperiod. Some insects are capable of perceiving polarized light and still others have color vision. For most adult insects, the largest concentration of photoreceptors is associated with the compound eyes, which are usually positioned on either side of the head (figure 10.14). Structurally compounds eyes are composed of thousands of photoreceptor cells and light sensory units referred to as **ommatidia** (singular = **ommatidium**). Each ommatidium has an outer cornea positioned over a longitudinal crystalline cone that is surrounded by pigment cells. The pigment cells contain a variety of pigments that can capture light of different wavelengths. Underneath each ommatidium is a sensory receptor, which can activate multiple neurons simultaneously.

Functionally, light is focused by the cornea through the crystalline cone to the pigment cells, which in turn undergo cellular changes based on the intensity of the light stimulus and wavelength (figure 10.15). Those changes are transduced to the sensory neurons that relay the information directly to the brain. Each ommatidium operates independently of the others; the brain receives sensory information from thousands of ommatidia, generating a composite image. Compared to those of humans, who rely on binocular vision, insects' compound eyes do not produce high-quality, crisp, clear images. However, the pigment cells recover quickly with each round of excitation evoked by light stimuli,

Figure 10.14. Compound eyes of insects are composed of thousands of individual facets, or ommatidia, that give the structures their distinctive look. Photo by Thomas Shahan available at http://bit.ly/1VlXLAN.

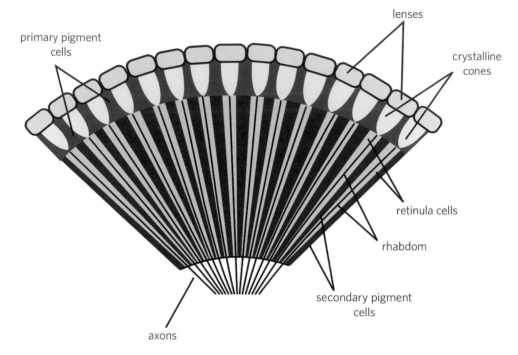

primary pigment
cells

lenses

crystalline
cones

retinula cells

rhabdom

secondary pigment
cells

axons

Figure 10.15. Cross section of an insect compound eye. Cartoon by Bugboy52.40 available at http://bit.ly/1VlYbaq.

permitting insects to have exceptional flicker vision. In other words, they can easily detect slow-moving objects in their environment. This is especially handy for recognizing stationary potential mates hiding out on vegetation or detecting predators that are stalking their insect prey! (Insect color vision: https://www.youtube.com/watch?v=MowLAMeX7x0.)

CHAPTER REVIEW

☀ Darwin and insect behavior

- In *Origin of Species,* Darwin first addressed one of the major challenges to his theories on evolution: whether instincts can be acquired and modified through natural selection. He believed they could be and were in fact as much influenced by natural selection as any anatomical feature of an animal. Today we know that the instincts referred to by Darwin are innate behaviors.

- One of the key links to entomology is that Darwin referred to several examples of insect innate behaviors, as well as complex interactions (sociality) in *Origin of Species,* helping to establish that insects do indeed display a range of behaviors previously thought to be reserved for higher animals. He went on to suggest that the difference between human behavior and intellect and that of other "higher" animals is a matter of degree, not kind.

- Darwin's writings also cast light on patterns of inheritance and the evolution of behavioral traits, the latter implying that primitive forms of complex behaviors should be evident in lower life forms. While the contention that insects are a lower life form is debatable, the concept of evolved behavioral complexity has strength when considering how similar insects are to primates: insects display rudimentary to complex forms of innate behaviors, communication, parental care, and social organization, as well as waging war and making slaves of their own kind.

- Darwin's intimations regarding animal behavior also contributed to the establishment of entomology as a discipline. His observations and theories regarding insect behavior helped spawn new research, launching several insect species into the role of model organisms for behavioral investigations. The trend has continued through today, with several insect species used to examine a number of questions related to behavioral evolutionary theory.

* **Ganglionic architecture: New age building designs or the key to insect neurological functions?**
  - Insects share in common with humans a central nervous system with an anteriorly located brain serving as the principal region for integration of sensory feedback, and nerve cells, or neurons, as the basic functional unit of the entire nervous system. There is also some overlap in the types of chemical messengers used by insects, humans, and other higher animals in relaying information throughout the body for sensing and responding to environmental changes and coordinating physiological systems. The remaining design features of the insect nervous system deviate substantially from our understanding of the human plan.
  - The brain best illustrates the significance of ganglia to insects. In fact, the insect brain is a ganglion. Correction—the brain represents the fusion of three, possibly four large ganglia. The structure is referred to as the supraesophageal ganglion, or dorsal ganglionic center. In simplest form, the brain is composed of the protocerebrum, deutocerebrum, and tritocerebrum—each region a separate ganglion that fused to form one large nerve center—all functionally distinct from each other.
  - The supraesophageal ganglion and subesophageal ganglion are physically in contact by two connections that permit electrical and chemical transmission between the two major ganglionic centers. In turn, the subesophageal ganglion functions to control feeding behaviors, including movements of all of the mouthparts. It also serves as a relay station for the neurons positioned anywhere else in the body, as it is physically connected to other nerve centers or mini brains via the ventral nerve cord. The ventral cord is located ventrally and represents a series of nerve connectives linking each ganglion with one positioned to the anterior and/or posterior. Two connections occur between each ganglion, thus forming a double ganglionic cord along the ventral side of the insect's body.
  - Smaller brains or ganglia occur outside the head. In other words, mini nerve centers are located in each segment of the thorax and abdomen in most insects. Each is a ganglion, although they do display size differences. Those in the thorax are composite ganglia, meaning that they represent multiple smaller ganglia that have fused together into a functional unit. Abdominal ganglia are not composite and therefore smaller. Each ganglion is connected to adjacent ganglia via the double ganglionic cord arrangement discussed earlier.
  - The nervous system of any insect is divided into functional systems, or divisions. Each system has defined roles in terms of physiological mechanisms or tissues they regulate and the sensory signals perceived and interpreted. The three divisions are the central, visceral, and peripheral nervous systems. Collectively, the three recognized divisions of the insect nervous system provide overall coordination for all physiological systems in the body, permitting an insect to respond to stimuli derived from the internal or external environment, within a time frame that maximizes the chances of not only surviving, but thriving in terrestrial and aquatic habitats.

* **The genetic basis for learning, memory, and innateness**
  - The capacity to perform any behavior, simple or complex, has genetic underpinnings. Of course this implies that insect behaviors are inherited, which is not totally correct. Some behaviors are indeed inheritable, as Darwin predicted over 150 years ago. Those are manifested as innate actions of insects. However, for other types of behaviors, specifically those that are learned, what is learned cannot be passed on from generation to generation via genetic transfer. No, learning must occur anew with each generation. That said, the capacity to learn is based on the DNA composition of the species and individual.
  - Learned behaviors also require a genetic blueprint for expression of the activity or response to stimuli. However, the behavior itself is learned as

a result of experience, trial and error, or even through social learning, in which other members of the hive or colony essentially teach it. The genetic component includes the ability to learn and what can be learned. Learning in insects is subject to evolution via natural selection, but only in instances in which heritable variation in individuals leads to differences in fitness.

- Short-term and long-term memories are features of insect learning, even to being stored in regions of the brain that display commonality to the primate condition. However, long-term memories can decay within a matter of minutes in many species. This is not really a shortcoming of the insect condition; rather, it reflects the genetic programming that corresponds to life cycles that typically are very short. In fact, learning is considered beneficial only to species that live long enough to reap the benefits.

Born to do it: Innate behaviors

- Innate behaviors are considered the least complicated of the behaviors insects can demonstrate. While certainly many are indeed simplistic compared with complex behaviors that are based on experiences or observations, not all innate behaviors are simple or quickly completed actions. Some are certainly detailed responses to environmental or other stimuli. With that said, of course insect innate behaviors appear quite simplistic, even mindless, when held against the complexities of human behaviors.

- Once oviposition is complete, that brings to an end the investment by either parent in the offspring. So learning behaviors by observation or from parental instruction does not happen with insects. This necessitates that juveniles immediately have the capacity to initiate behaviors essential for survival—like feeding, responding to the environment, and recognizing predators—at the moment of egg hatch. New innate behaviors are required as the insect matures: changing what it eats as it ages, seeking concealment to pupate, or initiating calling, courting, and copulation behaviors. Consequently, the need for innateness does not go away; rather, it changes over time.

- Innate behavior can be characterized as heritable, stereotypic, intrinsic, consummate, and possibly unchangeable. What this means is that the behaviors are genetically preprogrammed and passed from parents to offspring (heritable); performed the same way each time the same stimulus is encountered (stereotypic), which suggests that innate behaviors are inflexible regardless of experience (unchangeable), although there is evidence that some innate responses can be modified by experience; occur the same way whether the insects are raised in isolation or grouped together (intrinsic); and are expressed in their fullest form based on a particular stimulus with the first encounter (consummate).

- The best way to recognize innate behaviors is to observe insects in their natural environment, watching for the features that typify the broad classes of inherited behaviors: reflexes, kinesis, taxis, and fixed action patterns.

Learning in insects: There is no remedial class!

- When an organism changes its behavior as a result of experience, it is demonstrating learning. Modification of behavior through experience is a very simple definition of learning. Generally, the expectation is that such behavioral changes are adaptive, meaning that insects or other animals are making better choices or "decisions" through experience. The idea that insects are capable of decision-making is not universally accepted. In the view of some ethologists, insects are more akin to complex computers that respond to input (stimuli) in relatively predictable ways with every encounter of the same stimulus. To many others, experience demonstrates that insects not only learn but can also make informed choices.

- Insects must learn where they live, where and when to find food, and which other organisms to interact with, and how. In this respect, learning is of primary importance for dealing with aspects of time, place, and individuals. Because each of these features can change over time, the response of the insect needs to be adaptable, which, for the most part, cannot occur with innate behaviors.

So the capacity to learn is just as relevant to an insect as it is to humans.

- Learned behaviors have certain characteristics that make them distinct from those that are innate. For example, learning is not inherited. Of course, as we have already discussed, the capacity to learn is heritable. However, experiences of parents cannot be passed down to offspring genetically, not unless the behaviors represent variable traits in a population that are influenced by natural selection. Learned behaviors are extrinsic, in that they do not occur in insects that are isolated from others; they can be modified by experiences and trial and error, so learned behavior is adaptable; an insect may also become more proficient at a task with experience (progressive); and the behaviors are not fixed, as with innate behaviors, meaning they can change over time—they are permutable, even when encountering the same stimulus multiple times.

- The types of behaviors displayed by insects that share these features range from relatively simple learning (habituation and sensitization), to conditioned responses (classical conditioning/ associative learning and instrumental learning), to rudimentary forms of complex behavior (latent behavior). Each type of learning has a distinct purpose, so the capacity to demonstrate a particular form by an insect is critical to its survival; lacking the ability does not indicate remedial learning ability but reflects no need for the behavior.

How insects can tell it's raining and other environmental fun facts

- Insects are very good at "reading" the environment, being finely in tune with seasonal and even aseasonal changes. Their success as a group and as individual species is due in part to their extraordinary ability to sense what is occurring or will occur in the environment, which, in any habitat in any biogeographic location, is composed of a milieu of abiotic and biotic cues that must be interpreted so that appropriate responses can be made. For an insect residing in a temperate zone, the climatic conditions undergo sea-sonal change, and the signals are not static. Adding to that complexity, individual insects change over time, even if only by the molting associated with growth and development. Thus, the structures used by an insect to perceive its environment are modified simply because of aging.

- The typical arrangement is for the cuticle to be modified into sensory organs so that environmental cues in the form of chemicals, light, sound (vibration), touch, moisture, or heat make contact directly or indirectly with a sensory neuron or just a portion of a dendrite, which itself is encased by cuticle. Insect senses are highly acute, so detection can occur with minute signals, hence, minimal exposure of neurons is required to trigger a response that will be sent directly to the central nervous system. The basic sense organ just described is a cuticular sensillum and can be associated with almost any sensation (other than vision) and found in nearly every conceivable location on the body. Generally, sensilla are highly concentrated on mouthparts, antennae, footpads, cerci, styli, and even between body regions. Shape and location of sensory structures are modified for specific stimuli, so morphology of the sensory organs can be used to identify chemoreceptors, photoreceptors, mechanoreceptors, and sound receptors. Some can even sense changes in temperature, moisture, and air currents.

- Chemoreceptors can detect chemical signals that are airborne or that travel through aqueous solutions. Though many of these receptors are positioned on the mouthparts and are associated with ingestion of food, chemoreceptors are also located on antennae, as well as most any other region of the body, for detecting pheromones and an array of other chemical cues. Thus, insects can "smell" or "taste" with multiple body parts. The basic arrangement of a chemoreceptor is that of a cuticular sensillum with one or more pores located along the shaft.

- Mechanoreceptors show a similar appearance to those used for olfaction and taste but differ in that the cuticular shaft does not have a dendrite running through it. Instead, one or more sensory neurons are located at the base of the hair within the epidermis. Any type of bending or touching of the seta applies pressure to the sensory neuron, which in turn triggers generation of an action potential, again provided that a threshold is surpassed.

■ All adult insects possess Johnston's organs, a type of proprioceptor that detects movement of antennal flagella. Changes in antennal position can be attributed to a multitude of stimuli, and consequently these proprioceptors may record such features as flight speed, body position during flight or traversing through or across water, or sound detection. Even sound detection relies on mechanoreceptors in the form of chordotonal sensilla and tympana. Tympana are cuticular membranes that can detect vibrations. In some cases, the membranes work in conjunction with chordotonal sensilla by covering the receptors; vibrations of the membranes can in turn stimulate the sensilla.

■ Vision and light detection depends on photoreceptors. The receptors are located in the compound eye, ocelli, stemmata, along the exoskeleton, and, in some species or specific developmental stages, light is directly perceived by neurons in the brain. An array of different photoreceptors permits detection of light, objects, movement, and changes in light intensity, as well as photoperiod. Some insects are capable of perceiving polarized light and still others have color vision. For most adult insects, the largest concentration of photoreceptors is associated with the compound eyes, which are usually positioned on either side of the head.

MUSHROOM FARMING (SELF-TEST)

Level 1: Knowledge/Comprehension

1. Define the following terms:
   (a) innate behavior   (d) olfactory receptor
   (b) learned behavior   (e) ventral nerve cord
   (c) ganglionic center   (f) taxis

2. Describe the basic architecture of the insect brain.

3. Where are memories formed?

4. Describe the characteristics of innate and learned behaviors.

5. Explain how cuticular sensilla "sense" a chemical signal.

6. Describe the functions of the protocerebrum, deutocerebrum, and tritocerebrum.

Level 2: Application/Analysis

1. Explain the influence of Charles Darwin on insect behavior and entomology.

2. Describe how an insect can detect a vibrational noise.

3. Make a diagram comparing the morphology of a chemoreceptor used for olfaction with that of a trichoid sensilla used in mechanoreception.

4. Detail how associative learning likely operates in a foraging honeybee visiting flowers.

Level 3: Synthesis/Evaluation

1. Speculate on the effects on leg movements if the supraesophageal ganglion is removed from a cockroach.

2. If the same experiment is performed, but instead of head removal, the ventral nerve cord is ablated or cut behind the subesophageal ganglion, what would be the expected effect on leg movements? On feeding behavior?

3. Insects rely on photoreceptors to detect changes in the photoperiod, which in turn can program internal or biological clocks within the insects. Speculate on the biological clocks, in terms of what they are and where they are located, and also how photoperiodic cues can influence their functioning.

REFERENCES

Boakes, R. 1984. From Darwin to Behaviourism: Psychology and the Minds of Animals. Cambridge University Press, Cambridge, UK.

Chapman, R. F. 1998. The Insects: Structure and Function. 4th ed. Cambridge University Press, Cambridge, UK.

Darwin, C. 1909. The Origin of Species. Harvard Classics edition (C. W. Eliot, ed.). P. F. Collier & Son Company, New York, NY.

Dukas, R. 2007. Evolutionary biology of insect learning. *Annual Review of Entomology* 53:145-160.

Fahrbach, S. E. 2006. Structure of the mushroom bodies of the insect brain. *Annual Review of Entomology* 51:209-232.

Giurfa, M. 2007. Behavioral and neural analysis of associative learning in the honeybee: A taste from the magic well. *Journal of Comparative Physiology A* 183:737-744.

Heisenberg, M. 1998. What do the mushroom bodies do for the insect brain? An introduction. *Learning & Memory* 5:1-10.

Homberg, U., T. A. Christensen, and J. G. Hildebrand. 1989. The structure and function of the deuterocerebrum in insects. *Annual Review of Entomology* 34:477-501.

Howse, P. E. 1975. Brain structure and behaviour in insects. *Annual Review of Entomology* 20:359-379.

Isingrini, M., A. Lenoir, and P. Jaisson. 1985. Preimaginal learning as a basis of colony-brood recognition in the ant *Cataglyphis cursor. Proceedings of the National Academy of Sciences (USA)* 82:8545-8547.

Ito, K., K. Shinomiya, M. Ito, J. D. Armstrong, G. Boyan, V. Hartenstein, S. Harzsch, et al. 2014. A systematic nomenclature for the insect brain. *Neuron* 81:755-765.

Leadbeater, E., and L. Chittka. 2007. Social learning in insects—From miniature brains to consensus building. *Current Biology* 17:R703-R713.

Meller, V. H., and R. L. Davis. 1996. Biochemistry of insect learning: Lessons from bees and flies. *Insect Biochemistry and Molecular Biology* 26:327-335.

Meyer, J. R. 2006. Elements of behavior. http://cals.ncsu.edu /course/ent425/tutorial/Behavior/. Accessed February 8, 2015.

Papaj, D. R., and A. C. Lewis. 1993. Insect Learning: Ecological and Evolutionary Perspectives. Chapman and Hall, New York, NY.

Raubenheimer, D., and D. Tucker. 1997. Associate learning by locusts: Pairing of visual cues with consumption of protein and carbohydrate. *Animal Behaviour* 54:1449-1459.

Staddon, J. E. R. 1983. Adaptive Behavior and Learning. Cambridge University Press, Cambridge, MA.

Stopfer, M. 2014. Central processing in mushroom bodies. *Current Opinion in Insect Science* 6:99-103.

THE ENTOMOLOGIST BOOKSHELF (SUPPLEMENTAL READINGS)

Alcock, J. 2013. Animal Behavior: An Evolutionary Approach. 10th ed. Sinauer Associates, New York, NY.

Davis, R. L., and K.-A. Han. 1996. Neuroanatomy: Mushrooming mushroom bodies. *Current Biology* 6:146-148.

Heinrich, B. 2004. Bumblebee Economics. Rev. ed. Harvard University Press, Cambridge, MA.

Matthews, R. W., and J. R. Matthews. 2009. Insect Behavior. 2nd ed. Springer Sciences, New York, NY.

Mosqueiro, T. S., and R. Huerta. 2014. Computational models to understand decision making and pattern recognition in the insect brain. *Current Opinion in Insect Science* 6:80-85.

Snodgrass, R. E. 1993. Principles of Insect Morphology. Cornell University Press, Ithaca, NY.

Von Frisch, K. 1967. The Dance Language and Orientation of Bees. Belknap Press, Cambridge, MA.

Wigglesworth, V. B. 1972. The Principles of Insect Physiology. 7th ed. Chapman and Hall Publishers, London, UK.

Wilson, E. O. 2000. Sociobiology: The New Synthesis. 25th anniv. ed. Belknap Press, Cambridge, MA.

ADDITIONAL RESOURCES

Animal Behaviour
http://www.journals.elsevier.com/animal-behaviour/

Atlas of *Drosophila* brain
http://flybrain.neurobio.arizona.edu

Digital atlas of honeybee brain
http://www.neurobiologie.fu-berlin.de/beebrain/

Insect brain and nervous system
http://cronodon.com/BioTech/insect_nervous_systems .html

Insect nervous system
http://www.cals.ncsu.edu/course/ent425/tutorial/nerves .html

Insect vision
https://www.youtube.com/watch?v=TU6bgQnTil8

Interview with E. O. Wilson
http://yhoo.it/1JynoaL

Journal of Insect Behavior
http://link.springer.com/journal/10905

Tracking neurons in mushroom bodies
http://yhoo.it/1LXmXZF

View from bug-eyed camera
https://www.youtube.com/watch?v=C1mvq_3gVVE

11 Instant Messaging in the Insect World

Communication with Kin and Non-Kin

Be a two-way, not a one-way, communicator.

Dr. Phil McGraw
Dr. Phil's 6 Rules for Talking and Listening*

Insects are not selfish creatures. Well, most are not. When one member of a species locates a great food source, generally that individual is more than willing to share the information with his or her kin. This is also true when a nesting or an overwintering site is discovered, and sometimes even when the opposite sex has been found. Insects communicate bad news as well good. So a cry goes out in some form to let others know that predators are on the way, or are already here, which may be abundantly evident if the alarm is only sounded while the body is being consumed. Sharing such valuable information correctly implies that insects are animals that communicate with one another. Insects, like all advanced animals, need to be able to "talk" to members of their species, as well as to those of other species. How do they accomplish this?

Like you, insects communicate with members of the same species using a multitude of mechanisms, including visual, chemical, auditory, and even tactile (touch) forms. Unlike you, insects do not restrict "whom" they talk to, as their messages are conveyed to other insect species, as well as to other organisms. Would you believe that communication occurs with plants, bacteria, and even fungi? Well, it does. The messages used for interkingdom signaling are chemical cues. In fact, chemical communication within and between insect species is perhaps the most sophisticated and refined among all animal groups, reaching a crescendo with social insects. Important to the survival of all insects is the fact that the ability to communicate is not learned, as it is in humans. Rather, all forms of "talking" used by insects represent innate behaviors, meaning that each individual is genetically predisposed to engage in communication. Here we explore insect communication by examining the major forms used, discussing how and when each is employed, and briefly explaining the sensory systems necessary for two-way communication—that is, signaling and receiving. A discussion

---

*Dr. Phil of Oprah Winfrey fame and later his own show provides wisdom for effective communication; presumably it's meant exclusively for humans but somehow seems apropos for insects as well (http://www.oprah.com/relationships/Dr-Phils-Six-Rules-of-Talking-and-Listening).

of communication in social insects, particularly Hymenoptera, will also dispel the rumors that "gossip" originated in insect communities.

Key Concepts

- Communication is the key to every successful relationship
- The basics of insect communication
- Visual displays, camouflage, and mimicry
- Insect phonics: Auditory messages in the insect world
- Whiff this! Chemical communication
- Chemical dependency: On being a social insect
- Interspecies chemical communication

Communication is the key to every successful relationship

Insects *can* communicate and *need to* communicate. Both of these ideas may be shocking if you are still carrying the view that insects are mindless robots, lacking every appreciable attribute humans like to believe are ours alone. However, like you, insects have an absolute need to convey information to other members of the same species, sometimes relatives, sometimes not. In fact, the ability to communicate is not restricted to just **conspecifics,** or intraspecific communication. No, several species share information with other species (**allospecifics**), relying on interspecific communication to send or receive messages to and from other species of insects as well as to other organisms. Those other organisms include vertebrate animals, plants, bacteria, and fungi. It is entirely plausible that some creatures have been left off the list and may soon be discovered to be insect whisperers as well. I await with anticipation the day it is announced that cockroaches really do speak to penguins while they are dreaming* (figure 11.1).

Interkingdom signaling is quite common in the world of insects, but is not nearly so with humans. Yes, we can say that people talk to their pets, but do *they* in turn listen? Or do pets display nothing more than a conditioned response to a positive or negative reinforcer? When relying on nonverbal indicators, the only way to really tell whether the other animal is listening is through a change in behavior—that is, if the receiver displays an adaptive response to the communication signal, regardless in what form it is presented (i.e., acoustic, visual, chemical or tactile). My view is tainted by less-than-dynamic pet dogs that occupy my home as feedable rugs; listening is apparently not inherent with them, and so far, it is not a learned trait either! Other animals, like my teenage daughter, have an advanced capacity for communication, as evidenced by her ability to symbolically channel through nearly 1,100 texts in a thirty-day window![†] There is no evidence, however, that the recipients of those messages are engaged in true communication, for as far as I can tell, no perceivable adaptive behaviors have ever been displayed.

Effective communication is the key to any successful relationship. Any marriage guru would concur. Would such counselors expect that this adage applies to insects as well? Quite likely, no. Nor would they consider whether two-way communication is needed with creatures with six legs. The oversight is disturbing. By now you have likely come to accept that whatever insects do, they do well. So communication within species or outside their "friends and family network"[‡] is done with the same exceptional efficiency that is observed with food acquisition, digestion, reproduction, body defenses, or just about any other feature of insect natural history.

The question that now arises is, what do insects talk about? One can assume, just important stuff. Playful banter, telling jokes, or engaging in gossip probably are uniquely human, as they would appear to serve no purpose in the existence of insects. Of course, there is currently no empirical means to test the hypothesis that insects do not tell jokes. As mentioned earlier, the only way for us to recognize that insects do communicate is

---

*A reference to the glorious world of *Bloom County*, a cartoon strip (1980–1989) created by Berkeley Breathed and starring the affable Opus, a pudgy and often dim-witted penguin. During his sleep, the cockroach Milquetoast would frequently fill Opus's dreams with subliminal messages.

[†]I have come to recognize that my wife is a genius, if for no other reason than signing us up for unlimited texting. Can you hear me now!

[‡]Friends and Family is the trademarked cell phone network available through Verizon in North America in which you can add up to ten friends or family members to your network list that can be called, texted, and even yelled at for free, after you pay for it (the network plan). All insect communication plans are free.

Figure 11.1. What better example of interkingdom communication than a cockroach chatting with a penguin! *Bloom County* used with the permission of Berkeley Breathed and the Cartoonist Group. All rights reserved.

through observation of behavioral changes in the receiver. What has been detected is that insects convey information back and forth, and thus fulfill one (two-way communication) of Dr. Phil's recommendations for effective communication. Why the need for two-way communicators? Quite simply because insects need to share information related to attracting or locating mates (part of sexual selection); warning members of their own species (conspecifics) about impending danger (e.g., predators); establishing territories (it's a male thing); locating offspring; soliciting parental care (begging for food); advertising noxiousness to potential predators; or simply putting on a show so that insect bullies back down. As you can see, insects have a lot to talk about. There is a real need to share signals back and forth, since the messages are powerful. They convey information relevant to survival.

Recognizing that insects communicate depends on interpreting adaptive responses to nonverbal signals transmitted between insects. Sometimes the transmitter is not sending a message deliberately. Rather, a message is sent based on body morphology, such as shape or color patterns on wings. In those instances, the signal is clearly innate, since the insect has no control over its own anatomical design or body markings (figure 11.2). Other

Figure 11.2. Large milkweed bug, *Oncopeltus fasciatus* (Hemiptera) displaying aposematic coloration. Photo by Andrew C available at http://bit.ly/1MB4O6I.

forms of communication are also considered innate. Insect life cycles are too short to depend on learning to communicate. That said, experience most certainly influences what messages are shared with conspecifics. This

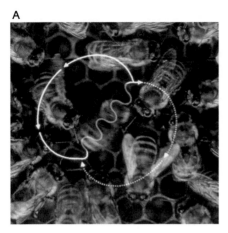

A

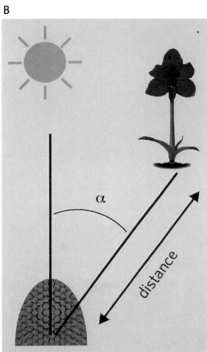

B

Figure 11.3. The waggle dance of the honeybee *Apis mellifera* (Hymenoptera: Apidae) is used to visually communicate information about where food is located. A waggle run oriented 45° to the right of "up" on the vertical comb (A) indicates a food source 45° to the right of the direction of the sun outside the hive (B). The abdomen of the dancer appears blurred because of the rapid motion from side to side. Photo by J. Tautz and M. Kleinhenz available at http://bit.ly/1LgToaMand PLoS Biol 2/7/2004: e216. http://dx.doi.org/10.1371/journal.pbio.0020216.

is especially true with social insects and has been intensively studied among many members of the social Hymenoptera (figure 11.3). Insects can only communicate using mechanisms they have the capacity to perform. So insects can't learn a new language, as humans can, but

can improve their innate communicative abilities. In the next section, we examine how insects use visual, acoustic, chemical, and tactile signals for intra- and interspecific communication. Keep in mind that the sensory structures discussed in chapter 10 are used for perceiving all forms of stimuli in the environment, including communication signals. Often these same structures are used for generating the signals. As we will soon see, this is especially true during acoustical communication.

The basics of insect communication

One feature of insect biology that permits these creatures to be so amazingly successful is their connection to the environment. They are attuned to changes that occur in their surroundings through perception of cues that signify a deviation from "normal" or previous conditions. The change itself may be barely perceivable, a disturbance so small that other organisms fail to recognize any deviation or simply ignore it as insignificant. Insects, however, not only recognize the array of signals trafficking through their environment, but also decipher the messages to know how and when to respond or, equally important, whether to ignore a stimulus. When environmental cues are sensed, signals are interpreted by the central nervous system, followed by an appropriate response, which, again, may include a "do nothing" approach (figure 11.4).

Of course, at other times, an immediate response may be warranted, whereby interpretation or integration is bypassed altogether. Chapter 10 focused on the portions of the insect nervous system involved in each facet, from sensory input to motor output. Those structural components are the same as those involved in insect communication. This is not surprising, when we view insect communication as part of an insect's ability to respond to changes in the environment. Visual signals like wing-flapping displays in butterflies, shrill singing from cicadas vibrating tymbals, or release and detection of sex pheromones by adults moths all constitute forms of sensory stimuli, then intermixed with a milieu of other environmental "noise" that insects must discern with precision. Failure to do so may result in a lost opportunity, such as finding a meal, making a love connection, or avoiding danger. Precise discernment of signal identification occurs through the sensory organs (sensory receptors) discussed in chapter 10. Each type is

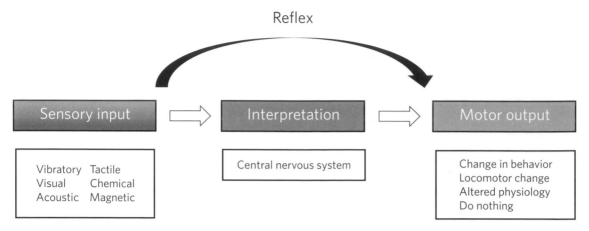

Figure 11.4. Typical pathway along which sensory cues from the environment are processed by an insect.

Table 11.1. Major types of sensory receptors used for detection of various communication signals used by insects. Acoustical and vibrational signals are grouped as one (acoustic) form of communication.

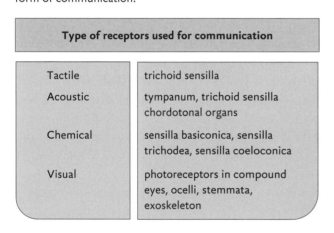

Type of receptors used for communication	
Tactile	trichoid sensilla
Acoustic	tympanum, trichoid sensilla chordotonal organs
Chemical	sensilla basiconica, sensilla trichodea, sensilla coeloconica
Visual	photoreceptors in compound eyes, ocelli, stemmata, exoskeleton

specific for the form of communication being employed (table 11.1). Thus sensory receptors are communication-signal specific.

The sensation of the need to "speak" (communicate) in insects is also associated with internal changes. In other words, an internal readiness must be achieved before some forms of communication can occur. This should sound familiar from our discussions of **fixed action patterns** in the last chapter. Specific hormonal titers need to be reached in the hemolymph; tissues, including sensory receptors, need to mature; or specific genes must be expressed before the signal can be generated. However, communication does not require a series of successive steps to be followed before the behavior is elicited, as with fixed action patterns. That said, an innate behavior could in fact be specifically used for communication.*

Such is the case with some species of migratory locusts, in which an alpha male attains a developmental status (internal readiness) that signifies "time to fly." The male in turn pounds a limb (leg) against the ground, creating a vibrational signal detected by conspecifics in the immediate vicinity, who, in turn, copy his action to pass on the message. The message is for individuals to take flight. This example is considered a **preprogrammed motor program,** or fixed action pattern, as well as vibrational communication. Many related examples occur with calling and courting, in which an internal threshold must be surpassed before communication behaviors like singing can occur or pheromones are synthesized and/or released. In other words, horniness in insects has a physiological basis!

Now that we understand that insects have a need to exchange information as part of their environmental sensing capabilities, the question of how they communicate must be addressed. We already have some idea of the major types of communication used: visual, acoustic, chemical, and tactile. One of the most common means of communicating involves mechanical stimuli in the form of vibrations. Where exactly does this fall among the forms of communication discussed so far? Here we need to make an adjustment to our categories. Many in-

---

*This may be somewhat confusing from the standpoint that all forms of communication are considered innate. However, some specific forms of innate behaviors that typically are not communication-related can be, in some specific instances.

Figure 11.5. Spiders immediately recognize that "lunch is ready" through mechanical stimuli produced from the struggling prey caught in the web. Photo by Schizoschaf available at http://bit.ly/1KOmuJe.

sects communicate by making or perceiving sound signals. In reality these are forms of vibrational disturbances that occur in the environment. Several other types of vibrational stimuli are not audible (like singing) but fall into the same category of vibrational communication. These signals include sound traveling through air and water, ripples on the surface of water, and vibrations along the surface of the ground or other substrates. The act of creating such a signal does not require specialized structures (although some do exist); almost any insect can generate this form of mechanical stimuli, and most, but not all, insect species can detect vibrational disturbances. This is also the basis for how most spiders recognize that prey have been trapped in webs (figure 11.5). The widespread occurrence of conveying information through vibrations among all arthropods, not just insects, points to this being the most primitive form of communication. As we'll discuss in a later section, that does not mean that all vibrational communication forms are less refined than other types of communication, especially when it comes to serenades (i.e., singing) by an insect choir.

Other forms of communication are not as cosmopolitan in the insect world, but certainly are not rare either. For example, visual communication is widespread across

the orders. Primitive forms include body design, markings, and coloration, which insects cannot modify (yet we will discuss some that can!). Others rely on body motion like wing flapping, dancing, body gyrations, and signaling via antennal movements. Some species rely on multiple methods, such as the bee waggle dance, which combines choreographed dance moves with beating of the wings, using honeycomb as the dance floor. Noise production occurs across several taxa, and each group seems to perfect its own method of sound production. Some insects rely on elastic membranes to sing, some rub their wings together or use other body parts, and some can blow air out of their body in a manner that is deemed pleasing to the receiver.

Some insects that depend on chemicals for communication can also release a signal from the body, but it is not always a pleasurable experience for the receiver. Chemical communication is considered the most common and sophisticated form of information exchange among insects. The signals themselves are often structural art forms, permitting specificity in the message yet rugged in design to withstand traveling through multiple forms of media (e.g., water, air), sometimes over incredibly long distances. One class of these compounds is

pheromones, which are used in intraspecific communication to tell many stories, from where to find a mate or food to warning of the presence of a predator to forecasting a change in climatic conditions. Social insects have a chemical vocabulary that rivals the language skills of many adult humans that I have encountered. (You can interpret that in whatever way you like.) Still other species of insects pass information back and forth via direct physical contact (tactile). This is especially true during the intimacy of courtship and copulation, in which a male and female are evaluating each other as potential mates and prospective parents. In such scenarios, tactile stimulation is also driven by (or drives) **positive feedback**, as the message conveyed by the mate can heighten the response of the receiver, and continues until insemination (or rejection) ensues. Some termites appear to use tactile communication without direct contact. Instead, they respond to work done by others, adding to nest construction without receiving instructions from nest mates on what to do next.

In the coming sections, we explore each of these in detail, focusing especially on examples that are common forms of each type of communication, as well as those that are truly unique to the world of insects. We will also explore how some forms of communication, namely chemical, can be exploited to facilitate information exchange between different species, including several non-insects.

Visual displays, camouflage, and mimicry

Looks do matter, especially when trying to impress. Obviously insect appearance is relevant to many phases of sexual selection. This extends also to several facets of their everyday lives in which messages are conveyed through visual patterns or displays. Communicating visually is very common with insects, and rightfully so; it is easy to do. Another way of looking at it is that visual communication is oftentimes energetically free: cellular energy investment is not required to simply look "stunning." Shapes, patterns, and colors are inherent and do not require an individual to do anything other than be seen (figure 11.6). The receiver of the message has the greater investment, since **photoreception** is required. Here is also where the visual signal can be modified

Figure 11.6. Insects display many different colors, shapes and patterns that can all be associated with visual communication: (A) female orchard swallowtail butterfly (Jillian Visser, http://bit.ly/1h3AiqK), (B) spicebush caterpillar (Michael Hodge, http://bit.ly /1OCl9Lo, (C) adult velvet ant (Mcevan, http://bit.ly/1JyCz3Q), and (D) adult buffalo treehopper (gbohne, http://bit.ly/1P1OWg3).

almost effortlessly, in that visual perception is influenced by light and wavelength intensity, contrast (imagine the difference in information from an insect that is green resting in a tree canopy versus one that is brilliant red), color of the object(s), distance from the "message," and whether the sender or receiver is in motion, or is visualizing with ultraviolet or polarized light.

Ultraviolet light has wavelengths between 400 nm to 10 nm, a range that is shorter than visible wavelengths but longer than X-rays. Many moths and butterflies have identification markings on the wings that only reveal the true species identity when seen in ultraviolet. Adult females of the cabbage butterfly, *Pieris rapae* (Lepidoptera: Pieridae) take UV signaling a step further by flapping the wings to create flash of light that draws the attention of any male in the area. This perhaps is also an example of sexual role reversal (see chapter 9) since males are more likely to expose themselves (display) to females, than the other way around.

Visual communication is not only easy, but also efficient. As already discussed, a single visual message can be interpreted in multiple ways simply as a result of the conditions of photoreception. The message is not distorted by wind, as occurs with chemical signals, and visual cues can be shared while the insect is on the move or at rest, as well as across long distances, limited only by the vision capabilities of the receiver. The downside is the lack of specificity, with the exception of ultraviolet signaling, in that any organism, friend or foe, that can perceive the visual signs of the insect now know where that individual is located. Another major limitation is that most visual messaging must occur during daylight. So if a bright coloration is used as warning, it is useless to fend off a predator at night (figure 11.7).

Bioluminescence is an exception to most of the rules just discussed. This method of communication relies on chemical production of light, usually during periods of calling by fireflies and lightning bugs (which, in chapter 9, were revealed as actually beetles). Light is produced when the enzyme luciferase catalyzes the oxidation of the light-producing pigment luciferin. Duration, frequency, and pattern of flashing using luciferin confer uniqueness among different insect species that bioluminesce. This method of visual communication is effective but is energetically expensive, works well only at night, and, though the light flashing patterns are species-specific, the code has been broken by several species that lure in unsuspecting prey by mimicking the flashing pattern. Since it is dark outside, the innocent victim cannot rely

Figure 11.7. A beetle larva (aka glow-worm) (Coleoptera: Lampyridae), communicating through the use of bioluminescence. Photo by Timo Newton-Syms available at http://bit.ly/1WEeR1P.

on other visual cues to recognize that it has been duped. Similarly, the bright flash of light can be seen clearly at night, exposing the beetles to potential attack by any predator active at night.

So far we have touched on different means of visual communication through the lenses of efficiencies and drawbacks. This leaves other forms of visual signals still unaccounted for. Insects exchange information via visual signals in at least five ways:

1. Color patterns and shapes
2. Displays
3. Aposematic coloration
4. Bioluminescence
5. Larval mimicry

We have already discussed conveying information through patterns, shapes, colors, and bioluminescence in some detail. So let's turn our attention next to visual displays. Insects can communicate by moving their bodies, or portions of them, to draw the attention of conspecifics and allospecifics. In chapter 9, we examined how some species of flies rely on dancing and body gyrations to win over the opposite sex. Similarly, wing flapping can be used for the same purpose. In other instances, wing movements may be quite sudden, to reveal markings that temporarily ward off attack. Eye spots on the wings of moths are an excellent example of this defensive tactic (figure 11.8). The idea is that the spots are kept hidden until just before a predator strikes; the sudden appearance of eyes resembling some sort of predator bird (e.g., an owl) or even snake evokes a momentary pause in attack as the would-be-predator deals with fear that it is about to be eaten. This strategy is referred to as the **startle response**, because in fact that is what happens to the attacker.

The waggle dance of honey bees is used exclusively for intraspecific communication. Worker bees returning to the hive from a day of foraging for nectar and pollen share information on location of the bounty using the bee version of MapQuest,* a dance. Karl von Frisch shared the Nobel Prize in Physiology or Medicine in 1973 for his ingenious work in deciphering the language of bees. Foragers perform the dance in front of other

Figure 11.8. Prominent eye spots on the hindwings of an adult io moth. Photo by Jeremy Johnson available at http://bit.ly /1P20Hmt.

workers on honeycomb; the angle of the body on the honeycomb, coupled with the number and direction of body turns, and wing buzzing, conveys to the others the location of the food source. Other foragers in turn follow the directions to the exact location directed by the messenger bee.

Several other species rely on displays as means to announce that they have arrived or to establish a territory. A male hissing cockroach (*Gromphadorhina portentosa*) uses **stilting**—a series of movements of lifting the body or just the abdomen up and down—to ward of competing males that enter his territory. Antennal movements can aid the defending of territory, often simply by moving them in an aggressive fashion. In some species, antennae are moved with the precision of a signal flagger aboard a naval ship.

Some species get their message across by being adorned in bright colors. Now this may seem identical to our earlier discussion, but it differs in one important aspect: the message is a warning to other species to back off. Color patterns of red, orange, and yellow, alone or in combination, as well as black and white, are universal in the animal kingdom in conveying warning, otherwise known as **aposematic coloration**. The insect is deliberately advertising its existence, sending the message to potential predators that there is a cost to considering them for dinner. The cost is associated with the fact that many of the insects displaying aposematic colors harbor toxins, venoms, or other noxious compounds, or

---

*MapQuest is a free online mapping service available for finding directions to almost any location in the United States.

Figure 11.9. Classic example of Müllerian mimicry: the monarch and viceroy butterflies. The markings of the adult monarch butterfly, *Danaus plexippus* (left) and adult viceroy, *Limenitis archippus* (right) reflect aposematic coloration of a toxic and noxious species, respectively. Photos by PiccoloNamek (monarch) and Derek Ramsey (viceroy) available at http://bit.ly/1JyIGoJ.

that they are capable of defending themselves by biting, stinging, or releasing potent compounds. Many insects belonging to the orders Hymenoptera, Hemiptera, Coleoptera, and Lepidoptera depend on aposematism.

Other species that take on the colors of insects that possess well-developed defenses are in actuality innocuous. Such insects are mimics; that is, they have evolved to share characteristics of another species or model. When an otherwise harmless insect evolves to mimic one that is harmful if eaten or messed with, the mimic displays what is called **Batesian mimicry**. The idea is that one species gains protection by taking on the morphological features of one that is more than capable of defending itself. **Müllerian mimicry**, on the other hand, occurs when multiple species, each being unpalatable or possessing defense capabilities, evolve very similar aposematic coloration. Predators will associate all species with the similar warning coloration and avoid them. The classic example from the world of insects is the monarch butterfly, *Danaus plexippus,* and the viceroy butterfly, *Limenitis archippus* (figure 11.9). Both display similar but not identical patterns of orange, black, and white on the fore and hindwings. Monarchs are deadly if consumed, having sequestered cardiac glycosides during larval feeding on the host plant. Viceroys are so noxious in taste that almost no predatory species can consume them. The story of the two butterflies may be in sharp contrast with what you have heard or read previously. When I was in college, viceroys were used as examples of Batesian mimicry, because the belief was that they had no means to defend themselves. But of course today we know that the previous view was incorrect and that viceroys can indeed defend themselves.

Bug bytes

How to avoid being eaten

https://www.youtube.com/watch?v=B2JdRPKYyTc

Mimicry can also be used as a form of camouflage. Yes, camouflage falls under the umbrella of visual communication (colors, patterns, and markings), as well as functioning as a form of defense. Hiding works to avoid predators. We will discuss camouflage defensive tactics more in chapter 12. Before ending our discussion of visual signals, however, we'll look at one more unique form of mimicry. **Larval mimicry** occurs in a few families (e.g., Papilionidae, Saturniidae) of Lepidoptera, in which images of the caterpillar are evident along the margins of the mesothoracic (fore) wings. When the wings are held open, a predator is directed toward the fringes of the wings rather than the body proper. If an attack occurs, damage will be minimized, since tattered wing margins do not prevent flying. Deception is used to protect the most defenseless portion of the body, strategies not unique to insects, as fish and birds also commonly employ this approach.

Quick check

When insects rely on camouflage for visual communication, with "whom" are they attempting to exchange information?

Table 11.2. Insect orders that can make and/or 'hear' various forms of sound. Vibrational communication is the most primitive of the methods and hence is widespread among the Class Insecta.

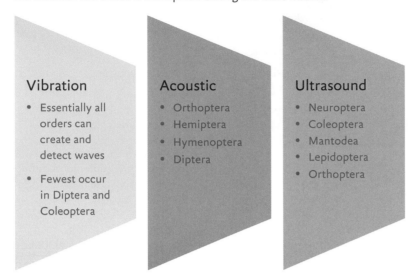

Vibration
- Essentially all orders can create and detect waves
- Fewest occur in Diptera and Coleoptera

Acoustic
- Orthoptera
- Hemiptera
- Hymenoptera
- Diptera

Ultrasound
- Neuroptera
- Coleoptera
- Mantodea
- Lepidoptera
- Orthoptera

Insect phonics: Auditory messages in the insect world

Many species of insects make sound to communicate with other animals. Far fewer insects have the capacity to hear (table 11.2). The form of communication being referenced is audible sound signals as opposed to nonaudible vibrational stimuli discussed earlier in the chapter. As with other forms of communication, insects use sound or noise production for a variety of functions. Sex always seems to top the list for any discussion of insect biology, and here again is another example. Male crickets, grasshoppers, cicadas, some beetles, even hissing cockroaches make noise, sometimes classified as "singing," to win the hearts of lonely females looking for, well . . . sex.* Noise is also used as a signal warning of imminent danger. Most insects are capable of distinguishing between the two. In the cases of attracting a mate, calling for aggregation, or signifying territorial boundaries, members of the same species are being summoned. Sound

produced for intraspecific communication tends to be highly patterned and frequently repeated (https://www.youtube.com/watch?v=oqys8lKsu4s).

By contrast, warning sounds or other acoustical signals used for interspecific communication are unpatterned and sporadic, rarely being repeated or, if so, not necessarily in the same sequence each time (http://ars.usda.gov/sp2UserFiles/person/3559/s2b1geoegeriei.wav). To the untrained ear of most humans, insect sounds are simply background environmental noise. Most of us may recognize the sounds, broadly, as cicadas singing or crickets chirping (**stridulation**), but likely not as patterned versus sporadic. It is not that we cannot make the distinction. Rather, it is more a lack of paying attention. As we have discussed, insects do not have the luxury of being passive or disinterested observers of environmental noise.

Acoustical communication is not as easy as using visual signals. The processes involved are always active, meaning cellular energy must be expended. Lots of energy is necessary for extended calling episodes that frequent the warm evenings during which cicada, cricket, or katydid choruses seem to last for hours. In most cases, specialized structures are required to produce sound, or at least common features such as wings or legs must be modified to permit the production of the desired sound signals. Likewise, for those species that can hear, special-

---

*Unlike online dating services like Match.com or Farmers Only.com used by humans, insects are not looking for long-term relationships. No, when an insect takes the time to communicate an interest in finding a mate, he or she has one thing on the mind, sex. Maybe two things, because eating the potential partner is often an option if sex seems out of the question.

ized structures are needed. As indicated earlier, not all species that create sound necessarily have an interest in the response of the receiver. In other words, all species that make audible noise do not have the necessary morphological apparatus to hear, and the absence of "ears" should not be considered a deficiency. There is no need, necessarily, for the sound emitter to perceive sensory feedback, particularly if a message is directed at allospecifics.

Acoustical performances by insects often lack sophistication, meaning the messages are relatively simplistic in design. However, this is not an indication that insects lack efficiency with acoustical communication. The sheer feat of singing for hours is an amazing accomplishment considering the energetics involved, that predators do not discover "loud" individuals frequently, and that any insect species has *so* much to say! Of course, if it is just the same song repeated over and over, well then, that is not so impressive. More to the point, acoustical signals can travel over long distances, in multiple directions. The signal carries specificity by default, in that most other insects do not have the ability to hear. Sound messages can be sent day or night. Some insects have found creative ways to amplify or change the wavelengths of sound signals, relying on subterranean burrows (mole crickets), plant material (grasshoppers), or other structures external to the body. By doing so, they give the impression that the sound was produced by an individual much larger than they really are. Size is important when attempting to establish a territory, scare off threatening organisms, or thwart other suitors for potential mates.

As we learned with other forms of communication, there are always trade-offs to counter the efficiencies of a particular method. Acoustical communication is energetically expensive in comparison to other methods for exchanging information, with perhaps the only exception being chemical communication. As we have already discussed, specialized structures are typically required to make noise and definitely to hear audible signals. Perhaps the most obvious downside is that noise production, especially if for an extended period, makes the insect conspicuous to just about every living thing in the surrounding environment, even long distances away. Augmenting this problem is that many vertebrate predators have exceptional depth perception when detecting sound, whereas insects typically do not. Thus, vertebrates can easily zero in on the location of the noisy insect. If this does occur, sound-producing insects are not defenseless—but that is a topic for chapter 12.

How do insects make noise that can be heard by others? We know some can sing, but singing is an act or behavior, not the actual method of sound production. Insects rely on multiple methods for producing sound.

1. Rubbing two body parts together
2. Pounding a body part(s) against a substrate
3. Blowing air through or over an opening
4. Using elastic membranes (tympana and tymbals)

Rubbing body parts together. Insects can make audible noise by using two or more body parts. This is frequently achieved by simply vibrating the wings against each other, creating the distinctive "buzzing" sound of flying insects. A buzzing frenzy is commonly associated with the bee waggle dance discussed earlier as a form of visual display. Stridulation is perhaps the most readily recognizable form of acoustical communication. The process can be accomplished by more than one means, but that of crickets and some grasshoppers is best known. In crickets, a **scraper** (a hardened edge) on one wing is rubbed across a **file** on another wing. Grasshoppers perform the same act with the exception that the scraper is located on a leg (metathoracic or hind leg). Passalid beetles rub a portion of the abdominal tergum across their hindwing to stridulate. In short, multiple species of beetles and weevils (Coleoptera), ants (Hymenoptera), assassin bugs (Hemiptera), and flies (Diptera) communicate via some form of stridulation (figure 11.10).

Pounding body part against substrate. Sound can be produced by hammering a substrate with a body part. Think about that statement for just a moment. Insects are small creatures. To create a sound in this manner that is audible to another organism requires at least one of two things: (1) an incredibly sensitive sense of hearing on the part of the listener, or (2) an incredible force in relation to body size to slam the anatomical structure in question such that it can be heard. Now consider that some insects do this with their head, like some moths (Lepidoptera) or sawflies (order Hymenoptera, suborder Symphyta). They represent nature's original head bangers! Other

Figure 11.10. Grasshopper in the act of stridulating or singing. The hindleg (metathoracic) is being rubbed across a hardened edge of the wing. Photo by G.-U. Tolkiehen available at http://bit.ly/1KG4dzR.

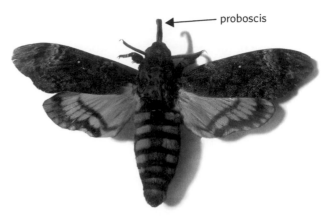

Figure 11.11. Adults of the death's head moth can hiss by blowing air through the extended proboscis. Photo by Jeffdelonge available at http://bit.ly/1KKYVA7.

insects use appendages or body parts other than the head, like grasshoppers thumbing a leg on the ground to gain attention of nearby conspecifics. Click beetles can make noise by inserting a spine on the prosternum into a groove on the mesosternum, which, once released, produces a loud "click" used to frighten predators. Clicking has the added benefit of propelling the insect into the air, useful as an escape response or to simply to right itself if flipped over (https://www.youtube.com/watch?v=0jXp9JAl7kU). In some species, the antennae are used to drum a surface. Vibrational feedback is used for information gathering, but acoustic signaling to other species may also result as part of courting or warning.

Blowing air through or over an opening. A loud audible sound or just a hiss can be created when air is forced out of the body through a small opening or, similarly, over an opening. One of the best examples of sound production by air blowing is demonstrated by adult death head moths (Lepidoptera: Sphingidae) (figure 11.11). When an adult is agitated, say by a potential predator, it lets out a loud squeak. Sound is created by air being forced from the pharynx through the coiled proboscis mouthparts. Similar acoustical signaling occurs with Madagascar hissing cockroaches (Blattodea: Blaberidae). Adults and older nymphs respond to tactile stimulation by hissing. The signal occurs when air stored in large tracheal trunks is forced out of spiracles along the abdomen. Hissing is also used by males establishing a territory, during courtship of females, and during

fighting with other males (https://www.youtube.com/watch?v=W7Efjt6UClg).

Using elastic membranes. Tympanic membranes are used by several insect groups (Lepidoptera, Hemiptera, Orthoptera, Coleoptera, and Neuroptera) to hear sound and, in some cases, to also make noise. The membranes represent thin areas of the exoskeleton that have several sensory organs (chordotonal sensilla) lying just below. Alternating contractions of muscles connected to the tympanic membrane creates sound. Contraction of the muscles literally pulls on the membrane, creating tension and vibration. A tympanum functions like a crude drum on which a thin layer of leather is stretched over a hollow tube, producing sound when a stick or similar object is brought down on the surface. All insects relying on tympanic membranes use them acoustically in a nearly identical fashion. What differs between the groups is where the membranes are positioned. Some grasshoppers have them located on the forelegs, while other members of the order Orthoptera have the membranes on the abdomen. The latter is also true for some species of moths, and in noctuid moths, the structures can be found on the mesothoracic sternum (figure 11.12). In cicadas, elastic cuticular membranes are referred to as tymbals and are positioned at various locations along the abdomen. The membranes function similarly to tympana, producing sound through distortion of membrane shape, which results as muscular tension is increased or decreased.

Figure 11.12. Male cicada in the throes of singing. The thorax appears raised to expose the sound-producing organs (tymbals) underneath. Photo by Jon Richfield available at http://bit.ly/1PJLfcQ.

Whiff this! Chemical communication

Quick check
Would visual or acoustical signaling be more conducive to females being called by males?

Chemical communication is regarded as the most widespread form of communication among all animals. It also viewed as the most ancient means of communication. Both of these evolutionary trends are indeed true with insects. In fact, insects have taken chemical communication to another level of complexity and efficiency, exceeding any comparison to other animals. The chemical signals used are designed for specific functions with the intent of conveying information to either friend or foe, in air or water. Such diversity of functions suggests that the chemical compositions of the messages are highly varied. In actuality, this is not a universal trait. Rather, chemical diversity is dependent on function.

Broadly speaking, chemicals used for communication that modify behavior of recipients are termed **semiochemicals**. Semiochemicals can be subdivided into two broad classes: **pheromones** and **allelochemicals**. The distinction between the chemicals is based on the intended audience of the signal. For example, pheromones are used for intraspecific communication. In other words, they are used for within-species messaging. By contrast, allelochemicals are used for interspecific communication, and many different organisms participate in chemically tweeting* insects. We begin here with pheromones (we'll address allelochemicals occur later in the chapter).

Pheromones, like all semiochemicals, are produced in **exocrine glands**. Exocrine glands are modified epithelial cells that release their products to the outside of the body, or secrete into lumens of visceral organs. Pheromones

---

*Tweeting is a form of communication associated with the social media platform Twitter. Twitter is a microblog site where messages, photos, and other attachments can be shared between broadcaster and receiver. The major caveat is the size of the message: no more than 140 characters per transmission. This is intended to keep the communications short and simple and, presumably, focused on important stuff, but the latter is definitely not true. Most insect communications can be thought of as tweets, because they are concise and to the point. They stray from the human version in that insect tweets probably have no vanity messages associated with them.

can be released directly from cuticular surfaces or directly from exocrine glands. In the case of the latter, the glands are typically referred to as **scent glands** and can be distributed almost anywhere on the insect's body. The pheromonal signals can occur as hydrocarbons in liquid form, used for physical (contact) communication, or be designed to travel as airborne (volatile) signals. Volatile pheromones traverse long distances, requiring chemical stability in an ever-changing and complex environment. Volatility is constrained by the size of the molecule, in that smaller molecules are more likely to become volatile than larger ones. In contrast, larger compounds afford more structural diversity and, therefore, specificity, which is important for making sure that the intended recipient is the only one who "hears" the message being sent.

Pheromone signal specificity depends predominantly on receptor-mediated responses. What this means is that the chemical signals physically bind to receptors on the dendritic membranes of the sensory receptor (cuticular sensilla) to trigger a behavioral response in the receiver. And a response will be generated, as the insect on the receiving end has no choice but to obey the message. What do pheromones ordinarily convey? Many messages, but usually "I am interested in sex" is at the top of the list. We will explore the types of pheromones a little later. For now, we focus on the chemical nature of pheromones: they can be volatile or contact signals, they are structurally stable owing to the environment in which they operate and long distances traveled, and the cues must bind to receptors to elicit a response. Though specificity is desired, many of the same chemical compounds are shared in the pheromone blends of different insects. The consequences of repetitive signals in the environment can be that the chemical code is broken by insects and other organisms intent on a free lunch. This natural form of espionage will be discussed in the section describing **kairomones**.

Pheromones can be broadly classified based on immediacy of response. Time-sensitive messages are referred to as **releaser pheromones**, while those geared toward future events are called **primer pheromones**. Releaser pheromones demand an immediate behavioral response in the recipient. Such pheromones convey messages of alarm, aggregation, sex attraction, mate recognition, and

oviposition stimulation. The recipient does not have a choice in whether it will respond; the chemical signal binds directly to an olfactory sensillum to trigger an innate behavioral response. By contrast, primer pheromones do not cause an immediate reaction in the receiver, nor is one expected. Rather, these signals work by conditioning the recipient for a behavioral or physiological action later in time. The time frame is usually longer than hours: a response may not occur for days, weeks, or even months later. Events associated with growth, development, and reproduction are most influenced by primer pheromones. There appears to be no relationship between the volatility of a pheromone and its status as a primer or releaser. For example, sex attractants can be either volatile or work via direct contact, but in either case, the chemical signal functions as a releaser pheromone.

Classification of pheromones also depends on function. Regardless of whether the pheromone operates as a primer or releaser, the intraspecific signals are grouped based on whether they are used for stimulating sexual behaviors (sex pheromones), finding food (trail pheromones), establishing territories or avoiding overcrowding (spacing pheromones), signaling warning (alarm pheromones), or promoting the formation of clusters or masses of individuals (aggregation pheromones), which is a behavior important to insects for a number of reasons, including mate-finding, defense, resource utilization, and seasonality. Among these categories, sex pheromones have been studied most extensively. Sex pheromones represent a complex blend of compounds rather than a single entity. The uniqueness or specificity of the signal is achieved through varying the exact composition and the proportions of the chemical constituents and components. Either sex may release a sex pheromone, but in many groups, the female is chiefly responsible for signaling. In turn, males display acute chemotaxic and anemotaxic* abilities in tracking the scent of the female suitor(s), a description best applied to volatile sex pheromones. Contact pheromones are generally not emitted

---

*A taxis is an innate response in which the body changes position in relation to a stimulus. A positive response (chemotaxis) is shown by male insects flying toward the source of sex pheromone. Flying upwind is a positive anemotaxis.

until the two sexes physically touch. The hydrocarbon blend of sex attractant lies just below the exoskeleton surface and is secreted with tactile stimulation. Because of the close proximity of the adults, a complex and volatile pheromone is not needed. Specificity is achieved through the behavioral interactions that preceded touch.

Sex pheromones also have practical purposes for humans. Now, before this gets your mind racing toward thoughts of ill repute, let me clarify: I refer to insect control measures. Since insects do not have the ability to deny the instructions of sex pheromones, the use of these behavior modifiers makes for ideal attractants to lure in harmful or destructive insects to traps. A wide range of insect traps, some laced with insecticides, some with sticky adhesives, and others with no escape routes, rely on the sex pheromone of one sex to draw in the opposite sex. Does the approach work? And how! Hundreds to thousands of individuals can be collected in a single trap, depending on the insect and design of the trap. In chapter 14 we explore this use of applied insect biology as part of efforts to control and manage insect populations.

Chemical dependency: On being a social insect

The examples of communication discussed thus far predominantly reflect the efforts of reclusive insects. That is not totally fair, but the vast majority of insects do live as solitary individuals. They only come together for short-term excursions, such as to temporarily share food or to unite as a couple and then quickly resume a life alone. Social insects are different beasts altogether. Their entire existence is spent in community. Part of the benefit of residing in a hive or colony is group—meaning cooperative—food collection, parental care, and defensive efforts. The key is cooperation, which can only occur if an efficient means of communication exists between all members of the community association.

The social language is not restricted to one type of modality; visual, acoustical, tactile, chemical, and even magnetic signals are used to convey messages between individuals in a hive or colony. Without a doubt, chemical communication is by far the most important means for exchange of information. This would lead to the immediate speculation that pheromones are the primary signals exchanged. All five types (alarm, sex, aggregation, spacing, and trail pheromones) play a role in the lives of social insects. In reality, most colonies of insects rely on specialty pheromones to carry out day-to-day functions. In particularly, maintaining social organization so that the hive or colony functions as a unit and avoids collapsing into chaos requires chemical control over group members. Similarly, to realize the benefits of living in a group, alarm pheromones must do more than simply convey warning; they must mobilize the worker force to fight with such aggression that the attacker(s) rethinks the decision to ever commit the act again. Sociality requires a complex language that is not needed for a solitary existence.

Among social insects, the social language of honey bees (*Apis mellifera*) (Hymenoptera: Apidae) has been most extensively studied. Several pheromones unique to the honey bee hive are produced to communicate at precise times in the development of the brood and workers. The timing of signal release coincides with specific tasks that must be accomplished. Thus, the emitter of the signal, whether it is the queen, adult workers, or even developing larvae, are programmed with the inherent awareness that a chemical message cannot be broadcast constantly and still have a meaningful effect. To yell out instructions at an inappropriate time will have the same results on a bee as it does on you: individuals habituate and orders are ignored, becoming nothing more than background noise. Here we briefly examine some of the specialty pheromones used by honey bees.

Alarm pheromone. The mixture of compounds that make up alarm pheromones are stored and released in the venom glands of workers. During an attack on an individual during foraging or on the colony proper, a worker honey bee responds by stinging. The alarm pheromone is released in the venom, which functions to recruit other workers to aid in defense. This type of chemical signaling is common among many of the social Hymenoptera, a feature you may have unfortunately encountered if ever smashing an ant or yellow jacket near the colony.

Nasanov pheromone. Nasanov pheromone functions in nonaggressive forms of swarming behavior. Workers leaving the existing colony to form a new one use this complex blend of compounds. Typically swarming occurs when multiple queens have been produced in response to an aging and dying queen, or when the hive has become too large to be sustained.

Queen's substance. Retinue pheromone, often termed queen's substance, is a blend of compounds produced by a newly mated queen once she returns to the colony. The pheromone attracts workers to the queen. She further entices them by offering droplets of the pheromone so that the workers will antennate and lick her, further solidifying the authority of the new queen. The queen controls the behavior of these workers through the queen's substance, which in turn recruits other workers in the hive to join in attending to the queen's desires. A gradual decrease in production of retinue pheromone by the queen is a signal to nursery workers that the old queen is dying and that new ones must be produced.

Brood pheromone. Larval forms of honey bees are capable of communicating to nursery (or nurse) bees. The latter are adult workers dedicated to feeding and caring for developing larvae. As the young near the completion of larval development, they emit a complex of ester compounds that trigger the nursery workers to cap the brood cells for pupation (figure 11.13). Some of the esters released from the larvae function to inhibit ovarian maturation in the workers. Similarly, esters in queen's substance also yield the same result on workers' capacity to reproduce. If a colony suffers from a prolonged absence of brood and a queen, the inhibitory effect is lifted and workers (all females) begin to mature ovarian tissue.

Other specialty pheromones exist not only in honey bee colonies but in other species of social insects as well. An exhaustive discussion of all types exceeds the scope of this textbook. However, the point is that pheromones are part of the special language used to communicate in a group setting. Based on the enormous success of social insects, the pheromone repertoire appears to be quite effective.

Interspecies chemical communication

Chemical signals that permit communication between insects of different species (interspecific communication) are termed allelochemicals. In reality, such chemical cues are not restricted to just exchanging information between insects, as other signalers include a wide range of organisms such as plants, bacteria, and fungi. Undoubtedly other organisms also participate, but they have yet to be caught in the act of talking to insects. Allelochemicals are far more diverse than pheromones. This is because a

Figure 11.13. Capped brood cells in a honey bee (*Apis mellifera*) colony. Photo by Abalg available in public domain at http://bit.ly /1KG5iYp.

Insect séance: Speaking to the dead

Insects can channel the dead. Well, not really. But what they can do is communicate with dead things. At first glance, you may find this very odd, even repugnant. However, when you consider two things, your view may change. First, you are attracted to dead and nonliving things as well. Consider most of the foods that humans eat. Dead, dead, and dead. Turkey or pie smell delicious when baking. The odors may even have you talking to food like Fat Bastard in the Austin Powers movies,* "Get in my belly!" (https://www.youtube.com/watch?v=nixR6wVa4HY). We also have a propensity as a species to talk to nonliving objects, more frequently than insects. Consider how people react to watching sports on television or how they jump off the couch to answer an inanimate cell phone.

Second, many insects functionally recycle dead animals in terrestrial ecosystems, a service that makes our lives much easier. What is the attraction? Animal remains are nutrient pantries that must be exploited immediately upon recognition. If insects delay, other organisms will undoubtedly not. As a consequence, carrion feeding and breeding insects are finely attuned to the chemical signals that signify death. Waves of insect colonizers arrive on an animal corpse, each vying for the chance to feed at the time most opportune for their own success. Flies in the family Calliphoridae, otherwise known as blow flies or bottle flies, are usually the first colonizers. Females can detect a decomposing body within minutes of death under the right conditions. What did the flies detect? The short answer is apneumones. A detailed answer is not available because the precise chemical signals that draw the attention of any **necrophilous** insect have not been fully deciphered. What is known is that ammonia and sulfur-rich compounds in complex make up the apneumones. Gases like methane, ammonia, putrescine, and cadaverine, in combination with other chemicals, function as strong communicators to flies. However, none of the gases alone are effective at serving as the chemical signature of death. Recent evidence indicates that bacteria residing on a corpse also emit chemical signals that attract blow flies. In this instance, the signal is a synomone, since both the fly and bacterium benefit: the fly finds a food source and an oviposition site, and the bacteria are transported to another location (dispersal) by the fly. Odors emanating from a corpse change over time, as a result of continuous chemical and physical breakdown of the corpse, as well as changes induced by feeding invertebrates and vertebrates. The net effect is that the chemical signals released, and hence the communication between carrion and insects, is unique to the conditions of each instance of death. Thus, the fauna attracted over time change in response to the chemical cues emitted and detected.

Some plants in the families Araceae and Rafflesiaceae get in on the act by producing sulfur-containing compounds that smell like rotting meat. The corpse or carrion flowers of these plants are responsible for odor emission, which draws in the same necrophagous flies attracted to a corpse. Flies feed on nectaries of the flowers while at the same time becoming covered in pollen. Here is yet another example of a synomone. A few species of plants even consume the insect that arrives. In those instances, the chemical signal is an allomone and represents the ultimate death signal for the fly.

---

*Fat Bastard is a villain turned softie in the Austin Powers movies created by and starring Mike Myers, formerly of NBC TV's *Saturday Night Live*. Mike Myers plays Fat Bastard, a large, red-bearded Scottish-accented antagonist with a penchant for eating everything in large quantities.

wide range of chemicals can be used for interspecific communication. Any chemical or complex of compounds that changes the behavior of the receiver, favorably or not, is considered an allelochemical, provided the receiver is not of the same species as the "emitter." Emitter, in this instance, does not have exactly the same meaning as we used for pheromones, in that the signal may not be volatile and is released by mechanisms such as a contact secretion, a spray, an injection, or in an airborne form.

Such a wide swath of chemicals also means that the origin, composition, receptor (or not), and mode of action of each chemical is unique from every other. This diversity in chemical properties is also in contrast to pheromones, which are much more uniform in composition. Two characteristics that are universal to nearly all allelochemicals are (1) their use in interspecies or allospecific communication, and (2) their synthesis by exocrine glands. Structurally, the exocrine glands are distinct from those that produce pheromones, in that the gland lumen is lined with cuticle. Yes, the same cuticle that we discussed in chapter 6 that makes up the major portion of the exoskeleton. A cuticular lining is a necessary prerequisite to avoid **autointoxication** for those species that produce defensive secretions that are nondiscriminate or that do not require a receptor to trigger responses in the receiver.

Why are some chemical signals considered allelochemicals when they do not share the two basic unifying features? All chemicals classified under this umbrella share the trait of being used for interspecific communication. Where they stray is in origin: not all the signals are produced by insects. Remember that interkingdom communication occurs, and thus organisms other than animals generate several allelochemicals. Short story, plants and microorganisms do not possess exocrine glands. Some allelochemicals are sequestered compounds such as cardiac glycosides, terpenoids, or alkaloids, obtained from plants used for food rather than synthesized de novo. Several sequestered compounds offer the advantage of already being in active form, a common feature of many allelochemicals, yet others must be activated from a precursor (e.g., some hymenopteran venoms) or require an enzymatic reaction like the hot sprays generated by bombardier beetles (Coleoptera: Carabidae).

Allelochemicals are categorized based on their effect on the receiver and, to a lesser extent, their effect on the emitter. We examine four of these chemical signals.

1. allomones
2. kairomones
3. synomones
4. apneumones

Allomones. Chemical substances produced and released by an individual of one species that is harmful to the receiver are called **allomones**. Generally, the receiver is always a species different from the emitter, but there are a few exceptions. The emitter may not benefit from the chemical release, but typically does, since allomones are most typically used in defense. Chemically, allomones are diverse, ranging from proteinaceous venoms used by parasitic Hymenoptera and predatory true bugs (salivary venoms), quinolic sprays or secretions, alkylpyrazine odors, and aldehyde repellants. Some stinging hair caterpillars possess hemorrhagic or neurotoxic venoms, while social Hymenoptera (ants, wasps, bees, hornets) produce venom cocktails composed of peptides, proteins, catecholamines, and an array of enzymes (https://www.youtube.com/watch?v=F4pEO93yeqE). Sequestered compounds (cardiac glycosides, terpenoids, or alkaloids) obtained from food sources also represent chemically diverse allomone signals (figure 11.14). Among the more entertaining spectacles in nature are the defensive activities of ground beetles in the genus *Chaenius* (Coleoptera: Carabidae). When attacked by predatory ants, the beetles emit a volatile 'sedative' from glands positioned near the anus. The ant is temporarily immobilized, allowing the would-be-prey an opportunity to escape. If the ant takes too long to recover from its slumber, it may be consumed by a predator. When death does occur, the scenario earns the phrase *fatal flatulence* for this allomone (figure 11.15).

Kairomones. Chemicals that are produced and released by one species and that evoke a positive or beneficial response in the recipient are referred to as **kairomones**. The emitter does not benefit from release of the signal, often in volatile form, but may be harmed if the receiver uses the chemical cue to track the releaser so that the originating species is consumed as food or used as a host for parasitic species. For some insects, kairomone signals detected by the receiver are pheromones produced by another species. Oviposition pheromones on the eggs of

Figure 11.14. An adult bombardier beetle (Coleoptera: Carabidae) that produces allomones in the form of host quinones. Photo by Peter Halasz available at http://bit.ly/1ss6x8X.

blow flies (Diptera: Calliphoridae) or sex pheromones emitted during mate attraction by any number of insects may be intercepted and decoded by predatory species. Adults of the yellow jacket *Vespula germanica* (Hymenoptera: Vespidae) are capable of using sex pheromones of the fruit fly *Ceratitis capitata* (Diptera: Tephritidae) as kairomones. In other instances, the chemical cues are derived from plants, frequently synthesized as secondary metabolites that attract the attention of potential herbivorous insects and other animals.

Synomones. When the chemical substance released generates a response in the receiver that also results in benefits to the emitter, the signal is called a **synomone**. The definition does not imply that the chemical substance itself benefits both the originator and recipient. Rather, the response of the receiver leads to a positive outcome for individuals of both species. Numerous examples can be found associated with plant herbivory. For example, a bark beetle (Coleoptera: Curculionidae) attack of pine tree species can result in the injured tree tissue releasing terpene compounds that serve as attractants to parasitic wasps that subsequently locate and parasitize the beetles. Thus, the wasp locates a host aided by plant-derived chemicals, and the tree experiences reduced herbivory since the wasps are **parasitoids** that ultimately kill the beetle hosts. A similar relationship is initiated when

Figure 11.15. Saddleback caterpillar (Lepidoptera: Limacodidae) with stinging hairs prominently on display. The urticating hairs are quite painful when embedded into human skin. Photo by Gerald Lenhard available at bugwood.org.

onions attacked by the onion maggot fly, *Delia* (*Hylemya*) *antiqua* (Diptera: Anthomyiidae) release a volatile compound, the same one that evokes tears in humans when slicing onions. The compound serves as a synomone for a braconid parasitoid, *Aphaerata pallipes* (Hymenoptera: Braconidae) that utilizes the fly larvae as hosts, thus benefiting the onion as well.

Apneumones. Chemical signals emitted from a nonliving object that trigger a response in the recipient is known as **apneumones**. The emitter may not be of biological origin, but usually is, and typically the first example that comes to mind is dead and decaying animals. The initial attraction of flies in the families Calliphoridae and Sarcophagidae to a dead body is believed to be triggered by odors emanating from decomposing tissues of a corpse, prompting such behaviors as adult feeding, mate-finding, and oviposition. In several species of social Hymenoptera that form well-organized colonies or nests, some workers have the task or ability, depending on your point of view, of detecting dead workers and removing their bodies from the colony. The workers are often identified as the undertaker caste, and in the case of some ant species, the dead are simply piled onto the colonies' trash heap.

CHAPTER REVIEW

❋ **Communication is the key to every successful relationship**

- Effective communication is the key to any successful relationship. Any marriage guru would concur. Would such counselors expect that this adage applies to insects as well? Quite likely, no. So communication within species or outside their "friends and family network" is done with the same exceptional efficiency that is observed with food acquisition, digestion, reproduction, body defenses, or just about any other feature of insect natural history. The question that now arises is, what do insects talk about? One can assume, just important stuff. Playful banter, telling jokes or engaging in gossip probably are uniquely human, as they would appear to serve no purpose in the existence of insects. Of course, there is currently no empirical means to test the hypothesis that insects do not tell jokes. The only way for us to

recognize that insects do communicate is through observation of behavioral changes in the receiver.

- Insects convey information back and forth, and thus rely on two-way communication. Why the need for two-way communicators? Quite simply because insects need to share information related to attracting or locating mates (part of sexual selection); warning members of their own species (conspecifics) about impending danger (e.g., predators); establishing territories (it's a male thing); locating offspring; soliciting parental care (begging for food); advertising noxiousness to potential predators; or simply putting on a show so that insect bullies back down.

- Recognizing that insects communicate depends on interpreting adaptive responses to nonverbal signals transmitted between insects. Sometimes the transmitter is not sending a message deliberately. Rather, a message is sent based on body morphology, such as shape or color patterns on wings. In those instances, the signal is clearly innate, since the insect has no control over its own anatomical design or body markings. Other forms of communication are also considered innate. Insect lifecycles are too short to depend on learning to communicate. That said, experience most certainly influences what messages are shared with conspecifics. This is especially true with social insects and has been intensively studied among many members of the social Hymenoptera. Insects can only communicate using mechanisms they have the capacity to perform. So insects can't learn a new language as humans can, but can improve their innate communicative abilities.

❋ **The basics of insect communication**

- One feature of insect biology that permits these creatures to be so amazingly successful is their connection to the environment. They are attuned to changes that occur in their surroundings through perception of cues that signify a deviation from "normal" or previous conditions. The change itself may be barely perceivable, a disturbance so small that other organisms fail to recognize any deviation or simply ignore it as insignificant. Insects, however, not only recognize the array of signals trafficking through their environment, but also decipher the messages to know how and when to respond or, equally important, whether to ignore a stimulus. When environmental cues are sensed, signals are interpreted by the central nervous system, followed by an appropriate response, which, again, may include a "do nothing" approach. Of course at other times, an immediate response maybe warranted whereby interpretation or integration is bypassed altogether.

- The sensation of needing to communicate in insects is also associated with internal changes. In other words, an internal readiness must be achieved before some forms of communication can occur. This should sound familiar from our discussions of fixed action patterns in the last chapter. Specific hormonal titers need to be reached in the hemolymph; tissues, including sensory receptors, need to mature; or specific genes must be expressed before the signal can be generated. However, communication does not require a series of successive steps to be followed before the behavior is elicited, as with fixed action patterns. That said, an innate behavior could in fact be specifically used for communication.

- The major types of communication used by insects include visual, acoustic, chemical, and tactile forms. One of the most common means of communicating involves mechanical stimuli in the form of vibrations. Where exactly does this fall among the forms of communication discussed so far? Here we need to make an adjustment to our categories. Many insects communicate by making or perceiving sound signals. In reality these are forms of vibrational disturbances that occur in the environment. Several other types of vibrational stimuli are not audible but fall into the same category of vibrational communication.

- Other forms of communication are not as cosmopolitan in the insect world, but certainly are not rare either. For example, visual communication is widespread across the orders. Noise production occurs across several taxon, and each group seems to perfect its own method of sound production. Some insects rely on tympanic membranes to sing, some rub their wings together or use other body parts, and some can blow air out of their body in a manner that is deemed pleasing to the receiver. Chemical communication is considered the most common and sophisticated form of information exchange among insects. The signals themselves are often structural art forms, permitting specificity in the message yet rugged in design to withstand traveling through multiple forms of media (e.g., water, air), sometimes over incredibly long distances. Still other species of insects pass information back and forth via direct physical contact (tactile). This is especially true during the intimacy of courtship and copulation, in which a male and female are evaluating each other as potential mates and prospective parents.

Visual displays, camouflage, and mimicry

- Communicating visually is very common with insects and rightfully so; it is easy to do. Another way of looking at it is that visual communication is oftentimes energetically free: cellular energy investment is not required to simply look "stunning." Shapes, patterns, and colors are inherent and do not require an individual to do anything other than be seen.

- Visual communication is not only easy, but also efficient. As already discussed, a single visual message can be interpreted in multiple ways simply as a result of the conditions of photoreception. The message is not distorted by wind, as occurs with chemical signals, and visual cues can be shared while the insect is on the move or at rest, as well as across long distances, limited only by the vision capabilities of the receiver. The downside is the lack of specificity, with the exception of ultraviolet signaling, in that any organism, friend or foe, that can perceive the visual signs of the insect now know where that individual is located. Another major limitation is that most visual messaging must occur during daylight.

- Insects exchange information via visual signals in at least five ways: through color patterns and shapes, displays, aposematic coloration, bioluminescence, and larval mimicry.

- Bioluminescence is an exception to most of the rules just discussed. This method of communication relies on chemical production of light, usually during periods of calling by fireflies and lightning bugs. This method of visual communication is effective but is energetically expensive, works well only at night, and, though the light flashing patterns are species-specific, the code has been broken by several species that lure in unsuspecting prey by mimicking the flashing pattern.

- Some species get their message across by being adorned in bright colors. Now this may seem identical to our earlier discussion, but it differs in one important aspect: the message is a warning to other species to back off. Color patterns of red, orange, and yellow, alone or in combination, as well as black and white, are universal in the animal kingdom in conveying warning, otherwise known as aposematic coloration. The insect is deliberately advertising its existence, sending the message to potential predators that there is a cost to considering them for dinner.

Insect phonics: Auditory messages in the insect world

- Many species of insects make sound to communicate with other animals. Far fewer insects have the capacity to hear. As with other forms of communication, insects use sound or noise production for a variety of functions. Noise is used as a signal warning of imminent danger. In the case of attracting a mate, calling for aggregation, or signifying territorial boundaries, members of the same species are being summoned. Sound produced for intraspecific communication tends to be highly patterned and frequently repeated. By contrast, warning sounds or other acoustical signals used for interspecific communication are unpatterned and sporadic, rarely being repeated or, if so, not necessarily in the same sequence each time.

- Acoustical communication is not as easy as using visual signals. The processes involved are always active, meaning cellular energy must be expended. In most cases, specialized structures are required to produce sound, or at least common features such as wings or legs must be modified to permit the production of the desired sound signals. Likewise, for those species that can hear, specialized structures are needed. As indicated earlier, not all species that create sound necessarily have an interest in the response of the receiver. In other words, all species that make audible noise do not have the necessary morphological apparatus to hear, and the absence of "ears" should not be considered a deficiency.

- Acoustical performances by insects often lack sophistication, meaning the messages are relatively simplistic in design. However, this is not an indication that insects lack efficiency with acoustical communication. The sheer feat of singing for hours is an amazing accomplishment considering the energetics involved, that predators do not discover "loud" individuals frequently, and that any insect species has so much to say! More to the point, acoustical signals can travel over long distances, in multiple directions. The signal carries specificity by default, in that most other insects do not have the ability to hear. Sound messages can be sent day or night.

- Insects can make noise that can be heard by other insects and organisms by a number of methods. We know some can sing, but singing is an act or behavior, not the actual method of sound production. Insects rely on multiple mechanisms for producing sound, including rubbing two body parts together, pounding a body part(s) against a substrate, blowing air through or over an opening, and using elastic membranes.

Whiff this! Chemical communication

- Chemical communication is regarded as the most widespread form of communication among all animals. It also viewed as the most ancient means of communication. Both of these evolutionary trends are indeed true with insects. In fact, insects have taken chemical communication to another level of complexity and efficiency, exceeding any comparison to other animals. The chemical signals used are designed for specific functions with the intent of conveying information to either friend or foe, in air or water. Such diversity of functions suggests that the chemical compositions of the messages are highly varied. In actuality, this is not a universal trait. Rather, chemical diversity is dependent on function. Broadly speaking, chemicals used for communication that modify behavior of recipients are termed semiochemicals. Semiochemicals can be subdivided into two broad classes: pheromones and allelochemicals. The distinction between the chemicals is based on the intended audience of the signal.

- Pheromones, like all semiochemicals, are produced in exocrine glands. Exocrine glands are modified epithelial cells that release their products to the outside of the body, or secrete into lumens of visceral organs. Pheromones can be released directly from cuticular surfaces or directly from exocrine glands. In the case of the latter, the glands are typically referred to as scent glands and can be distributed almost anywhere on the insect's body. The pheromonal signals can occur as hydrocarbons in liquid form, used for physical (contact) communication, or be designed to travel as airborne (volatile) signals. Volatile pheromones traverse long distances, requiring chemical stability in an ever-changing and complex environment. Volatility is constrained by the size of the molecule, in that smaller molecules are more likely to become volatile than larger ones.

- Pheromone signal specificity depends predominantly on receptor-mediated responses. What this means is that the chemical signals physically bind to receptors on the dendritic membranes of the sensory receptor to trigger a behavioral response in the receiver. And a response will be generated, as the insect on the receiving end has no choice but to obey the message.

- Pheromones can be broadly classified based on immediacy of response. Time sensitive messages are referred to as releaser pheromones, while those geared toward future events are called primer pheromones. Releaser pheromones demand an immediate behavioral response in the recipient. Such pheromones convey messages of alarm, aggregation, sex attraction, mate recognition, and oviposition stimulation. The recipient does not have a choice in whether it will respond; the chemical signal binds directly to an olfactory sensillum to trigger an innate behavioral response. By contrast, primer pheromones do not cause an immediate reaction in the receiver, nor is one expected.

- Classification of pheromones also depends on function. Regardless of whether the pheromone operates as a primer or releaser, the intraspecific signals are grouped based on whether they are used for stimulating sexual behaviors (sex pheromones), finding food (trail pheromones), establishing territories or avoiding overcrowding (spacing pheromones), signaling warning (alarm pheromones), or promoting the formation of clusters or masses of individuals (aggregation pheromones), which is a behavior important to insects for a number of reasons, including mate-finding, defense, resource utilization, and seasonality.

Chemical dependency: On being a social insect

- Social insects' entire existence is spent in community. Part of the benefit of residing in a hive or colony is group—meaning cooperative—food collection, parental care, and defensive efforts. The key is cooperation, which can only occur if an efficient means of communication exists between all members of the community association. The social language is not restricted to one type of modality; visual, acoustical, tactile, chemical, and even magnetic signals are used to convey messages between individuals in a hive or colony. Without a doubt, chemical communication is by far the most important means for exchange of information.

- Among social insects, the social language of honey bees (*Apis mellifera*) has been most extensively studied. Several pheromones unique to the honey bee hive are produced to communicate at precise times in the development of the brood and workers. The timing of signal release coincides with specific tasks that must be accomplished. Thus, the emitter of the signal, whether it is the queen, adult workers, or even developing larvae, are programmed with the inherent awareness that a chemical message cannot be broadcast constantly and still have a meaningful effect.

* **Interspecies chemical communication**
  - Chemical signals that permit communication between insects of different species are termed allelochemicals. In reality, such chemical cues are not restricted to just exchanging information between insects, as other signalers include a wide range of organisms such as plants, bacteria, and fungi. Allelochemicals are far more diverse than pheromones. This is because a wide range of chemicals can be used for interspecific communication.
  - This diversity in chemical properties is also in contrast to pheromones, which are much more uniform in composition. Two characteristics that are universal to nearly all allelochemicals are (1) their use in interspecies or allospecific communication and (2) their synthesis by exocrine glands. Structurally, the exocrine glands are distinct from those that produce pheromones, in that the gland lumen is lined with cuticle.
  - Allelochemicals are categorized based on the effect on the receiver and, to a lesser extent, their effect on the emitter. These chemical signals include allomones, kairomones, synomones, and apneumones.

MUSHROOM FARMING (SELF-TEST)

Level 1: Knowledge/Comprehension

1. Define the following terms:
   - (a) pheromone
   - (b) allelochemical
   - (c) kairomone
   - (d) intraspecific communication
   - (e) aposematic coloration
   - (f) social language

2. How can cicadas' singing be distinguished between mating songs and warning?

3. Describe the difference in function between pheromones and allelochemicals.

4. Explain the differences between Batesian and Müllerian mimicry.

5. Explain why visual communication is considered easier for the emitter than for the receiver.

Level 2: Application/Analysis

1. Describe how the same chemical signal can be considered a pheromone and a kairomone.

2. Explain why social insects have an absolute need for a sophisticated chemical vocabulary.

3. What are the benefits for an insect to communicating with non-insect organisms?

Level 3: Synthesis/Evaluation

1. What are the potential trade-offs for adult Lepidoptera that rely on larval mimicry?

2. Speculate how closely related species of insects that rely on similar blends of pheromones or that sing nearly identical mating calls are capable of achieving specificity despite similarity.

REFERENCES

Aak, A., G. K. Knudsen, and A. Soleng. 2010. Wind tunnel behavioural response and field trapping of the blowfly *Calliphora vicina*. *Medical and Veterinary Entomology* 24:250-257.

Berenbaum, M. R. 1995. Bugs in the System. Addison-Wesley, New York, NY.

Birch, M. C. Chemical communication in pine bark beetles: The interactions among pine bark beetles, their tree hosts, microorganisms, and associated insects for a system superbly suited for studying the subtlety and diversity of olfactory communication. *American Scientist* 66:409-419.

Cardé, R. T, and J. G. Millar. 2011. Advances in Insect Chemical Ecology. Cambridge University Press, Cambridge, UK.

Cocroft, B., and R. L. Rodríguez. 2005. The behavioral ecology of insect vibrational communication. *BioScience* 55:323-334.

Eisner, T. 2003. For Love of Insects. The Belknap Press of Harvard University Press, Cambridge, MA.

Gerhardt, H. C., and F. Huber. 2002. Acoustic Communication in Insects and Anurans. University of Chicago Press, Chicago, IL.

Greenfield, M. D. 2002. Signalers and Receivers: Mechanisms and Evolution of Arthropod Communication. Oxford University Press, New York, NY.

Hedwig, B. 2013. Insect Hearing and Acoustic Communication. Springer, London, UK.

Hölldobler, B., and E. O. Wilson. 1990. The Ants. Harvard University Press, Cambridge, MA.

Ma, Q., A. Fonseca, W. Liu, A. T. Fields, M. L. Pimsler, A. F. Spindola, A. M. Tarone, T. L. Crippen, J. K. Tomberlin, and T. K. Wood. 2012. *Proteus mirabilis* interkingdom swarming signals attract blow flies. *ISME Journal* 6:1356-1366.

Michelsen, A., F. Fink, M. Gogala, and D. Trane. 1982. Plants as transmission channels for insect vibrational songs. *Behavioral Ecology and Sociobiology* 11:267-281.

Osorio, D., and M. Vorobyev. 2008. A review of the evolution of animal colour vision and visual communication signals. *Vision Research* 48:2042-2051.

Schiestl, F. P., F. Steinbrunner, C. Schulz, S. von Reub, W. Francke, C. Weymuth, and A. Leuchtmann. 2006. Evolution of "pollinator"-attracting signals in fungi. *Biology Letters* 2:401-404.

Slessor, K. N., M. Winston, and Y. Le Conte. 2005. Pheromone communication in the honeybee (*Apis mellifera* L.). *Journal of Chemical Ecology* 31:2731-2745.

Steiger, S., and J. Stökl. 2014. The role of sexual selection in the evolution of chemical signals in insects. *Insects* 5: 423-438; doi:10.3390/insects5020423.

Thomas, J. A., J. J. Knapp, T. Akino, S. Gerty, S. Wakamura, D. J. Simcox, J. C. Wardlaw, and G. W. Elmes. 2002. Parasitoid secretions provoke ant warfare. *Nature* 417:505-506.

Vickers, N. J., T. A. Christensen, H. Mustaparta, and T. C. Baker. 1991. Chemical communication in heliothine moths. III. Flight behavior of male *Helicoverpa zea* and *Heliothis virescens* in response to varying rations of intra- and interspecific sex pheromones components. *Journal of Comparative Physiology A* 169:275-280.

Wajnberg, E., and S. Colazza (eds.). 2013. Chemical Ecology of Insect Parasitoids. Wiley-Blackwell, West Sussex, UK.

Wilson, E. O. 1971. The Insect Societies. The Belknap Press of Harvard University Press, Cambridge, MA.

THE ENTOMOLOGIST BOOKSHELF (SUPPLEMENTAL READINGS)

Bell, W. J., and R. T. Cardé (eds.). 1984. Chemical Ecology of Insects. Chapman and Hall, New York, NY.

Bradbury, J. W., and S. L. Vehrencamp. 1998. Principles of Animal Communication. Sinauer Associates, Sunderland, MA.

Claridge, M. F. 1985. Acoustic signals of the Homoptera: Behavior, taxonomy, and evolution. *Annual Review of Entomology* 30:297-318.

Hölldobler, B., and E. O. Wilson. 1990. The Ants. Harvard University Press, Cambridge, MA.

Lloyd, J. E. 1983. Bioluminescence and communication in insects. *Annual Review of Entomology* 28:131-160.

Romoser, W. S., and J. G. Stoffolano, Jr. 1998. The Science of Entomology. 4th ed. WCB-McGraw-Hill, New York, NY.

Rowe, C., and T. Guilford. 1999. The evolution of multimodal warning displays. *Evolutionary Ecology* 13:655-671.

Wyatt, T. D. 2014. Pheromones and Animal Behavior. Cambridge University Press, Cambridge, UK.

ADDITIONAL RESOURCES

Bug bytes insect sound library
http://ars.usda.gov/pandp/docs.htm?docid=10919

Cicada mania
http://www.cicadamania.com/audio/

Insect communication
http://www.cals.ncsu.edu/course/ent425/tutorial/Communication/index.html

Insect communications
http://www.hometrainingtools.com/a/insect-communications

Pherobase: Database of insect pheromones
http://www.pherobase.com/

Pheromone traps and lures
http://www.gemplers.com/pheromone-lures

Tactile communication
http://www.cals.ncsu.edu/course/ent425/library/tutorials/behavior/tactile.html

Vibrational communication in insects and spiders
http://www.mapoflife.org/topics/topic_584_Vibrational-communication-in-insects-and-spiders/

12 Small but Fortified

Insects Are Not Defenseless

You must not fear death, my lads; defy him, and you drive him into the enemy's ranks.

Napoleon Bonaparte
Emperor of France[*]

For animals that rarely reach two inches in length and more often are shy of an inch, each day is filled with challenges of finding food and avoiding being food for others. The daily grind for insects entails contending with those that are bigger, with the ability to engulf them whole or in parts, or those that are small enough to stealthily sneak up without detection and slip past the exoskeleton to extract the nutrients inside. Human society preaches that size does not matter. Try existing as an insect and see if that idiom is still true! The small size of insects makes living in terrestrial and aquatic environments a 24/7 ordeal, with life and death hanging in the balance. A multitude of predators and parasites view insects as food pantries. In practical terms, insects are bountiful stores of nutriment, containing on average five- to tenfold more nutrients within the confines of the exoskeleton than most any other food option available to predators and parasites. Nature is filled with such unwelcome diners, seeking insects to fill their gustatory needs. Predators are almost always larger and quite resourceful at capturing many insects for their dining pleasure. Likewise, parasitic species are very efficient in exploiting insects, although none would be classified as large. In fact, the vast majority of parasites are microorganisms that are smaller than any cell in the insect body. Only those species that are also insects rival the size of the host.

Chapter 12 examines the defensive strategies, behaviors, and physical and chemical defenses employed by different insect species to thwart unwanted attack. Some of the defenses rely on the various forms of communication discussed in chapter 11, and these will be reexamined here in relation to their use toward allospecifics and cannibalistic relatives. Defenses associated with microorganisms and other types of parasites that attempt to invade the intracellular environment, or that successfully break through the first line of defense to take up residence inside the host, will be discussed with an eye on the different habitats occupied by insects. As we are about to learn, despite

---

[*]Napoleon Bonaparte served as a military general and emperor of France (1804-1814, 1815) during the French Revolution and associated wars. He is regarded as one of the finest military minds the world has ever known.

their small size, insects are quite feisty and possess an extraordinary arsenal of weapons to defend themselves!

Key Concepts
* What are insects afraid of? Predatory and parasitic threats
* Hide and seek: The use of camouflage and mimicry to stay alive
* Behavioral tactics to combat predators and parasites
* Chemicals to the rescue: Allelochemicals
* Keep out! The role of the exoskeleton in protection from parasites
* Oh those wonderful hemocytes!

What are insects afraid of? Predatory and parasitic threats

Insects are unbelievably efficient at basically every task. Nearly all of us would point to their proficiency in reproduction as being in the top one or two categories that insects are really, really good at. A degree in entomology is not essential to recognize that insects are everywhere; we examined this idea in chapter 1 and chapter 9. Most of us are also aware that all insects do not look the same. Many varieties (species) of six-legged creatures inhabit the nooks and crannies of every ecosystem, with the only exception being marine environments, and even there, it is simply a matter of knowing where to look for them. We encountered this topic previously as species richness. So what do abundance and species richness have to do with insect defenses? And how do these ideas connect to what insects are scared of? Simple: the vast abundance of insects on the planet—estimated at over twenty quintillion individuals on any given day—means that they are a ready food source for a wide range of animals, including other insects. Even some plants show predatory tendencies for insects. The small size of juvenile and adult insects in relation to other animals undoubtedly provides a sense of confidence to most predatory species to pursue these creatures for a dining experience (figure 12.1). Of course prey size would have no bearing on attractiveness to plants. So then why are insects targeted (figure 12.2)? Is it just a matter of abundance and, to a lesser extent, size? Although numbers are important in terms of sustained sustenance, a very

Figure 12.1. A crab spider has subdued a robber fly (Diptera), showing that possession of wings is not always enough to escape predation. Photo by gbohne available at http://bit.ly/1KC7PSm.

Figure 12.2. A fly (Diptera) moments before becoming a meal for a predatory Venus fly trap. Photo by Conrad Erb, Chemical Heritage Foundation available at http://bit.ly/1VmzY3P.

important feature that is largely neglected in these types of discussions is the nutritional value of raw insects.

In chapter 8 we developed an understanding that insects obey neural commands that regulate what and when they should eat. The result is that nondomesticated insects are healthy and physically fit. For a predator, a meal of maggots, grubs, termites, and so on is not just slimy and satisfying,* it also a healthy choice over many other alternatives. Most insects offer an abundance of the basic macromolecules needed by all animals, some plants, and a number of microorganisms. In fact, the blood (**hemolymph**) concentrations of amino acids, carbohydrates, organic acids, and even some fats (di- and triglycerides) are commonly severalfold higher than in most other animals, including us. Why are insects so nutrient concentrated? The basic reason is associated with body pressure. Macromolecules function as solutes dissolved within the hemolymph to create osmotic pressure. In this capacity, the nutrients are more correctly referred to as **osmolytes**. The osmotic pressure of hemolymph helps maintain body shape, particularly of soft-bodied insects. Obviously the nutrients are also used in cell functions, like building molecules and membranes and serving as cellular fuels. By storing many macromolecules in the hemolymph, they can thus be used for the dual purpose of maintaining shape. This was a long-winded way of saying that insects are concentrated sources of nutrients, making them especially good food sources for predators and parasites.

So what are insects afraid of? Obviously, of being eaten! Insects are a primary option as a food source or to serve as host to a large number of parasites and predators. Insects are great nutrient source; they are smaller than many predators, making them appear easy to subdue; and they are the most abundant animals on the planet, meaning that insects are a reliable food source. Because insects are reliable and seasonally predictable, predators and parasites can evolve to utilize specific species. Perhaps we need to pause for a moment to recognize that this is one of the few times in our study of insects that we have really seen that these marvelous creatures can be vulnerable. Is there anything insects can do to balance the attack? The answer must be yes, otherwise world domination would not have been possible.

Insects rely on several strategies to overcome predators and parasites. The precise methods and mechanisms

are not identical, as there are fundamental differences between the parasites and predators that pose threats to insects. We will examine those differences shortly. For now, we can broadly classify the defenses of insects into four categories.

1. Morphological defenses
2. Behavioral defenses
3. Physiological (or chemical) defenses
4. The exoskeleton

The remainder of this chapter dives into these defense strategies and mechanisms within the context of antipredator and antiparasite mechanisms. We also need to add another category of threat to insects in the form of **parasitoids**. Parasitoids are insects that are specialized parasites that always end the association with the insect host by killing it. Parasitoids often evoke unique defense mechanisms in their insect hosts, and thus will be addressed as a subsection of antiparasite mechanisms (figure 12.3).

Before we begin our exploration of defenses, we need to delineate the distinctions between predators and parasites. A large number of animals—vertebrates and invertebrates—prey on insects. All predators of insects share some basic characteristics. They are typically bigger than their insect prey, they consume many prey individuals at a time, and they are often generalists

Figure 12.3. Caterpillar of the tobacco hornworm, *Manduca sexta* (Lepidoptera: Sphingidae) parasitized by a parasitic wasp. After the wasp larvae finish feeding on their host, they form pupae on the outside of the body. Photo by Stsmith and available at http://bit.ly/1WtSuK3.

---

*A reference to the phrase "slimy yet satisfying," an adage used in the Disney movie *The Lion King* (1994) wherein all sorts of insects are collected from rotting logs and then consumed by a meerkat, warthog, and lion cub (https://www.youtube.com/watch?v=HqREvb2VTjw).

(figure 12.4), meaning the predator is not picky about which insect species or stages of development they consume. Predatory insects may be much closer in size to their prey, but not always. In some cases, insects have a wide range of predatory species that target them. This is typical of many herbivorous species that feed on innocuous plant sources. The same is true of carrion-feeding insects that are rich sources of protein. However, other species may have a limited number of predators that seek them out. This occurs with species that live in concealment and consequently are difficult to detect and/or reach. It also occurs with species with unusual lifecycles, which can (but not always) make them unpredictable in timing and location. Those predators that feed almost exclusively on a few or a single insect species are termed specialists.

Parasites of insects differ from predators in that they usually are much smaller than the insect being targeted. The target is referred to as a host. A parasite needs the host for survival (table 12.1). The insect host is therefore not killed by the parasite; rather a portion of its nutrients is consumed so that the parasite can complete development and, in the case of microorganisms, reproduce within the host. To achieve these developmental goals, the parasite often must suppress the host's immune

Figure 12.4. Insects are frequently the preferred food choice of many large, nondiscriminating vertebrate predators. Photo by Alpsdake available at http://bit.ly/1LKvrYm.

Table 12.1. Types of common metazoan (animal) and microbial parasites that attack insects. Many species of insects use other insects as hosts. When the parasite always kills the host, the attacker is classified as a parasitoid.

Metazoan

- [ ] Nematodes
- [ ] Non-insect arthropods (i.e., mites, isopods)
- [ ] Parasitic wasps (order Hymenoptera)
- [ ] Parasitic flies (order Diptera)
- [ ] Twisted wing parasities (order Strepsiptera)
- [ ] Nest or space parasites (order Coleoptera, Diptera, and others)

Microbial

- [ ] Bacteria
- [ ] Protozoa
- [ ] Fungi
- [ ] Viruses

Figure 12.5. An adult wasp (Hymenoptera) parasitized by the fungus *Cordyceps* sp. Fungal hyphae are the spear-like projections radiating from the body. Photo by Erich Vallery and available at bugwood.org.

system to escape detection and combat internal defenses (figure 12.5). It is this facet of the relationship that may be the ultimate demise of the insect, because it becomes vulnerable to attack by other types of parasites. The original parasite itself does not deliberately try to kill the host, because that is not in the parasite's own self-interest. In fact, a "good" host may be suitable to sustain multiple generations of some parasites. Parasites utilize only one individual for sustenance, whereas a predator feeds on multiple individuals.

As mentioned earlier, parasitoids are specialized parasites that basically share the same characteristics as your average, everyday, run-of-the-mill parasite, with the exception that the host is always killed. Parasitoid relationships with their hosts are amazing studies of symbiosis. Many are parasitic wasps that use venoms, viruses, and specialized cells to subdue, paralyze, and manipulate the host so that it becomes a nutrient factory for the offspring of the wasps. Based on this brief description, it sounds like the insect hosts are defenseless. But the situation is anything but one-sided. In the next sections we examine the mechanisms that insects use to defend themselves from the plethora of creatures that want to eat them or reside inside them. Many of the strategies are designed to avoid detection in the first place. However, several rely on the insect taking matters into its own tarsi, using active defensive methods to thwart any and all

attackers, regardless of size and numbers. Let the hunger games begin!

Quick check

What are the differences between a predator, a parasite, and a parasitoid that feeds on insects?

Hide and seek: The use of camouflage and mimicry to stay alive

Insects use their bodies for defense. This is not to say that they will sacrifice themselves for the good of others. In fact, the opposite is more likely to be the case, where a sibling is nudged into the path of an oncoming predator to save numero uno, aka him or her insect self. Sacrifice for others implies emotional attachment, which there is absolutely no evidence that insects have. Of course, some social species do display group protection, especially for the greater good of the colony. In reality such actions are still relatively self-serving because of the genetic relatedness of all individuals in the hive or colony (a topic more suited for behavioral defense strategies that we'll examine in the next section). So, insects rely on morphological defenses to combat threats from predators. This is true to a lesser extent with fending off potential parasites and parasitoids. In any scenario, the strategies are generally passive, since there is not too much an individual can do to modify its appearance.

The most common approach is to evolve morphological features (appendages, coloration, shapes, and patterns) that permit evasion from attack in the first place. In other words, species utilize camouflage to hide in plain sight. However, this is not the only approach to morphological defenses, as some depend on stealing the skin of another creature or purposefully standing out in the crowd. Insect use of anatomical features for defenses can be classified as crypsis, mimesis, aposematic coloration, mimicry, and physical protection. Some of these morphological defenses were discussed in the last chapter as forms of visual communication. It should be apparent that the message is intended for allospecifics, with the goal, most of the time, of not being found by the receiver. As we delve into different examples of each form of morphological defense, it will also become apparent that overlap exists between crypsis, mimesis, and mimicry. Globally, the former two strategies are types of mimicry. Traditionally, however, each is treated as a separate type of defense and so we will do so here.

Crypsis is a defense adaptation as well as a form of visual communication. Crypsis or cryptic coloration relies on the insect displaying body coloration that allows it to blend into its surroundings. Many grasshoppers and katydids (Orthoptera) are colored green, gray, or brown, or a combination of shades, so that if they remain motionless, detection is nearly impossible when they are resting on a tree leaf, the bark of a tree, or in a grassy meadow. Obviously if the insect strays from its natural surroundings or seasonality results in a change in leaf coloration, the defense mechanism is no longer effective. This means that the insect does not migrate far from its defensive position, or is most active at night (nocturnal), when it cannot be seen (figure 12.6). Some insects are multi-colored or display patterned coloration that permits concealment, while others depend on countershading (the dorsal surface is lighter or darker than the sternum) or patterns that disrupt the lines of the body.

An alternative to relying on cryptic coloration alone for the purposes of camouflage is to resemble the shape of a natural object in the environment, a tactic called **mimesis**. The insect's body or portions of it may be shaped like a leaf, stick, thorn, berries, or even bird poop. The key for this strategy to work is for the insect to appear as an object that a predator or parasitoid would have no interest in. Some leaf insects have evolved to not only take on the appearance of a leaf from their preferred host plant, but also the margins of the abdomen are concave, as if a caterpillar had consumed a portion of the "leaf" (figure 12.7). Members of the hemipteran family Membracidae display some of the most elaborate body plans to hide in plain sight, from thorns and berries dangling above the thorax to translucent abdomens that give the appearance of plant leaf venation (figure 12.8). The eye spots used to evoke a **startle response** (see chapter 11 for more details) in would-be predators are also a form of mimesis, since the spots mimic the eyes of predatory bird species, particularly owls. Mimesis can occur with adults or larvae, and in most cases (other than startle response), as long as the insect does not move quickly, it remains elusive to detection by most predators and parasites (figure 12.9).

In contrast to most morphological defense strategies, in which avoidance is achieved through hiding, **aposematic coloration** is a blatant visual scream into the environment. Bright coloration in the shades of red, yellow, and orange, interspersed with black and white, is used to warn others of a payload of nastiness. The insect in

Figure 12.6. A katydid (Orthoptera) displays crypsis (the green coloration of the body), which camouflages the body when in a tree canopy, but not so much when "hiding" in a flower. Photo by David Burgess available at http://bit.ly/1OCvc37.

Figure 12.7. A leaf insect (Phasmatodea) displaying mimesis. It really looks like a leaf! Photo by McKay Savage available at http://bit.ly/1FDFqxg.

Figure 12.8. Insects in the family Membracidae (Hemiptera) show remarkable forms of mimesis, as this adult appears to have thorns and berries for a prothorax. Photo by Sergio Monteiro available at http://bit.ly/1OCZ9Qx.

Figure 12.9. The eye spots of an io moth (Lepidoptera: Saturniidae) are used to generate a startle response in a would-be attacker, which is not expecting to encounter the "eyes" of predatory bird species. Photo by Patrick Coin available at http://bit.ly/1FE39xp.

question is potentially noxious, toxic, or has the capacity to bite or sting. For this strategy to be effective, it is important for the insect to announce its presence for all to see. Colors used for aposematism are universal throughout the animal kingdom, and generally predators and parasites have an innate capacity to recognize and then avoid those displaying aposematic coloration. Failure to avoid such species can have deadly consequences (figure 12.10).

Some insects take on the coloration of species that possess well-developed defenses, when in actuality they are innocuous. Such insects are mimics; that is, they have evolved to share characteristics of another species or model. When an otherwise harmless insect evolves to mimic one that is harmful if eaten or provoked, the mimic displays **Batesian mimicry** (figure 12.11). The defenseless species gains protection by taking on the morphological features of another that is more than

Figure 12.10. Box elder bugs rely on aposematic coloration to indicate that they are the trifecta of badness: they release repugnant odors, are noxious if consumed, and bite. Photo by Bruce Marlin available at http://bit.ly/1MVQMPW.

capable of defending itself. Numerous species of insects from several orders have evolved this type of defense strategy. It has the added benefit that predators are not likely, or at least not many, to take the gamble to learn to distinguish between those that are innocuous and those that are toxic. One dining mistake by a predator may indeed be its last lunch.

By contrast, **Müllerian mimicry** occurs when multiple species, each being unpalatable or possessing defense capabilities, evolve very similar aposematic coloration. The approach is nature's version of a mass marketing campaign. Predators will make the same associations for all species with the similar warning coloration, and avoid them. There are numerous examples of such mimicry, including the often-cited (for instance, see chapter 11) monarch butterfly, *Danaus plexippus,* and the viceroy butterfly, *Limenitis archippus.* Both display similar but not identical patterns of orange, black, and white on the fore- and hindwings.

Mimicry can also occur without involving aposematic coloration; the mimic appears as another insect that is

Figure 12.11. A bee fly (Diptera: Bombyliidae) displaying Batesian mimicry. The fly's defense is to look like a bee. Photo by John Tan available at http://bit.ly/1LYhTEb.

equipped with aggressive or noxious defenses, or takes on an appearance to deceive potential attackers. Larval mimicry is an example of the latter, in which some species of moths and butterflies (Lepidoptera) showcase images of the caterpillar at the fringes of the forewings to divert attention away from the body. Crypsis and mimesis are technically also forms of mimicry. The external surface of many insects is adorned with an array of structures that confer some physical protection. The structures can range from hardened sclerites that function as body armor to more elaborate structures like spines, bristles, horns, and hairs. Sclerites are very effective at blocking ovipositor penetration by parasitic wasps, even more so when the soft cuticle of intersegmental membranes (conjunctiva) are physically covered by the sclerite plates. Similarly, heavily sclerotized regions of the exoskeleton do not necessarily prevent attack by predators but may serve to inhibit mandible penetration. The same can be said for cuticular extensions in the form of spikes, bristles, and hairs: they function as a deterrent during an actual attempt to bite down on the prey. A bird, frog, or toad is not likely interested in finishing the meal after getting a mouthful of spikes or hairs. The structures prevent a female parasitoid from inserting the ovipositor or from planting the legs for oviposition penetration or drilling. This is especially true with some parasitic wasps that rely on concealed hosts. Other structures like mandibles and stingers are used as well, and will be discussed later, as a function of aggressive or defensive behaviors.

Some species do in fact modify their appearance over time. The most passive form is through absorption and release of water vapor, which changes light reflection from the exoskeleton, thereby resulting in different body colors depending on relative humidity. Some species simply surrender the body part to the attacker. Scales covering the wings of moths and butterflies can detach in the mouth of a predator, allowing the adult insect to fly away. Other species notch this approach up to the level of **autotomy**, in which the actual appendage releases from a weak point in the exoskeleton. Depending on the species and age of the insect, it is possible to replace the lost appendage during the next molt. Still other forms of physical protection are much more aggressive. An extreme example occurs with some West African assassin bugs (Hemiptera: Reduviidae) that coat the dorsal surfaces of

the thorax and abdomen with sticky secretions. The bugs then add sticks, vegetation, and even cast-off skins of other insects to the adhesive to camouflage their bodies as a pile of debris. Many other examples exist of insects covering their bodies with objects from the environment to either hide or deceive.

Behavioral tactics to combat predators and parasites

If anatomical weapons fail to deter predators and parasitoids, or such defenses are not the answer for fending off attack, more active measures are needed. From an energetics perspective, the next level of defense involves behavioral tactics. Such mechanisms generally involve insect movement, either a change of body position or a call to motion. The point is that behavioral defenses are not passive, as are most examples of morphological defenses, yet are not necessarily as expensive as a long-term investment in a physiological or chemical defense adaptation. This example is not meant to give the impression of an either/or type of defense strategy. Most insects utilize multiple modalities to protect themselves from predators, parasites, and parasitoids. Based on the type and severity of attack, an insect may need to move through a series of defensive tactics, or alternatively, display the most severe (strongest) defense immediately.

This seems to be a good place to address an obvious question that has not yet been raised: How do insects recognize predators and parasitoids? The short answer is, through sensory perception. Visual recognition using photoreceptors in the compound eyes is the most common means for detection. Predators approaching from any direction will displace air, causing distortion of mechanical receptors and giving themselves away regardless of which end (or side) of the insect they attack. In some instances, chemical detection occurs through olfactory receptors. This is particularly common with social insects that use alarm pheromones for warning and also as an indication that an attack has already occurred.

Unfortunately for some species, recognition does not happen until tactile information is received. In other words, a stealthy predator or parasite may not be perceived until the act of aggression has been initiated. Tactile recognition is most likely to be associated with nonmobile stages of development, especially those stages engaged in some form of metabolic suppression (**aestivation**, **diapause**), or with eggs and pupae. The latter examples really have no opportunity to use behavioral defense strategics.

Before proceeding, yes, parasites have been omitted. Generally, microorganism detection occurs only after parasites have successfully entered the body cavity. So detection may not occur via sensory perception. Instead, first recognition of a foreign entity is by cells (**hemocytes**) associated with insect immune responses. Hemocytes carry out a number of defensive functions within circulating fluids (hemolymph, interstitial), including detection of certain types of non-self (foreign) cells, compounds, or particles, and then can relay that information to the rest of the body via chemical signal molecules (more details on that in the last section of the chapter).

Regardless of the means of recognition, insects frequently use behavioral responses to ward off or evade predators and parasitoids. This is especially true if an initial morphological strategy failed to keep the attackers at bay. The most common behavioral defenses include **thanosis**, construction of protective enclosures, escape responses, aggressive behaviors, **reflex bleeding**, and cooperative or group behavior. The mechanics of several of these behaviors are intuitively obvious, so not as much explanation is needed. Others are unique, or the name does not reveal what the insect does to defend itself. For example, thanosis is when the insect plays dead. This strategy is not unique to the world of insects. In fact, opossums are well known for this characteristic defense. The behavior only works for potential prey if the predator cannot detect the hunted when motionless or if an insect is likely to be rejected if already dead. Insect parasitoids are not as easily fooled, and will typically probe their food with the ovipositor, relying on chemosensory receptors at the tip to assess host quality rather than perceived behavioral activity. Some insects preemptively

Figure 12.12. Protective case of a caddisfly larva (Trichoptera) made of pebbles and sand. It offers the added benefit of blending in to the environment. Photo by Ashley Pond V available at http://bit.ly/1MC9Klz.

construct their own version of a fort in the form of tunnels or casings, or conceal themselves in another creature to afford protection, while others may seek shelter only once they are detected by a potential assailant (figure 12.12).

More active strategies include escape responses or maneuvers to flee an attacker that is close by or that has already made an unsuccessful attempt at eating or parasitizing the insect. For adults with wings, taking flight is the obvious first choice of escape. However, this strategy may not be immediately available, as the flight muscles must be prepared to initiate wing flapping—the process requires a minimum temperature threshold—and if the attacker strikes at the wings first, they may be too damaged to permit lift. Insects like grasshoppers, crickets, katydids, fleas, and some beetles have enlarged femurs associated with the metathoracic legs to permit jumping as an evasive maneuver (figure 12.13). If the jump alone is not sufficient to escape the predator, the prey may remain elusive by initiating flight. A jumping defense can

Figure 12.13. The enlarged femur of the metathoracic leg permits grasshoppers and other orthopterans to use jumping as an escape defense from predators. Photo by Victordk available at http://bit.ly/1NYel9t.

also be employed without legs, as occurs with a cheese skipper (a fly maggot in the family Piophilidae) that grabs its anus with its mouth to create muscular tension; upon release, the insect is flung into the air. Springtails (not an insect, but closely related) achieve a similar feat using the equivalent of a spring-loaded appendage (**ferculum** that fits into the **tenaculum**) to escape by springing into the air. Many insects simply jump from a higher location to land lower in the tree canopy, to another elevated locale, or to reach the ground. For these species, jumping permits fleeing immediate threats, but may in turn place them further in harm's way since they are landing into an area unknown to them.

In aquatic environs, swimming is equivalent to using wings to evade would-be attackers. Jumping while in water is physiologically very difficult because of the viscosity of water and the small mass of insects. However, some species (dragonflies in the order Odonata) make up for this shortcoming by relying on jet propulsion to escape from attackers. Movement is achieved in a manner similar to that of cephalopod mollusks, in that water is forcibly ejected from the body via a specialized rectal chamber in the anus. Short, quick bursts of movement are achieved, sufficient to permit escape in a cloud of sediment to find refuge under a submerged rock or log, or even longer-term swimming movement (https://www.youtube.com/watch?v=cEgZL32HSxo).

Speaking of the anus, glands near the aforementioned opening in some beetle species release a secretion that temporarily breaks the hydrogen bonds of water. When the bonds re-form, water is pushed forward, permitting the insects to jet across the water surface. A similar feat can be performed by hemipteran adults that carry their long mouthparts under the body with the opening pointed toward the posterior end; salivary secretions achieve the same effect on hydrogen bonds of water molecules. Water striders (Hemiptera) simply outrun any predator, shooting across the water surface in short bursts and in somewhat zigzag patterns (figure 12.14).

A few insect species respond to attack by reflex bleeding. Blood (hemolymph) is released from specific points along the body, most commonly at the legs. The preyed-upon insect is able to control the bleeding by increasing hemolymph pressure where the attacker is positioned. Hemolymph contains the toxin isopropyl methoxy pyr-

Figure 12.14. A water strider perched atop the surface of a stream, ready to shoot across the water if threatened. Photo by Tim Vickers available at http://bit.ly/1Vn4Scr.

azine, a compound that is toxic to most predators (the details are more suited for our discussion of chemical defense strategies). When reflex bleeding works well, the hemolymph coats the mandibles of the predator and then quickly begins to thicken (akin to clotting), so that the mouthparts become frozen in place. In the short term, the prey can escape while the predator deals with the unexpected problem. If the predator cannot remove the dried hemolymph, it will starve to death, or be consumed by a predator of its own.

Most behavior tactics are focused on the victim, meaning the insect prey, escaping from the attacker. However, many insect species are not passive to attack. They do not flee, but instead stand put to defend themselves. Others actually go on the offensive and chase after the predator. This is especially true for species that can bite and sting, even more so when the attacked can depend on siblings to "have its back." For these species, being attacked just seems to tick them off! Praying mantids (Mantodea) are classic examples of aggressive defenders: adults will turn to face a predator, fending it off with raptorial legs and biting with their powerful mandibles (figure 12.15). The "taking a stand" strategy makes sense for these insects because they are slow clumsy fliers, and running away is not an option due to the modified prothoracic legs. Many species will attempt to bite an

Figure 12.15. Raptorial legs of a praying mantis (Mantodea). The cuticular spikes on the legs make them a formidable weapon for fighting off predators and also for capturing prey. Photo by Bj. Schoenmakers available at http://bit.ly/1VjGX2S.

attacker, rather than turning tail. A forceful bite from chewing mandibles is an effective deterrent when an insect is facing only one challenger of comparable size. When the attacker is 'huge' in relation to the prey, or multiple individuals participate in a gang attack, mandibular defense becomes as futile attempt at survival. An extreme scenario occurs with some earwigs (Dermaptera), in which the mother will defend her brood of eggs to the death against a multitude of predators, some that are much larger, and others that attack en masse. The adult female is armed with only a modest pair of mandibles and pincer-shaped **cerci** that offer admirable but often insufficient defense against an army of foraging ants, a monster-sized venomous scorpion or spider, or small but menacing pseudo-scorpions (https://www.youtube.com/watch?v=V6eAYITila4; https://www.youtube.com/watch?v=g5Qz58jNlgE).

Aggression is also a defense against parasitoids. This is especially evident when parasitic wasps attempt to parasitize mobile stages of a particular host insect. For example, upon recognition that a female wasp is present, caterpillars and fly maggots initiate violent thrusts of the body to disrupt oviposition by the parasitoids. Oftentimes this movement is effective at deterring parasitoids, sometimes even causing injury to the wasps. To a lesser extent, the pupal stages of some beetle species will contract the abdomen, creating enough agitation to frustrate a wasp trying to insert the ovipositor. The balance sheet likely still favors the wasp that is patient, since the pupa has a limited source of nutriment and ATP to continue the muscular contractions.

Group defense is a strategy common with eusocial insects and also some that are solitary. Nonsocial species may form clusters or aggregations that permit group displays (visual and behavioral), as well as cooperative chemical defenses. The aggregations are often the result of pheromonal signaling, and as we discussed in chapter 10, since the receiver has no choice but to obey the commands of the message, thousands of individuals may make up the assemblage. Imagine the visual image of ten to twenty thousand orange and black ladybird beetles mass-accumulated on a single tree (figure 12.16). Large assemblages of necrophagous fly larvae on **carrion**, termed **maggot masses**, are believed to confer some protection against both predators and parasites. How the fly maggot masses form is not completely understood, but it is believed that large aggregations aid in predator avoidance in at least two ways. One is associated with **spatial partitioning**, in that the larvae positioned in the center of the mass are most protected from predation whereas those to the periphery are most susceptible. In mixed-species maggot masses, the species that is most heat tolerant typically resides in the center of the mass where temperatures are the highest, and thus competing species are driven to the periphery as a result of their thermal tolerance capacity.

The second aspect to maggot defense is associated with the unique ability of necrophagous fly larvae to

Figure 12.16. Aggregation of ladybird beetles ("lady bugs") (Coleoptera: Chrysomelidae). In some cases, literally thousands of individuals can cluster in a single location, providing a sea of aposematism as well as reflex bleeding. Photo by Singer Ron available at http://bit.ly/1KGAuH3.

Figure 12.17. Carnage following a wasp attack of a honey bee colony. The bees defend the nest cooperatively, fighting valiantly to the death to protect their queen and home. Photo by Qypchak available at http://bit.ly/1MCbmC3.

demonstrate **heterothermy**; that is, the flies constituting the larval aggregations can generate heat. Heat production is thus a major feature of maggot masses. The internal temperatures are influenced by a number of factors, including species composition, age of larvae, larval density and/or volume, and microbial activity in the mass. Maggot mass temperatures directly influence the metabolic rate of the flies. The net effect is that the higher the temperature (up to a species-specific point), the faster the development of the fly, and consequently the less time exposed to predators and parasites.

Cooperative defense is universal among social insects—to assault one individual is akin to attacking the entire hive or colony. At the most basic level, the first level of behavioral defense is the construction of the nest. All workers aid in the building processes, and the resulting nest confers protection to all members of the society, no matter what age or standing in the hierarchy. Individuals that are attacked by predators or parasitoids generally display an eye-for-an-eye response. In other words, they aggressively defend themselves, relying on chewing mandibles to ferociously bite the attacker. Social Hymenoptera also possess an **ovipositor** modified into a sting apparatus that permits stinging behavior. When necessary, the sting can deliver a potent venom cocktail that can minimally evoke pain, potentially incapacitate through paralysis, and, in some cases, elicit death (we'll

have a more thorough examination of salivary toxins and venoms when we discuss chemical defenses). Release of venom also mobilizes siblings to come to the aid of their sister* in distress. Alarm pheromones are stored in the venom, so injection into a predator sends the volatile chemical signals into the air. Depending on how close the alarm pheromone release is in relation to the location of the colony, a few to tens of individuals can be quickly summoned to protect the individual. Again, since an attack on one is considered the same as violating the whole (hive), the mobilized will fight back aggressively (stinging and biting), and repeatedly. For the demise of the initial aggressor, aka predator or parasitoid, the colony workers are prepared to fight till the death (figure 12.17).

Beyond the text

Reflex bleeding is an unusual defense, but may have its origins in another form of defense utilized by insects. Speculate how reflex bleeding may have evolved.

---

*Why only sisters? In the case of social Hymenoptera, the entire colony, or nearly so, comprises daughters of the queen. The few males that exist are there for mating purposes only, and thus they have no role in hive protection. The situation for termites (Isoptera) is substantially different, in that workers and soldiers can be male or female.

Figure 12.18. An array of stinging hair caterpillars: (A) io moth caterpillar, *Automeris* sp. (Alan Rockefeller, http://bit.ly/1WurZEh), (B) stinging rose caterpillar, *Parasa indetermina* (Megan McCarthy, http://bit.ly/1P2vfoa), (C) larva of *Lonomia obliqua* (Rodrigomorante, http://bit.ly/1MCc2r2), and (D) saddleback caterpillar, *Sibine stimulea* (thatredhead4, http://bit.ly/1Gdng0k).

Chemicals to the rescue: Allelochemicals

Of all the defenses displayed by insects, those involving chemicals may be the most familiar to humans—not by the names of the chemicals or the modes of action that deter us and other animals from engaging in further hostilities toward the six-legged soldiers. No, what we have is a basic understanding that certain insects will sting or bite us, leaving their mark in the form of welts, swelling (inflammation), pain, sometimes intense agony, and, in extreme scenarios, long-term injury and even death. The bite or sting alone does not generate these powerful responses. Rather, it is the injection of saliva or venom that produces the desired goal of thwarting the attacker. Each fluid contains a wealth of compounds that aid in the chemical defense of the supposedly vulnerable insect. Many of these venomous fluids have evolved to elicit an immediate and lasting impression, so that the aggressor does not repeat the activity. Aiding this defensive strategy is aposematic coloration; the beasts that bite and sting are often adorned in the bright warning colors and patterns described earlier as biological advertisement for trouble if messed with. Toxins can also be released in the form of secretions via glands that empty to the outside of the exoskeleton or by contact with hairs that sting. The latter form of toxin delivery is quite common with a number of caterpillars, which display a dazzling array of colors and hair patterns that amounts to a toxic fashion show (figure 12.18).

The use of toxins for defense is most well developed in true bugs (Hemiptera) and aculeate Hymenoptera. The latter group includes, essentially, all eusocial members of the order Hymenoptera, meaning ants, wasps, hornets, and bees, which possess a stinger (or sting apparatus) (figure 12.19). A cocktail of toxins and enzymes are produced in modified exocrine glands (venom glands) that produce and store venom. Most of the active components are small peptides and proteins that function by binding to cellular receptors associated with target cells, frequently inducing nonregulated opening of ion channels, which in turn can cause pain and paralysis.

Venom toxins produced by some ant and wasp species are among the most deadly of any animals toward humans (see the Fly Spot on page 321 for more details). Several types of enzymes—most commonly lipases and

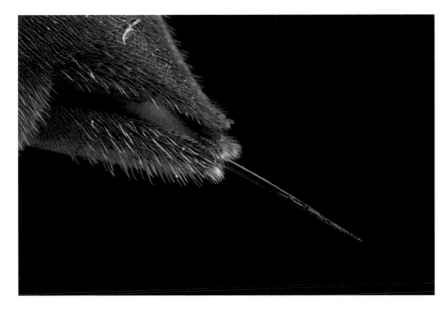

Figure 12.19. The sting or stinger of a wasp, *Polistes carolina*. A stinger typifies the aculeate Hymenoptera and represents both a physical and chemical deterrent to predation. Photo by Insects Unlocked available at http://bit.ly/1jrNAPm.

some carbohydrases—are also present in the venoms, and all function more or less in a nondiscriminate manner. This means that they do not require attachment to a receptor to carry out their function(s), and the enzymes do not show specificity for the type of cell affected. The destructive action of venom enzymes is most commonly responsible for inflammation responses that occur around the point of sting entry (figure 12.20). Ants are also capable of delivering toxins, albeit usually in the form of weak acids, through saliva injection. So when species like fire ants protect themselves or go on the offensive, they offer the dual threat of venom by mouth and stinger (figure 12.21). By contrast, some predatory true bugs, namely those with common names like assassin, pirate, or ambush bugs, pump highly toxic salivary venoms into prey and attackers alike.

In some instances, potent neurotoxins are injected that induce localized to systemic paralysis. Powerful digestive enzymes are also injected that nondiscriminately induce cellular lysis, which in turn triggers pain and inflammation. An array of carbohydrases, lipases, and proteases functions to cleave or hydrolyze various components of cellular membranes or soluble macromolecules. The bite of some bugs, such as those from the family Belostomatidae (water bugs or toe biters) or Red-

uviidae (assassin bugs), are known to cause an intense throbbing pain in a hand, foot, or anywhere else bitten, which can last for hours. Perhaps the most painful scenario is not a natural defense at all, but instead one devised by man. Assassin bugs featured in an old medieval torture technique would be starved for several days, and then placed in a pit with human prisoners. Hundreds of bugs would attack, injecting salivary venom into the prisoners. Enzymes in the saliva would slowly but effectively digest soft tissues of their human prey, eventually causing the tissues to slide right off the bone. The pain had to have been unimaginable (table 12.2).

Chemical defense strategies are not limited to just toxins. Many other compounds are in the arsenal of weapons available to insects, including some that are noxious (i.e., taste horrible and often cause **reverse peristalsis**); some that are irritants (such as sprays from bombardier beetles or secretions from some beetles, including during reflex bleeding); others that repel or are repugnant (the delightful odors of stink bugs); and several that are not even produced by insects. Those not produced by insects are derived from food sources and stored (**sequestered**) in the insect, so that if by chance the insect is consumed, it will likely be the attacker's last meal. The example of monarch butterflies discussed earlier (with

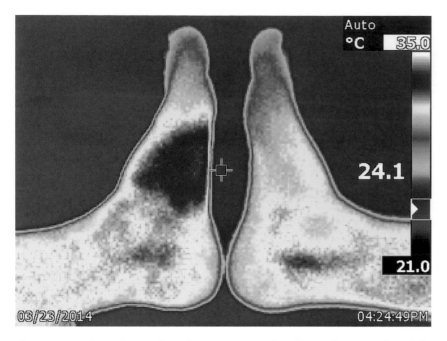

Figure 12.20. Thermal scan of two feet to compare the effects of an insect bite (left) with no bite (right). Inflammation is associated with area in red. Image by Volkan Yuksel available at http://bit.ly/1QJhg55.

Figure 12.21. Venom of the ant *Ectatomma tuberculatum* is one of the most lethal for humans of any insect venom known. Photo by Alex Wild available at http://bit.ly/1WusPB4.

aposematic coloration) reflects a defensive strategy relying on accumulation of toxic host plant compounds (cardiac glycosides) in their own tissues during herbivory. Collectively, all these chemicals are allelochemicals since they involve messages being shared across taxa. Specifi-

cally, the chemical agents described are **allomones**. As we learned in the last chapter, allomones are chemical signals that benefit the emitter at the expense of the receiver. Dying after venom injection is a clear example of a non-benefit to the receiver. An examination of allomones used

Table 12.2. Examples of the types of digestive enzymes present in the salivary venoms of predatory true bugs (Order Hemiptera).

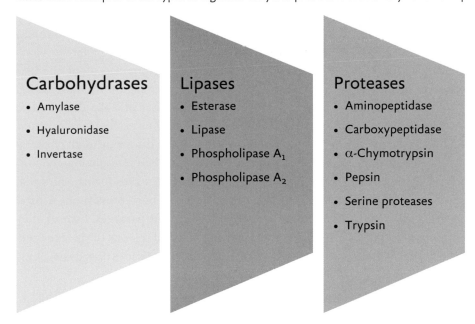

Carbohydrases
- Amylase
- Hyaluronidase
- Invertase

Lipases
- Esterase
- Lipase
- Phospholipase A_1
- Phospholipase A_2

Proteases
- Aminopeptidase
- Carboxypeptidase
- α-Chymotrypsin
- Pepsin
- Serine proteases
- Trypsin

by insects in total certainly gives the impression that the idea of chemical warfare may have originated with insects. If it did not, then insects have been among the most adept at adaption of a concept for their own personal use.

The chemical defense system of bombardier beetles (Coleoptera: Carabidae) deserves special consideration. Why? Because it's just plain cool! As alluded to earlier, bombardier beetles rely on chemical sprays for defense. But this simple description does not give a full picture of what they do. The central theme of the defense plan is to spray hot quinones out of openings near the anus, with the added feature of a loud pop resulting from microexplosions occurring in the glands. Many forms of quinones function as skin irritants to a wide range of animals. Heat the compounds up to near boiling (of water) now you have a memorable experience! How does the spray work? Two glands near the tip of the anus store a mixture of enzymes: catalases and peroxidases. In a separate reservoir, an aqueous solution of hydrogen peroxide and hydroquinones are stored until needed. When the beetle is threatened or physically attacked, the valve to the reservoir opens, permitting the aqueous solution to mix with the enzymes. A multistep exothermic chemical reaction occurs that ultimately produces a boiling hot liquid composed of 1,4-benzoquinone and gases. Gas buildup in a closed container means that internal pressure elevates. Once the external valves of the glands open, a pulse of hot quinone fluid sprays out. This can be repeated several times a second and can continue as long as new reactants are released into the glands. The beetle is protected from its own secretions by closing the valve to the reservoir immediately after mixing of reactants, and by the thick-walled design of each gland. The recipient of the hot irritating quinone spray immediately ceases the assault on the beetle and typically does not make another attempt. Interestingly, most of these ground beetles show no signs of aposematism. Consequently, the first encounter with bombardier beetles by a predator appears to be out of sheer ignorance regarding what is about to occur.

Bug bytes
Bombardier beetles
http://yhoo.it/1P2xFTP

The final types of chemicals used in defense do not fall into any of the categories mentioned so far. Instead, they function as communication signals to prime for other types of defenses, namely behavioral or chemical.

The chemical signals are pheromones and kairomones. Details of their roles in insect communication are discussed in chapter 11, and so will not be repeated here. What is important to note is that pheromones are vital for conveying alarm when others are threatened, promoting aggregation for a display defense, or mobilizing the forces of social insects to defend the hive or colony. We already know what happens when you rile the nest; a mass attack involving biting and stinging occurs. Woe to the attacker that makes this mistake! Kairomonal signaling works similarly, but generally involves only solitary insects. In these instances, volatile chemical signals are released from an organism during predatory or parasitic attack. A different species detects the compound and then migrates toward the releaser. Once the emitter has been discovered, the insect version of a superhero assaults the villain, which in this case is the attacker of the emitter. Are you still with me?

Here is an example that demonstrates how kairomones function in defense. An adult moth known as the corn earworm, *Helicoverpa zea* (Lepidoptera: Noctuidae) deposits an egg mass on a corn plant. Following egg hatch, the larvae (caterpillars) begin to feed on the reproductive tissues of the corn. Eventually, the young caterpillars embed themselves in the corn ear. As they feed, the corn plant responds by producing secondary compounds that are released into the air. If the chemicals are detected by the parasitic wasp *Microplitis croceipes* (Hymenoptera: Braconidae), the compounds have officially become kairomones. The wasp can then follow the odor plume back to the emitter (the corn plant), to locate the caterpillars. In turn, female wasps will parasitize the herbivore. In this example, immediate salvation is not achieved for the corn plant, since the wasp's strategy is to keep the host alive for a period of time. Eventually, however, the caterpillar will die. Thus, kairomonal signaling is a long-term strategy that does not generate instantaneous results like most chemical defense strategies.

Keep out! The role of the exoskeleton in protection from parasites

The exoskeleton is an amazing structure in terms of everything it does for an insect. At the very least, it can be considered a dynamic barrier to the environment, keeping at bay most living and nonliving entities that would otherwise pass into the nutritious internal world of an insect's body. The design of the exoskeleton is very effective in regulating what moves in and out of the body. By necessity, the structure cannot be completely impervious. For example, openings (**spiracles**) along the thorax and abdomen permit gas exchange, the mouth intakes food and water, the anus releases metabolic and digestive wastes, and the gonopore permits exchange of gametes. The point is that the insect skin cannot be totally impregnable, and because it is not, there are obvious regions that can serve as potential entry sites for parasites that wish to utilize the insect as a host. Sometimes natural openings are not required by the parasites. Instead, they penetrate through less fortified regions of the exoskeleton, such as intersegmental membranes or **ecdysial** cleavage lines.

Still other organisms, like several species of fungi or protozoa, rely on hyphae or polar filaments, respectively, to ram, drill, or digest their way through even the most sclerotized (hardened) portion of the exoskeleton. And some are opportunistic, taking advantage of a wound site in the integument to gain passage into the insect's body. It is safe to say that in most cases, microorganisms are the main parasites that target the exoskeleton as the port of entry into their desired host. This answers, in part, a question that perhaps has crossed your mind, particularly after reading about how insects are not passive to attack: How do microorganisms so easily or effectively cross the exoskeleton? Generally they are not recognized until physically making contact with an individual. In fact, recognition of non-self usually does not happen until after the parasite or its products encounter living cells or circulating fluids. That basically means the microorganisms are not detected before reaching the epidermis, which is buried under several layers of nonliving cuticle (chapter 6 reviews the layers and functions of the exoskeleton). This seems to imply that some parasites penetrate deep into the insect's body before meeting with any type of resistance. As we are about to see, the exoskeleton presents a formidable defense even to stealth-like microorganisms. So the scorecard is not balanced in favor of the parasites; rather an intense evolutionary battle is at play between insects and the microorganisms that wish to use them as hosts. Eventually equilibrium is reached in which neither has the overall upper hand in the relationship,

Insects are not passive when it comes to protecting themselves. Nowhere is this better illustrated than with the chemical defense strategies employed by several groups. In many instances, insects are so "confident" in their ability to ward off an attack that they go on the offensive. This is especially true with the aculeate Hymenoptera, meaning members of the suborder Apocrita, which possess a sting apparatus and are well known to us for their ability to inject painful venoms. But stinging is not the only method of toxin delivery. Other insects rely on biting, stinging hairs or even secretions to introduce venoms or toxins to potential attackers. Most of these chemical weapons are designed to trigger short-term pain, lasting just long enough to permit escape. However, for a small but significant number of insects, the toxic agents of defense are highly potent, resulting in long-term injury and even death.

Why do some insects produce compounds that are so much more potent than others? The key is in effective and efficient defensive strategies. Chemical compounds, particularly those that are amino acid-based, are energetically expensive to produce. Thus, to maximize a toxin's effectiveness, it is ideal that a little go a long way. For predatory species, intense and painful interactions with prey as a result of stinging, biting, or secretion are more likely to be remembered. In other words, venom or toxin potency equates to long-term memory.

High potency toxins made to ward off potential predators can have deadly consequences for us. In the truest sense, insect venoms did not evolve to drive off humans; other species are their primary targets. As a result, humans have not adapted to the effects of insect venoms. This is evident in the simplest form of deadly venom interactions, **anaphylaxis**. Anaphylaxis is an acute systematic allergic reaction to the components of insect venoms or secretions. The constituents are not deadly in terms of their mode of action toward us. Instead, individual components may represent a unique allergen to our immune system that in turn triggers a hypersensitive reaction that can cause mild to major health problems. Death can and frequently does result. You likely know someone who carries injectable epinephrine in a pen syringe with them for use in case of anaphylaxis.

In other instances, insect fluids do indeed contain toxins that are lethal. These venoms and secretions are designed to be destructive, either through modifying specific signal transduction pathways or evoking cellular chaos by nondiscriminate modes of action. Fortunately, not many species are capable of producing such toxins, and very few occur in North America. Of those insects that synthesize the deadliest toxins, the vast majority are members of the order Hymenoptera. For example, the ant species *Ectatomma tuberculatum* and *Pogonomyrmex maricopa* (family Formicidae) produce the most lethal toxins of any known insect (**Figure 12.21**). Others are highly toxic as a result of group attack, such as occurs with the red imported fire ant *Solenopsis invicta* or the Oriental wasp *Vespa mandarina* (family Vespidae). The latter delivers the most toxic payload of venom of any animal on the planet when taking into account that the wasp attacks as a large group.

Lethality is not restricted to one insect group, as there are examples of beetles (Coleoptera), bugs (Hemiptera), and even caterpillars (Lepidoptera) that generate potent and deadly toxins. One of the most deadly is a stinging hair caterpillar (*Lonomia oblique*) found in South America. Toxins are released by tactile stimulation (touch) and evoke an immediate painful response in humans. The injected venom is hemorrhagic, meaning that diffuse and uncontrolled bleeding occurs. A sufficiently high dose will result in death. Adding to the nastiness of the caterpillar, as well as a few other species, is that they do not advertise their toxicity via aposematic coloration. This is in sharp contrast to most toxic species, which are readily evident by their distinctive bright colors and banding patterns. Those that do not telegraph their existence usually attempt to hide by relying on some form of camouflage—avoiding conflict appears to be their method. Others, however, offer no warning and do not hide. They are just plain sadistic!

but at any given moment, one or the other may have a slight advantage.

How does the exoskeleton provide protection from parasites and even some predators? In the simplest form, a hardened, heavily sclerotized cuticle is nearly impenetrable by any potential invader or hungry predator. Again, these are the regions other than the vulnerable areas just mentioned. That said, if all of the insect's natural body openings are closed, and the body is positioned so that the armored sclerite plates cover the openings and intersegmental membranes, a locked-down fortress is created. Few creatures will be able to breach this simple yet amazingly effective defense. But some inevitably do! Just think how difficult it is to be an insect living in an aquatic environment or within burrows of a subterranean habitat. Attack can come from all angles, almost constantly; the slightest mistake in assuming a defensive position is all that is needed for a microscopic organism to slip past and in. What can an insect do at that point?

The next line of defense, once the exoskeleton barrier has been penetrated, is to rely on enzymes and, in some cases, circulating cells called hemocytes. Several types of enzymes are present throughout the cuticular layers, many functioning to produce various components in the layers or contributing to chemical reactions that tan (**melanize**) or harden (sclerotize) the exoskeleton. These enzymes—or more often, intermediate products from the chemical reactions they mediate—are toxic to many organisms, including microorganisms and several insect larvae. Phenolic compounds are just one example of these protective intermediaries. An array of proteases contributes to the immune defenses of the insect cuticle, largely by targeting membrane-bound proteins in cell walls of microorganisms. One class of these enzymes is the cathepsins, all of which are classified as serine proteases. Mixed-function oxidases similarly function in cuticular defense but more in the role of detoxification of chemicals that have crossed the exoskeleton or that were secreted by parasites once past the first line of defense.

Cellular defenses are especially critical to recognizing and removing parasites that enter the insect's body. Hemocytes have the principal role in this regard, but generally only after the invaders reach the hemolymph-filled body cavity (those details are presented in the next section). Some are believed to be associated with the epidermis and perhaps innermost layers of the cuticle. Oenocytes or oenocytoids are a type of hemocyte that has a defined role in synthesizing structural cuticular lipids and, in some insects, also in the production of the enzyme phenoloxidase, which functions in defensive reactions. The reactions yield phenolic compounds that are toxic to most microorganisms and parasitic insects.

Oh those wonderful hemocytes!

When all else fails, turn to cellular defenses. Of course this refers to what happens if the parasites have not been deterred by the time they reach hemolymph and living tissues. Bacterial cells and other types of microorganisms will almost immediately be identified as non-self once infesting the insect blood (hemolymph). Proteins that circulate in insect hemolymph, called lectins, bind to carbohydrates in the cell walls of the bacteria. Lectin-bound cells are now flagged by circulating defense cells, hemocytes, to remove the invaders, usually by engulfing the foreign cell. This cellular consumption is termed **phagocytosis**. Plasmatocytes are the dominant type of hemocytes involved in phagocytosis and function when the numbers of microorganisms are low. Bacterial cells, fungal spores, or protozoans can be effectively engulfed and then digested once inside the plasmatocytes.

What happens if a significant infection occurs? In other words, what if the numbers of microorganisms become excessively high, so that phagocytosis alone will not be able to rid the insect of the infection. The typical response is mobilization of multiple hemocytes, namely granulocytes (or granular cells) and plasmatocytes, to form nodules. **Nodule formation** entails the bacteria or other foreign entity being surrounded by granulocytes, trapping them in pockets of hemolymph. In some insects, the cells release phenoloxidases to trigger melanization reactions (i.e., production of phenolic compounds), which in turn begin to lyse the bacterial cell walls. In other species the enzymes are already present in the hemolymph, albeit in an inactive form. Eventually, plasmatocytes are summoned to the sites so that they can form a layer of cells surrounding granulocyte-bacterial complexes. The end result is a nodule that has been formed by layers of plasmatocytes to the outside, microorganisms to the inside, and granulocytes sandwiched in between (table 12.3).

Table 12.3. Examples of cellular responses to non-self within the body of an insect. The hemocytes involved in any response are typically order specific, so the listing given may not be applicable to all groups of insects.

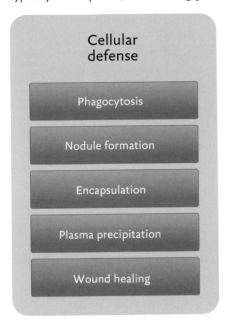

Cellular defense	Hemocytes involved
Phagocytosis	Plasmatocytes, granular cells
Nodule formation	Plasmatocytes, granular cells, oenocytes
Encapsulation	Plasmatocytes, granular cells, oenocytes
Plasma precipitation	Plasmatocytes, granular cells
Wound healing	Plasmatocytes, granular cells, spherule cells

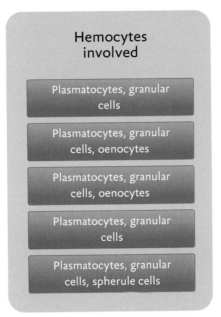

When the invasion involves organisms that are much bigger than microorganisms, such as parasitic insect larvae, the cellular response is associated with **encapsulation**. Larger attackers require a more rapid mobilization of hemocytes and usually a higher recruitment response, meaning more cells are needed. Encapsulation works by granulocytes recognizing an insect or nematodes as non-self in the hemocoel, and then immediately degranulating. **Degranulation** is the process of releasing granules and other intracellular components onto the surface of the parasites. The granules serve as recruitment molecules, drawing the attention of other hemocytes (plasmatocytes) to the invaders. Plasmatocytes arrive and then begin to attach to the parasites. These cells will flatten and spread, forming continuous membranous sheets around the insect larvae or nematodes. Several layers of plasmatocytes will form around the parasites. Typically, phenoloxidase is released to stimulate melanization. The result is formation of a hardened outer layer of the capsule. If the enzyme is released inside the layers, then the toxic phenolic compounds produced will likely kill the parasites. However, if melanization occurs exclusively in the outer layers of the capsule, death of the parasites is more typically the result of suffocation or starvation.

Numerous proteins produced by fat body tissues also aid in the removal of bacteria from the hemolymph. Among the most prominent are antibacterial proteins, which often are species-specific in terms of size, amino acid composition, and type of bacteria targeted. Cecropins, manducins, sarcotoxins, and hemolins are just some of the most common antibacterial proteins produced by insects. These compounds generally bind to the surface of bacterial cells to induce lysis. Other bactericidal proteins are in the form of enzymes. Phenoloxidases and cathepsins found in the hemolymph have functions similar to those described in the cuticle, but differ in chemical structure. **Lysozymes** are produced in response to non-self-recognition. These enzymes digest bacterial cell walls, either before or after antibacterial proteins attack the foreign cells. In some insects, lysozyme concentrations increase severalfold during the process of molting, apparently as a preemptive defense strategy. Molting is when the insect is most susceptible to predation and parasitism, since the new cuticle begins as a soft, non-tanned and non-sclerotized structure. Obviously, in such a state, the exoskeleton affords none of the protections we have discussed. Thus, the importance of hemocytes and hemolymph proteins elevates during a molt.

CHAPTER REVIEW

✴ **What are insects afraid of? Predatory and parasitic threats**

- Many varieties of six-legged creatures inhabit the nooks and crannies of every ecosystem, with the only exception being marine environments, and even there, it is simply a matter of knowing where to look for them. What do abundance and species richness have to do with insect defenses? The vast abundance of insects on the planet—estimated at over twenty quintillion individuals on any given day—means that they are a ready food source for a wide range of animals, including other insects. Even some plants show predatory tendencies for insects. The small size of juvenile and adult insects in relation to other animals undoubtedly provides a sense of confidence to most predatory species to pursue these creatures for a dining experience. Although numbers are important in terms of sustained sustenance, a very important feature that is largely neglected in these types of discussions is the nutritional value of raw insects.

- Insects are a primary option as a food source or to serve as host to a large number of parasites and predators. Insects are great nutrient source; they are smaller than many predators, making them appear easy to subdue; and they are the most abundant animals on the planet, meaning that insects are a reliable food source. Because insects are reliable and seasonally predictable, predators and parasites can evolve to utilize specific species.

- Insects rely on several strategies to overcome predators and parasites. The precise methods and mechanisms are not identical, as there are fundamental differences between the parasites and predators that pose threats to insects. The defenses of insects can be broadly classified into four categories: morphological defenses, behavioral defenses, physiological (or chemical) defenses, and the exoskeleton.

- Predators and parasites of insect can easily be delineated based on physical and behavioral characteristics. All predators of insects share some basic characteristics. They are typically bigger than their insect prey, they consume many prey individuals at a time, and they are often generalists. Parasites of insects differ from predators in that they usually are much smaller than the insect being targeted. The target is referred to as a host. A parasite needs the host for survival. The insect host is therefore not killed by the parasite; rather a portion of its nutrients is consumed so that the parasite can complete development and, in the case of microorganisms, reproduce within the host. Parasites utilize only one individual for sustenance, whereas a predator feeds on multiple individuals. Another category that feeds on insects are parasitoids. These specialized parasites share the same characteristics as your average, everyday, run-of-the-mill parasite, with the exception that the host is always killed.

✴ **Hide and seek: The use of camouflage and mimicry to stay alive**

- Insects rely on morphological defenses to combat threats from predators. This is true to a lesser extent with fending off potential parasites and parasitoids. In any scenario, the strategies are generally passive, since there is not too much an individual can do to modify its appearance. The most common approach is to evolve morphological features that permit evasion from attack in the first place. In other words, species utilize camouflage to hide in plain sight.

- Insect use of anatomical features for defenses can be classified as crypsis, mimesis, aposematic coloration, mimicry, and physical protection. Some of these morphological defenses are forms of visual communication. It should be apparent that the message is intended for allospecifics, with the goal, most of the time, of not being found by the receiver. Some of these defenses do overlap, as occurs with crypsis, mimesis and mimicry.

- In contrast to most morphological defense strategies, in which avoidance is achieved through hiding, aposematism is a blatant visual scream into the environment. Bright coloration in the shades of red, yellow, and orange, interspersed with black and white, is used to warn others of a payload of nastiness. The insect in question is potentially noxious, toxic, or has the capacity to bite or sting. For this strategy to be effective, it is important for the insect to announce its presence for all to see. Colors used for aposematism are universal throughout the animal kingdom, and generally predators and parasites have an innate capacity to recognize and then avoid those displaying aposematic coloration.

- Some insects take on the coloration of species that possess well-developed defenses, when in actuality they are innocuous. Such insects are mimics; that is, they have evolved to share characteristics of another species or model. When an otherwise harmless insect evolves to mimic one that is harmful if eaten or provoked, the mimic displays Batesian mimicry. By contrast, Müllerian mimicry occurs when multiple species, each being unpalatable or possessing defense capabilities, evolve very similar aposematic coloration. Predators will make the same association for all species with the similar warning coloration, and avoid them.

Behavioral tactics to combat predators and parasites

- If anatomical weapons fail to deter predators and parasitoids, or such defenses are not the answer for fending off attack, more active measures are needed. From an energetics perspective, the next level of defense involves behavioral tactics. Such mechanisms generally involve insect movement, either a change of body position or a call to motion. The point is that behavioral defenses are not passive, as are most examples of morphological defenses, yet are not necessarily as expensive as a long-term investment in a physiological or chemical defense adaptation.

- The most common behavioral defenses include thanosis, construction of protective enclosures, escape responses, aggressive behaviors, reflex bleeding, and cooperative or group behavior. The mechanics of several of these behaviors are intuitively obvious. Others are unique or the name does not reveal what the insect does to

defend itself. For example, thanosis is when the insect plays dead. The behavior only works for potential prey if the predator cannot detect the hunted when motionless or if an insect is likely to be rejected if already dead. Insect parasitoids are not as easily fooled, and will typically probe their food with the ovipositor, relying on chemosensory receptors at the tip to assess host quality rather than perceived behavioral activity. Some insects preemptively construct their own version of a fort in the form of tunnels or casings, or conceal themselves in another creature to afford protection, while others may seek shelter only once detected by a potential assailant.

- More active strategies include escape responses or maneuvers to flee an attacker that is close by or that has already made an unsuccessful attempt at eating or parasitizing the insect. For adults with wings, taking flight is the obvious first choice of escape. Insects like grasshoppers, crickets, katydids, fleas, and some beetles have enlarged femurs associated with the metathoracic legs to permit jumping as an evasive maneuver. If the jump alone is not sufficient to escape the predator, the prey may remain elusive by initiating flight. A jumping defense can also be employed without legs, as occurs with a cheese skipper that grabs its anus with its mouth to create muscular tension; upon release, the insect is flung into the air. Many insects simply jump from a higher location to land lower in the tree canopy, to another elevated locale, or to reach the ground.

- In aquatic environs, swimming is equivalent to using wings to evade would-be attackers. Jumping while in water is physiologically very difficult because of the viscosity of water and the small mass of insects. However, some species make up for this shortcoming by relying on jet propulsion to escape from attackers. Movement is achieved in a manner similar to that of cephalopod mollusks, in that water is forcibly ejected from the body via a specialized rectal chamber in the anus.

- A few insect species respond to attack by reflex bleeding. Hemolymph is released from specific points along the body, most commonly at the legs. The preyed-upon insect is able to control the bleeding by increasing hemolymph pressure where the attacker is positioned. When reflex bleeding works well, the hemolymph coats the mandibles of the predator and then quickly begins to thicken, so that the mouthparts become frozen in place.

- Most behavior tactics are focused on the victim, meaning the insect prey, escaping from the attacker. However, many insect species are not passive to attack. They do not flee, but instead stand put to defend themselves. Others actually go on the offensive and chase after the predator. This is especially true for species that can bite and sting, even more so when the attacked can depend on siblings to "have its back," as in the case of cooperative or group defense.

Chemicals to the rescue: Allelochemicals

- Of all the defenses displayed by insects, those involving chemicals may be the most familiar to humans. We have is a basic understanding that certain insects will sting or bite us, leaving their mark in the form of welts, swelling, pain, sometimes intense agony, and, in extreme scenarios, long-term injury and even death. The bite or sting alone does not generate these powerful responses. Rather, it is the injection of saliva or venom that produces the desired goal of thwarting the attacker. Each fluid contains a wealth of compounds that aid in the chemical defense of the supposedly vulnerable insect. Many of these venomous fluids have evolved to elicit an immediate and lasting impression, so that the aggressor does not repeat the activity. Aiding this defensive strategy is aposematic coloration; the beasts that bite and sting are often adorned in the bright warning colors and patterns described earlier as biological advertisement for trouble if messed with.

- The use of toxins for defense is most well developed in true bugs and Hymenoptera. Most of the active components are small peptides and proteins that function by binding to cellular receptors associated with target cells, frequently inducing nonregulated opening of ion channels, which in turn can cause pain and paralysis. Venom toxins produced by some ant and wasp species are among the most deadly of any animals toward humans. Several types of enzymes—most commonly lipases and some carbohydrases—are also present in the venoms, and all function more or less in a nondiscriminate manner.

- Some predatory true bugs, namely those with common names like assassin, pirate or ambush bugs, pump highly toxic salivary venoms into prey and attackers alike. In some instances, potent neurotoxins are injected that induce localized to systemic paralysis. Powerful digestive enzymes are also injected that nondiscriminately induce cellular lysis, which in turn triggers pain and inflammation. An array of carbohydrases, lipases, and proteases functions to cleave or hydrolyze various components of cellular membranes or soluble macromolecules.

- Chemical defense strategies are not limited to just toxins. Many other compounds are in the arsenal of weapons available to insects, including some that are noxious, some that are irritants, others that repel or are repugnant, and several that are not even produced by an insect. Those not produced by insects are derived from food sources and stored in the insect, so that if by chance the insect is consumed, it will likely be the attacker's last meal. The example of monarch butterflies discussed earlier reflects a defensive strategy relying on accumulation of toxic host plant compounds in their own tissues during herbivory.

- The final types of chemicals used in defense do not fall into any of the categories mentioned so far. Instead, they function as communication signals to prime for other types of defenses, namely behavioral or chemical. The chemical signals are pheromones and kairomones.

Keep out! The role of the exoskeleton in protection from parasites

- The exoskeleton is an amazing structure in terms of everything it does for an insect. The design of the exoskeleton is very effective in regulating what moves in and out of the body. By necessity, the structure cannot be completely impervious. Openings along the thorax and abdomen permit gas exchange, the mouth intakes food and water, the anus releases metabolic and digestive wastes, and the gonopore permits exchange of gametes. The point is that the insect skin cannot be totally impregnable, and because it is not, there are obvious regions that can serve as potential entry sites for parasites that wish to utilize the insect as a host. Sometimes natural openings are not required by the parasites. Still other organisms, like several species of fungi or protozoa, rely on hyphae or polar filaments, respectively, to ram, drill, or digest their way through even the most sclerotized portion of the exoskeleton. And some are opportunistic, taking advantage of a wound site in the integument to gain passage into the insect's body.

- Generally parasites are not recognized until physically making contact with an individual. In fact, recognition of non-self usually does not happen until after the parasite or its products encounter living cells or circulating fluids. That basically means the microorganisms are not detected before reaching the epidermis, which is buried under several layers of nonliving cuticle.

- How does the exoskeleton provide protection from parasites and even some predators? In the simplest form, a hardened, heavily sclerotized cuticle is nearly impenetrable by any potential invader or hungry predator. Again, these are the regions other than the vulnerable areas just mentioned. That said, if all of the insect's natural body openings are closed, and the body is positioned so that the armored sclerite plates cover the openings and intersegmental membranes, a locked-down fortress is created. Few creatures will be able to breach this simple yet amazingly effective defense.

- The next line of defense once, the exoskeleton barrier has been penetrated, is to rely on enzymes and, in some cases, circulating cells called hemocytes. Several types of enzymes are present throughout the cuticular layers, many functioning to produce various components in the layers or contributing to chemical reactions that tan or harden the exoskeleton. Cellular defenses are especially critical to recognizing and removing parasites that enter the insect's body. Hemocytes have the principal role in this regard, but generally only after the invaders reach the hemolymph-filled body cavity.

Oh those wonderful hemocytes!

- Bacterial cells and other types of microorganisms will almost immediately be identified as non-self once infesting the hemolymph. Proteins that circulate in insect hemolymph, called lectins, bind to carbohydrates in the cell walls of the bacteria. Lectin-bound cells are now flagged by circulating defense cells, hemocytes, to remove the invaders, usually by engulfing the foreign cell. This cellular consumption is termed phagocytosis. Plasmatocytes are the dominant type of hemocytes involved in phagocytosis and function when the numbers of microorganisms are low.

- What happens if a significant infection occurs? In other words, what if the numbers of microorganisms become excessively high so that phagocytosis alone will not be able to rid the insect of the infection. The typical response is mobilization of multiple hemocytes, namely granulocytes and plasmatocytes, to form nodules. Nodule formation entails the bacteria or other foreign entity being surrounded by granulocytes, trapping them in pockets of hemolymph.

- When the invasion involves organisms that are much bigger than microorganisms, such as parasitic insect larvae, the cellular response is associated with encapsulation. Larger attackers require a more rapid mobilization of hemocytes and usually a higher recruitment response, meaning more cells are needed. Encapsulation works by granulocytes recognizing an insect or nematodes as non-self in the hemocoel, and then immediately degranulating. The granules serve as recruitment molecules, drawing the attention of other hemocytes to the invaders. Plasmatocytes arrive and then begin to attach to the parasites. These cells will flatten and spread, forming continuous membranous sheets around the insect larvae or nematodes.

- Numerous proteins produced by fat body tissues also aid in the removal of bacteria from the hemolymph. Among the most prominent are antibacterial proteins, which often are species-specific in terms of size, amino acid composition, and type of bacteria targeted. Other bactericidal proteins are in the form of enzymes.

MUSHROOM FARMING (SELF-TEST)

Level 1: Knowledge/Comprehension

1. Define the following terms:
   (a) allomone   (d) sequestration
   (b) mimesis   (e) hemocyte
   (c) thanosis   (f) autotomy

2. Describe the morphological defenses used by insects to ward off predators.

3. What types of behaviors are used by insects for defense?

4. Explain how insects use cooperation or group behavior to defend against attack by predators.

5. Provide examples of how insects use chemical defense strategies for protection from predation and parasitism.

6. Discuss how components of the exoskeleton deter entry of microorganisms into the insect's body.

7. Describe how hemocytes combat bacterial infections in the hemolymph.

Level 2: Application/Analysis

1. Explain how crypsis, mimesis, and mimicry are essentially identical forms of defense used by some insects.

2. Describe how the cellular defense responses of insects differ if a protozoan infestation occurs versus a parasitic wasp depositing a clutch of ten eggs in the hemolymph.

3. Discuss how kairomonal signaling works as part of the defense strategy of some organisms subjected to insect herbivory.

4. Why are some insect venoms more potent or painful than others?

Level 3: Synthesis/Evaluation

1. For insects residing within concealed locations like underground burrows, what is/are the most likely defense strategies used? Explain why.

2. Speculate on why some highly noxious or toxic insects do not advertise aposematic coloration and almost seem to relish being attacked.

3. Explain when a pheromonal signal of one insect may serve as a defensive kairomonal signal for another.

REFERENCES

Ashida, M., and P. T. Brey. 1995. Role of the integument in insect defense: Pro-phenol oxidase cascade in the cuticular matrix. *Proceedings of the National Academy of Sciences (USA)* 92:10698-10702.

Brandt, M., and D. Mahsberg. 2002. Bugs with backpacks: The function of nymphal camouflage in the West African assassin bugs: *Paredocla* and *Acanthiaspis* spp. *Animal Behaviour* 63:277-284.

Brossut, R. 1983. Allomonal secretions in cockroaches. *Journal of Chemical Ecology* 9:143-158.

Cardé, R. T., and J. G. Millar. 2011. Advances in Insect Chemical Ecology. Reissue ed. Cambridge University Press, Cambridge, UK.

Chapman, R. F. 1998. The Insects: Structure and Function. 4th ed. Cambridge University Press, Cambridge, UK.

Clarke, T. E., and R. J. Clem. 2003. Insect defenses against virus infection: The role of apoptosis. *International Reviews of Immunology* 22:401-424.

Dean, J., D. J. Aneshansley, H. E. Edgerton, and T. Eisner. 1990. Defensive spray of the bombardier beetle: A biological pulse jet. *Science* 248:1219-1221.

Eisner, T. 2003. For Love of Insects. Belknap Press, Cambridge, MA.

Gross, P. 1993. Insect behavioral and morphological defenses against parasitoids. *Annual Review of Entomology* 38:251-273.

Gullan, P. J., and P. S. Cranston. 2014. The Insects: An Outline of Entomology. 5th ed. Wiley-Blackwell, West Sussex, UK.

Gupta, A. P. 2009. Insect Hemocytes: Development, Forms, Functions, and Techniques. Cambridge University Press, Cambridge, UK.

Lavine, M. D., and M. R. Strand. 2002. Insect hemocytes and their role in immunity. *Insect Biochemistry and Molecular Biology* 32:1295–1309.

Lockwood, J. A. 2009. Six-Legged Soldiers: Using Insects as Weapons of War. Oxford University Press, New York, NY.

Meyer, J. 2006. Insect defenses. http://www.cals.ncsu.edu/course/ent425/tutorial/Ecology/defense.html. Accessed March 31, 2015.

Mill, P. J., and R. S. Packard. 1975. Jet-propulsion in anisopteran dragonfly larvae. *Journal of Comparative Physiology* 97:329–338.

Moore, B. P., and W. V. Brown. 1981. Identification of warning odour components, bitter principles, and antifeedants in an aposematic beetle: *Metriorrhyncus rhipidius* (Coleoptera: Lycidae). *Insect Biochemistry* 11:493–499.

Nishida, R., and H. Fukami. 1990. Sequestration of distasteful compounds by some pharmacophagous insects. *Journal of Chemical Ecology* 16:151–164.

Ortiz-Urquiza, A., and N. O. Keyhani. 2013. Action on the surface: Entomopathogenic fungi versus the insect cuticle. *Insects* 4:357–374.

Palma, M. S. 2006. Insect venom peptides. In: The Handbook of Biologically Active Peptides (A. Kastin, ed.), pp. 409–416. Academic Press, San Diego, CA.

Prestwich, G. 1983. The chemical defenses of termites. *Scientific American* 249:78–97.

Rivers, D. B., C. Thompson, and R. Brogan. 2011. Physiological trade-offs of forming maggot masses by necrophagous flies on vertebrate carrion. *Bulletin of Entomological Research* 101:599–611.

Schmidt, J. O., S. Yamane, M. Matsuura, and C. K. Starr. 1986. Hornet venoms: Lethalities and lethal capacities. *Toxicon* 24:950–954.

Sword, G. A., P. D. Lorch, and D. T. Gwynne. 2005. Insect behaviour: Migratory bands give crickets protection. *Nature* 433:703. doi:10.1038/433703a.

Waldbauer, G. 2012. How Not to Be Eaten: The Insects Fight Back. University of California Press, Berkeley, CA.

THE ENTOMOLOGIST BOOKSHELF (SUPPLEMENTAL READINGS)

Blum, M. S. 1981. Chemical Defenses of Arthropods. Academic Press, New York, NY.

Eisner, T. M., M. Eisner, and M. Siegler. 2007. Secret Weapons: Defenses of Insects, Spiders, Scorpions, and Other Many-Legged Creatures. Belknap Press, Cambridge, MA.

Evans, D. L., and J. O. Schmidt. 1990. Insect Defenses: Adaptive Mechanisms and Strategies of Prey and Predators. State University of New York Press, Albany, NY.

Quicke, D. L. 1997. Parasitic Wasps. Chapman and Hall, New York, NY.

Witz, B. W. 1989. Antipredator mechanisms in arthropods: A twenty year literature survey. *Florida Entomologist* 73:71–99.

ADDITIONAL RESOURCES

Animal venom research international
http://usavri.org

Butterfly defense mechanisms
http://sciencelearn.org.nz/Science-Stories/Butterflies/Butterfly-defence-mechanisms

Chemical communication
http://www.cals.ncsu.edu/course/ent425/tutorial/Communication/chemcomm.html

Chemical composition of insect venoms
http://www.compoundchem.com/2014/08/28/insectvenoms/

Insect arcade game
http://www.hackedarcadegames.com/game/11160/Insect-Defence.html

Most toxic insect venoms
http://entnemdept.ufl.edu/walker/ufbir/chapters/chapter_23.shtml

13 Life on the Edge

Coping with Stress

Conditions are seldom ideal, and if one waits long enough for ideal conditions one is just making excuses.

Dr. Bernd Heinrich
A Year in the Maine Woods (1995)

Stress is not unique to the human condition. Oh, we may believe it is, or that our lives are near chaos most of the time. Certainly for some there is an element of truth to that statement. For most of us, however, the stress we face daily is not quite the same as it is for other animals, insects included. Most humans in the United States live a comparatively sheltered existence, literally and figuratively. By contrast, insects are small, nutritious, and live outdoors. This is a recipe for stress. Small and nutritious means interest is piqued in every conceivable type of predator, parasite, and parasitoid seeking insects as prey or hosts. The hunters are relentless, keeping the hunted on high alert 24/7. For the hunted, lowering their guard for even a moment can be the difference between life and death.

The most common forms of stressors are associated with the nonliving environment. Seasonal change represents extraordinary challenges for insects regardless of their geographic region or localized habitat. Seasonality can range from the predictable seasons of temperate zones, rainy seasons with sustained heavy precipitation, and long-term drought to periods of extreme high or low temperatures. Such changes require that insect residents be attuned to environmental signals that forecast impending climatic conditions and then respond appropriately. The responses are adaptive and occur before harsh weather arrives. Thus, insects essentially possess antistress genetic programs to cope with seasonal stress and hence promote survival. Many environmental stresses are not predictable, such as uncharacteristic conditions for a particular season. Although the absolute changes are not necessarily extreme in themselves, aseasonal disturbances tend to be sudden, and so individuals in the path of change are not typically prepared. Consequently, physiological stress can occur that, if experienced for prolonged periods, may induce injury and even death.

Chapter 13 explores common forms of abiotic and biotic stresses that challenge the lives of insects. We will explore how insects detect impending climatic change through environmental cues, as well as the behavioral, morphological, and physiological adaptations they rely on to cope with stress. We will also delve into the genetic programs that regulate seasonality in insects, including defined periods of metabolic suppression such as hibernation, diapause, aestivation, and quiescence. What will become apparent is that insects have excellent stress management plans in place to deal with almost any of life's daily threats.

Key Concepts
- Talk about stressed: 24/7, 365 days a year
- Dealing with stress on a typical day: General stress responses
- Environmental tokens tell the tale of impending changes
- Seasonality and insect life history traits
- Genetic regulation of seasonal survival
- Coping with the unknown: Aseasonality

Talk about stressed: 24/7, 365 days a year

It is a cold blustery day in late November somewhere in the northeast corner of the United States. The autumn sky has turned dark with thick smoke-colored clouds occupying every position that rays of sunlight could potentially pass. The air stings with small, frozen water crystals barely visible yet ever felt. A gust of wind rips across your exposed cheeks causing tears to form in response to a brisk environment that you wish to escape. Escape for you is indeed an option. Almost any of us would surrender to the natural assault by moving indoors. A warm artificial "nest" awaits, stocked with an excess of supplies that permit warmth despite the weather outside, food that satisfies gustatory needs and pleasures, sufficient space to house the immediate brood and then some, and a multitude of devices facilitating communication and entertainment, nearly all depending on a power source other than your own. Harsh environmental conditions like those of winter pose little stress to the average person living in the United States, with perhaps the notable exceptions of attempting to drive through snow drifts or across ice, or overcoming the accumulation of girth while binge-watching your favorite shows on Netflix.* (Just to be clear, snowboarding down a long tall pine is not a naturally occurring stress; it is an entertainment-related stress and an optional indulgence for those with an intense need to break something.†)

The situation is substantially different for most other animals, but especially for **ectotherms** like insects. Northern regions of temperate zones are cold and dry during the long duration of winter. Insects have limited ability to regulate their own body temperatures. Their body temperature is completely at the mercy of prevailing environmental temperatures, a condition referred to as being **poikilothermic**. Obviously, if temperatures plunge below zero or climb to excessive highs, an insect's body temperature will follow suit. In the absence of preventive measures, a poikilothermic animal will experience thermal stress during seasonal changes or periods of unpredictable weather.

This is in contrast to you, an **endotherm** that will maintain essentially the same body temperature regardless of climatic conditions. Of course, if we are exposed to brutal temperatures long enough, our bodies cannot keep up. But that almost never occurs, because when we have had enough of the weather, inside we go. Insects do not have the option of removing themselves from nature to take a seat by the comforting hearth, waiting for ideal conditions to return (figure 13.1). No, they must contend with prevailing conditions, whether related to the abiotic or biotic features of the environment, at all times. In other words, stress avoidance is not an on/off mechanism for them. Rather, they rely on mechanisms of stress management the entire duration of their lives.

What is evident so far is that insects live on high alert 24/7 (figure 13.2). Two questions come to mind, and the first has two parts: Why do insects face stress so frequently, and what exactly is meant by "stress"? Let's start with the idea of stress. The term stress typically conjures up negative connotations, because after all, we have been taught that to be stressed is a bad thing for us. Humans strive to avoid stress, almost at all costs. Experts appear on television to provide advice on removing stress

---

*Netflix is an Internet provider of on-demand programming, including original sitcoms and series. Viewers can watch their favorite programs any time they want, which has led to the phenomenon of binge-watching entire series in one setting.

†Outdoor sports, with their attendant injuries, are practiced by humans by choice, not out of necessity. This is a conversation that a colleague of mine and I engage in on a regular basis, as we discuss his latest injuries from rollerblading, distance cycling, backpacking, walking (long story), or snowboarding, the latter being a non-Olympic experience for him that yielded a multifractured arm bone after mistiming a ramp jump.

Figure 13.1. Insects do not have the option of retreating from the cold of winter to nestle next to a warm fire. Instead they must become resistant to low temperatures or perish. Photo by Robbie Sproule available at http://bit.ly/1BpzyQZ.

from our lives. Courses can be taken in college to learn how to cope with and manage stress. Psychoanalysts are hired so that we can unburden ourselves of stress. You can even recruit a life coach to help you learn to live stress-free. No such luck if six-legged.

Stress can be defined in two ways, as a condition and as a coping mechanism. When viewed as a condition, stress results when demands are placed on an individual that causes deviation from "normal." Normal refers to the physiological state of **homeostasis**, in which internal systems are functioning at ideal or optimal levels. We will examine this concept more in the next section, but suffice to say, rarely is an animal able to maintain homeostatic conditions easily; life is filled with challenges that throw the body out of balance. Stress can also be thought of as the response, usually in the form of a physiological mechanism(s), to adverse conditions. If the response syndrome is working properly, stress itself does not necessarily lead to negative consequences. Instead, the individual recognizes stimuli termed **stressors** that are associated with adversity and then responds in a healthy fashion to cope with or overcome the unfavorable situation. It is a normal process in life. Stress is also inevitable when living in the natural environment. Under this

view, stress as a response is adaptive and necessary. The human view of stress avoidance is actually not natural, but of course, we also live under artificial circumstances.

What about the second question? Why are insects stressed so often? Our example of winter temperatures makes sense, but are insects really taxed all the time? Think about the duration of winter in regions of the United States that experience a true winter, where temperatures drop significantly to the point that freezing is possible for more than just a few days and snow accumulations are measured in several inches to feet annually. In these areas, winter conditions typically last four to six months; northernmost regions may push the envelope even longer. As a consequence, insects that overwinter must contend with the threat of cold or, more correctly, low temperatures daily until the season changes.

Cold temperatures are not the only stress encountered during winter. Corresponding drops in moisture or relative humidity occur as well, meaning that the air is dry. Stated another way, low relative humidity favors evaporation of liquid water. You know this concept from a familiar example: freezer burn that occurs with meats not wrapped properly or left in a household freezer too long. Insects, then, are faced with the threat of body

Figure 13.2. Hypothetical stress alert levels for insects dealing with abiotic or biotic stressors. Insects are on high to severe alert 24/7. The levels are modified from those used by the US Department of Homeland Security as advisory levels regarding terrorist threats.

water loss in cold, dry environments. Compounding this potential problem is the fact that insects are small-bodied. This creates a physiological challenge at any temperature, not just during winter, to avoid desiccation, because the amount of surface area is very high in relation to the volume of their bodies. Surface area increases at a square root, whereas volume increases at a cube root. Thus, if body size is small, volume decreases at a faster rate than surface area. What this means to an insect is that there is an enormous amount of surface area available for water to evaporate from in comparison to the volume or mass of the body (figure 13.3). The threat of desiccation is very real in a terrestrial environment, no matter what the ambient temperature.

As if that is not enough stress, food is typically not available during winter. So whatever nutrient needs the insect may have they must be fulfilled from the reserves on hand. Thus, metabolic use of nutrient reserves must be judicious over the entire seasonal window to ensure that enough remains to resume development once favorable conditions return. Accurate food accounting depends on internal clocks that are regulated or set by external environmental cues. We have discussed this before and will throughout the chapter, but it cannot be emphasized enough: insects must be able to sense changes in their environments to make appropriate adjustments in life cycles, behavior, physiology, and morphology (interpreting environmental cues or tokens and the concept of internal clocks [**biological or circadian clocks**] will be examined later in this chapter).

We have described so far the environmental stresses associated with seasonality. Seasonal change is more or less predictable. So much so that we even mark on our calendars when spring starts, then summer, fall, and winter. An animal synchronized with environmental signals can plan accordingly and be prepared for or avoid harsh weather altogether. (For example, my wife is so attuned with the environment that she can anticipate the arrival of fall with a preemptive shopping spree that permits her to change out the entire summer wardrobe before long-sleeved and sweater weather arrives.) Unfortunately, impending climatic stress is not always predictable. Any abiotic component of the environment can change without warning. When it does, an insect is exposed to a stressful situation, the severity of which is dependent on the factor that has changed, how quickly a departure from the previous condition has occurred, and the difficulty of the resulting new conditions. For example, in the case of an unexpected temperature change, a rapid rise in ambient temperature can be more severe than a subsequent drop. The rate at which temperatures decline can be the difference between causing one form of low temperature injury or another (i.e., **chilling injury** vs. **cold shock**). Similarly, the final temperature reached

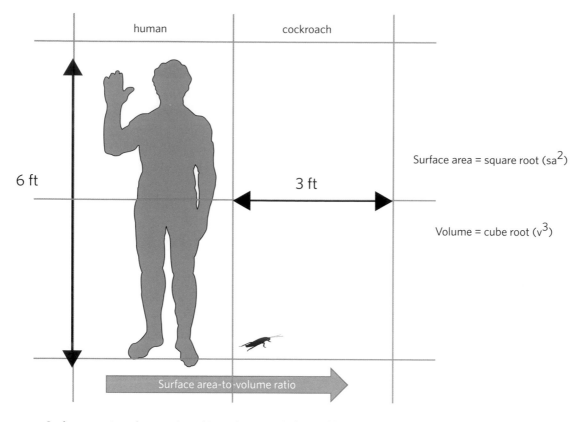

human   cockroach

6 ft

3 ft

Surface area = square root (sa^2)

Volume = cube root (v^3)

Surface area-to-volume ratio

Surface area-to-volume ratio = skin surface area/volume of body

Figure 13.3. The surface area-to-volume relationship between an insect and adult human male. Since volume increases or decreases at a faster rate than surface area, a small animal like an insect has a much larger surface area-to-volume ratio than a comparatively enormous beast like us. Image of a cockroach by Jan Gillbank (http://bit.ly/1Aumcb4) and scale image of a human by Samsara (http://bit.ly/1LINU4j).

will also influence the level of stress elicited. Temperatures above freezing are less severe in their effect in aseasonal situations than those below freezing. Table 13.1 provides examples of aseasonal stressors than commonly threaten insect survival.

Not all stressors in the lives of insects are abiotic, derived from climatic change or weather-related events. A whole array of predators, parasites, and pathogens targets insects as a food source. In chapter 12 we examined why insects are food favorites of each of these organisms. It basically comes down to three things: size, abundance, and nutrition. Insects are small-bodied; most predators are much larger and consequently they pick on the smaller-sized prey. We also know that insects are in extraordinary abundance in comparison to all other animals and organisms. So basically there are lots of them

to eat. The fluids and tissues of insects are also loaded with nutrients, and at far higher concentrations than most any other organisms. Thus, they are well worthwhile to eat. The net effect is that other creatures are constantly seeking out insects as food, and in turn, the "food" must be on constant guard to prevent itself from being eaten or parasitized. In other words, the biotic portion of the environment majorly contributes to the high stress of insects.

In the coming sections, we explore in more detail the types of stresses experienced by insects and how they cope. Stress responses are vastly different for seasonal versus aseasonal stress, and the same is true with biotic stressors like predators and parasites. That said, insects are also excellent at relying on a series of general stress responses to deal with a variety of in-

Table 13.1. Common examples of stressors in the environments insects occupy. Abiotic stressors are non-living components of the environment while biotic are living entities.

Abiotic	Biotic
☐ Temperature (chilling injury, cold or heat shock)	☐ Predators, parasites, and parasitoids
☐ Ultraviolet radiation	☐ Nutrient depletion
☐ Drought or dehydration	☐ Overcrowding
☐ Anhydrobiosis (aquatic species exposed to air)	☐ Competition
☐ Chemical or metal toxicants	☐ Nucleic acid mutation
☐ Overhydration or precipitation	☐ Injury

sults and thus display efficiency even when faced with adversity.

Quick check

Does stress always have to imply negative consequences to an insect?

Dealing with stress on a typical day: General stress responses

Stress happens—deal with it. This is the reality of any organism residing in a natural environment. Stress avoidance is not a practical approach to dealing with daily challenges and threats. As a consequence, animals, insects included, have evolved general stress responses to cope with a normal existence. What this means is that insects rely on a generic system of detecting and responding to daily challenges as part of their everyday lives. This is in contrast to the very specific mechanisms used to prepare and cope with severe and/or long-term adversity, such as seasonal change. Seasonality and genetic programs conferring protection from stress are the subjects of subsequent sections; here, the focus is on your average, everyday, run-of-the-mill stress response. General stress responses, also called **general adaptation syndrome**, in insects (as well as in most any animal) have three phases.

1. Detecting stressors, which occurs via sensory perception (see chapters 10 and 11).
2. Responding to stress, which, for an insect, means fleeing, escaping, or defending the body via behavioral and/or physiological mechanisms.
3. Succumbing to stress, an indication that the response mechanisms cannot keep pace with the stressors. The individual may be driven to exhaustion because the stress responses have been "on" for a prolonged period. Death may be the end result.

General stress responses are activated upon detection of a stimulus. Insects generally rely on two broad categories of stress responses: those that are behavioral and those involving physiological mechanisms. Examples of behavioral stress responses include simply leaving the immediate area where the stressor was detected. Mobile insects have a multitude of options in this regard, from jumping, running, swimming, or flying to calmly walking away. Others may engage the stress-causing agent head-on by relying on any number of defensive responses like those detailed in chapter 12. Biting and stinging are effective remedies to biological stress. However, it is possible that the defense strategy or escape plan does not deter the stressor, whether of abiotic or biotic origin (figure 13.4). As a consequence, if the stressed insect continues to maintain the response activity, eventually ATP depletion will ensue, resulting in exhaustion and possibly the ultimate demise of the individual. The take home message is general stress responses are not always effective; they are situation dependent. We will learn later in the chapter that escape is not restricted to general stress responses, as several species rely on migration as the answer to seasonality.

Figure 13.4. Stinging is an excellent way to cope with stress, for the bee that is! Photo by Hadi available at http://bit.ly/29hk4L3.

Most often when general stress responses are discussed, the focus is on the physiological responses of the organism. Insects have been model organisms in this regard, with particular attention given to how they cope with stresses associated with temperature, overcrowding, and noxious or toxic compounds. The two dominant components of the physiological stress response are hormones and proteins. Hormone involvement appears to be primarily associated with stressor detection. Immediately after perceiving that stress is imminent, the brain or associated endocrine tissues produce neurohormones in the form of biogenic amines (e.g., octopamine and dopamine), juvenile hormone (multiple versions of this hormone exist), or several other possibilities. While the hormones can trigger an array of tissue and cellular responses, a common path is to arouse the insect (table 13.2). In other words, the insect's version of fight-or-flight is stimulated by mobilizing energy reserves to elevate metabolic rate. Why is this necessary? (figure 13.5). Well, if you consider the escape or defense responses mentioned earlier, those activities are highly energy demanding, and flight or leg muscles must be prepared for sustained mobility.

Table 13.2. Examples of physiological agents associated with the general stress response of insects. Most function to either stimulate the insect version of 'fight-or-flight' or serve to protect cellular proteins from oxidative damage during stress.

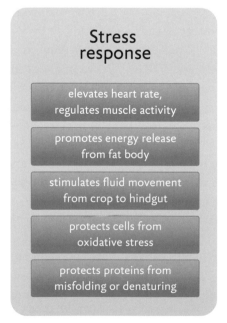

Stress modulators	Stress response
octopamine	elevates heart rate, regulates muscle activity
adipokinetic hormone	promotes energy release from fat body
diuretic hormone-I	stimulates fluid movement from crop to hindgut
vitellogenin	protects cells from oxidative stress
heat shock proteins	protects proteins from misfolding or denaturing

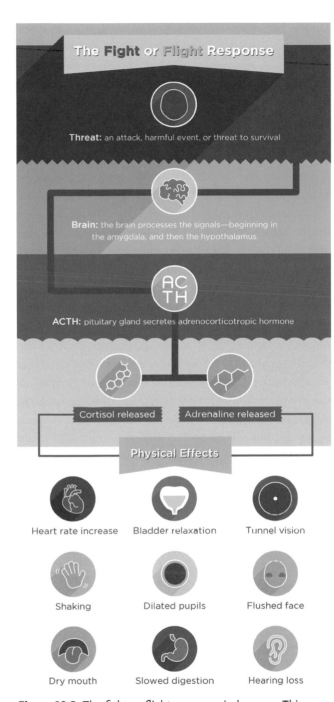

Figure 13.5. The fight-or-flight response in humans. This common stress response in humans, dealing with immediate threats, real or imagined, actively recruits participation of multiple neural and hormonal pathways. Insects do have comparable stress responses that involve hormones like octopamine and dopamine, which can stimulate mobilization of energy reserves and increase circulation. Image by Jnkfood available at http://bit.ly /1AxPmWE.

Several proteins are produced in response to various stressors to protect cells or their constituents. The proteins are globally referred to as stress proteins, but are probably best known as **heat shock proteins or HSPs.** "Stress" better describes what the proteins are produced for, and when, as they are synthesized in response to a much larger range of stimuli than only under heat shock conditions. What is heat shock? It is a condition in which temperatures unexpectedly and quickly rise from ambient. Cold shock also occurs. Both situations represent types of aseasonal stress that will be examined a little later in the chapter.

Stress proteins were originally identified during experiments associated with heat shock, and hence the name has stuck for all forms of stress. In practice, stress proteins also include a large group of antioxidant proteins that are stimulated under some stress conditions that do not involve HSPs. In general, stress proteins are produced in response to both abiotic and biotic stresses, including temperature, chemicals, ultraviolet radiation, overcrowding, parasitism, infection, food shortage, and injury. The proteins are associated with folding and unfolding of other proteins, not always during times of stress, to essentially stabilize intracellular proteins. Those proteins may be membrane-bound or essential components of metabolic pathways that occur within the cytosol of a cell. It is important to note that though stress proteins aid in the survival of the cell, their production comes at a cost. Any type of protein is energetically expensive to synthesize and requires a carbon backbone. Producing them is therefore at the expense of energy and carbon stores. The longer the stress proteins are needed within the cell, the more of these vital materials are consumed. Ultimately, stress protein synthesis robs the cells of producing critical macromolecules needed for synthesis and other functions during nonstress conditions.

General stress responses are not limited in scope. Behavioral and physiological stress responses occur with common stressors associated with the daily environment and may also be activated when aseasonal or seasonal weather changes occur. However, the general responses are not designed to provide long-term protection. As already mentioned, they are metabolically expensive. So other types of stress responses are needed when dealing with adversity for a prolonged period. Stress that can be anticipated, so that adaptive preparations can be made, is less energetically demanding over the long haul for insects and provides more substantial protection during times of extreme conditions, such as an extended window of drought or cold temperatures.

Environmental tokens tell the tale of impending changes

The evolutionary success of insects as a whole has hinged in part on their ability to interact with and in the environment. While that may not reveal much at first glance, a deeper look shows that insects are excellent at perceiving what is occurring around them. Chapter 10 and chapter 11 explored the sensory structures and mechanisms employed by these creatures for detecting what is going on outside their bodies, so that they can interpret the meaning of the stimuli and then, if need be, respond in an appropriate way. Common sense would indicate that the responses are adaptive, functioning to ensure survival of the individual or group.

By extension, this same form of oneness with the environment is needed for dealing with stress. The various forms of stressors, regardless of whether abiotic or biotic, evoke many of the types of behaviors and communication signals we examined in chapters 11 and 12. Consequently, sensory perception of stress stimuli relies on the same sensory pathways described for auditory, mechanical, visual, chemical, and tactile sensation. Stimuli are stimuli, perception is perception; it is all the same in the environment. It is just a matter of degree and timing of the stimuli that differs. Right? Yes and no. Yes, many different environmental signals are perceived via the same mechanisms, because some stimuli trigger multiple events.

For example, visual recognition of a predator may promote communication signals to warn conspecifics, while at the same time evoking escape behaviors as part of the defense strategy, which simultaneously is also aimed at stress avoidance. However, detection of future climatic change associated with seasonality may rely on environmental cues that have nothing to do with the immediate day-to-day activities of an insect. So, no, it is not just a matter of degree and timing that differs. Case in point is an insect residing in temperate regions that perceives autumn is arriving. Autumn months in many

regions begins a transition from warm days and nights, long days of sunshine, and relatively high humidity with precipitation in the form of rain, to cooler temperatures, a gradual decrease in day length, and an eventual drop in relative humidity until dry conditions are achieved. By the onset of winter, the difference between these abiotic features and those of summer is drastic. The same set of life history characters is not sufficient to deal with the environmental stressors to come, meaning during the seasonal transitions. If the insect waits to respond once the stress arrives, it is too late. They will most surely die. So what is needed is a way for insects to anticipate when the changes (the impending stressors) are coming and to prepare long in advance.

Hopefully you can see the difference that has emerged between stress associated with seasonality and other forms of threats; the latter requires sensory perception of the immediate stressor while the former depends on detecting cues that foretell future events. But what can possibly permit natural fortune-telling? The answer is, the abiotic environment. Or, that is a partial answer. Most abiotic features of the environment change from summer to winter, but some begin to change prior to the onset of harsh conditions. With these features, the changes are gradual and do not adversely affect insects. As a result, such abiotic characters can serve as cues that signal climatic change is coming. These signals are termed **environmental tokens**.

Environmental tokens can range from those as apparent as absolute temperatures or amount of day length (**photoperiod**) to less-apparent tokens like surpassing threshold levels of relative humidity or atmospheric pressure, or the relationship between daytime and nighttime temperatures (thermoperiod). The tokens are not just a set point, but also represent levels that have deviated from the ideal conditions experienced by the insects previously and in which development and reproduction thrived. They thus are used as anticipatory cues that signify two important details of the future: (1) conditions favoring growth, development, and reproduction are coming to an end; and (2) unfavorable conditions are approaching. We already have an understanding that environmental tokens forecast early enough to allow insects and other organisms to prepare for seasonal change. Now the question is, how do these cues accomplish this task?

The best-studied examples involving insects are associated with photoperiod. Most insects respond to changes in day length to adjust some feature of reproduction, migration, or dormancy. So it is no surprise that many also rely on photoperiod as an important environmental token to program seasonal adaptations. The short days that characterize late summer and the beginning of autumn are very reliable indicators of winter's encroachment. The use of photoperiod as a signal of seasonal environmental change depends on insects having the ability to (1) perceive changes in day length and (2) count the number of short days experienced. Day length appears to be recognized as either long or short. The precise definition varies depending on insect and location. For example, several species of flesh flies (Diptera: Sarcophagidae) in North America perceive a long day as sunshine greater than 13.5 hours in a twenty-four-hour cycle (figure 13.6). Even relatively low light intensity can be detected by photoreceptors and thus count toward a long day.

To use photoperiod, an insect must have the ability to keep track of the number of long days experienced. Thus, the insect must have some means to count and store the number of long days experienced before a threshold has been reached to stimulate developmental and physiological pathways that prepare the individual for seasonal change. The latter is an incredibly important safeguard that prevents the insect from responding to aseasonal changes that are only temporary. Once a seasonal program like dormancy is started, it runs course until completion. If a diapause cycle has been triggered, completion means the arrival of spring, which is characterized by a unique set of environmental tokens. Dependency on specific seasonal cues ensures the insect does not initiate a seasonal adaptation too soon or terminate the program too early.

In some cases, environmental tokens are used for more routine events than stress adaptations. Internal or biological clocks regulate several daily, even seasonal, physiological or behavioral activities. The clocks are a series of genes in neuronal cells that function in timekeeping. When a precise time in a twenty-four-hour window is reached, the genes express neuropeptides and other modulator peptides that start or stop cellular

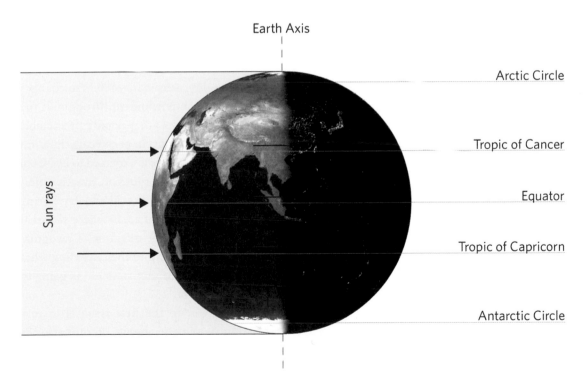

Figure 13.6. The Earth's daily rotation on its axis leads to a precise photoperiod depending on your location on the globe. Location matters because of the tilt of the planet on its own axis. Illustration by Przemyslaw Idkiewicz at http://bit.ly/1HMh2a2.

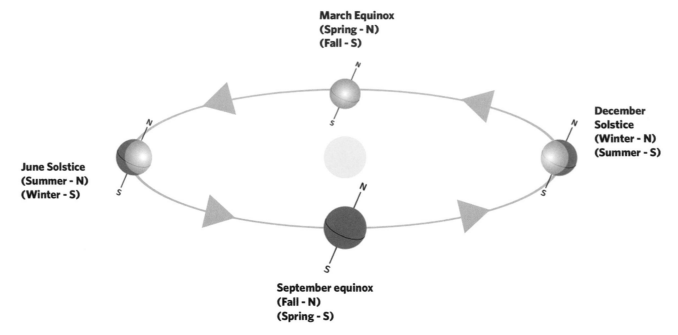

Figure 13.7. The movement of the Earth around the sun over the course of a year results in changing seasons. Photoperiod is the most reliable environmental token used by insects and other organisms for predicting seasonal change. Image by Colivine available at http://bit.ly/1dyr3Os.

events. Typically, events tied to daily cycles are termed **circadian rhythms**. Timing by the clocks is regulated or set by environmental tokens associated with rotation of the Earth on its axis (figure 13.7). In contrast, events on annual cycles respond to tokens that occur as a result of the Earth's movement around the sun. As you might have surmised, some regions of the planet are less likely to respond to the environmental cues discussed so far because of the tilt in the Earth's axis. For example, photoperiod and temperature change very little along the equator. Similarly, aquatic insects residing in deep benthic regions or inhabitants of polar environs respond to unique signals, different from those utilized by terrestrial insects occupying temperate ecosystems. The point is that one size does not fit all, even with regard to responding to environmental stimuli.

Bug bytes

Monarch seasonal migration

https://www.youtube.com/watch?v=yIFB9reAkwU

Seasonality and insect life history traits

It is time to put your new knowledge of environmental tokens to use. We will use photoperiod—the major example discussed in the last section—as the focal point for examining seasonality. Seasonality refers to changes that occur in the insect's environment throughout the course of a year. Such changes are generally predictable in terms of timing (when they occur each year), duration (e.g., summer in the United States is expected to last from June 21 to September 21), and the types of climatic differences that will occur (i.e., winter is cold and dry, spring is cool and wet). Insects living in areas where seasonal change occurs are faced with the dilemma of recognizing when the change is coming and then deciding what to do once they know that the environment is going to become drastically different.

Obviously the answer to the first issue is to rely on tokens that signify climatic change. Photoperiod is by far the most reliable for predicting seasonality. For most insects living in temperate zones, short days forecast the arrival of autumn and then winter. As we dis-

Figure 13.8. Young caterpillar of the gatekeeper butterfly, *Pyronia tithonus* (Lepidoptera: Nymphalidae) in winter diapause. Photo by Gilles Saint Martin (http://bit.ly/1AuoMy2).

cussed in the last section, an insect must be able to distinguish a short day from a long one, and then count the number of short days experienced. When a critical threshold number has been surpassed, the insect will initiate its seasonal program. In the example of approaching winter, insects do not remain active, because the conditions are too harsh for survival and food is not available. As a consequence, the seasonal program is often one of dormancy. Dormancy is known by different names (i.e., hibernation, **diapause, aestivation** and **quiescence**), and the various forms are not the same. However, all forms of insect dormancy do share in common sustained periods of reduced physiology manifested as suppressed metabolic activity and a halt in development. Diapause is essentially the insect equivalent to vertebrate hibernation, in which individuals respond to reduced photoperiod by entering a developmental arrest that will be maintained until a separate and sustained environmental token indicates that favorable conditions have returned. Arrival of spring, with the associated seasonal cues, initiates the stop signal for diapause. We will return to a discussion of winter dormancy in a moment.

For comparison, metabolic arrest during warmer conditions or rainy seasons is termed aestivation, which differs from diapause and is characterized by unique life history traits. Quiescence is also a form of reduced activity in response to weather, but differs from aestivation and diapause in that the insects are not locked into a program. Instead, once temperatures climb to more favorable levels, individuals can immediately resume activity (figure 13.8). Quiescence is quite common to seasonal transitions associated with autumn and spring, and sometimes with aseasonal conditions. Some insects also use it as an adaptation to more severe weather like winter, but not for the entire individual. Say what? Several species of blow flies (Diptera: Calliphoridae) spend the winter as adults in quiescence, yet their gonads are in diapause. Thus, if a winter day in January is uncharacteristically warm, the adults can resume activity, perhaps even being lucky enough to find some juicy dead animal to feed on. However, mating and egg laying cannot occur, which is a safeguard against ovipositing progeny that are not adapted for cold, low humidity, and an absence of food (even the juicy animal will freeze) once typical winter conditions return.

Winter dormancy results in dramatic alterations in an individual before the onset of harsh conditions. What types of changes are necessary? The most significant change is preparation for cold. Low temperatures, often approaching or dipping below freezing, typify winter. The life history characteristics favored during summer are not useful for contending with cold. So if the goal is to survive winter—and it is—the diapause program must prepare insects for the impending low temperatures. This is frequently achieved via acquiring **cold hardiness** (or, cold hardening), in which several **cryoprotectant** compounds are synthesized; body water content is reduced to concentrate solutes, thereby lowering the **supercooling point*** of body fluids; and membrane lipids are changed to permit stretching in case of ice crystal formation and expansion.

Several insects acquire cold hardiness through a **rapid cold hardening** response, in which resistance to the adverse effects of low temperatures is acquired quickly from a brief exposure to nonlethal low temperatures before exposure to more extreme conditions. The protection is short-lived if the insect returns to warmer temperatures, but is long lasting if associated with a seasonal program like diapause. Some species also experience **thermal hysteresis**, a phenomenon in which antifreeze proteins attach to ice that has already formed in body fluids; protein attachment inhibits further growth of the ice crystals. Complementing these adaptations to low temperature is the synthesis of small heat shock proteins that function to protect intracellular proteins. Those produced in response to or in anticipation of low temperatures tend to be of lower molecular mass than the corresponding proteins produced during times of heat stress.

Several of the adaptions mentioned thus far deal with ice prevention. Why are insects so concerned with frozen water? Ice is a major concern not just for insects, but for all organisms that must contend with subzero temperatures. Ice can be deadly when it occurs inside the body of an animal. The crystals that form as temperatures fall below freezing have sharp edges that can easily slice through membranes as they are carried along by the circulatory system. If present inside a cell, the growing crystal stretches membranes, causing materials to flux in both directions or, worse, degrade the nucleus and/or burst the cell. Even the initial event of ice formation—**sublimation**—can create localized problems for cells and macromolecules in the immediate vicinity. Sublimation refers to the physical change in state of liquid water to a solid (ice). At the precise moment of the phase shift, a small but significant temperature burst occurs as energy is released. Cells close by may experience injury from the temperature shock, which often is manifested as leaky membranes. It is worth noting that the temperature at which sublimation occurs is the supercooling point mentioned earlier.

Enhanced synthesis of cuticular hydrocarbons occurs with some species to prepare for diapause. The extra layers help create a watertight structure, thereby reducing water loss to the desiccating air of winter. The reverse is also true; additional hydrocarbons prevent water from penetrating the exoskeleton. Why would this be important? Consider that during diapause, the insect is completely immobile. If dormancy occurs in the soil, then individuals can be subject to immersion in pools of water and saturated soil as snowpack melts (figure 13.9). Hydrocarbons thus function as waterproofing.

Figure 13.9. The need for waterproofing as a seasonal adaptation is evident after snow melts. Standing water in early spring can last for several weeks, which will lead to the quick demise of an unprepared insect. Photo by David Anstiss available at http://bit.ly/1AyrjXu.

---

*The supercooling point (SCP) of a fluid is the temperature at which ice crystal formation can (but not necessarily will) occur. Many insects can influence the SCP by reducing the water content of body fluids, thereby concentrating solutes, which in turn lowers the SCP. Species that spend the winter frozen may alternatively decrease the solute composition of fluids to elevate the SCP, with the goal of promoting ice formation.

The long duration of winter also means the absence of food, at least for most insects, for four to eight months, depending on location. When food becomes limited, development cannot be sustained. Consequently, one adaptation of the diapause program is to enter an arrested state of development. A key feature of this developmental arrest is suppressed intermediary and respiratory metabolism. Actually it is more correct to say that insects shift to an alternative form of metabolism that is best adapted for winter dormancy. Preparation for the dynamic metabolism of diapause begins prior to developmental arrest, by the individual acquiring food and then converting the absorbed nutrients into metabolic reserves like glycogen, fat (triacylglycerols), or proteins (hexameric in structure).

Proteins often are used for storing amino acids. In other words, the protein is a nutrient sink of amino acids, which in turn can be used as metabolic reserves or a source of cryoprotective compounds during diapause. Metabolic reserves accumulate primarily in fat body tissue to be used throughout the duration of winter. Interestingly, most insects studied do not rely on the same metabolic fuel for the entirety of diapause. Early onset diapause may depend on carbohydrate or lipid, switch to lipid during midwinter, and then shift to amino acids or protein in late diapause. The switch to the final metabolic reserve usually signifies that diapause-terminating conditions are expected relatively soon, since the final fuel is limited in supply. Mistiming, or an unexpectedly long winter, can result in starvation.

All seasonal changes do not involve major physiological adaptation per se. Some preparations for dealing with adversity are as simple as seeking out refuge or vacationing in a warmer climate. For example, some species may burrow into soil, below the frost line, to ride out the harsh conditions. This type of strategy often requires an accumulation of metabolic reserves, but not always, if a food source like tree roots can be tapped into. Shelter-seeking has the additional advantage of not necessarily requiring cold hardening or morphological changes, both of which are energetically expensive to perform. There is even evidence of some species displaying winter **synanthropy**, meaning that the critters have come to depend on arti-

Figure 13.10. The familiar face of a winter guest. Adult brown marmorated stink bugs, *Halyomorpha halys* (Hemiptera: Pentatomidae) commonly move indoors prior to the onset of winter along the east coast of the United States, sometimes accumulating in masses of hundreds to thousands. Photo by Joltthecoat (http://bit.ly/1FDRRG6).

ficial dwellings constructed by humans to circumvent the need for other forms of more natural winter adaptation.

One has to look no further than the mid-Atlantic region of the United States to find an excellent example of synanthropy in the form of the brown marmorated stink bug, *Halyomorpha halys* (Hemiptera: Pentatomidae) (figure 13.10). This invasive pest was accidentally introduced into the eastern United States sometime around 1998 and has become quite adept at spending the winter in homes and other dwellings in quiescence, often congregated in aggregations of hundreds to thousands! A true dormancy has not been observed for this species when residing in artificial structures. In fact, it commonly becomes active throughout winter, taking on its usual behavior of dive-bombing table lamps or

Figure 13.11. Overwintering monarch butterflies, *Danaus plexippus* (Lepidoptera, family Nymphalidae) after a long migration to central California. Photo by Michael Baird available at http://bit.ly/1FimT3W.

patrolling the faces of individuals (meaning people) who mistakenly fall asleep at night.

The idea of heading to a warm climate to wait out winter is not a novel concept to humans (especially those who qualify for AARP membership).* Many insects follow suit by relying on seasonal migration to leave before impending inclement weather arrives. A classic example is that of the monarch butterfly, *Danaus plexippus* (Lepidoptera: Nymphalidae), which responds to photoperiodic cues signaling autumn's return by migrating thousands of miles south. Western populations of monarchs found in the United States migrate to the Pacific coast, overwintering together on the Monterey peninsula. Those originating from the east coast to the Midwest travel a different path, heading to parts of Central America just south or west of Belize, or taking up residence in the Sierra Madre of Mexico (figure 13.11). It is there they will spend the rest of their lives. A return northern migration occurs in late spring, only in this case, multiple new generations participate to repopulate areas of the United States abandoned before winter. Technically, the butterflies are in diapause, because their gonads are in a state of arrested development. A

Figure 13.12. A stand of milkweed plants in a meadow. The highly toxic plants are a dining delicacy for the winter migration of the monarch butterfly (*Danaus plexippus*) and giant milkweed bug (*Oncopeltus fasciatus*). Photo by Pookie Fugglestein available at http://bit.ly/1Ks5TeT.

similar migratory diapause, albeit greatly reduced in terms of distance covered, occurs with the large milkweed bug, *Oncopeltus fasciatus* (Hemiptera: Lygaeidae). Interestingly, both insects feed on the highly poisonous milkweed plant during juvenile development (figure 13.12).

---

*A reference to "old" people, as in, seasoned for fifty-five years or more.

Beyond the text
If ice can be so devastating in cells and body fluids, why do some insects purposely promote freezing solid as their strategy for winter survival?

Genetic regulation of seasonal survival

Insects have an incredible arsenal of seasonal adaptations, which are part of highly evolved genetic programs that serve to promote species survival when enduring long-lasting stress. As we just discussed, the array of physiological, morphological, and behavioral changes are precisely timed to be in place before the onset of adverse weather, regardless of whether climatic harshness takes the form of extreme temperatures, precipitation, or drought. Keep in mind that insects are not only excellent at being prepared for adversity, but they can also anticipate the duration of the climatic conditions to allocate energy and nutrient usage to match the timing, and then modify their life history traits at the end of dormancy to begin the next chapter in life. How are all these necessary changes coordinated? Yes, it is true that photoperiod is the major environmental cue used to activate seasonal programs and adjust internal clocks. But the question now is, what mechanisms regulate the seasonal programs and the internal biological clocks?

We can address the question of coordination by breaking down the entire process of seasonality into individual components. For example, we already know that insects are responding to an environmental signal, in this case photoperiod. Photoperiod must be interpreted daily to obtain two important pieces of information: (1) is the

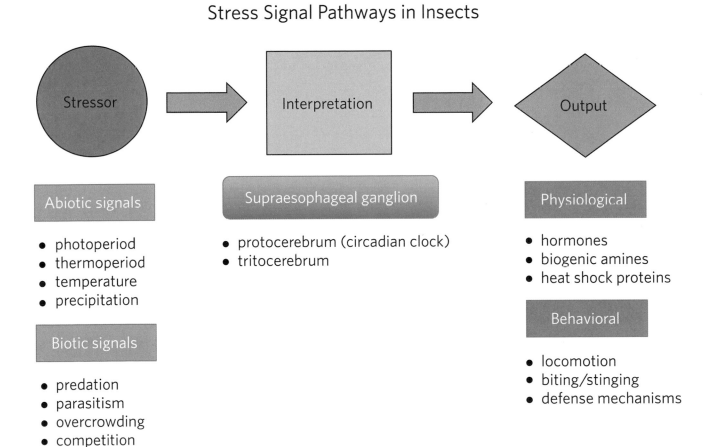

Figure 13.13. Stress signal pathways in insects involve any number of abiotic or biotic stressors. Seasonal changes are perceived primarily via photoperiodism, with the output manifested through hormonal signals to trigger behavioral, physiological, and even morphological adaptations.

current day recognized as "long" or "short," and (2) how many long- or short-day signals were experienced over a critical window of time. In simplest terms, a timer and counter are needed for photoperiodism. The environmental information that is deciphered must then lead to an output signal that sets an appropriate response mechanism into motion. In the case of responding to seasonal change, one or more hormones are involved, which in turn can activate or terminate a cascade of cellular, physiological and behavioral activities throughout an individual. Thus, a broad view of seasonal program coordination moves from environmental signal to interpretation to output (figure 13.13).

The intimate details of the pathways involved in seasonal programs are not necessary for our understanding of insect adaptations. However, it is worth noting that multiple hormones are typically involved in preparing, programming, maintaining, and terminating a particular seasonal program. For example, most insects rely on a diapause-inducing hormone, as well as juvenile hormone, and then a separate terminating hormone to initiate resumption of development. In the vinegar fly, *Drosophila melanogaster* (Diptera: Drosophilidae), the titers of three groups of hormones—juvenile hormones, ecdysteroids, and insulin-like peptides—are linked to environmental signals that in turn dictate the induction or avoidance of ovarian diapause. The internal clocks associated with photoperiodism are referred to as circadian clocks. The original name was associated with clock genes that anticipate daily environmental changes. It is now understood that these same clocks are involved in seasonal programs as well.

Multiple genes have been identified that regulate clock functioning and that respond to clock signals to produce the output mentioned earlier. For example, the clock genes called period (per) and timeless (tim) provide negative input, while cycle (cyc) is a positive regulator. In essence the genes provide increase/decrease or on/off signals to the circadian clocks. Similarly, other clock genes function as the timers and counters of photoperiodic signals. Multiple genes are involved in receiving and interpreting environmental tokens, as well as in generation of the outcome signals. Thus, there is complexity and multiple checkpoints along the pathways that encompass seasonal programs. As there should be, since

commitment to enter or terminate dormancy like diapause should be based on as much information as possible. Why? Any of the seasonal programs results in dramatic changes in the life history traits of a given species, and ultimately, their very survival depends on them.

Coping with the unknown: Aseasonality

So far we have spent the bulk of our time focused on how insects handle the challenges of seasonal stress. Certainly the many seasonal adaptations employed are impressive, and afford the major advantage of being predictable. Life as a whole, however, is filled with the unpredictable, especially so for organisms that live 24/7 in the natural environment. Most stress experienced by an insect is associated with unpredictable changes in the abiotic environment. Aseasonal stressors arrive suddenly, evoking dramatic shifts in climatic and environmental conditions, and therefore do not include anticipation. As a consequence, photoperiodism and other environmental tokens are not factors in terms of preparing for aseasonal changes, because preparation is not a feature of this type of stress. In practical terms, this means that the outcome response is generated only after the stressor has been experienced. Stress responses are therefore more akin to emergency responders, functioning either to minimize the damage or to provide first aid, since injury already occurred. In this context, aseasonal stress responses are short-term phenotypic changes that come about on a case-by-case basis. This is in sharp contrast to seasonal adaptations, which represent evolutionary adjustments in life history traits.

Inevitably, stressful conditions will be experienced when living in the environment. Even when climatic conditions appear to be stable, like during midsummer, short-term chaos is just around the corner. For example, the bright sunshine of a July summer day feels delightful to us when lying on the beach or next to a pool. However, when our skin is unprotected from ultraviolet radiation, short- and long-term consequences can prevail. The short-lived sunburn may be an outward indication of much more serious damage to the epithelial cells of the skin, perhaps leading to DNA damage or, worse, some form of skin cancer. Insects are just as sensitive, and in fact, because of their large surface area-to-volume ratio, are in many ways more susceptible than humans. Other

Stealing winter

The ever-efficient insect has evolved to maximize winter survival by decreasing the overall energetics of its seasonal strategies. In other words, they steal winter protection from someone else. The "someone" is usually another insect but can also be plants. One example is the strategy used by the parasitic wasp *Nasonia vitripennis* (Hymenoptera: Pteromalidae). Female wasps lay their eggs on pupae of many species of flies. The fly hosts all share in common that the pupal stage is not naked, but instead encased by a hardened shell (actually the exoskeleton of the last larval stage), termed the **puparium**. Many of these flies also have a similar winter seasonal program that involves responding to short-day photoperiod in mid- to late August to enter diapause in the pupal stage by September. The fly hosts are loaded with cryoprotective compounds like glycerol, trehalose, and alanine, possess extra layers of hydrocarbon inside the puparia, and bury themselves a few inches into the topsoil, protected from fluctuating air temperature. The sly parasitic wasp delays the onset of its seasonal program until later into autumn, locating fly hosts already in dormancy. This alone is a slick adaption because it involves recognizing fly species that overwinter in diapause as pupae. Wasp larvae that then feed on dormant hosts directly acquire premade cryoprotectants that they in turn use as their own. By late October, the offspring halt development as last instar larvae, committing to diapause within the waterproofed host until the spring thaw. The thicker puparia also serve to buffer changes in ambient air temperatures for those that become exposed directly to air, permitting slower cooling rates for the wasp larvae inside. The wasp thus avoids injury associated with rapid changes in air temperature. Once favorable conditions return in spring, the fly parasites resume development and eventually emerge from the protective host puparia in late spring. By that time, a new generation of fly hosts is already available, just waiting to be parasitized by egg-bearing females. The winter strategy of *N. vitripennis* is ingenious, since the wasp has essentially evolved a survival strategy of letting another insect do all the work!

environmental stressors include drought, over hydration, ion imbalance, and exposure to chemical toxicants or pollutants.

By far the most significant aseasonal stressor is temperature. This obviously sounds very similar to our discussion of seasonality, and to degree, there is a great deal of commonality. The differences in this case are (1) the lack of preparedness (unlike what occurs with seasonal adaptations), and (2) the absolute temperatures encountered. When we examined dormancy associated with winter, subzero temperatures were a primary threat that insects must contend with if they hope to survive. Thus a major feature of diapause is cold hardening. The situation is much different with aseasonal change, in which the real threat comes from the degree of change in temperature that occurs over a short window, not the absolute temperature experienced. For example, sudden drops in temperature, in which ambient air temperatures plummet by twenty to thirty degrees in just a few hours, can cause **cold shock** even though conditions remain above freezing. The opposite situation can also occur with rapid temperature elevations (**heat shock**). In either scenario, temperature related injury and death are possible outcomes, not because of the actual final temperature experienced, but instead because of the internal turmoil the insect faces as a result of the sudden and dramatic changes. Therefore, when temperature functions as an aseasonal stressor, whether injury will occur—and if so, its ultimate severity—depends on the degree of change that occurs (meaning from beginning to end temperature), the rate at which the temperature elevates or drops, and the absolute temperature if it is near an upper or lower survivorship threshold, the **critical thermal minima** or **maxima** for an individual species. Exposure

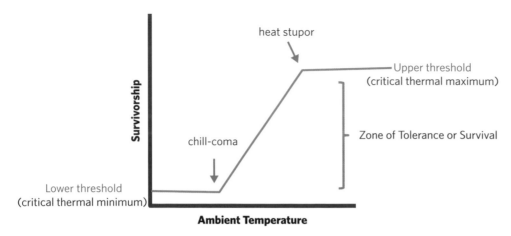

Figure 13.14. The relationship between environmental temperatures and survivorship for an insect. Since insects are poikilotherms, their body temperatures change directly with ambient conditions. Each species has a temperature zone in which they can survive; temperatures outside that range will lead to death. Prior to reaching critical thermal temperatures, a physiological arrest (chill-coma or heat stupor) will occur.

to temperatures at or below the critical thermal minimum will result in death. The same is true for the critical thermal maximum. Of course for both of those conditions, the onset of the death depends on the length of exposure (figure 13.14).

If aseasonal stress is inevitable, then living sounds futile for an insect! Perhaps on paper it seems that way, but we know the reality: that is, insects are doing very well for themselves. So this obviously means that they have coping mechanisms to deal with even the unpredictable. The main stress response used to deal with temperature stressors are heat shock proteins. Yes, the same proteins we have discussed a few times in the chapter already, including with seasonal changes. Several different types of heat shock proteins have been identified in a plethora of insect species. In general, HSPs are named according to their molecular weight, with the higher mass proteins being most associated with high temperature stress. The ubiquity of HSPs in a wide range of stress situations is a reflection of their ability to confer protection to cells from stress-induced damage. Protection is manifested through serving as chaperone proteins to intracellular proteins, thereby helping them resist denaturing conditions. In some insect species, HSPs are believed to facilitate resolubilization of proteins denatured by thermal stress, in essence raising them from the dead. A **rapid heat hardening** response, nearly identical to rapid cold hardening, occurs in several species in which a brief exposure to a nonlethal high temperature stimulates the production of several HSPs, which in turn protect intracellular proteins from a subsequent lethal aseasonal elevated temperature change.

In reality, heat shock proteins operate in concert with several other of the general stress responses discussed earlier in this chapter. Regardless of which responses are used, none are called into action until after the insect encounters the stressor. What this means in practical terms is that stress occurs before a response can be mounted, and sometimes the stress responses will not be sufficient to counter the effects. The best-case scenario is that a short-term injury will be experienced but will not alter the insect's ability to reproduce. At other times, the stressed insect is not as lucky, as aseasonal stress evokes more serious injuries, such as inability to reproduce, and thus fitness is lowered. Obviously the ultimate fate is that death results, and as we have discussed earlier in the text, death has a profound influence on the ability to reproduce.

CHAPTER REVIEW

❋ **Talk about stressed: 24/7, 365 days a year**
- Insects do not have the option of removing themselves from nature to take a seat by the comforting hearth, waiting for ideal conditions to

return. They must contend with prevailing conditions, whether related to the abiotic or biotic features of the environment, at all times. In other words, stress avoidance is not an on/off mechanism for them. Rather, insects rely on mechanisms of stress management the entire duration of their lives.

- Stress can be defined in two ways, as a condition and as a coping mechanism. When viewed as a condition, stress results when demands are placed on an individual that causes deviation from "normal." Normal refers to the physiological state of homeostasis, in which internal systems are functioning at ideal or optimal levels. Stress can also be thought of as the response, usually in the form of a physiological mechanism(s), to adverse conditions. If the response syndrome is working properly, stress itself does not necessarily lead to negative consequences.

- Stress is also inevitable when living in the natural environment. Under this view, stress as a response is adaptive and necessary. Stress is associated with seasonal change, where conditions that typify winter include the arrival of cold temperatures, low humidity, and an absence of food. In some areas, these harsh conditions may last from four to eight months, meaning that the insects present must contend with stress on a daily basis. Unfortunately, impending climatic stress is not always predictable. Any abiotic component of the environment can change without warning. When it does, an insect is exposed to a stressful situation, the severity of which is dependent on the factor that has changed, how quickly a departure from the previous condition has occurred, and the difficulty of the resulting new conditions.

- Not all stressors in the lives of insects are derived from climatic change or weather-related events. A whole array of predators, parasites, and pathogens target insects as a food source.

Dealing with stress on a typical day: General stress responses

- Stress avoidance is not a practical approach to dealing with daily challenges and threats. As a consequence, animals, insects included, have evolved general stress responses to cope with a normal existence. What this means is that insects rely on a generic system of detecting and responding to daily challenges as part of their everyday lives. This is in contrast to the very specific mechanisms used to prepare and cope with severe and/or long-term adversity such as seasonal change.

- General stress responses, also called general adaptation syndrome, in insects (as well as in most any animal) have three phases: (1) detecting stressors, (2) responding to stress, and succumbing to stress.

- General stress responses are activated upon detection of a stimulus. Insects generally rely on two broad categories of stress responses, those that are behavioral and those involving physiological mechanisms.

- Most often when general stress responses are discussed, the focus is on the physiological responses of the organism. Insects have been model organisms in this regard, with particular attention given to how they cope with stresses associated with temperature, overcrowding, and noxious or toxic compounds. The two dominant components of the physiological stress response are hormones and proteins.

Environmental tokens tell the tale of impending changes

- The evolutionary success of insects as a whole has hinged in part on their ability to interact with and in the environment. By extension, this same form of oneness with the environment is needed for dealing with stress. The various forms of stressors, regardless of whether abiotic or biotic, evoke many of the types of behaviors and communication signals we examined in chapters 11 and 12. Consequently, sensory perception of stress stimuli relies on the same sensory pathways described for auditory, mechanical, visual, chemical, and tactile sensation.

- Most abiotic features of the environment change from summer to winter, but some begin to change prior to the onset of harsh conditions.

With these features, the changes are gradual and do not adversely affect insects. As a result, such abiotic characters can serve as cues that signal climatic change is coming. These signals are termed environmental tokens.

- Environmental tokens can range from those as apparent as absolute temperatures or amount of day length (photoperiod), to less-apparent tokens like surpassing threshold levels of relative humidity or atmospheric pressure, or the relationship between daytime and nighttime temperatures (thermoperiod). The tokens are not just a set point, but also represent levels that have deviated from the ideal conditions experienced by the insects previously and in which development and reproduction thrived. They thus are used as anticipatory cues that signify two important details of the future: (1) conditions favoring growth, development and reproduction are coming to an end; and (2) unfavorable conditions are approaching.

- In some cases, environmental tokens are used for more routine events than stress adaptations. Internal or biological clocks regulate several daily, even seasonal, physiological or behavioral activities. The clocks are a series of genes in neuronal cells that function in timekeeping. When a precise time in a twenty-four-hour window is reached, the genes express neuropeptides and other modulator peptides that start or stop cellular events. Typically, events tied to daily cycles are termed circadian rhythms. Timing by the clocks is regulated or set by environmental tokens associated with rotation of the Earth on its axis. In contrast, events on annual cycles respond to tokens that occur as a result of the Earth's movement around the sun.

Seasonality and insect life history traits

- Seasonality refers to changes that occur in the insect's environment throughout the course of a year. Such changes are generally predictable in terms of timing, duration, and in the types of climatic differences that will occur. Insects living in areas where seasonal change occurs are faced with the dilemma of recognizing when the change is coming and then deciding what to do once they know that the environment is going to become drastically different. Obviously the answer to the first issue is to rely on tokens that signify climatic change. Photoperiod is by far the most reliable for predicting seasonality.

- Dormancy is known by different names (i.e., hibernation, diapause, aestivation and quiescence) and the various forms are not the same. However, all forms of insect dormancy do share in common sustained periods of reduced physiology manifested as suppressed metabolic activity and a halt in development. Diapause is essentially the insect equivalent to vertebrate hibernation, in which individuals respond to reduced photoperiod by entering a developmental arrest that will be maintained until a separate and sustained environmental token indicates that favorable conditions have returned.

- Winter dormancy results in dramatic alterations in an individual before the onset of harsh conditions. The most significant change is preparation for cold. Low temperatures, often approaching or dipping below freezing, typify winter. The life history characteristics favored during summer are not useful for contending with cold. So if the goal is to survive winter—and it is—the diapause program must prepare insects for the impending low temperatures. This is frequently achieved via acquiring cold hardiness, in which several cryoprotective compounds are synthesized; body water content is reduced to concentrate solutes, thereby lowering the supercooling point of body fluids; and membrane lipids are changed to permit stretching in case of ice crystal formation and expansion.

- Several of the adaptions mentioned thus far deal with ice prevention. Why are insects so concerned with frozen water? Ice is a major concern not just for insects, but for all organisms that must contend with subzero temperatures. Ice can be deadly when it occurs inside the body of an animal.

- All seasonal changes do not involve major physiological adaptation per se. Some preparations for dealing with adversity are as simple as seeking out refuge or vacationing in a warmer climate. For example, some species may burrow into soil, below the frost line, to ride out the harsh conditions. Others may engage in migration from the area.

Genetic regulation of seasonal survival

- Insects have an incredible arsenal of seasonal adaptations, which are part of highly evolved genetic programs that serve to promote species survival when enduring long-lasting stress. How are all of these necessary changes coordinated? In simplest terms, a timer and counter are needed for photoperiodism. The environmental information that is deciphered must then lead to an output signal that sets an appropriate response mechanism into motion. In the case of responding to seasonal change, one or more hormones are involved, which in turn can activate or terminate a cascade of cellular, physiological, and behavioral activities throughout an individual. Thus, a broad view of seasonal program coordination moves from environmental signal to interpretation to output.

- Most insects rely on a diapause-inducing hormone, as well as juvenile hormone, and then a separate terminating hormone to initiate resumption of development. The internal clocks associated with photoperiodism are referred to as circadian clocks. The original name was associated with clock genes that anticipate daily environmental changes. It is now understood that these same clocks are involved in seasonal programs as well. Multiple genes have been identified that regulate clock functioning and that respond to clock signals to produce the output mentioned earlier. In essence the genes provide increase/decrease or on/off signals to the circadian clocks. Similarly, other clock genes function as the timers and counters of photoperiodic signals. Multiple genes are involved in receiving and interpreting environmental tokens, as well as in generation of the outcome signals.

Coping with the unknown: Aseasonality

- Most stress experienced by an insect is associated with unpredictable changes in the abiotic environment. Aseasonal stressors arrive suddenly, evoking dramatic shifts in climatic and environ-

mental conditions, and therefore do not include anticipation. As a consequence, photoperiodism and other environmental tokens are not factors in terms of preparing for aseasonal changes, because preparation is not a feature of this type of stress. In practical terms, this means that the outcome response is generated only after the stressor has been experienced.

- By far the most significant aseasonal stressor is temperature. This obviously sounds very similar to our discussion of seasonality, and to degree, there is a great deal of commonality. The differences in this case are (1) the lack of preparedness (unlike what occurs with seasonal adaptations), and (2) the absolute temperatures encountered. During dormancy associated with winter, subzero temperatures are a primary threat that insects must contend with if they hope to survive. Thus a major feature of diapause is cold hardening. The situation is much different with aseasonal change, in which the real threat comes from the degree of change in temperature that occurs over a short window, not the absolute temperature experienced.

- The main stress response used to deal with temperature stressors are heat shock proteins. Several different types of heat shock proteins have been identified in a plethora of insect species. In general HSPs are named according to their molecular weight, with the higher mass proteins being most associated with high temperature stress. The ubiquity of HSPs in a wide range of stress situations is a reflection of their ability to confer protection to cells from stress-induced damage.

MUSHROOM FARMING (SELF-TEST)

Level 1: Knowledge/Comprehension

1. Define the following terms:

   (a) stress   (d) environmental token

   (b) stressor   (e) diapause

   (c) heat shock   (f) seasonality

2. How do insects perceive stress?

3. Describe how the general stress response of insects operates.

4. What is the difference between seasonal versus aseasonal climatic change?

5. Dormancy is a very broad term when used in the context of insect biology. Describe some of the various forms of dormancy used by insects.

6. Explain how internal clocks respond to environmental tokens.

Level 2: Application/Analysis

1. Explain why more than one form of dormancy is necessary for insect survival.

2. Photoperiodism is central to seasonal programs. Describe how insects use photoperiodic signals to prepare for seasonal change.

3. Discuss which heat shock proteins (in general) would be synthesized in response to a sudden drop in temperature and describe their functions to protect cells from injury.

Level 3: Synthesis/Evaluation

1. Speculate on how circadian clocks operate even during several days of continuously cloudy and rainy conditions.

2. If the clock genes timeless (tim) and period (per) were ablated (meaning destroyed without otherwise harming the insect), what would happen to the functioning of the circadian clock?

REFERENCES

Brackenbury, J. 1994. Insects: Life Cycles and the Seasons. Blandford Press, London, UK.

Danks, H. V. 2004. Seasonal adaptations in Arctic insects. *Integrated Comparative Biology* 44:85-94.

Danks, H. V. 2005. Key themes in the study of seasonal adaptations in insects I. Patterns of cold hardiness. *Applied Entomology and Zoology* 40:199-211.

Danks, H. V. 2006. Key themes in the study of seasonal adaptations in insects II. Life-cycle patterns. *Applied Entomology and Zoology* 41:1-13.

Denlinger, D. L. 2002. Regulations of diapause. *Annual Review of Entomology* 47:93-122.

Denlinger, D. L., and R. E. Lee Jr. 2010. Low Temperature Biology of Insects. Cambridge University Press, Cambridge, UK.

Duman, J. G. 2001. Antifreeze and ice nucleator proteins in terrestrial arthropods. *Annual Review of Physiology* 63:327-357.

Emerson, K. J., W. E. Bradshaw, and C. M. Holzapfel. 2009. Complications of complexity: Integrating environmental, genetic and hormonal control of insect diapause. *Trends in Genetics* 25:217-225.

Even, N., J.-M. Devaud, and A. B. Barron. 2012. General stress responses in the honey bee. *Insects* 3:1271-1298. doi:10.3390/insects3041271.

Hahn, D. A., and D. L. Denlinger. 2007. Meeting the energetic demands of insect diapause: Nutrient storage and utilization. *Journal of Insect Physiology* 53:760-773.

Heinrich, B. 1995. A Year in the Maine Woods. Da Capo Press, Boston, MA.

Heinrich, B. 2013. The Hot-Blooded Insects: Strategies and Mechanisms of Thermoregulation. Springer, New York, NY.

Ikeno, T., S. I. Tanaka, H. Numata, and S. G. Goto. 2010. Photoperiodic diapause under the control of circadian clock genes in an insect. *BMC Biology* 8:116. http://www.biomedcentral.com/1741-7007/8/116.

King, A. M., and T. H. MacRae. 2015. Insect heat shock proteins during stress and diapause. *Annual Review of Entomology* 60:59-75.

Korsloot, A., C. A. M. van Gestel, and N. M. van Straalen. 2004. Environmental Stress and Cellular Responses in Arthropods. CRC Press, Boca Raton, FL.

Koštál, V. 2006. Eco-physiological phases of insect diapause. *Journal of Insect Physiology* 52:113-127.

Leather, S. R., K. F. A. Walters, and J. S. Bale. 1996. The Ecology of Insect Overwintering. Cambridge University Press, Cambridge, UK.

Nelson, R. J., D. L. Denlinger, and D. E. Somers. 2010. Photoperiodism: The Biological Calendar. Oxford University Press, New York, NY.

Saunders, D. S., R. D. Lewis, and G. R. Warman. 2004. Photoperiodic induction of diapause: Opening the black box. *Physiological Entomology* 29:1-15.

Sørensen, J. G., C. J. Vermeulen, G. Flik, and V. Loeschcke. 2009. Stress specific correlated responses in fat content, Hsp70 and dopamine levels in *Drosophila melanogaster* selected for resistance to environmental stress. *Journal of Insect Physiology* 55:700-706.

Waldbauer, G. 1996. Insects through the Seasons. Harvard University Press, Cambridge, MA.

Zhao, L., and W. A. Jones. 2012. Expression of heat shock protein genes in insect stress responses. *Invertebrate Survival Journal* 9:93-101.

THE ENTOMOLOGIST BOOKSHELF (SUPPLEMENTAL READINGS)

Bale, J. S. 2002. Insects and low temperatures: From molecular biology to distributions and abundance. *Philosophical Transactions of the Royal Society of London B* 357:849–862.

Block, W. 1996. Cold or drought—the lesser of two evils for terrestrial arthropods? *European Journal of Entomology* 93:325–339.

Hahn, D. A., and D. L. Denlinger. 2011. Energetics of insect diapause. *Annual Review of Entomology* 56:103–121.

Heinrich, B. 2009. Winter World: The Ingenuity of Animal Survival. Harper Perennial, New York, NY.

Hochachka, P. W., and G. N. Somero. 1984. Biochemical Adaptation. Princeton University Press, Princeton, NJ.

Hoffman, A. A., and P. A. Parsons. 1991. Evolutionary Genetics and Environmental Stress. Oxford University Press, New York, NY.

Pimentel, D. 1994. Insect population responses to environmental stress and pollutants. *Environmental Reviews* 2:1–15.

Selye, H. 1956. The Stress of Life. 2nd ed. McGraw-Hill Publishers, New York, NY.

Storey, K. B. 2004. Functional Metabolism: Regulation and Adaptation. Wiley-Liss, Hoboken, NJ.

Tauber, M. J., C. A. Tauber, and S. Masaki. 1985. Seasonal Adaptations of Insects. Oxford University Press, New York, NY.

ADDITIONAL RESOURCES

Byrd Polar and Climate Research Center
http://bpcrc.osu.edu/

Climate change affects insect distribution
http://www.natureworldnews.com/articles/6135/20140221/climate-change-affects-insect-distribution.htm

Diapause in insects
http://www.as.wvu.edu/biology/bio21site/Diapause%20in%20insects.pdf

Honey ant adaptations
http://education.nationalgeographic.com/media/reference/assets/honey-ant-adaptations-1.pdf

How insects survive the long, cold winter
http://www.hys.org/news/2011-02-insects-survive-cold-winter.html

Popular Science Blog
Do insects get stressed? http://www.popsci.com/do-insects-get-stressed

Science Daily: For many insects, winter survival is in the genes
http://www.sciencedaily.com/releases/2007/05/070530124251.htm

14 Revenge of the Humans

Insects as the Hunted

The struggle between man and insects began long before the dawn of civilization, has continued without cessation to the present time, and will continue, no doubt, as long as the human race endures. It is due to the fact that both men and certain insect species constantly want the same things at the same time.

S. A. Forbes (1915)*

If human beings are to continue to exist, they must first gain mastery over insects.

Dr. L. O. Howard (1911)†

Insects have the major disadvantage of not being cute and cuddly like bear cubs advertising fabric softener and extra-strength toilet paper. The end result is that humans rarely find these animals to be endearing creatures. Couple this poor marketing effort by the class Insecta with the nasty tendency of some species to want what we want, to attack us, or to invade our homes or other property, and you have a recipe for confrontation. A war for existence ensues, which may seem overly dramatic, but is often the true motivation for either member of the battle.

As we know from earlier chapters, insects are amazingly efficient at nearly everything they do, including eating our food and living as parasites on us or on animals we care about. They usually have the upper hand in the relationship. However, insects do not go unscathed. When they attack us, humans fight back, frequently developing ingenious (e.g., novel chemical weapons), and at times not so ingenious (e.g., fire), means for taking the battle to those with six legs. Chapter 14 explores the struggle between insects and man by examining under what circumstances insects achieve the status of enemy: When are insects truly pests in our lives and when should they simply be ignored, albeit cautiously? Methods of insect control are also discussed, ranging from the context of historical efforts to suppress pest populations to today's approach, in which the idea of "control" has been replaced with that of management. As we will learn, the latter practice is an attitude that attempts a more holistic view, so that the health of the environment is not compromised for the sake of trying to achieve the unrealistic goals of the past, namely, eradication of insect pests.

Key Concepts

- A few bad apples: Insects that are pests
- Insect control before the advent of electricity

---

*The Insect, the Farmer, the Teacher, the Citizen, and the State, Illinois Laboratory of Natural History, pamphlet.

†The House Fly, Disease Carrier: An Account of Its Dangerous Activities and of the Means of Destroying It, F. A. Stokes Company, New York, NY.

- The golden age of killing: Insecticides and death
- *Silent Spring* and the end to the eradication dream
- Sustainability and management: A new way of proceeding
- Towers of death, scents of love, and recombinant weapons: Tools of the twenty-first century

A few bad apples: Insects that are pests

A criminal case in the United States requires that the prosecution prove the elements of the case, in essence the guilt of the accused, beyond a reasonable doubt. Civil cases are less demanding, requiring that responsibility or guilt of the defendant is established by a mere preponderance of evidence, meaning by more than 51%. Using either definition to make the case, insects as a whole—the defendants—are not guilty of being harmful toward humans. In fact, the overwhelming majority can be classified either as beneficial to the human condition or as benign. The case for innocence was developed in chapter 3, in which the many beneficial and cultural influences of insects on our lives were established. The case for the prosecution (i.e., public opinion), condemning all insects, was presented in chapter 4. While there are definitely some species that can be classified as "pests," less than 1% of all described species fall into this category.

Our earlier discussions of pests required insects to meet or exceed certain thresholds, which are, for the most part, subjectively determined. In essence, pest status comes down to economics and tolerance. The economic side is obvious; an insect that causes damage that lowers profits of a crop is deemed a pest. A no-brainer! The same can be said of a beast that feeds on the wood frame of a home or building. That tangible damage can be measured in terms of economic loss. When we first explored the topic of pest status in chapter 4, insects were characterized based on whether and/or when they approached or exceeded critical damage thresholds, specifically **economic injury levels** and **economic thresholds**. Insect-induced injury or damage is modeled with respect to anticipated crop yield, profit realized from selling the crop, and costs associated with attempting to control the insect populations.

As we see, a number of factors influence the thresholds. Consequently, pest status for an insect can vary from year to year, during the season, and from one location to another. Two points can be made about insects viewed as pests in agricultural scenarios: (1) the process of establishing the thresholds is complex and combines analytics with some subjectivity, and (2) the same insect will not necessarily be considered a pest all the time. The latter is particularly important to understand. Insects are indeed opportunistic. So if conditions favor them eating our food, they most certainly will. However, if insects arrive in a situation that does not favor them, they will do very little damage to the cropping system. In other words, just because they exist is not reason to attempt to kill them! Later in the chapter we will see how this view has become an integral part of modern agricultural philosophies, in which integrated management and sustainability are centerpieces to dealing with insect pests.

Humans have no tolerance for some insect species. It may be a matter of situation or location. For example, those that vector disease are considered pests at any population density. We value human life above all else, and do not use thresholds to establish acceptable levels of injury or death. The mere presence of, say, a mosquito that vectors a pathogen such as those causing malaria, yellow fever, encephalitis, or related diseases elevates the insect to pest status. Economic injury levels and economic thresholds do not apply, because we cannot and do not place a monetary value on human life. If we did use such thresholds, they would be equal to one another and set at zero (figure 14.1). Similarly, species that invade our homes or habitation, meaning anywhere we are at the moment, are deemed unacceptable. This attitude is not universal, but certainly it is the case for many. For people who despise insects in their space, any and all insects need to be killed; people often do it themselves. However, if they possess no tolerance at all, then a professional exterminator may be called. In either scenario, economics does not play into the decision of when an insect achieves pest status. They exist, therefore they are (pests)! We apply the term **aesthetic injury level (AIL)** to such situations. The individual consumer makes the call on when an insect is a pest.

My home is a classic example of the variability that exists among individuals in terms of insect acceptance or tolerance. This past winter, as has occurred over the last several years, my home became an abode for overwintering brown marmorated stink bugs, *Halyomorpha*

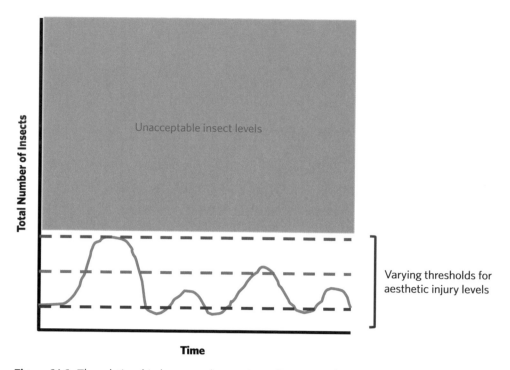

Figure 14.1. The relationship between the number of insects and aesthetic injury levels. The threshold is consumer-specific, so the precise injury level is determined by an individual's tolerance of a specific insect species. The bottom (red) line reflects an individual with virtually no tolerance of insects, while the top (green) line demonstrates a more entomologically accepting person. The absolute values are arbitrary, since anyone's opinion or tolerance of insects can change daily.

halys (Hemiptera: Pentatomidae). Usually they are lethargic throughout most of the winter, but occasionally the stink bugs take flight around the room, resting on the ceiling, windows, or my daughter's hair. If I discover one or more, I usually pick them up bare-handed and throw them outside (alternatively, I may add a step in which the stink bug is flown past my youngest daughter's face or food before being removed from the premises). My wife usually reacts more intensely, either with aggression (at them or me) or by unceremoniously flushing a stink bug wrapped in several layers of toilet paper down the commode. You may have heard it when my youngest daughter came across one of the critters, for the sound meets the combined shrill decibel level of a rock concert when Bruno Mars* first takes the stage (http://bit.ly/1SZ0pOz). The point is that pest status in such cases is completely arbitrary and totally depends on the tolerance of each individual.

---

*Bruno Mars, born Peter Gene Hernandez, is a singer, songwriter, record producer, and actor, and is one of the best-known and successful male vocalists in the world.

The aesthetic injury level *is* a real measure of pest status; it simply has no economic dependency and is individually based. Sometimes, the AIL is tied to individuals who have an extreme dislike or even paranoia with regard to creatures with six legs. Dislike or fear may be manifested as **entomophobia**, a condition in which the individual cannot be around insects at all. If fear is the driving force, then the person may become physically ill in the presence of insects. Others may be more severely affected and suffer from **delusional parasitosis**. Such individuals show an irrational and often uncontrollable fear toward insects that is so severe that it can be physically and emotionally debilitating for them. They cannot handle insects near them, so pest status is achieved before the insect even enters their homes.

Our discussions so far have focused on defining when an insect becomes a pest. The question that also needs to be addressed is, why do they become pests? To a degree, the answer has been looked at throughout this discussion and seems intuitively obvious. The knee-jerk response is, because they can. Not totally satisfying, but

Table 14.1. Important invasive insect species found in the United States. Data from the USDA (http://1.usa.gov/1LO6Y1v).

Invasive Insect Species

Order	Scientific name	Common name
Hymenoptera	*Apis mellifera scutellata*	Africanized honey bee
Coleoptera	*Anoplophora glabripennis*	Asian long-horned beetle
Lepidoptera	*Cactoblastis cactorum*	cactus moth
Diptera	*Aedes albopictus*	Asian tiger mosquito
Lepidoptera	*Lymantria dispar*	European gypsy moth
Hemiptera	*Megacopta cribraria*	kudzu bug
Coleoptera	*Pectinophora gossypiella*	pink bollworm
Diptera	*Bactrocera dorsalis*	Oriental fruit fly
Coleoptera	*Agrilus planipennis*	emerald ash borer
Hemiptera	*Bemisia tabaci*	silverleaf whitefly

not altogether wrong either. Insects are opportunistic, so when presented with a situation of abundantly available nutrients, they will most definitely partake. When might this occur? For one, if an insect is introduced into an area it is not native to, the potential exists for it to flourish. The insect in this case is referred to as an **exotic** or **invasive species**. Basically, the scenario that plays out is an insect is accidentally introduced into another region or another country (table 14.1). If a suitable food source is present, the insect begins its ordinary habit of eating. Since the insect is not native to the area, the predators, parasites, and pathogens keeping it in check in its native land are absent. Voila—the opportunity exists for the insect population to explode.

Similarly, if a nonnative plant is introduced into an area, native insects that were not pests previously might achieve pest status because they feed on and ultimately damage profit gain from the introduced plant. Agricultural practices in the United States and Europe entice some species to become pests. The practice of **monoculture**, with cropping systems in which large areas of land are planted in the same crop year after year, makes food availability predictable for insects that become drawn into becoming agricultural pests. Likewise, dairy, beef, swine, and poultry production is concentrated in specific regions in the United States and by type of animal (e.g., the east coast of the United States contains the majority of poultry producers, the Midwest is where most hogs are raised) (figure 14.2). This too facilitates insects to show an interest in the concentrated source of nutrients (e.g., the livestock, their feed, and their poo) associated with animal production. In every example given, humans are the key link to bringing insects and these opportunities together, since the accidental or intentional introductions were brought about by our actions. Likewise, agriculture is not a natural use of land. We are pest enablers.

What we see from these examples is that some insects are considered pests for a wide variety of reasons. So when is an insect a pest? The answer is, it depends on the situation. And because the context for achieving the status of pest is variable, so, too, are the approaches and methods for controlling insects. What works in a wheat field is not applicable to an apple orchard or in a restaurant, and so forth. In the coming sections, we examine the biggest task facing those who study insects professionally: that is, how to manage insect populations so that we can coexist. The strategies and philosophies have changed dramatically over time, from simply trying to

Figure 14.2. Large, fully automated hog farm in Lyons, Georgia. Such facilities draw the attention of insects just looking for opportunities to become pests! Photo by Jeff Vangua, USDA Natural Resources Conservation Service.

survive to the belief that eradicating all insects was possible to today, in which sustainability is our mantra.

Quick check

If a common house fly enters your kitchen, is it a pest?

Insect control before the advent of electricity

"When I was kid, we used to control insects by picking them off plants with our teeth, removing their wings and legs with our incisors, and then spitting 'em out on the street. They never messed with my food again!" That walk down memory lane took a disturbing turn, quick. Yet it still serves to remind us of how stories of what conditions were like in the olden times always seem to present a vision of times much tougher than today. I know any story my father told me had those familiar elements: walking to school uphill both ways, in blizzards or firestorms, all the while being chased by bears and unicorns. The only things missing from those tales were facts and relevancy!

In the case of insect control, it is generally true that the methods available in the past were much more challenging than those before us today. Why, you ask? The answer has several layers. For example, consider the discussion from the last section, regarding why insects become pests in the first place. None of the factors would really have been much of an influence in a nonelectric, nonindustrialized world. Undoubtedly some insect species occasionally hitched a ride on a ship to be introduced to a new region. This in turn would have set up the potential to wreak havoc in the new land. But travel was comparatively slow by today's standards, trade was limited, farms were small and individualized (family farms), and monoculture was not the norm. So at the very least, the introduction and spread of an invasive species would have been a slow process. That said, once beachheaded on a new shore, the beast would have easily gone undetected, and even once discovered, there would be little a family could do to rid themselves of the pest. The Internet did not exist for a Google search of insect remedies, nor were there pest control companies to call, or even dashing entomologists to consult about the pest problem. Large corporate farms were nonexistent prior to the mid- to late twentieth century. Food was basically raised on each family's small parcel of land. Knowledge of insect pests was typically passed down from one generation to another in the form of oral history. The same was true for what should be done to control the unwanted guests. Thus, discovering and controlling insect pests fell completely on the shoulders of the family or individual. Hence the saying, it takes a village to control an insect. Actually, no one says that, ever.

Consider life on a family farm in Kansas during the early part of the nineteenth century (figure 14.3). The farm was the key to the family's existence. Typically, livestock in the form of cattle, chickens, and possibly swine were raised to provide meat, eggs, and milk (nonpasteurized), as well as other household items such as leather, soap, candles, and bone meal. Corn and wheat were grown as row crops. Mechanized machinery had not yet been invented to work the land (figure 14.4). Rather, this task was accomplished with horse or oxen hitched to a plow. Seeds were hand-sown and covered with dirt with a shovel, foot, or hands. Being a farmer was labor intensive—a truly 24/7 job. Taking a day off was not an option, because the farm provided essentially all of the food for the household for the year. The produce and meat produced also served as currency to exchange for other needed goods with neighbors or in a nearby town.

Now consider what must have crossed the minds of family members looking up into the July sky to

Figure 14.3. A diorama of a mid- to late-eighteenth-century farm in the northeastern United States. The depiction speaks a thousand words to the degree of labor required for a family to survive on the homestead. Original work by Theodore Pitmann, photo by Daderot (http://bit.ly/1Lyjy86).

Figure 14.4. Painting (*Oxen Plowing*, 1875) by Rosa Bonheur depicting the difficult work of tilling a field using oxen and a small plow. All aspects of farming were a challenge, include methods of insect control. Photo by Sotheby's of New York (http://bit.ly/1fJQnSb).

see it blackened with clouds moving toward the farm (figure 14.5). A few minutes later a distinctive roar could be heard, not of thunder or the familiar train locomotive signaling a tornado's impending arrival, but instead the horrifying symphony of insect wings beating in unison. Swarms of migratory locusts composed of too many individuals to count—but well past a million—roamed the prairies and grasslands of the Great Plains and Midwest.

Their arrival meant a path of destruction in their wake. The swarms consumed all vegetation—corn, wheat, sorghum, beans, grasses, and even weeds—in their path. For the locusts, survival meant reaping the harvest of other's toil. For the family, the farm was ruined, maybe only temporarily but possibly permanently. The locusts thrived on this existence. In contrast, the Kansas farm family was potentially driven from the land, or worse, died.

Figure 14.5. Painting from 1884 by Alfred Brehm depicting the futile battle between man and great swarms of locusts attacking en masse. Early methods of insect control literally involved beating insects with whatever objects could be found. Photo in public domain at http://bit.ly/1CyQ1mn.

We know how the story turns out. Man "conquered" the land, not just in Kansas but also throughout the United States and beyond. In many ways, this victory depended on meeting the challenge penned by L. O. Howard (1911), of gaining "mastery over insects." Even before the advent of synthetic chemical insecticides, tractors and planes equipped with sprayers, and experts trained in insect death (aka exterminators), man fought back (figure 14.6). We defended our crops, livestock, and homes from insect invasions. Farmers used whatever was available to them to fight insects, devel-

oping control strategies mostly via trial and error. So how did they combat an enemy like swarms of migratory locusts? For that matter, what did they do to deal with lesser pests that did not attack en masse like the locusts?

To say that they used any available method is by no means an overstatement. Locust storms set off panic on the farm. As the insects landed, family members would attack using brooms, blankets, pots, pans, or anything else to beat the locusts into submission (death). This method killed hundreds, offering little resistance to clouds of millions. The practice probably gave the farmers a sense of fighting back, but they had to know it would be futile. In severe infestations, when it was obvious that massive destruction was inevitable, the crops of one or more farms would be set ablaze. The hope was that the intense heat of the fire would subdue the locusts. But most often the result was triggering wildfires that spread uncontrollably over hundreds to thousands of acres. Using fire seems like a ridiculous method of pest control to us now. However, keep in mind that it was an act of desperation for people who saw their future on the verge of being wiped out (figure 14.7).

Fire is an extreme example of insect control for a severe insect outbreak. More typical forms of insect control involved handpicking or physically removing each individual pest discovered in the fields or on livestock. Obviously this was a very labor-intensive approach that only proved successful when the pest population numbers were was low. It should be noted that this approach is still used today, especially around the home with non-aggressive pests on garden plants and ornamentals, or those that move indoors. Survival on a small farm one hundred plus years ago required that individuals become resourceful. This was evident in the development, again by trial and error, of compounds and mixtures serving as early forms of homemade **insecticides**. Some were derived from plants (botanicals) and others from household materials:

1. Botanical insecticides (plant extracts)
2. Herbs and oils (from plants and livestock)
3. Soaps (from animal fats and plant extracts)
4. Fire ash (literally, the ash from a fire, which is highly alkaline)

Figure 14.6. Portrait of Leland Ossian Howard, an entomologist instrumental in promoting the insecticide revolution in the United States. Photo available at http://bit.ly/1e8z7Fi, author unknown.

Figure 14.7. Fire was often used as a last resort to combat insect pests like locusts that were too numerous to physically battle by hand. Photo by Jonas Dovydenas at http://bit.ly/1QVJFcK.

5. Paris green (an inorganic compound that was highly toxic to everything)
6. Chemicals (sulfur, arsenic, and lead)

The concoctions were applied directly to the crop, fruit, vegetable, or livestock, through physically painting exposed surfaces or topically applying powders, particularly in the case of ash and chemical compounds, onto leaves or skin. If death of the insect occurred, then the substance became a weapon in the arsenal against all insects, not just the one tested. If the desired result was not achieved, then the dose was increased to see if that allowed victory. Often, the mixture was just as destructive to the crop as it was to the insect, so the net effect could be a total loss of the treated plant. At other times the result was far worse, as the substance employed was toxic to the family. In the case of arsenic and lead, this would not have been immediately obvious, so applications most likely continued for months or longer, and only after prolonged exposure did it become apparent that one or more family members were seriously ill. Even then, no one probably suspected that the source of the ailment was the insecticide concoction being used on his or her food. And, because arsenic was commonly used to treat body lice, the compound was applied directly to the skin, greatly accelerating the toxic effects to the louse and human!

Early forms of insect control shared several features that would make them impractical and unsafe for use today. All tended to be nonspecific, so on the one hand they were used broadly toward most any insect deemed a pest, but on the other they offered limited efficacy in reducing populations. Those that were especially toxic for insects had similar effects on the targeted crops or livestock, which, in turn, posed serious health problems to humans. The methods of application lacked precision and were labor intensive. And though the concoctions were relatively cheap to make, no one had any real idea of what they did to insects, how much to use, or when to apply them in relation to insect activity. Despite these limitations, the practice of using such remedies continued well into the twentieth century.

This period, from the beginning of man's plight with insects until roughly the 1940s, has become known as the pre-insecticide era. In reality, the label is misleading, because any compound used to kill insects is technically an insecticide, and as we have just discussed, many different mixtures were used to combat pestiferous species. However, modern references tend to focus on synthetic organic (i.e., containing carbon molecules) compounds for the definition of insecticides. By contrast, prior to World War II, insecticides were predominantly inorganic in composition. The notable exceptions were the botanical insecticides, in particular **pyrethrum,** which is extracted from dried *Chrysanthemum* flowers and various by-products from manufacturing processes (figure 14.8). Yes, the latter implies that industrial waste products were applied to crops, cattle, and people as a

Figure 14.8. Flowers of one type of *Chrysanthemum*. The dried flowers are the source of pyrethrum, from which insecticidal toxins called pyrethrins are extracted. Photo by Digigalos at http://bit.ly/1LBMEn4.

means to remove insects or prevent future damage. As we will see in the next section, methods of insect control via synthetic chemicals made a dramatic leap with the onset of World War II.

The golden age of killing: Insecticides and death

The dawn of the twentieth century ushered in a revolution with respect to man's relationship with insects. Up until then, the window of time called the pre-insecticide era was largely the playhouse for all things with six legs. Insects that preyed on the human condition met with little resistance, at least not enough to detract insect activity by any substantial measure. However, the situation changed radically and rather suddenly with the marvelous discovery of insecticides (by the modern definition). That statement can be broadened to include all forms of

Figure 14.9. US soldier demonstrating hand-spraying of DDT. The insecticide provided a decisive advantage for Allied forces during World War II in preventing the spread of typhus (lice) and malaria and dengue fever (mosquitoes). Photo available through the Centers for Disease Control and Prevention (http://bit.ly/1GCrLBr).

biocidal agricultural compounds, more commonly referred to as **pesticides**, which include such compounds as herbicides (targeting plants), fungicides (fungi), rodenticides (rodents), nematocides (nematodes, a type of worm), acaricides (mites and ticks), and, yes, insecticides. A chemical to kill just about anything found on a farm and anywhere else was being developed.

Far and away the most numerous in type and in use were insecticides. What brought about this revolution in insect control? The answer most often cited is the discovery of DDT or dichlorodiphenyltrichloroethane. This insecticide may be as responsible for the Allied victory in World War II as any other weapon or strategy employed. However, there were clearly earlier events that laid the foundation for the development of DDT and other compounds (figure 14.9). For instance, the beginning of the twentieth century was characterized by an

incredible rise in industrial manufacturing throughout the world. This included the emerging automobile industry in the United States that, importantly, relied on the development of a broad number of synthetic products. The point is that chemical research and development exploded during this period.

The novel compounds created were in turn tested for efficacy in a wide range of applications. Not surprisingly, many were found to be biocidal, especially toward insects. These early synthetic insecticides included phenolic compounds (chlorophenols, nitrophenols), petroleum-based oils, naphthalene, and creosote. All are highly insecticidal but none are fit for long-term environmental exposure because of their lack of specificity and their toxicity for many nontarget organisms. Just try to imagine treating food with creosote, the same tarlike, noxious-smelling material used to coat railroad ties to prevent wood rot.

Also contributing to the chemical revolution was World War I. The so-called Great War introduced the world to large-scale use of chemicals as strategic offensive weapons. Blistering agents and neurological compounds were developed into weaponry with one purpose: to injure and kill human soldiers. Ignoring the morality of such weapons, their production demonstrated that synthetic chemicals could be manufactured in large-scale cheaply and quickly. Correspondingly, relatively simple delivery systems were developed so that the chemicals could reach their intended targets. This was the underlying foundation needed to launch the development of a new class of synthetic insecticides.

In 1939, the Swiss chemist Paul Hermann Müller discovered that DDT displayed insecticidal activity toward several medically important insects. Among these pests were mosquitoes that vectored malaria and yellow fever, and body lice, which transmitted typhus (figure 14.10). The compound was used by the United States and other Allied forces to effectively delouse troops and civilians during the European campaign of World War II and to dramatically reduce the incidence of malaria and dengue fever in the Pacific theatre. DDT provided a decisive advantage for Allied forces throughout the war and is credited with reducing the effects of body lice to such an extent that World War II was the first war in which the casualties of typhus were less than those resulting from

dichlorodiphenyltrichloroethane

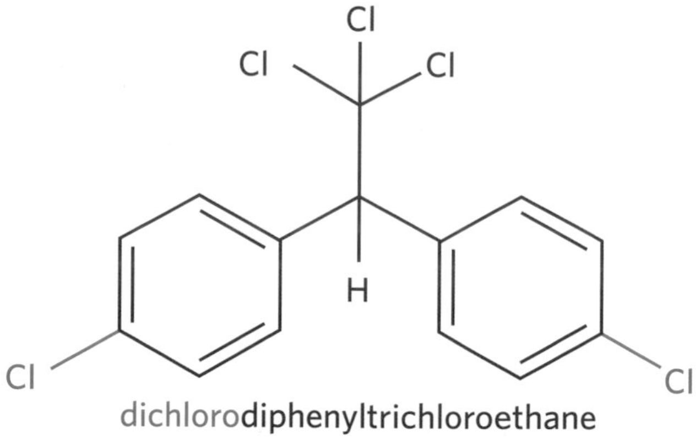

Figure 14.11. Chemical structure of DDT. The ring structures contribute to its ability to pass across biological membranes, and the chlorine side groups allow it to anchor in nervous tissues and to force open ion channels. Image by Leyo at http://bit.ly/1RGz4gG.

Figure 14.10. A memorable if not effective marketing poster for DDT. Image available via the Otis Historical Archives of the National Museum of Health and Medicine at http://bit.ly/1IjYYq3.

battle wounds (see discussion in chapter 2 of Napoleon Bonaparte's failed attempt to conquer Russia). For his efforts, Müller received the 1948 Nobel Prize in Physiology or Medicine. After the war, DDT and several other structurally similar compounds were introduced into the arsenal of weapons used to aid farmers in controlling agricultural pests and to rid homes of unwanted guests (with six legs, that is). The result was almost instantaneous success; pest populations were quickly reduced. So much so that some entomologists actually believed, and to an extent feared, that those insects could not only be controlled but eradicated. What they feared was that, once insects were wiped out, entomologists would no longer be needed.

Bug bytes

Goodbye Mr. Cockroach

http://bit.ly/1FeAlo4

The success of DDT and related compounds is quite astonishing in terms of how quickly they achieved insect control. In fact, within less than a decade, man went from trying to keep up with insects to believing that insects could be annihilated. Obviously the latter attitude was misguided, but it does raise several questions about the class of insecticides being produced, particularly when trying to understand why those compounds were so much better at killing insects than anything that had preceded them: What exactly was different in the new compounds from the old-style insecticides? What do the insecticides do to insects to control them? Why are they so effective in insect control? and, Where did they go? This last question addresses the fact that we do not use DDT any longer in the United States, nor several of the compounds that were considered so effective for about a decade during the 1940s and 1950s. Why not? Let's tackle the four questions in the order presented.

Synthetic insecticides like DDT are not naturally occurring chemicals. Instead, they are synthesized in a laboratory and typically have multiple carbon atoms making up their molecule, often in the form of ring structures, with side groups like chlorine or sulfur (figure 14.11). Carbon-containing compounds are referred to as organic, hence the new wave of insecticides being synthesized after World War II were synthetic organic compounds. Again, this is in contrast to insecticides prior to the discovery of DDT, which were predominantly inorganic (lacking carbon). So the defining characteristic of post-World War II insecticides is the presence of carbon. Does carbon make that much difference to killing insects? It seems the answer must be yes, because they did (and do) in fact have very high kill rates. They often are also broad in spectrum, meaning that the compounds are effective against a fairly large array of pest species (and non-pest species).

This leads us to our second question, how do these insecticides kill their targets? Organic-based compounds have the advantage that they can interact with other organic compounds. In other words, similar compounds can permit unique binding or movement of materials across or through organic structures. Living organisms are composed of organic compounds—proteins, lipids, nucleic acids, and carbohydrates—and, consequently, are uniquely sensitive to synthetic organic insecticides. A compound like DDT has carbon formed into an arrangement similar to that of lipids located in membranes, particularly those associated with the nervous system. So the insecticide is able to move easily across biological membranes to reach the target tissues, where it can then embed itself. The physical binding of the insecticide alone, or in combination with its side groups like chlorine or sulfur, can cause those tissues to become "leaky." This basically occurs by ion channels being forced open, unable to close. As a result, the cell no longer can control what moves in or out. When this occur with excitable membranes—those associated with neurons and muscles—the cells can be overstimulated and driven to the point of exhaustion, or they are prevented from being able to function properly. Eventually the cells will be thrown into chaos and likely die. If done long enough or with the "right" target tissues (e.g., nervous and muscle tissues), the insect can become injured, paralyzed, or dies. This is just a broad overview of how many insecticides work, but it does give a general idea of why organic-based compounds kill differently than inorganic insecticides.

Can the unique modes of action of organic insecticides also account for why they were so effective during the golden age, answering our third question? To an extent, an argument can be made to support this view. Insecticidal compounds that target nervous tissue tend to be effective toward many species of insects—again, the idea of broad spectrum referred to earlier. Thus, one compound can achieve satisfactory control against a range of pest insects. This was indeed considered a good

trait for an insecticide, particularly in terms of cost-effectiveness. However, these new compounds were also valued for the ability to be easily and cheaply produced, and they were not labor intensive to apply. In fact, the insecticides could be applied to any situation with the appropriate-sized sprayer: simple handheld sprayers for home use; backpack sprayers for lawns, trees, or localized pest outbreaks; and tractor or aerial sprayers when large acreage needed treatment (figure 14.12). Application technology evolved almost as rapidly as new insecticides. Obviously, insect control in the 1950s was far superior to the practices on that Kansas farm in 1820. Insect pests were being controlled quickly and cheaply. Correspondingly, agricultural production achieved yields never before realized. Yet by the early 1970s, none of these miraculous compounds were still in use in the United States. What happened?

One clear outcome of having weapons to fight back with is that the definition of "pest" tends to become larger in scope. We as a species, then and now, view any creature that interferes with what we want as a pest, and pests need to be controlled. To that end, a massive arsenal of pesticides was developed. The production of novel tools to fight the insect horde shifted the balance of power from man simply trying to survive the insect attacks to seeing that insect populations could be reduced to the eventual overconfidence that we could defeat and annihilate insects as a whole. That final solution obviously never happened. In fact, it was never an achievable or even reasonable goal, but the process did contribute to the modern approaches used today for insect management. Most significant was the fact that during the golden age of developing and using insecticides, no one paid any attention, at least not initially, to whether there were any side effects or consequences to pouring biocidal compounds into the environment as a whole. It turns out that there were, and the public started to take notice. One individual, Rachel Carson, created quite a stir with the publication of *Silent Spring* in 1962. This seminal text forever changed the landscape of insecticide use in the United States and ultimately led to reshaping agricultural practices. So the answer to where did all of the insecticides go can be summed up in two words: *Silent Spring*.

Figure 14.12. Backpack sprayers were commonly used during World War II by Allied soldiers to dispense DDT for mosquito control. Often the insecticide formulation was mixed with petroleum extracts or kerosene. Photo available through the Otis Historical Archives at the National Museum of Health and Medicine (http://bit.ly/1TY5iuG).

Silent Spring and the end to the eradication dream

By the 1950s, synthetic organic insecticides, coupled with the remarkable action of other pesticides, produced an agricultural boom in the United States. Insect pests of agriculture were being controlled, and food production was at an all-time high in terms of yield. The cost of control was cheap, the insecticides were broad spectrum—effective against a wide range of pests, not just those associated with agriculture—and everyone, especially chemical companies, was realizing substantial profits. Fewer insects, more food, and more money! Why would anyone not be happy during the golden age? The romance with pesticides began to fade as it became apparent that synthetic pesticides were not without problems. What kinds of problems? Remember from our discussion in the last section that pesticides, insecticides especially, evolved rather rapidly. DDT is the best example of a synthetic compound that one day was discovered to be highly effective against important pest insects in laboratory testing, and the next (metaphorically speaking) was

being deployed in insect control efforts. What did not occur, which does today, is testing the compounds for unwanted effects, such as determining whether the insecticide is toxic for us or other plants and animals.

At first, probably, no one cared much if there were side effects because DDT performed miraculously during World War II. Similarly, after the war, the insecticide was used with great success in agricultural applications, targeting household pests like cockroaches, and was the main weapon in the United States Department of Agriculture's assault on red imported fire ants (*Solenopsis invicta*) that were spreading throughout the southeast. The latter example is especially relevant to the downfall of insecticides, for the USDA's 1957 fire ant eradication program caught public attention, which eventually turned to an outcry. At the time, fire ants were contained within a few states but spreading fast. The USDA strategy was an aggressive plan relying on aerial spraying of DDT mixed with fuel oil. Public and private lands were sprayed, repeatedly, and without approval of the landowners. Many of those property owners filed suit to have the spraying stopped. They were unsuccessful in their bid, but the case was significant because (1) it reached the US Supreme Court, and (2) the court ruled that the landowners could file for injunctions in the future if environmental damage from the spraying was demonstrated. The case was significant for another very important reason: it gained the attention of a US Fish and Wildlife employee, who also happened to be an environmentalist and writer. That employee was Rachel Carson.

Carson had been expressing concern about the environmental safety of all pesticides since the 1940s. She was primarily worried that no one really had any idea whether the biocides were safe or were creating harmful side effects for nontarget organisms. In 1958, Carson received a letter from a close friend detailing the death of birds on her land that she believed was the result of aerial spraying of DDT. The letter fueled Carson to begin delving further into the effects of synthetic pesticides on the environment. She spent the next four years intensely researching the topic, uncovering case after case detailing health problems and deaths in humans and wildlife believed to be linked to multiple chemical compounds. This research would be the basis for her seminal book, *Silent Spring,* in 1962.

It is important to note that the examples that Carson and others reported implied correlative or associative effects, and not cause and effect relationships. What this means is that the environmental problems were anecdotally observed to occur after treatment with a particular compound, but direct proof was not available, at least at the time, to show that DDT or any other compound was definitively the cause of the environmental and health problems. This is very similar to the situation that existed with trying to blame tobacco companies for incidences of various cancers from cigarette smoking; the cause and effect data was not available for years. A lack of direct proof did not deter Carson, who butted heads with government officials, heads of chemical companies, and, to an extent, even some of her own sympathizers, as she challenged them to make changes to the way we use pesticides. In her book, Carson asked whether these chemicals (pesticides) were safe, claiming that the companies producing the compounds deliberately misled the public concerning the safety of DDT and other synthetic chemicals, and that government officials simply turned a blind eye. *Silent Spring* caused the public to take notice,

and in turn they demanded answers to Carson's questions. And that is what killed the dream of insect eradication. Okay, admittedly, several more steps were taken between the publishing of Carson's book and the eventual changes that occurred with the use of agricultural biocides, but the time frame for change was relatively short. This serves as an impressive example of what can occur when the public is informed and holds public officials accountable (figure 14.13).

Carson could never have imagined that her efforts would lead to a national environmental movement. In 1969, the National Environmental Policy Act was adopted that led to the eventual creation of the US Environmental Protection Agency (EPA). The act established the requirement that all chemicals be registered with the federal government before they could be used in the environment or in places where humans would be exposed. Mandatory testing was required of the compounds to

Figure 14.13. Photo of Rachel Carson, whose book *Silent Spring* laid the foundation for an environmental movement in the United States that included a mandate for responsible use of insecticides. Photo by Cornischong at http://bit.ly/1dnZpCz.

Pesticides are great tools to help manage problems with insects in almost any scenario. This may seem like a puzzling statement in an age that is emphasizing "going green," preferring organically grown foods, and attempting to minimize agents that negatively affect human health and the well-being of the environment as a whole. The reality is that pesticides, insecticides specifically, are not really a threat to achieving such goals. Say what? Okay, perhaps some context is needed to convince you. When insecticides are used *properly*, they provide a safe and effective weapon to help reduce pest populations. Obviously the key to that statement is when used *properly*. It is also important to note that proper use of any pesticide is not just the responsibility of commercial growers and pesticide companies; it is yours too. How am I supposed to know how to correctly use an insecticide? Easy. By reading the label.

All pesticides sold in the United States possess labels that contain the relevant information needed for proper use and handling, including safety considerations and directions for storage (http://npic.orst.edu /health/readlabel.html). In fact, labeling guidelines are federally regulated. So failure to provide the required information on a pesticide label is in violation of federal law. So, too, is the requirement to follow the instructions on the label. In other words, a homeowner who does not use an insecticide according to the instructions provided on the label is violating federal law. More importantly, that individual may be exposing him or herself or other people or organisms in the immediate vicinity to a potential health hazard. And, oh yes, the original goal of trying to control the pest probably was not accomplished either.

The take-home message is, read the label before you use an insecticide, or any type of chemical for that matter. I know this is often a foreign concept to many of us. After all, how many times have we started to build, assemble, or program something without any regard for the instructions? That may work for your DVR, but do not use that approach with any form of biocide.

This brings us to the question of what types of information are typically found on an insecticide label? Quite a few actually! For example, information is provided on which pests the insecticide was developed for, how to use it, what the active and inactive (inert) ingredients are, safety concerns, how to store and dispose of the insecticide, and what to do in case of emergency. In terms of actually getting down to the task of treating the insect, the insecticide label provides everything you need to know:

- A list of pest insects that the insecticide is registered to control. Registered with who? The EPA.
- How much should be used.
- Where the insecticide can be applied (indoors/outdoors, types of vegetation, etc.).
- How to apply the insecticide.
- If potential food has been treated, how long after application before it is safe to eat.
- If a home or building has been treated, how long before you, pets, or other animals can reenter.
- The frequency of application if more than one treatment is needed.

Really, if you follow directions, the insecticide is likely to provide the desired outcome with little to no problems. However, if you ignore the instructions, like spraying for ants in your kitchen with an insecticide designed for outdoor use only on caterpillars, then do not be surprised if the result is not to your liking. Such actions are not only improper use of pesticide; they represent potential threats to your health and are illegal.

demonstrate that they actually did what they claimed and whether they demonstrated any mammalian (like, to us) toxicity, and also to ensure that each had minimal to no effect on other nontarget organisms. Today, the EPA, in conjunction with the USDA, also stipulates that appropriate dosages for each insecticide targeting particular insect pests be determined, and that to use these compounds, individuals must complete a certification process. Carson's *Silent Spring* started a movement that, if nothing else, has led to responsible use of pesticides in all applications. Pesticides were not eliminated, but today their use in agricultural systems and household applications is simply one choice among many. Equally important is that insecticides are used only when necessary rather any time a potential pest is observed.

Sustainability and management: A new way of proceeding

By the end of the 1960s, public discontent with pesticides had reached a crescendo, demanding change in the narrow-minded attitude that had dominated the previous two decades, during which synthetic chemicals were considered the solution to all our pest problems. This view extended to insect pests. No longer was it acceptable to apply the insecticide-only approach to insect control. In fact, the very notion that control was the goal was being dismissed. As we discussed earlier, the early success of insecticides bred the idea that insects could be controlled through killing and that eradication was achievable. What finally became apparent was that insects really could not be controlled, at least not for very long. The killing and eradication tactics of the insecticide era were deemed impractical, unattainable, and not environmentally safe. Consideration of safety included human health, as DDT had awakened the public to the potential hazards to us as well as wildlife.

A new approach to insects as pests was emerging that has continued on into the twenty-first century. The idea of **pest management** evolved, in which managing populations of insects was the strategy. A key adjustment to this new philosophical approach was accepting that insect control cannot be achieved and that coexistence or tolerance is absolutely essential. The concept of pest management is really based on common sense. It is a philosophy with four basic elements:

1. *Pest management should be selective for the pest of interest.* This immediately stands out in contrast to the insecticide era, during which the chemical weapons were desired to be broad spectrum. Lack of selectivity increases the likelihood that resistance can develop in one or more species quickly and that nontarget organisms will be affected. By contrast, selective methods focus on a specific pest while minimizing exposure of nontarget organisms.

2. *Pest management should be comprehensive for the production system or living environment.* In other words, the strategy should take into account the entire field, household, or other area containing the pest, being mindful of other inputs and outputs into the overall system.

3. *Pest management should be compatible with agricultural or ecological principles.* Obviously it defeats the purpose of treating a crop with an insecticide if the synthetic chemical applied is harmful to the plant or otherwise lowers the yield or value of the crop. This also extends to insecticide treatment of insects affecting livestock and us. For example, it makes no sense to use an insecticide in a home to reduce fleas if the chemical stains the walls or carpet, or has a foul-smelling residual odor. Likewise, the strategy should pose no health risks to humans, animals, or other nontarget organisms.

4. *We must be tolerant of potential pest species within economically acceptable limits.* This idea goes back to a much earlier discussion in which we attempted to define pest status. If the insect is not significantly affecting profits, then no action is required. Similarly, acceptance of insects that really are not pests at all but have been deemed so through aesthetic injury levels (AIL) is part of the tolerance attitude.

This last point is without a doubt the most difficult to fully implement, even today, for all the reasons outlined when discussing AIL. Nonetheless, pest management is a sound, responsible approach to dealing with harmful and destructive insect species that minimizes the harmful effects on the environment and the organisms residing within it.

While the practices of pest management are refreshing compared with the practices of the bygone "golden age," one basic question may still appear unanswered. If pest control through killing was the strategy during the middle part of the twentieth century, what is today's alternative? The main objective of pest management today is to reduce pest status, which arguably was the same objective as during the insecticide era. The difference is that killing is but one approach of many used today to reduce the insect population. Multiple, integrated strategies are applied to a given pest situation. This approach is known as **integrated pest management** (IPM), and it relies on understanding the biology of the insect so that a comprehensive strategy is developed to reduce pest status. It should be noted before proceeding that IPM works best with agricultural systems and is tougher to apply toward medically important species in which the EIL and ET are equally positioned at zero. For those species, a level of tolerance is not applicable. That said, ecologically sound principles can still be adhered to when combating such pests. Integrated pest management relies on multiple strategies to reduce the pest status for a given insect. Broadly, there are four strategies for agricultural systems:

1. Adopting a do-nothing strategy
2. Reducing the pest population density
3. Reducing the susceptibility of the crop or livestock to the pest
4. Reducing the pest numbers *and* changing the susceptibility of crop or livestock

Again, remember that multiple strategies can be developed within these categories based on the biology of the insect and the cropping system. The do-nothing strategy represents the idea of tolerance. As long as the insect is not harmful or is causing only minimal damage, nothing needs to be done about it, recognizing that there is no cost benefit to applying control measures with the pest population at such low levels. Reducing the pest population can be done in multiple ways, including using synthetic insecticides, releasing natural enemies in the form of predators and parasitoids, and relying on traps to collect one or both sexes. Pest numbers can also be reduced by developing methods that diminish the insect's ability to reproduce. A classic example is the sterile male technique, in which sterile males are produced in the laboratory and then released en masse into targeted areas. The

sterile insects compete with naturally occurring males to mate with females. This approach works only for species in which the females mate but once in their lifetime.

Another nonlethal strategy is to reduce the carrying capacity of the environment. Making conditions unfavorable for the pest will cause it to leave, fail to reproduce, or conceivably die by indirect means. Arguably one way to alter the carrying capacity of the environment is to alter the susceptibility of the crop or livestock being targeted by the pest. The most common way to accomplish this task is through selective breeding programs, in which desired characteristics, such as insect resistance, are produced by crossbreeding different cultivars or populations. The process can be relatively drawn out depending on the life cycle of the plant or animal. Another important consideration is that the insect resistance traits (i.e., those that confer decreased susceptibility) must not compromise other features of the organism that are highly valued from an agricultural standpoint. For example, where is the value in creating insect-free corn that is neither nutritious or tastes good?

The nuances of each strategy vary depending on the type of damage caused by a particular insect pest, the life cycle of the insect, the behaviors of each life stage, the seasonal occurrence of the insect and the frequency of the crop, and overall pest numbers, as well as the rate at which the population is increasing. Of course all these factors are influenced by seasonal conditions. So any variation in the abiotic environment, like temperature and precipitation, can drastically modify the growth characteristics of the insect and cropping system. Consequently, a great deal of information is needed about the pest and the crop to develop appropriate pest management strategies. As would be expected, the pest system (meaning the environment in which the pest exists) is complex and thus decision making in terms of strategies is not easy. So sometimes a particular plan works well, and at other times, it does not. Rarely, however, does an integrated strategy fail to achieve some level of pest status reduction. This is largely because today pest management strategies are developed via computer software, which rely on recommendation algorithms. In other words, strategies are created based on inputs in the form of biological and economic information to yield outputs— pest management strategies—for the given conditions.

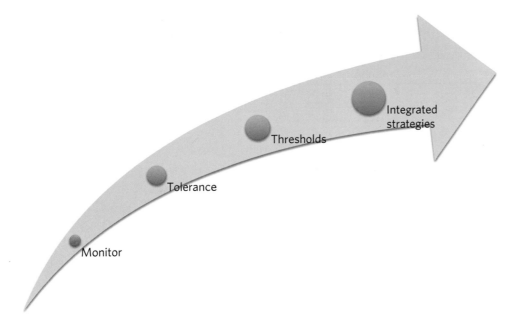

Figure 14.14. Key facets to integrated management of a potential insect pest. The insect population must be monitored to determine when it is approaching or exceeding action or economic thresholds. When the insect density is below levels that require action, tolerance or "do nothing" is practiced. When action is needed, multiple integrated strategies are developed based on the biology of the insect and cropping system, while not compromising agricultural principles or environmental health.

The end result is an informed decision-making process to reduce pest populations using the most economically practical and environmentally conscious approaches. Integrated pest management attempts to achieve pest reduction in a sustainable manner that promotes tolerance and a quality environment (figure 14.14).

Beyond the text

What type of pest management strategies should be used for cockroaches and termites that invade a home?

Towers of death, scents of love, and recombinant weapons: Tools of the twenty-first century

Up to this point, most of the focus on reducing insect populations has been on chemical-based methods. We know from prior discussions that chemicals have been convenient to obtain, cheap to produce, and relatively simple to apply. Of course there is plenty of evidence to show that naturally occurring and synthetic chemicals have limitations and, at times, pose serious health risks. As a consequence, alternative forms of insect control or

management techniques are always being sought. Many have been introduced over the years, some short-lived, while others are proving to be valuable additions to the tools used to manage pest populations. Among those with staying power, a few really stand out as novel or intriguing weapons, such as those that play on the behavior of specific pests, pheromones used for seeking love, and those created by recombinant DNA technology.

In the case of modulating behavior, mosquito females are targeted through the use of traps that emit gas. These so-called towers of death* work because adult mosquitoes locate a mammalian host (like us) by attraction to either a chemical, such as carbon dioxide or octenol, or visual stimuli, such as light of specific wavelengths, or a combination of the two. Of the two, chemical attraction is most commonly used to manipulate mosquito behavior. Warm-blooded animals release carbon dioxide as they exhale, and the rate of emission increases with heat and physical activity. A mosquito follows the gas plume back

---

*Actually, I may be the only person who calls these traps towers of death. The name is sort of catchy, so I do not intend to stop.

Figure 14.15. An example of a carbon dioxide insect trap. It works by burning propane gas to generate carbon dioxide. The gas is attractive to insects such as biting midges and mosquitoes, which fly into the trap but cannot escape. Photo by Michael Hardman at http://bit.ly/1fS27lS.

Figure 14.16. An example of an insect pheromone trap. The trap contains the sex pheromone of an adult female (a moth, in this case) with the goal of luring in males. Males that enter the trap cannot escape. Photo by Beentree at http://bit.ly/1Nnz9Vx.

to the source using sensory receptors located on the antennae. Carbon dioxide traps operate by possessing either a carbon dioxide tank, for direct gas release, or a propane tank, identical to one on a gas grill, that can be lit so that a small flame burns inside the trap, thereby releasing carbon dioxide (figure 14.15). Female mosquitoes are drawn to the released gas. Eventually, some of the mosquitoes fly inside the trap to the source. Once inside, the mosquitoes may be vacuumed into a catch net, dunked in liquid, or exposed to a fan powerful enough to pull the insect in so that they cannot escape. Variations in trap design also include heat emission as a further attractant, as well as chemical lures designed to draw in either sex of mosquito. The traps are most effective when placed in close proximity to where humans live. This of course means that the enemy must be close to us for the trap to work, which almost gives the impression that we are the bait.

Sex attractants are the bases for luring in unsuspecting and horny male insects to their demise. The traps are baited with sex pheromones, derived usually from the female, with the intent of enticing males to approach and enter the trap. Once inside, the insects get trapped in a sticky adhesive, which holds them in place until death. Aggregation pheromones function in a similar way, with the notable difference that males, and not females, pro-

duce the chemical lures. Such pheromones function to call males together so that they collectively form a visual display that females cannot resist. Unfortunately for the males in this case, the odor is being emitted from a trap that no self-respecting female would consider entering and the males get stuck permanently on the inside (figure 14.16). A clever version of the pheromone trap is one in which males are lured to a trap by female sex attractant; they enter, only to be covered with a powdered form of the female's scent, and then are free to escape. Why? So that males will chase each other around, rather than females, believing falsely that they are in hot pursuit of a female. In the end, everyone is frustrated and few to none actually reproduce.

Unlike the simplicity of traps, the use of recombinant DNA technology offers a complex yet highly selective

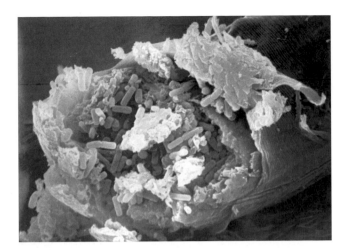

Figure 14.17. Vegetative rods of *Bacillus thuringiensis* within the gut of the nematode *Caenorhabditis elegans*. The bacterium's protein toxins are biocidal to several important agricultural and medically important insects. Photo by Joe Lange at http://bit.ly /1rUdRdD.

means to reduce pest populations or to change the susceptibility of plants and animals. Basically, the technique is used to combine DNA from different sources so that the genetic material can be expressed in a specific organism. Sometimes the goal is to get the host organism to produce a protein of interest for other uses, such as to synthesize human insulin in bacterial cells. However, in the case of pest management, a common approach is to introduce a toxin gene from another organism into a plant's genome with the goal of conferring protection against insect herbivores. A classic example is the introduction of insecticidal toxin genes from the soil bacterium *Bacillus thuringiensis* (Bt) into corn and cotton. There are actually many varieties of Bt, each producing a unique type of insecticidal protein that targets beetles, caterpillars, or mosquitoes (figure 14.17).

The goal of recombinant DNA technology is to introduce a species-specific insecticidal gene into the plant that targets a unique pest of corn or cotton or any other type of crop. Further manipulation of the plant's genome permits regulation of which tissues produce the toxins and when. This slows the rate at which insect resistance to the toxins might develop and also minimizes exposure of nontarget organisms. Several different sources (e.g., spiders, insects, scorpions) of insecticidal toxins have been tested for possible manipulation in this fashion.

Likewise, unique compounds, such as **insect growth regulators** that mimic the action of insect hormones, show promise for similar inclusion into plant tissues. Most examined to date interfere with the molting process. This also represents a potential problem, since all arthropods and a few related other animal phyla depend on similar hormones to molt. To date, such problems are more theoretical than practical, because introduction of recombinant plants into the environment has been slow to occur.

CHAPTER REVIEW

✦ A few bad apples: Insects that are pests

- While some insects can definitely be classified as "pests," less than 1% of all described species fall into this category. In essence, pest status comes down to economics and tolerance.

- Insect pests can be characterized based on whether and/or when they approach or exceed critical damage thresholds, specifically economic injury levels and economic thresholds. Insect-induced injury or damage is modeled with respect to anticipated crop yield, profit realized from selling the crop, and costs associated with attempting to control the insect populations. As we see, number of factors influence the thresholds. Consequently, pest status for an insect can vary from year to year, during the season, and from one location to another. Two points can be made about insects viewed as pests in agricultural scenarios: (1) the process of establishing the thresholds is complex and combines analytics with some subjectivity, and (2) the same insect will not necessarily be considered a pest all the time.

- We value human life above all else, and do not use thresholds to establish acceptable levels of injury or death. The mere presence of, say, a mosquito that vectors a pathogen such as those causing malaria, yellow fever, encephalitis, or related diseases elevates the insect to pest status. Economic injury levels and economic thresholds do not apply, because we cannot and do not place a monetary value on human life. If we did use such thresholds, they would be equal to one another and set at zero.

- Species that invade our homes or habitation, meaning anywhere we are at the moment, are deemed unacceptable. This attitude is not universal, but certainly it is the case for many. For people who despise insects in their space, any and all insects need to be killed. They exist, therefore they are pests! We apply the term aesthetic injury level (AIL) to such situations. The individual consumer makes the call on when an insect is a pest.

- The question that also needs to be addressed is, why do insects become pests? Insects are opportunistic, so when presented with a situation of abundantly available nutrients, they will most definitely partake. When might this occur? For one, if an insect is introduced into an area it is not native to, the potential exists for it to flourish (as an exotic or invasive species). Basically, the scenario that plays out is an insect is accidentally introduced into another region or another country. Since the insect is not native to the area, the predators, parasites, and pathogens keeping it in check in its native land are absent. The same can happen if a plant species is introduced to a new area; insects that were not pests before now have a new opportunity.

Insect control before the advent of electricity

- In the case of insect control, it is generally true that the methods available in the past were much more challenging than those before us today. Why, you ask? The answer has several layers. The Internet did not exist for a Google search of insect remedies, nor were there pest control companies to call, or even dashing entomologists to consult about the pest problem. Large corporate farms were nonexistent prior to the mid- to late twentieth century. Food was basically raised on each family's small parcel of land. Knowledge of insect pests was typically passed down from one generation to another in the form of oral history. The same was true for what should be done to control the unwanted guests. Thus, discovering and controlling insect pests fell completely on the shoulders of the family or individual.

- Survival on a small farm one hundred plus years ago required that individuals become resourceful. This was evident in the development, again by trial and error, of compounds and mixtures serving as early forms of homemade insecticides. Some were derived from plant and animals, and others from household materials; herbs and oils (from plants and livestock); soaps (from animal fats and plant extracts); fire ash (literally the ash from a fire, which is very alkaline); Paris green (an inorganic compound that was highly toxic to everything); and an array of potent chemicals (sulfur, arsenic, and lead).

- The concoctions were applied directly to the crop, fruit, vegetable, or livestock, through physically painting exposed surfaces, or topically applying powders, particularly in the case of ash and chemical compounds, onto leaves or skin. If death of the insect occurred, then the substance became a weapon in the arsenal against all insects, not just the one tested. If the desired result was not achieved, then the dose was increased to see if that allowed victory. Often, the mixture was just as destructive to the crop as it was to the insect, so the net effect could be a total loss of the treated plant. At other times the result was far worse, as the substance employed was toxic to the family.

The golden age of killing: Insecticides and death

- The dawn of the twentieth century ushered in a revolution with respect to man's relationship with insects. Up until then, the window of time called the pre-insecticide era was largely the playhouse for all things with six legs. However, the situation changed radically and rather suddenly with the marvelous discovery of insecticides. That statement can be broadened to include all forms of biocidal agricultural compounds, more commonly referred to as pesticides, which include such compounds as herbicides, fungicides, rodenticides, nematocides, acaricides, and insecticides. Far and away the most numerous in type and in use were insecticides.

- What brought about this revolution in insect control? The answer most often cited is the discovery of DDT or dichlorodiphenyltrichloro-ethane. In 1939, the Swiss chemist Paul Hermann Müller discovered that DDT displayed insecticidal activity toward several medically important insects. Among these pests were mosquitoes that vectored malaria and yellow fever, and body lice, which transmitted typhus. DDT provided a decisive advantage for Allied forces throughout the war and is credited with reducing the effects of body lice to such an extent that World War II was the first war in which the casualties of typhus were less than those resulting from battle wounds. After the war, DDT and several other structurally similar compounds were introduced into the arsenal of weapons used to aid farmers in controlling agricultural pests and to rid homes of unwanted guests. The result was almost instantaneous success; pest populations were quickly reduced.

- The success of DDT and related compounds is quite astonishing in terms of how quickly they achieved insect control. In fact, within less than a decade, man went from trying to keep up with insects to believing that insects could be annihilated. Obviously the latter attitude was misguided, but it does raise several questions about the class of insecticides being produced, particularly when trying to understand why those compounds were so much better at killing insects than anything that had preceded them. The defining characteristic of post-World War II insecticides is the presence of carbon. Does carbon make that much difference to killing insects? It seems the answer must be yes, because they did in fact have very high kill rates. They often are also broad in spectrum, meaning that the compounds are effective against a fairly large array of pest species (and non-pest species). Organic-based compounds have the advantage that they can interact with other organic compounds. In other words, similar compounds can permit unique binding or movement of materials across or through organic structures. A compound like DDT has carbon formed into an arrangement similar to that of lipids located in membranes, particularly those associated with the nervous system. So the insecticide is able to move easily across biological membranes to reach the target tissues, where it can then embed itself.

- Insecticidal compounds that target nervous tissue tend to be effective toward many species of insects—again the idea of broad spectrum referred to earlier. Thus, one compound can achieve satisfactory control against a range of pest insects. This was indeed considered a good trait for an insecticide, particularly in terms of cost-effectiveness. However, these new compounds were also valued for the ability to be easily and cheaply produced, and they were not labor intensive to apply. In fact, the insecticides could be applied to any situation with the appropriate-sized sprayer: simple handheld sprayers for home use; backpack sprayers for lawns, trees, or localized pest outbreaks; and tractor or aerial sprayers when large acreage needed treatment. Application technology evolved almost as rapidly as new insecticides.

- One clear outcome of having weapons to fight back with is that the definition of "pest" tends to become larger in scope. We as a species, then and now, view any creature that interferes with what we want as a pest, and pests need to be controlled. To that end, a massive arsenal of pesticides was developed. The production of novel tools to fight the insect horde shifted the balance of power from man simply trying to survive the insect attacks to seeing that insect populations could be reduced to the eventual overconfidence that we could defeat and annihilate insects as a whole. That final solution obviously never happened.

Silent Spring and the end to the eradication dream

- By the 1950s, synthetic organic insecticides, coupled with the remarkable action of other pesticides, produced an agricultural boom in the United States. Insect pests of agriculture were being controlled, and food production was at an all-time high in terms of yield. The cost of control was cheap, the insecticides were broad spectrum—effective against a wide range of pests, not just those associated with agriculture—and everyone, especially chemical companies, was realizing substantial profits. The romance with pesticides began to fade as it became apparent that synthetic pesticides were not without problems. What kinds of problems? Unwanted effects were becoming increasingly apparent, affecting the health and even survival of humans and other nontarget organisms.

- Since the 1940s, a US Fish and Wildlife employee named Rachel Carson, who also happened to be an environmentalist and writer, had been expressing concern about the environmental safety of all pesticides. She was primarily worried that no one really had any idea whether the biocides were safe or were creating harmful side effects for nontarget organisms. After encouragement from a close friend, Carson began delving further into the effects of synthetic pesticides on the environment. She spent four years intensely researching the topic, uncovering case after case detailing health problems and deaths in humans and wildlife believed to be linked to multiple chemical compounds. This research would be the basis for her seminal book *Silent Spring* in 1962. In her book, Carson asked whether these chemicals (pesticides) were safe, claiming that the companies producing the compounds deliberately misled the public concerning the safety of DDT and other synthetic chemicals, and that government officials simply turned a blind eye. *Silent Spring* caused the public to take notice, and in turn they demanded answers to Carson's questions. And that is what killed the dream of insect eradication.

- In 1969, the Environmental Protection Act was adopted that led to the creation of the US Environmental Protection Agency (EPA). The act established the requirement that all chemicals be registered with the federal government before they could be used in the environment or in places where humans would be exposed. Mandatory testing was required of the compounds to demonstrate that they actually did what they claimed and whether they demonstrated any

mammalian (like, to us) toxicity, and also to ensure that each had minimal to no effect on other nontarget organisms. Today, the EPA, in conjunction with the USDA, also stipulates that appropriate dosages for each insecticide targeting particular insect pests be determined, and that to use these compounds, individuals must complete a certification process.

Sustainability and management: A new way of proceeding

- The early success of insecticides bred the idea that insects could be controlled through killing and that eradication was achievable. What finally became apparent was that insects really could not be controlled, at least not for very long. The killing and eradication tactics of the insecticide era were deemed impractical, unattainable, and not environmentally safe. Consideration of safety included human health, as DDT had awakened the public to the potential hazards to us as well as wildlife. A new approach to insects as pests was emerging that has continued on into the twenty-first century. The idea of pest management evolved, in which managing populations of insects was the strategy. A key adjustment to this new philosophical approach was accepting that insect control cannot be achieved and that coexistence or tolerance is absolutely essential.

- The concept of pest management is a philosophy with four basic elements. Pest management should be selective for the pest of interest, pest management should be comprehensive for the production system or living environment, pest management should be compatible with agricultural or ecological principles, and we must be tolerant of potential pest species within economically acceptable limits.

- The main objective of pest management today is to reduce pest status, which arguably was the same as it was during the insecticide era. The difference is that killing is but one approach of many used today to reduce the insect population. Multiple, integrated strategies are applied to a given pest situation. This approach is known as integrated pest management, and it relies on understanding the biology of the insect so that a comprehensive strategy is developed to reduce pest status.

- Integrated pest management relies on multiple strategies to reduce the pest status for a given insect. Broadly, there are four strategies for agricultural systems: do nothing; reduce the pest population density; reduce the susceptibility of the crop or livestock to the pest; and a combination of reducing the pest numbers and changing the susceptibility of crop or livestock. Multiple strategies can be developed within these categories based on the biology of the insect and the cropping system.

- The nuances of each strategy vary depending on the type of damage caused by a particular insect pest, the life cycle of the insect, the behaviors of each life stage, the seasonal occurrence of the insect and the frequency of the crop, and overall pest numbers, as well as the rate at which the population is increasing. Of course all these factors are influenced by seasonal conditions. So any variation in the abiotic environment, like temperature and precipitation, can drastically modify the growth characteristics of the insect and cropping system. Consequently, a great deal of information is needed about the pest and the crop to develop appropriate pest management strategies.

✦ Towers of death, scents of love, and recombinant weapons: Tools of the twenty-first century

- Most strategies focused on reducing insect populations have been based on chemical-based methods. Chemicals have been convenient to obtain, cheap to produce, and relatively simple to apply. Of course there is plenty of evidence to show that naturally occurring and synthetic chemicals have limitations, and at times, pose serious health risks. As a consequence, alternative forms of insect control or management techniques are always being sought. Many have been introduced over the years, some short-lived, while others are proving to be valuable additions to the tools used to manage pest populations. Among those with staying power, a few really stand out as novel or intriguing weapons, such as those that play on the behavior of specific pests, pheromones used for seeking love, and those created by recombinant DNA technology.

- In the case of modulating behavior, mosquito females are targeted through the use of traps that emit gas. The traps work because adult mosquitoes locate a mammalian host by attraction to either a chemical, such as carbon dioxide or octenol, or visual stimuli such as light of specific wavelengths, or a combination of the two. Of the two, chemical attraction is most commonly used to manipulate mosquito behavior.

- Sex attractants are the bases for luring in unsuspecting and horny male insects to their demise. The traps are baited with sex pheromones, derived usually from the female, with the intent of enticing males to approach and enter the trap. Once inside, the insects get trapped in a sticky adhesive, which holds them in place until death.

- Unlike the simplicity of traps, the use of recombinant DNA technology offers a complex yet highly selective means to reduce pest populations or to change the susceptibility of plants and animals. Basically, the technique is used to combine DNA from different sources so that the genetic material can be expressed in a specific organism. In the case of pest management, a common approach is to introduce a toxin gene from another organism into a plant's genome with the goal of conferring protection against insect herbivores.

MUSHROOM FARMING (SELF-TEST)

Level 1: Knowledge/Comprehension

1. Define the following terms:
   - (a) pest management
   - (b) pesticide
   - (c) insecticide
   - (d) aesthetic injury level
   - (e) pre-insecticide era
   - (f) *Silent Spring*

2. Prior to the use of synthetic chemical insecticides, what were the main methods of insect control?

3. What were the limitations of the pre-insecticide era methods of insect control?

4. How did the discovery of DDT and related insecticides affect the strategies for dealing with insect pests of crops?

5. Describe the basic elements of pest management.

6. Provide examples of modern methods used to reduce pest status that do not utilize synthetic insecticides.

Level 2: Application/Analysis

1. Explain the significance of the 1939 discovery that DDT was highly insecticidal in in terms of the outcome of World War II.

2. Describe the contributions of Rachel Carson that led to the end of the golden age of insecticides.

3. Explain the bases for how most insect traps work to lure in adult male insects.

Level 3: Synthesis/Evaluation

1. Discuss how modern insecticides can be designed to be species-specific, overcoming one of the major limitations of the past.

2. Speculate on whether there ever could be a scenario in which a do-nothing strategy would be appropriate in dealing with a medically important insect of humans.

REFERENCES

Cardé, R. T., and A. K. Minks. 1995. Control of moth pests by mating disruption: Success and constraints. *Annual Review of Entomology* 40:559-585.

Carson, R. 1962. Silent Spring. Houghton Mifflin, Boston, MA.

Flint, M. L., and R. van den Bosch. 2013. Introduction to Integrated Pest Management. Springer, London, UK.

Forbes, S. A. 1915. The Insect, the Farmer, the Teacher, the Citizen, and the State. Illinois Laboratory of Natural History, pamphlet. Urbana, IL.

Gilbert, L. I., and S. A. Gill. 2010. Insect Control: Biological and Synthetic Agents. Academic Press, San Diego, CA.

Gullan, P. J., and P. S. Cranston. 2014. The Insects: An Outline of Entomology. 5th ed. Wiley-Blackwell, West Sussex, UK.

Howard, L. O. 1911. The House Fly, Disease Carrier: An Account of Its Dangerous Activities and of the Means of Destroying It. F.A. Stokes Company, New York, NY.

Liebhold, A. M., and P. C. Tobin. 2008. Population ecology of insect invasions and their management. *Annual Review of Entomology* 53:387-408.

Naranjo, S. E., C.-C. Chu, and T. J. Henneberry. 1996. Economic injury levels for *Bemisia tabaci* (Homoptera: Aleyrodidae) in cotton: Impact of crop price, control costs, and efficacy of control. *Crop Protection* 15:779-788.

Oerke, E.-C. 2006. Crop losses to pests. *Journal of Agriculture Science* 144:31-43.

Pedigo, L. P., and M. Rice. 2006. Entomology and Pest Management. 5th ed. Pearson/Prentice Hall, Upper Saddle River, NJ.

Penn State Pesticide Education Program. 2015. Why You Need to Know about Reading a Pesticide Label. http://bit.ly/1Ll7R1i. Accessed June 17, 2015.

Stefferud, A. 1952. Insects: The Yearbook of Agriculture. United States Government Printing Office, Washington, DC.

Unsworth, J. 2010. History of Pesticide Use. http://bit.ly/1dB8vMR. Accessed June 11, 2015.

Witzgall, P., P. Kirsch, and A. Cork. 2010. Sex pheromones and their impact on pest management. *Journal of Chemical Ecology* 36:9737. doi:10.1007/s10886-009- 9737-y.

Yu, S. J. 2014. The Toxicology and Biochemistry of Insecticides. 2nd ed. CRC Press, Boca Raton, FL.

THE ENTOMOLOGIST BOOKSHELF (SUPPLEMENTAL READINGS)

Bradley, F. M., B. W. Ellis, and D. L. Martin. 2010. The Organic Gardner's Handbook of Natural Pest and Disease Control: A Complete Guide to Maintaining a Healthy Garden and Yard the Earth-Friendly Way. Rodale Books, New York, NY.

Frishman, A. M., and P. J. Bello. 2013. The Cockroach Combat Manual II. AuthorHouse, Bloomington, IN.

Holmes, D. 2006. Pick the right insecticide for the pest. OCALA.com. http://www.ocala.com/article/20061111/BIGSUNHOMES/211110315. Accessed June 1, 2015.

Horn, D. J. 1988. Ecological Approach to Pest Management. Guilford Press, New York, NY.

Metcalf, C. I., W. P. Flint, and R. L. Metcalf. 1962. Destructive and Useful Insects: Their Habits and Control. 4th ed. McGraw-Hill Book Company, New York, NY.

Raupp, M. J., J. A. Davidson, C. S. Koehler, C. S. Sadof, and K. Reichelderfer. 1989. Economic and aesthetic injury levels and thresholds for insect pests of ornamental plants. *Florida Entomologist* 72:403-407.

Sweetman, H. L. 1958. The Principles of Biological Control. W.C. Brown Company, Dubuque, IA.

ADDITIONAL RESOURCES

History of pesticide use
http://people.oregonstate.edu/~muirp/pesthist.htm

How to make your own insecticide soap
http://www.hortmag.com/weekly-tips/pests-diseases/mix-your-own-insecticidal-soap-for-garden-pests

Insecticides
http://www.epa.gov/caddis/ssr_ins_int.html

Insecticide tips for farmers
http://www.agphd.com/ag-phd-newsletter/2015/05/19/insecticide-tips/

Insect soaps and detergents http://www.ext.colostate.edu/pubs/insect/05547.html

Integrated pest management
http://www.epa.gov/agriculture/tipm.html#How%20do%20IPM

National Pesticide Information Center
http://npic.orst.edu/ingred/ptype/insecticide.html

Pesticide buying guide
http://www.lowes.com/projects/lawn-and-garden/pesticide-buying-guide/project

Pesticide education and assessment program
http://pesticide.umd.edu/

Pesticides
http://www.niehs.nih.gov/health/topics/agents/pesticides/

Pheromone traps—using sex as bait
http://www.thenakedscientists.com/HTML/interviews/interview/911/

15 Forensic Entomology

Insects as Tools in Legal Investigations

That we here highly resolve that these dead shall not have died in vain.

Abraham Lincoln
16th President of the United States*

We think of the animals that do the important work of redistributing the stuff of life as scavengers, and we may admire and appreciate them for providing their necessary "service" as nature's undertakers.

Dr. Bernd Heinrich
Life Everlasting: The Animal Way of Death (2013)

All crimes do not involve death. For that matter, all legal issues are not criminal issues. However, homicides, or really just the presence of a corpse, draw the undivided attention of insects. Actually, us too! Humans are captivated by death, especially when foul play is suspected. Despite the reality that death by murder, abuse, or negligence represents some of the worst actions that one human can bestow on another, people are drawn to it. We watch the news intently, read novels about crime and murder, and have made crime shows the top-rated programming on television in the United States. So it should come as little surprise that our number one nemesis—insects—has the same interests that we do. Okay, insect attraction to human or other animal remains is motivated not by the macabre (like it is for us), but instead as part of foraging behavior. Yes, several insect species use dead animals (feces too) as their primary food source. By understanding which insects are attracted to the dead, when they will arrive, and how long they take to complete development under varying environmental conditions, we can use these species as useful pieces of physical evidence in investigations of suspicious or unexplained deaths.

Using aspects of insect biology in this deductive manner is the basis for forensic entomology, the branch of forensic science and subfield of entomology that deals with issues in which insect and arthropod biology intersect with the judicial system. For most people, this immediately jogs memories from crime shows, in which fly maggots found on a body reveal "secrets" about a homicide that, when presented to the accused, cause him or her to instantly provide a detailed courtroom confession. That is not exactly what occurs in real life, but insects can be quite useful in uncovering infor-

---

*An excerpt from the Gettysburg Address, a ten-sentence speech delivered by President Lincoln on Thursday, November 19, 1863, as part of the dedication of the Soldiers' National Cemetery in Gettysburg, Pennsylvania. It is considered one of the best-known speeches in American history (http://bit.ly/1bFJewr).

mation about the time, location, and climatic conditions associated with death of humans and other animals. Of course the reality is that forensic entomology is much broader than just death investigations, and also includes legal aspects of insects associated with food (stored product entomology), dwellings (urban entomology), and matters of national security. This chapter introduces the three branches of forensic entomology by detailing the types of legal matters associated with each, examines examples of insects that can serve as physical evidence or even be the culprits in legal matters, and delves into the biology of the insects involved to understand why they become associated with human habitation, stored food, or colonized corpse.

Key Concepts

- Murder, termites, and weevils: The many faces of forensic entomology
- There's a fly in my soup: Should I sue?
- Home invasion: Matters for urban entomology
- Maggots, murder, and men
- The fly who loved me: Myiasis and cases of neglect
- Maggots on crack: Agents of toxicology

Murder, termites, and weevils: The many faces of forensic entomology

Insects can provide us with amazing insight when it comes to legal matters. Take a suspicious death, for example. Some species have the potential to be silent witnesses to murders and other crimes. For this statement to make sense, you first need to understand what insects do when they discover a dead body. In a terrestrial environment in which a human corpse is exposed to the elements, certain species of insects can arrive just minutes after death. They are not gawkers the way humans are when driving past a severe car accident on the interstate. No, these insects are there for a purpose, to feed on the remains. Your immediate reaction is probably "gross" or 'disgusting.' After deeper thought, you may even be offended that six-legged infiltrators have violated the sanctity of the dead (figure 15.1). Yet the reality is that these creatures mean no disrespect. In fact, we need them to serve as nature's undertakers.

The insects in question are generally referred to as scavengers, but more specifically they are **necrophagous**,

Figure 15.1. Egyptian mummy from the Late Period of dynastic Egypt. Mummification was a process used specifically to prevent necrophagous insects from defiling the dead. Photo by Gerard Ducher at the National Museum of Alexandria, Egypt (http://bit.ly /1egPUpn).

Figure 15.2. Adults of *Chrysomya megacephala* (Diptera: Calliphoridae) are attracted to fresh carrion as well as human remains. Several species of blow flies feed on the dead, but they do not all prefer the same types of carcasses or stages of decomposition. Photo by Toby Hudson at http://bit.ly/1MIsqFr.

meaning they feed on the dead. The broader term **saprophagous** refers to insects that consume any type of dead organic material (i.e., if it is dead, it's delicious, whether animal, plant, or anything else), whereas necrophagous is usually confined to those that consume decomposing animals. Several species of insects are attracted to **carrion**, a term for the carcass of a dead animal at any stage of decay (figure 15.2). We are an exception, and refer to our dead as, well, dead, and also as the deceased, a corpse, a cadaver, or sometimes as a set of remains. The name does not matter so much as the fact that a body is available for a select group of animals that prefer the dead as a food resource.

Why a dead animal as opposed to one that is living? Well someone has to do it! Actually, this is literally true. Animals (and all organisms) are dying daily. In the natural environment, the remains begin to decompose exactly where the animal died. There are no road crews in nature to scoop up a carcass and place it out of sight and out of mind. The only way a body is removed is through decomposition and recycling. Insects play a major role in these processes, in which their feeding activity can remove all the soft tissue of an animal in just a matter of days to weeks, depending on climatic conditions and other factors like the size of the corpse and its location. And they are "happy" to do so. Well, I have no idea if this is true, but flies and other insects find carrion an excel-

lent source of protein. In the case of necrophagous flies, both adults and juveniles feed on a carcass to obtain protein as well as other nutrients. Those that they cannot use are liberated into the soil and aquatic habitats. Without this recycling activity, bodies would accumulate rather quickly, the smell would be unbearable, and the threat of disease would escalate. So with this in mind, be sure to say "thank you" to the next blow fly that perches on your shoulder when reclining outside!

Great, the underlying message is that without flies the world would stink. More or less, yes, you could say that. So what do recycling insects have to do with forensic entomology? It is the recycling activity of insects that make them so useful for solving crimes. For example, depending on the season, geographic location, and a series of other abiotic and biotic influences, the species of insects that arrive on a body and their rates of development are relatively predictable. These features of the life history strategies of several species of necrophagous insects are the bases for using insects in investigations of suspicious deaths or homicides and fall under the umbrella of medicolegal or **medicocriminal entomology**. The latter is becoming the accepted name for the branch of forensic entomology in which arthropod evidence is used in criminal cases, frequently those associated with violent acts.

By far the most important insects in medicocriminal entomology are necrophagous flies belonging to the families Calliphoridae (e.g., blow flies and bottle flies) and Sarcophagidae (flesh flies). We will discuss them in more detail later in the chapter, but the key to their importance lies in the fact that the adults are among the first colonizers of a body and that their offspring (larvae) are completely dependent on the corpse for nutrition and climatic conditions for development. Several other flies, beetles, aquatic insects, and even some terrestrial arthropods can be useful tools in criminal investigations, but again, not nearly as useful as necrophagous flies. That said, each of these critters offers unique insight to an investigator when a body is found outdoors or in an unusual location that only they can reach. Being small has its advantages! We will also see that necrophagous insects offer utility beyond the grave, meaning that death is not a requirement for them to be witnesses. Insects often are important physical evidence in cases of negligence and abuse, and not just toward humans, as we exercise wicked behavior toward pets and wildlife as well.

Medicocriminal entomology is the field most often referred to when mentioning forensic entomology, but the field is subdivided into three distinct areas: **urban entomology**, **stored product entomology**, and medicocriminal entomology. Urban entomology is predominantly focused on insects that interact with humans in residential or commercial settings, including the properties associated with these facilities. The field is not defined by geographic location—municipalities versus rural—as the term "urban" implies. Instead, it focuses on insects that invade human habitation, especially when those species become matters of legal concern, such as with civil or criminal cases. Stored product entomology deals with insect infestation of food and food products and the disputes that result from the presence of insects, their body parts, or obvious damage from their activity in foodstuffs. No matter how fond you are of insects, few people will tolerate food that has been infested with insects. The thought of insect **frass** (otherwise known as excrement) in an energy bar or breakfast cereal, or beetle parts floating to the top of a pot of boiling pasta tends to gross the average person out, and may lead to civil suits against the manufacturer, food distributer, or grocer. But as we will soon learn, just because you discover insect bits in your food does not mean any law has been broken.

There's a fly in my soup: Should I sue?

How is it possible that insects, live or dead, or pieces of their wings or legs, or even their poop, in my food is not a violation of some law, code, regulation, or Boy Scout creed? It seems impossible to imagine in this day and age that any food can be considered "perfectly"* suited for sale in the United States, knowing that it possibly contains live insects. We will address how this is possible in a moment. But be forewarned, if you find the notion of insects in food disturbing, then you really do not want to know what *else* is okay to eat. As for "bugs" in food, the mere presence of insects or fragments of them in certain foods is not necessarily an indication that

anything is wrong. Contrary to popular belief, you cannot sue (file a civil or criminal suit) simply because you have a mouthful of beetles after taking a bite of cereal. That might be overly dramatic, but the point is that certain criteria must be met with regard to insects (as well as other organisms) in food before it can become a legal matter. To make that determination, two important pieces of information must be known: (1) when is it all right for insects and their paraphernalia to be in food, and (2) how did they get in the food in the first place?

The answer to the "when" question is relatively straightforward. The US Food and Drug Administration establishes guidelines for all food types sold by retailers and grocers and at markets (including open air and farmer's) and restaurants. Those regulations address a wide range of topics, including **defect action levels (DALs)**. Food defect action levels are defined as the levels of natural or unavoidable defects in foods that present no health hazards for humans. In other words, during the harvesting, processing, and packaging of certain food types, the food may incur some damage or impurities that in no way affects the overall healthfulness or nutritional value of the food item. Insects are considered a food defect or impurity. Depending on the food type or how it will be used, the food defect action level may permit an acceptable level of live or dead insects to be present, or portions of their bodies, or some level of insect damage (figure 15.3). For some foods, no insects are permitted. For others, a broader DAL is used. Again, it comes down to food type and usage. A complete list of defect action levels can be found on the FDA's website at http://1.usa.gov/1gSCTzA. If you believe your food is "pure" when purchased, then I do not recommend visiting the website.

The role of forensic entomology in cases of insects in food is really to determine whether a DAL has been violated, and if so, how did the insect get into the food in the first place. Answering the "how" question is important for establishing who is responsible for the violation, meaning for the presence of insects in the food, which, in turn, determines who is named as the defendant in a civil or criminal case. None of this matters from a legal perspective unless, again, the DAL has been exceeded for a particular food item, and then a consumer decides to take legal action. To do so, the consumer (the plaintiff) must demonstrate how they were injured from consuming the

---

*Perfectly is not in reference to the food, as in fact, the presence of insects is referred to as a 'defect' by the Food and Drug Administration. But a defect does not mean automatically that the food is unsafe or unfit for human consumption.

In the summer of 2011, television networks across the United States broadcast an extraordinary event: the murder trial of Casey Anthony. Murder is not a unique event in the United States. So the fact that Casey was charged with first-degree murder does not account for the sensationalism of the case. Even the fact that she was accused of killing her own child—two-year-old Caylee Anthony—could not totally explain how Casey became an overnight, albeit notorious, celebrity. The case and trial were bizarre from the outset. Was it because reporters covered the trial like they were at the red carpet for the Oscars? Regularly scheduled television programs were interrupted so that daily, sometimes hourly, updates could be provided to the wanting public. Murder had not been this popular since the O. J. Simpson trial over a decade earlier. Did the attention derive from the stars of forensic science who were called upon as expert witnesses for both the prosecution and defense? Could the attraction have been related to the immediate family of Casey? After all, her own mother and father switched sides during the

trial, her father and brother were accused of incest, and both parents testified that a terrible accident took place in the home but neither would elaborate. No, despite how spellbinding each of these events was to the public, it was Casey herself who really created the uniqueness of this case.

The tragic story began in June of 2008, when Casey left her parents' house, where she lived, with her daughter Caylee. Casey returned, alone, thirty-one days later. Prior to her reappearance, Casey chatted frequently with her mother (Cindy) by phone. The grandmother, according to her own testimony, constantly asked to see Caylee, but each time was rebuffed with stories explaining why her granddaughter was unavailable to visit: the explanations ranged from the young toddler being with a nanny, which was later found to be a lie, to spending time at amusement parks in the Orlando (Florida) area to just hanging out at the beach. The two-year-old apparently had a very active lifestyle. Two weeks after Casey and Caylee departed from their home, the grandparents were notified that Casey's car had been found abandoned and was now impounded at a local tow yard. When Casey's father (George) went to pick up the car, he discovered a horrible odor associated with the trunk, which he and the tow yard assistant both described as smelling like death. One would assume that Casey's father, as

a retired police detective, knew that distinctive smell very well.

Casey later testified that she had not seen her daughter since essentially the day she left her parents' home, and she believed that the nanny (again, fictional) had kidnapped her daughter. Interestingly, pictures and videos surfaced of the grieving mother partying at clubs in the area during the period of time she stated that her daughter was missing. Casey also led detectives to Universal Studios, where she claimed to be employed, but in actuality was not. During the police investigation, four hundred pieces of evidence were collected from Casey's car, including several items in a trash bag, some insects, and hair later determined to be Caylee's. From George and Cindy's house a computer was confiscated after it was determined that several Internet searches had been performed involving neck-breaking and the making and using of chloroform. Finally in mid-December of 2008, the skeletal remains of Caylee were discovered in a wooded area not far from the grandparent's home. The body was reduced to bone, cartilage, and hair, with a piece of duct tape attached to the skull, and all enclosed within a trash bug.

The district attorney was convinced that Casey was responsible for Caylee's death, but needed to prove this beyond a reasonable doubt to the jury. A "who's who" of expert forensic witnesses was called upon to provide forensic analyses on behalf of the prosecution. However, with each seemingly damning testimony against Casey, the defense countered with experts that equally swayed the jury. The trial essentially reached a stalemate. What tipped the scales? Forensic entomology! Okay, others may disagree about how influential the entomological evidence was to the jury. But what can be stated with confidence is that the testimony of opposing forensic entomologists came toward the end of the trial and did seem to be the piece needed to generate a verdict.

The prosecution acted first, by hiring Dr. Neal Haskell to analyze insects found in the trunk of Casey's car and at the separate location where Caylee's skeletal remains were discovered. Haskell determined that two types of insects were present in the car: several adults of *Megaselia scalaris*, a type of fly interested in dry remains, and a few beetle larvae belonging to the genus *Dermestes*. As Haskell pointed out in his expert report, the beetles and flies both prefer human tissues that are nearly devoid of moisture, in other words, in advanced stages of decay. He was convinced, then, that the little girl's body had decayed in the trunk of the car. Haskell then turned his attention to the insects associated with the skeletal remains. The entomological evidence included empty puparia of some early colonizing blow flies, two species of flies that prefer body liquids, and several more *M. scalaris*. His conclusion was that the body had decayed in the trunk, and was then transferred to the woods near the Anthonys' home, which accounted for *M. scalaris* being discovered in both locations.

In response to the prosecution's witness, the defense hired Dr. Timothy Huntington to evaluate the insect evidence in the trunk of the car. Huntington testified that the flies and beetles were "unremarkable," since both are known to feed on trash, which was plentiful in the trunk of the car. He also indicated that his own research had shown that when human surrogates—pigs—are allowed to decompose in a car trunk, hundreds to thousands of blow flies enter the trunk and are present outside the car. By contrast, blow flies were not found in Casey Anthony's car. Huntington also explained that copious amounts of body fluids should also be present if the body decomposed to the degree implied by Haskell. Again, such evidence was not evident in Casey's car. Therefore, in Huntington's opinion, there was no entomological evidence to support the contention that a body had decomposed in the trunk of the car owned by Casey Anthony.

In the end, the jury accepted Huntington's analyses over Haskell's. Subsequently, Casey Anthony was found not guilty of first-degree murder, aggravated manslaughter, or aggravated child abuse. Several questions remain unresolved with this case. One question you may have is, why was Huntington asked to evaluate only insects in the car and not those in the woods? There is a simple explanation: the car was the only piece of evidence that was directly linked to Casey. Another question is, how could Haskell conclude the body decomposed in the trunk first if early colonizers were present only with the skeletal remains? A very good question, and one that only Dr. Haskell can answer. It is entirely plausible that Caylee's body was never in the trunk at all. Her body might have been placed in the woods from the outset. Casey was exonerated of all major charges brought against her and cannot be tried again for Caylee's death. If she is truly innocent, than the real culprit is presumably still at large. The chief medical examiner for the case, Dr. Jan Garavaglia—yes, Dr. G from the Discovery Health Channel—has claimed that several items, including insects from the trunk of the car, were not allowed into evidence for the trial. The details of that entomological evidence may tell a much different story from the one presented in court. Only time will tell whether the true story of Caylee Anthony's death will be uncovered.

insects or their parts *and* show that the defendant was responsible for the injury. Here is where a forensic entomologist comes into play. An individual trained specifically in stored product entomology would need to assess the contaminated food to determine which species of insect is or was present, the degree of contamination (entire insects, body parts, or evidence of activity such as feeding or webbing), and how and when the food became contaminated. The latter is important because it establishes who is responsible for the insect defects. For example, it may be determined that the identified insect is only present during a specific growth phase of a crop. Thus the insect was introduced during the harvest or processing phase of food production. By contrast, if a live insect specimen is collected, depending on its development stage, it may be determined that the insect entered the food prior to purchase. The responsibility then shifts toward the food distributer or the grocery chain itself. Establishing when and how the insect entered the food product is essential for the plaintiff to prove his or her case. Table 15.1 lists some of the most common stored product and urban pests found in the United States.

Whenever there is the possibility that monetary gain can be derived from any process, someone will invent a way to beat the system. Or at least make an attempt. Such is the case with insects in food, wherein an individual makes a fraudulent claim against a grocer, convenience store, or, more commonly, a restaurant. More than one restaurant chain has faced accusations of an insect being discovered in a pizza, sandwich, or other food item. This form of fraud is most common with restaurants that deliver,

Figure 15.3. Larva of the Indian meal moth, *Plodia interpunctella* (Lepidoptera: Pyralidae) infesting chocolate sprinkles. The presence of this insect does not automatically make it a legal matter; a defect action level usually must be exceeded. Photo by Pudding4brains at http://bit.ly/1lma5QN.

Table 15.1. Common insects that are the basis for legal problems dealt with in urban entomology. The ones marked in red can become health problems for humans.

Blattodea—cockroaches. Pests of food or habitation

Coleoptera—beetles and weevils. Food or wood

Diptera—flies, gnats, mosquitoes. Food or habitation

Hemiptera—true bugs. Habitation

Hymenoptera—ants, bees, wasps. Food, wood, or habitation

Isoptera—termites. Habitation

Siphonaptera—fleas. Habitation

because of course after the driver leaves, the insects can be added, and the accusations can begin. Usually the story goes something like, "I took a bite of my [insert food item here], only to see a half-eaten bug staring back at me!" Perhaps receiving insect-tainted food does occur frequently. Perhaps. However, it is interesting to note that whenever cockroaches are involved, no one ever finds a partially eaten one. No, they miraculously are always discovered whole! Obviously, the role of a forensic entomologist is to determine the identity of the insect and the circumstances that could account for the critter being in the food. As you might imagine, it is the restaurant chain that most often seeks out the forensic expert, not the plaintiff.

Home invasion: Matters for urban entomology

Insects, like relatives, can be unwelcome guests in our homes. They (insects and human relatives) share several characteristics, as they typically were not invited, consume our food, and do not leave as easily as they arrived (http://bit.ly/1TjR6YY). Unfortunately, it is legal to use biocides on only one form of uninvited guest (Did I say that out loud?). Urban entomology is the subdiscipline of entomology that is concerned with the unwelcome insects that invade human habitation or the human environment. This broadly encompasses anywhere that insect habitats and human habitats intersect in terms of homes, buildings, and the surrounding environment. The term "urban" does not indicate the location of the insect infestation, but does distinguish the topic from, say, insects typically associated with agriculture. That said, some species that normally inhabit feedlots, barns, or other types of livestock facilities, and then in turn invade human space, are considered subjects of urban entomology (figure 15.4).

Case in point are varieties of flies (e.g., house fly, *Musca domestica*; face fly, *M. autumnalis*; and stable fly, *Stomoxys calcitrans*) that breed in livestock manure and take up residence in nearby homes. Does the presence of insects in the home or other dwellings immediately mark a legal situation? In other words, should a forensic entomologist be called simply because a homeowner discovers a fly, cockroach, or other insect in the house? No. Here is where urban and stored product entomology share similarities. Both are subfields of general entomology as well as specialized areas of forensic entomology. The mere

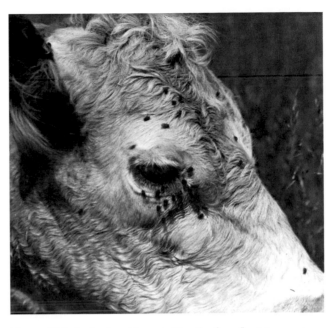

Figure 15.4. Agricultural insects like the face fly, *Musca autumnalis* (Diptera: Muscidae) become an issue for urban entomology when they invade human habitation. Photo provided by Clemson University, USDA Cooperative Extension Slide Series, Bugwood.org.

presence of insects in food or human habitation is an entomological issue, but not necessarily from the perspective of forensic entomology. For the insects to be of legal importance, either a critical threshold level must be surpassed—like with the defect action levels discussed in the last section—or a specific species must be present that represents potential harm. Let's briefly examine both of these criteria to understand when urban entomology falls under the umbrella of forensic entomology.

We know from earlier discussions that insects achieve pest status based on reaching certain arbitrary threshold levels. In some cases the thresholds are based on economic considerations, in others an insect becomes a pest through solely subjective determinations, known as aesthetic injury levels. Individual tolerance levels of insects dictate pest status. So a single cockroach in the home may be considered a pest by one person's standards, yet fifteen is not even an annoyance for someone else (and the house probably looks straight out of a TLC*

---

*TLC (The Learning Channel) airs several reality television shows including *Hoarding: Buried Alive*, http://bit.ly/1Q0WW3e.

Figure 15.5. Honey bee (*Apis mellifera*) infestation of a home and a homeowner's attempt at preventing them from entering the living room. Photo by Timothy Haley, USDA Forest Service, Bugwood.org.

hoarders show). Either way, both scenarios represent topics for urban entomology, but not for forensic entomology. Economic injury levels (EIL) and economic thresholds (ET) are more empirical in defining when an insect is a pest. Unfortunately these thresholds have very limited utility in urban entomology because insect populations are not predictable in human habitation, as they can be with agricultural pests, and are not easily sampled in most any type of building. For example, it is very difficult trying to estimate the population levels of cockroaches that hide during the day or termites that have tunneled into the wood frame of a home. On the other hand, some insects are potentially harmful, either directly to us (like medically important insects) or indirectly, by being highly destructive to a valuable commodity such as a home, building, or other structure (figure 15.5). In either scenario, no level of damage or injury is acceptable. Thus the EIL = ET = 0.

This brings us to the question of when exactly are urban pests forensically important? The most common example is when disputes arise between a homeowner or a business that have contracted a pest control company to deal with a pest situation such as outlined earlier. Legal action may be pursued because the plaintiff feels that inadequate or inappropriate control measures were used by the pest control company. This typically implies that the insects in question are still present or there is evidence of recent activity. An expert in urban entomology is tasked with determining if any insects are or were present, identifying any specimens collected, and, if insects are indeed present, determining whether they represent a new or a reoccurring infestation. Obviously a reoccurring infestation supports the contention of the plaintiff. It is also especially important when dealing with cases involving wood-destroying insects like termites or carpenter ants. The economic damage can easily be measured in thousands of dollars as a result of such pests, particularly if proper treatment has not been used when initially discovered in a home or other building. Again, keep in mind with wood-infesting insects that there is not a tolerable level in a home. So failure of a hired gun (aka a pest control company) to rid them from a premise magnifies the damage for the homeowner. Sometimes the issues overlap with stored product entomology, in that the food in the home, usually in cabinets or a pantry, are discovered to harbor insects. As discussed in the previous section, the forensic entomologist must determine when the insects gained entry. In this case it must be assessed whether the insects invaded the food once in the home, or whether the invasion occurred prior to purchase. If the latter is true, then the issue becomes a stored product entomology problem, and no longer an urban entomology one.

When health issues arise as a result of the action of a pest control company, the situation may escalate from a civil to a criminal case. Two common examples are the misuse of pesticides, which in turn adversely affects the health of individuals in a home or building, and failure to remove a nest of stinging or biting insects that produce venom. It is the venom that causes the health issues, usually in the form of **anaphylaxis**. Anaphylaxis is a condition in which an individual's immune system reacts to components in the venom through a powerful and systemic response that can cause massive releases of histamine, followed by inflammation. When this occurs in tissues associated with airways, breathing becomes labored and difficult since the air passages are constricted. This in turn stimulates accelerated heart rate to compensate for the diminishing levels of oxygen in blood. For some individuals, a severe anaphylactic response can lead to death. If this condition could have been avoided by proper treatment of the venom-packing insects, a case of criminal negligence against the pest control company can be made. Similarly, severe health problems or even death resulting from misuse of insecticides elevates a case to criminal negligence for the human damage inflicted and for violating federal laws in terms of pesticide use (see chapter 14 for a discussion of proper use of insecticides). Yet another criminal situation occurs when failure to properly treat a home for wood-boring or destroying insects results in the structure collapsing on the occupants. Though this is an extreme example, it has in fact happened.

Some insects pose unique challenges for urban entomology in the United States. One, the red imported fire ant, *Solenopsis invicta* (Hymenoptera: Formicidae) has been a formidable nemesis for over half a century, while a second, the bed bug, *Cimex lectularius* (Hemiptera: Cimicidae) is experiencing a resurgence in North America that really has not generated sentiments of nostalgia. What makes these two pests so special for urban and forensic entomology? Fire ants ordinarily live outdoors, and evoke terror in other living creatures that disturb their mounds or simply get in the way. The problem with these ants is that they often move indoors, invading homes, businesses, and even health care facilities. Once inside, the ants attack any creature that attempts to stop them. Remarkably—or maybe more correctly, expectedly—they even compete with patients for intravenous fluids by literally biting into IV lines to steal the nutritious liquids. There are several documented examples throughout the southern United States of definitive cases in which fire ants have directly caused the death of a patient in a health care facility. Because of their incredible ability to attack and colonize in force, as well as to protect the colony, it is essentially impossible to eradicate red imported fire ants from an urban or rural environment. In essence they have transcended control efforts, and most pest control companies in heavily infested areas provide no guarantees in terms of ant control or removal. Rather, treatment is applied in an effort to manage populations. You can see how this immediately influences forensic entomology in two ways: (1) a traditional civil suit against a pest control company changes based on the wording of the agreed-upon contract, and (2) the ants may be responsible for human deaths! And a homicide caused by any six-plus-legged

Figure 15.6. An adult bed bug, *Cimex lectularius* (Hemiptera: Cimicidae) doing what we dread: blood-feeding on a human host. Photo from the Centers for Disease Control and Prevention (http://bit.ly/1Ji0vx7).

creature becomes a matter for medicocriminal entomology.

The old bedtime adage, "Don't let the bed bugs bite" is back. Who says that to a child before sending them off to bed? It is just as comforting as "Happy nightmares, hope some creature from under your bed doesn't blood-feed on you during the night!" Regardless of the parenting skills of some people, the reality is that bed bugs are back. What prompted their return, and should you be concerned? Yes, you should be concerned, from the standpoint that it is never good to be a blood meal for any creature. Similar to mosquitoes, bed bugs are attracted to carbon dioxide. During the day, adults and juveniles (termed **nymphs**) hide in a dark secluded location, like along a bed frame, lampshade, or behind a picture frame just above the headboard. At night, however, while you participate in deep sleep, snoring loudly and expelling carbon dioxide, the bugs come out of hiding. They follow the trail of gas emission until they find your warm body. Bed bugs typically then crawl under the sheets to locate the ideal spot to insert their piercing-sucking mouthparts (figure 15.6). This is commonly along exposed areas of the belly or legs. Their bite is gentle, thereby ensuring that you do not wake to discover the unwelcome diners. Once full, they retreat to digest their meal and possibly have sex.

Fortunately, *C. lectularius* is not a vector for disease. That said, I have yet to find anyone who says, "Well, let 'em eat, then." Over the last ten to fifteen years, this pest of the past has returned with a vengeance, not just in the United States, but also in several countries in Europe and also Australia. What changed? No one knows for sure, but some blame the development of insecticide resistance as the leading cause, while others cite decreased use of broad-spectrum insecticides as a major contributing factor, as well as increased travel abroad, particularly by college and university students. One thing that is well known is that most people despise bed bugs, and if they are discovered in a hotel, luggage, or any other location, the idea of civil litigation is probably entertained to some degree. The reality is that a case can really be made only if the accused was aware of a bed bug problem and chose not to do anything about it. The burden of proof again falls on the plaintiff.

Quick check

Who should be consulted if a face fly enters your house? If the fly lays eggs on lunchmeat exposed on the counter? If insect eggs were already present when opening the package of food?

Maggots, murder, and men

How does that philosophical question go: If a human falls dead in the woods and no one is around, does it make a sound? Maybe that's not exactly right, because sound doesn't matter when it comes to the dead. They smell, and as a corpse decomposes, it becomes increasingly ripe (odiferous). The odors of human death pique the interest of many insects. For that matter, the remains of any warm-blooded vertebrate animal, large or small, inside or out, draws the undivided attention of several species of insects. This is especially true of several families of flies. But what is the link between animal death and such insects? One word: protein. We have already established that certain fly species feed exclusively, or nearly so, on carrion. What they seek more than anything is protein. For a female fly, animal tissues are rich sources of protein that are essential for producing or provisioning eggs. In fact, protein is so important that some males use their keen sense of smell to locate a freshly dead carcass and then advertise this skill through pheromone release to attract

"the ladies." If they are impressed, he may mate with one or more females.

Competition is intense for use of all types of corpses, because they are nutritionally rich but finite sources of nutrients and their occurrence (**ephemeral**) and location (**patchy**) are unpredictable. So when a death event occurs, a whole lot of creatures take notice and quickly mobilize for action. If you are necrophagous and are slow to detect or locate carrion, you lose (http://bit.ly/1FApslX). Competition is so intense on animal remains that it has shaped several aspects of the life history characteristics of insects that depend on the dead for survival. Nowhere is this more on display than with the reproductive strategies demonstrated by necrophagous flies. One group, the blow flies (family Calliphoridae), is recognized as first responders to death. In other words, many species are early colonizers. This generally means that adults are especially good at detecting and finding human or other animal remains. But even within this early pattern of colonization, there are several adaptations that permit the various species to decrease competition with each other (figure 15.7).

For example, some blow fly species prefer large versus smaller remains for laying eggs; others show preferences in terms of whether the body is in full sun, shade, or partial shade; and some are influenced by location on the body. The latter is manifested by those that prefer to deposit eggs (**oviposition**) in concealed locales like natural body openings instead of exposed on skin, hair, fur, or plumage (feathers). Many will lay eggs in exposed wounds, while others only deposit eggs on very moist surfaces such as in saturated soil (or carpet) under a body, or on clothing, hair, fur, or plumage soaked in exuded body fluids. Even the eggs and larvae are adapted for life in a competitive world, as eggs are "designed" to hatch quickly so that the resulting larvae can begin a period of rapid and continuous feeding. They also feed cooperatively in large aggregations or **maggot masses** that maximize the pace at which larval development can proceed. Maggot masses are unique microhabitats that permit larvae to thrive in carrion communities. Importantly, the life history characteristics of these flies and the subtle nuances that allow each to use a corpse slightly differently contribute to **resource partitioning**. Resource partitioning is the use of a resource like carrion by more than one species but

Figure 15.7. Several species of blow flies (Diptera: Calliphoridae) are early colonizers on a corpse, depositing eggs within just minutes of death. Photo by Susan Ellis, Bugwood.org.

Figure 15.8. Flesh fly species from the family Sarcophagidae display resource partitioning by larvipositing instead of laying eggs, arriving slightly later than blow flies and preferring different sized carcasses. Photo by Susan Ellis, Bugwood.org.

in different ways. It is a key feature of all necrophagous insects utilizing a dead animal as a primary food source.

Several other insects utilize animal remains either because they, too, are necrophagous or because other op-

portunities exist to capitalize on another animal's misfortune (it died) (figure 15.8). Flesh fly species from the family Sarcophagidae can arrive early or later during physical decay of an animal and reduce the competition with blow flies by preferring small carrion instead of larger; depositing larvae (larviposition) instead of eggs so the offspring hit the ground running, so to speak; and the young of some species are predatory on other species, which is a great way to get an edge on the competition. Some fly species (families Phoridae and Piophilidae) (figure 15.9) do not lay eggs until the body is very dry, long after blow flies and flesh flies no longer have an interest in the body.

Many species of beetles (Coleoptera) arrive early but after the initial wave of fly colonizers so that they can feed on both the corpse and the fly eggs. And as with flies, several species of beetles (families Dermestidae and Cleridae) prefer very dry remains, when the tissues are leathery and moisture is absent (figure 15.10). Once the body is nothing more than a pile of bone, cartilage, and hair, insect interest fades but is not totally absent. A few moths (Lepidoptera) feed on hair and remnants of clothing, some parasitic wasps (Hymenoptera) specialize on fly puparia scattered about the remains, and a few beetle species can be found searching for scraps on the bone. It is evident that multiple groups of insects utilize a decomposing body, and they arrive in waves. By special-

izing on different stages of decay, many organisms can successfully utilize the same resource. It is the ultimate in sharing, albeit competitively, in nature.

Since our interest is in murder, how does the activity of necrophagous insects tie in to criminal investigations? There are two key take-home points from our examination of necrophagous insects. One is that, as a body decomposes, each stage of decay is attractive to specific groups of insects. In essence, "who" will arrive and when they will arrive to a human corpse is relatively predictable. The second point is that though several insects will show up, in reality the number of necrophagous species that will colonize a body in a particular region of the United States is relatively small. In some cases ten to twenty species of flies may be present, represented by thousands of individuals, usually in larval form. Fortunately, the defining anatomical characteristics of each species make their identification straightforward. Now, to put this information in context, if a human corpse is discovered outdoors, covered by moving masses of insect larvae, it is possible to gain useful information based solely on entomological evidence. For example, by knowing when an insect colonizes a body, you can use the oldest-age insect collected on a corpse to estimate the length of time the body was available for colonization. Of course the other piece of information needed is how long that species takes to

Figure 15.9. Flies in the family Piophilidae prefer to oviposit on animal remains that are very dry, long after blow flies and flesh flies are no longer interested in the corpse. Photo by Susan Ellis, Bugwood.org.

Figure 15.10. Several beetle species compete with flies for use of carrion and also are predators of fly eggs and young larvae. Photo by Susan Ellis, Bugwood.org.

develop under the conditions in which the body was discovered. Most of that information can be found in the research literature, and if not, developmental experiments can be conducted relatively quickly in the laboratory.

Species identity is also useful for determining whether a body was moved to a location. This is because some fly species, for example, are restricted by geographic location, like west of the Mississippi River, or in the southeastern United States, because they cannot survive winter. More subtle distributions occur between rural and urban environments. The latter is associated with large metropolitan areas in which asphalt beltways surround and separate the city from the suburbs. In fact, if you examine all the life history characteristics discussed earlier that give the flies a competitive advantage on carrion, each of those provides useful clues for reconstructing the events surrounding a discovered body. It then becomes a matter of deductive sleuthing to match the insect clues with other physical evidence.

The advent of crime shows on television revealing how forensic science can be applied to police investigations has caused a massive upsurge in public interest. But it has also served to inform criminals of how to better cover their tracks. In the case of forensic entomology, bodies resulting from accidental or intentional homicides are disposed of in an effort to prevent insects from reaching the body, and thus serving as witnesses to the crime. It usually does not work. Some flies are adapted to burrow for buried bodies. Many flies and some beetles will eventually colonize a corpse hidden behind a wall, wrapped in a blanket, placed in an appliance, trash can, or trunk of a car (see Fly Spot 15.1 for more details). Dismembering a body accelerates fly colonization; coating with household chemicals often does nothing more than delay the inevitable feeding by insects; and dumping into aquatic environment simply shifts which insects are attracted. Bodies have been placed in almost any natural or manmade location imaginable to escape detection. Rarely does it stop necrophagous insects from making the discovery. Even burning or burying a body does not necessarily prevent insect colonization. Though humans are clever and devious, especially in the ways we manipulate each other, insects are very serious about their protein, meaning that which is in you. What most people do not comprehend is that insects will remain in hot pursuit of a protein meal, even if it kills them.

Bug bytes
Therapeutic myiasis
http://yhoo.it/1Jsv3bd

The fly who loved me: Myiasis and cases of neglect

Some necrophagous insects are infatuated with you. So much so that they literally will not wait until death to dine with you . . . err, on you. These species are adapted for locating open wounds or other decaying tissues on a living animal, including us. Wounds are made up of varying sized layers of dead tissues that give off the same odors as an entire corpse. Obviously, the larger the wound, the stronger the smell, which results in a greater attraction to necrophagous flies. Once discovered, flies lay eggs, the eggs eclose, and young fly larvae begin to feed on the dead tissue. Many questions are likely racing through your head at the moment, but one that probably stands out is, how does anyone just let a fly lay eggs on them? And then not remove the eggs or maggots? That's where forensic entomology comes in. Individuals most likely to become infested with fly larvae are ones that have been neglected, abused, or are incapacitated. In short, they generally are relying on someone else to take care of them. The presence of flies in any form may not be noticed, or if they are, often the individual is not physically able to prevent the flies from ovipositing or to remove the eggs or larvae once present. If the eggs or larvae persist for an extended time, then it is clear that the primary caregiver is neglecting the child or adult.

Fly infestations of living or dead tissues represent a parasitic condition called **myiasis**. Multiple forms of this condition inflict humans and other animals, but by far the one of most concern to forensic entomology is **facultative myiasis** (table 15.2). Facultative myiasis results when species of necrophagous flies detect the scent of dead tissues associated with a wound or other health issues (like diabetic ulcers), lay eggs, and the feeding larvae infest the tissue. If left unchecked, the larvae may consume all of the necrotic tissues and then switch to feeding on adjacent live tissue. The disease can progress

Table 15.2. Types of myiasis. Facultative myiasis is by far the most severe form because body is not adapted to fend off effectively these opportunistic parasites.

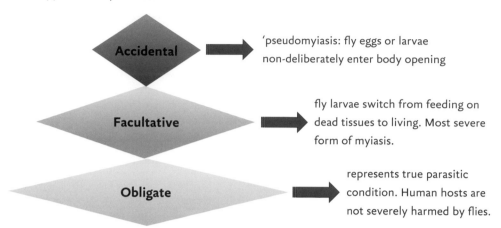

Accidental → 'pseudomyiasis: fly eggs or larvae non-deliberately enter body opening

Facultative → fly larvae switch from feeding on dead tissues to living. Most severe form of myiasis.

Obligate → represents true parasitic condition. Human hosts are not severely harmed by flies.

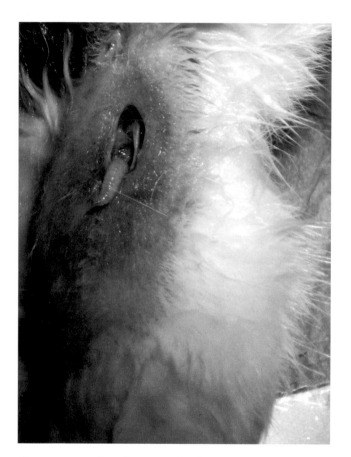

Figure 15.11. Fly infestation (facultative myiasis) of a flesh wound on a cat leg. Photo by Uwe Gille at http://bit.ly/1JiP605.

relatively quickly at that point, causing severe damage, even death, depending on the location of the initial infection (figure 15.11). Facultative myiasis is a matter for medicocriminal entomology because of the factors (abuse) that lead to fly oviposition and the failure to detect or stop the parasitic condition (neglect). This form of myiasis is actually on the rise in the United States because of at least two factors: (1) an increase in the number of individuals diagnosed with diabetes and displaying diabetic ulcers on feet and toes, and (2) an increasingly aging population. The factors are linked, as type 2 diabetes often is diagnosed in middle-aged or older individuals, but the disease has manifested itself much earlier in life. As a consequence, more and more individuals rely on caregivers to aid them in daily routines, including with hygiene and medical care. Relying on others, particularly nonrelatives, is a classic recipe for elevated instances of elder neglect and abuse.

Fly infestations are also associated with poor hygiene. For instance, failure to regularly change soiled diapers or clothing will lead to the attraction of some fly species that specialize on feces and urine. In most instances, the soiled diaper must be present long enough to be detected by flies living outdoors, for them to find a way inside, locate the diapers, lay eggs, and for the eggs to eclose. In other words, fly infestation did not occur overnight when a child may have had an "accident" in his or her sleep. Instead, insect colonization of the individual was clearly an outcome of neglect, since a parent, guardian, or caregiver did not provide appropriate supervision to prevent such poor hygiene practices. The situation can become more severe when a small child left alone, or an older individual unable to care for him- or herself, suffers from **retroinvasion**, in which the initial colonization of a soiled diaper over time leads to the fly maggots penetrating the anal or genital openings. Such parasitic infestations are very painful and also difficult to treat. A similar situation may occur in a nursing home or other health care facility, where failure to properly clean wounds or even catheters can result in facultative myiasis. As mentioned earlier, incidents of such issues are expected to rise with the increasingly aging population of the United States.

Myiasis is not restricted to humans. Facultative myiasis is often an indication of neglect and abuse of pets, domesticated animals, and even wildlife. As might be imagined, when animals are living outdoors, it is much more difficult to establish whether fly infestations are caused by accidental wounds, wounds inflicted by other animals, or abuse or neglect. Typically, however, instances of neglect and abuse go hand in hand with malnutrition and poor sanitary conditions. Forming a complete picture of the conditions that led to myiasis requires examining the living environment of the animal(s) in question, not just relying on entomological evidence. The severity of the crime obviously increases if abuse- or neglect-induced myiasis ultimately leads to death of the pet or other animals. Unfortunately, there are wide discrepancies among states in terms of penalties for such egregious acts. In some cases, severe forms of abuse and neglect may result in nothing more than a monetary fine.

Beyond the text

Why would facultative myiasis be more damaging to a human host than obligate myiasis?

Maggots on crack: Agents of toxicology

Do insects get high? An odd question, without a doubt, but yet it has merit in proper context. Consider that necrophagous insects, especially maggots, collectively consume most of a corpse. Their dining activity preferentially targets soft tissues and body fluids. Such tissues just happen to be the primary locations where most drugs, prescription or otherwise, accumulate after being ingested, injected, snorted, or smoked. On circumstantial evidence alone, we can say with confidence that maggots ingest drugs if the corpse has (obviously while the person was still living). If you are what you eat,

Figure 15.12. Empty fly puparia at a crime scene can reveal many details about the case, such as the presence of parasitic wasps, as evidenced by the emergence holes, or detection of prescription or illicit drugs that have become locked into the exoskeleton of the fly. Photo by USDA ARS Photo Unit, USDA Agricultural Research Service, Bugwood.org.

then some maggots are toxicological receptacles. What types of drugs have been detected in fly larvae? Every conceivable compound, from cocaine, heroin, and amphetamines to prescription medications such as the psychoactive compounds used for learning and behavioral disorders and more severe conditions like bipolar disorder and schizophrenia has been extracted from fly larvae feeding on human remains.

The importance of discovering such drugs in insect tissues is that some are retained in the insect body and may also influence aspects of insect development. The former implies that fly maggots can be used as surrogates to detect chemical compounds that were in the deceased in cases where insufficient tissues remain for direct toxicological analysis. There is ample experimental evidence to show that both illicit drugs and prescription medications can be found in fly tissues even after a human corpse has been reduced to hair, bone, and cartilage. In fact, some drug metabolites are retained in the last larval skin used to form the **puparium**, where the compounds can remain locked in place and unaltered for months to years (figure 15.12). The downside is that flies appear to be useful only for quantitative analyses; that is, the drugs or their metabolites can be detected but not quantified. Thus, it would not be possible to determine whether drugs found in the flies, and hence in the corpse, were at high enough concentrations to cause death. That said, the mere presence of a particular drug might provide potential leads for the investigators to find other forms of evidence.

Consumption of drugs or medications can alter fly development. For instance, cocaine has been shown to accelerate larval development. Heroin causes similar increases in developmental rates, yet extends the duration of stages within the puparia. The drug can even cause some species to be much larger than when heroin was not in their diet. This implies that heroin triggers fly larvae to overeat, essentially allowing them to ignore the **satiety** or fullness set point to continue eating. From a purely biological standpoint, overweight or beyond-optimal-size larvae is something that never occurs naturally with flies or any other wild (i.e., nondomesticated) insect.

The effect of some drugs is just the opposite of cocaine and heroin, functioning to depress one or more stages of development. Eating is not inhibited; it simply takes longer to consume enough food to molt to the next stage of development, and high doses can be lethal to larvae or pupae. Regardless of whether an ingested compound is an accelerant or depressant, if it alters fly development in any way, it conceivably compromises the utility of that species for use in calculating the postmortem interval (PMI). Remember from our earlier discussions that insects can aid in determining how long a body must have been present for oviposition to occur (period of colonization). Such calculations are possible only if you can accurately estimate the length of fly development. A drug that speeds up or slows down the rate of fly development, especially if investigators don't know that drugs are present, will cause under- or overestimations of the PMI. Depending on the chemical compound, the introduced error may cause the estimated PMI to be off by over a day, which is more than enough to corroborate or refute a suspect's alibi. Several compounds, such as those used to treat behavioral or learning issues (especially those associated with the autism spectrum disorders), have not been tested for their effects on fly development. Complicating the situation even more is that toxicological analyses can take weeks to complete and many of the new prescription drugs are not part of standard screening procedures unless suspected to be present. Consequently, estimations of the postmortem interval based on fly development may suffer from inherent error caused by

drug influences that were undetected for any number of reasons.

The toxicological utility of fly larvae extends to human tissues exposed to explosives or bullet fragments. How so? When an explosive device is detonated, it leaves behind chemical residues on anything close by, which unfortunately includes humans and other animals. In instances of terrorism, bomb-makers tend to have a specific modus operandi for the chemical composition of their devices; if the residues can be located and analyzed, the terrorist organization responsible for the egregious act may be identified. A similar situation occurs with gunshots, in which the resulting residues have a defined chemical signature. Gunshot residues (GSRs) contain spheroid particles that are easily detectable to the trained eye. They also are composed of unique levels of three elements: barium (Ba), lead (Pb), and antimony (Sb). Explosive and gunshot residues accumulate on skin surfaces and also on mucus membranes, such as found in the eyes, nose, and mouth. On a freshly dead corpse, the residues are readily identifiable. However, a badly decomposed body or one that has been consumed by insects and other creatures may leave behind no evidence of the residues. Here is where insects become useful to law enforcement once again. Insects, namely early colonizing flies, feed on surfaces where the residues accumulate. In situations in which foul play is suspected, the maggots can be ground up and analyzed for the chemical compounds that signify an explosive device was present, if not responsible for the death of the individual. Using flies in such toxicological applications works only for those species that arrive early or in cases where the residues are present when the insects gain access for colonization. Late colonizers or those that feed below the skin surface are not useful in such applications. Similarly, if the body is buried or disposed of in water, the residues are likely removed before insect feeding occurs. Nonetheless, the ability to detect bomb residues in fly larvae essentially elevates maggots to the status of agents of national defense.

CHAPTER REVIEW

✴ **Murder, termites, and weevils: The many faces of forensic entomology**

  ■ Insects can provide us with amazing insight when it comes to legal matters. Take a suspicious death for example. In a terrestrial environment in which a human corpse is exposed to the elements, certain species of insects can arrive just minutes after death. They are not gawkers the way humans are when driving past a severe car accident on the interstate. No, these insects are there for a purpose, to feed on the remains. In fact, we need them to serve as nature's undertakers. The insects in question are generally referred to as scavengers but more specifically they are necrophagous, meaning they feed on the dead. The broader term saprophagous refers to insects that consume of any type of dead organic material, whereas necrophagous is usually confined to those that consume decomposing animals. Several species of insects are attracted to carrion, a term for the carcass of a dead animal at any stage of decay.

- It is the recycling activity of insects that make them so useful for solving crimes. For example, depending on the season, geographic location, and a series of other abiotic and biotic influences, the species of insects that arrive on a body and their rates of development are relatively predictable. These features of the life history strategies of several species of necrophagous insects are the bases for using insects in investigations of suspicious deaths or homicides and fall under the umbrella of medicolegal or medicocriminal entomology. By far the most important insects in medicocriminal entomology are necrophagous flies belonging to the families Calliphoridae (e.g., blow flies and bottle flies) and Sarcophagidae (flesh flies).

- Medicocriminal entomology is the field most often referred to when mentioning forensic entomology, but the field is subdivided into three distinct areas: urban entomology, stored product entomology, and medicocriminal entomology. Urban entomology is predominantly focused on insects that interact with humans in residential or commercial settings, including the properties associated with these facilities. The field is not defined by geographic location—municipalities versus rural—as the term "urban" implies.

Instead, it focuses on insects that invade human habitation, especially when those species become matters of legal concern, such as with civil or criminal cases. Stored product entomology deals with insect infestation of food and food products and the disputes that result from the presence of insects, their body parts, or obvious damage from their activity in foodstuffs. No matter how fond you are of insects, few people will tolerate food that has been infested with insects.

There's a fly in my soup: Should I sue?

- The presence of insects or fragments of them in certain foods is not necessarily an indication that anything is wrong. Contrary to popular belief, you cannot sue simply because you have a mouthful of beetles after taking a bite of cereal. That might be overly dramatic, but the point is that certain criteria must be met with regard to insects (as well as other organisms) in food before it can become a legal matter. The US Food and Drug Administration establishes guidelines for all food types sold by retailers and grocers and at markets and restaurants. Those regulations address a wide range of topics, including defect action levels. Food defect action levels are defined as the levels of natural or unavoidable defects in foods that present no health hazards for humans. Insects are considered a food defect or impurity. For some foods, no insects are permitted. For others, a broader DAL is used. Again, it comes down to food type and usage.

- The role of forensic entomology in cases of insects in food is really to determine whether a DAL has been violated, and if so, how did the insect get into the food in the first place. Answering the "how" question is important for establishing who is responsible for the violation and determining the defendant in a civil or criminal case. None of this matters from a legal perspective unless, again, the DAL has been exceeded for a particular food item, and then a consumer decides to take legal action. To do so, the consumer (the plaintiff) must demonstrate how they were injured from consuming the insects or their parts *and* show that the defendant was responsible for the injury. Here is where a forensic entomologist comes into play. An individual trained specifically in stored product entomology would need to access the contaminated food to determine which species of insect is or was present, the degree of contamination, and how and when the food became contaminated. The latter is important because it establishes who is responsible for the insect defects.

- Whenever there is the possibility that monetary gain can be derived from any process, someone will invent a way to beat the system. Or at least make an attempt. Such is the case with insects in food, wherein an individual makes a fraudulent claim against a grocer, convenience store, or, more commonly, a restaurant. As you might imagine, it is the restaurant chain that most often seeks out the forensic expert, not the plaintiff.

Home invasion: Matters for urban entomology

- Urban entomology is the subdiscipline of entomology that is concerned with the unwelcome insects that invade human habitation or the human environment. This broadly encompasses anywhere that insect habitats and human habitats intersect in terms of homes, buildings, and the surrounding environment. The term "urban" does not indicate the location of the insect infestation, but does distinguish the topic from, say, insects typically associated with agriculture. That said, some species that normally inhabit feedlots, barns, or other types of livestock facilities, and then in turn invade human space, are considered subjects of urban entomology. Does the presence of insects in the home or other dwellings immediately mark a legal situation? No. Here is where urban and stored product entomology share similarities. Both are subfields of general entomology as well as specialized areas of forensic entomology. For the insects to be of legal importance, either a critical threshold level must be surpassed—like with the defect action levels discussed in the last section—or a specific species must be present that represents potential harm.

- Insects achieve pest status based on reaching certain arbitrary threshold levels. In some cases the thresholds are based on economic considerations, in others an insect becomes a pest through solely to subjective determinations. Individual tolerance levels of insects dictate pest status. So a single cockroach in the home may be considered a pest by one person's standards, yet fifteen is not even an annoyance for someone else. Either way, both scenarios represent topics for urban entomology, but not for forensic entomology. Economic injury levels (EIL) and economic thresholds (ET) are more empirical in defining when an insect is a pest. Unfortunately these thresholds have very limited utility in urban entomology because insect populations are not predictable in human habitation, as they can be with agricultural pests, and are not easily sampled in most any type of building.

- When exactly are urban pests forensically important? The most common example is when disputes arise between a homeowner or a business that have contracted a pest control company to deal with a pest situation such as outlined earlier. Legal action may be pursued because the plaintiff feels that inadequate or inappropriate control measures were used by the pest control company. This typically implies that the insects in question are still present or there is evidence of recent activity. An expert in urban entomology is tasked with determining if any insects are or were present, identifying any specimens collected, and if insects are indeed present, determining whether they represent a new or a reoccurring infestation. Obviously a reoccurring infestation supports the contention of the plaintiff. It is also especially important when dealing with cases involving wood-destroying insects like termites or carpenter ants. The economic damage can easily be measured in thousands of dollars as a result of such pests, particularly if proper treatment has not been used when initially discovered in a home or other building.

- When health issues arise as a result of the action of a pest control company, the situation may escalate from a civil to a criminal case. Two common examples are the misuse of pesticides, which in turn adversely affects the health of individuals in a home or building, and failure to remove a nest of stinging or biting insects that produce venom.

- Bed bugs are back. What prompted their return, and should you be concerned? Yes, you should be concerned, from the standpoint that it is never good to be a blood meal for any creature. Over the last ten to fifteen years, this pest of the past has returned with a vengeance, not just in the United States, but also in several countries in Europe and also Australia. What changed? No one knows for sure, but some blame the development of insecticide resistance as the leading cause, while others cite decreased use of broad-spectrum insecticides as a major contributing factor, as well as increased travel abroad, particularly by college and university students. One thing that is well known is that most people despise bed bugs, and if they are discovered in a hotel, luggage, or any other location, the idea of civil litigation is probably entertained to some degree. The reality is that a case can really be made only if the accused was aware of a bed bug problem and chose not to do anything about it. The burden of proof again falls on the plaintiff.

Maggots, murder, and men

- The odors of human death pique the interest of many insects. For that matter, the remains of any warm-blooded vertebrate animal, large or small, inside or out, draws the undivided attention of several species of insects. This is especially true of several families of flies. But what is the link between animal death and such insects? One word: protein. We have already established that certain fly species feed exclusively, or nearly so, on carrion. What they seek more than anything is protein. For a female fly, animal tissues are rich sources of protein that are essential for producing or provisioning eggs. In fact, protein is so important that some males use their keen sense of smell to locate a freshly dead carcass and then advertise this skill through pheromone release to attract "the ladies." If they are impressed, he may

mate with one or more females. Competition is intense for use of all types of corpses, because they are nutritionally rich but finite sources of nutrients, and their occurrence and location are unpredictable.

- One group, the blow flies, is recognized as first responders to death. In other words, many species are early colonizers. This generally means that adults are especially good at detecting and finding human or other animal remains. But even within this early pattern of colonization, there are several adaptations that permit the various species to decrease competition with each other. Some blow fly species prefer large versus smaller remains for laying eggs; others show preferences in terms of whether the body is in full sun, shade, or partial shade; and some are influenced by location on the body. The latter is manifested by those that prefer to deposit eggs in concealed locales like natural body openings instead of exposed on skin, hair, fur, or plumage (feathers). Many will lay eggs in exposed wounds, while others only deposit eggs on very moist surfaces such in saturated soil (or carpet) under a body, or on clothing, hair, fur, or plumage soaked in exuded body fluids. Even the eggs and larvae are adapted for life in a competitive world, as eggs are "designed" to hatch quickly so that the resulting larvae can begin a period of rapid and continuous feeding. They also feed cooperatively in large aggregations or maggot masses that maximize the pace at which larval development can proceed.

- Since our interest is in murder, how does the activity of necrophagous insects tie into criminal investigations? There are two key take-home points from our examination of necrophagous insects. One is that, as a body decomposes, each stage of decay is attractive to specific groups of insects. In essence, "who" will arrive and when they will arrive to a human corpse is relatively predictable. The second point is that though several insects will show up, in reality the number of necrophagous species that will colonize a body in a particular region of the United States is relatively small.

- If a human corpse is discovered outdoors, covered by moving masses of insect larvae, it is possible to gain useful information based solely on entomological evidence. For example, by knowing when an insect colonizes a body, you can use the oldest-age insect collected on a corpse to estimate the length of time the body was available for colonization. Of course the other piece of information needed is how long that species takes to develop under the conditions in which the body was discovered. Most of that information can be found in the research literature, and if not, developmental experiments can be conducted relatively quickly in the laboratory. Species identity is also useful for determining whether a body was moved to a location. This is because some fly species, for example, are restricted by geographic location, like west of the Mississippi River, or in the southeastern United States, because they cannot survive winter. The life history characteristics that give flies a competitive advantage on carrion provide useful clues for reconstructing the events surrounding a discovered body.

The fly who loved me: Myiasis and cases of neglect

- Fly infestations of living or dead tissues represent a parasitic condition called myiasis. Multiple forms of this condition inflict humans and other animals, but by far the one of most concern to forensic entomology is facultative myiasis. Facultative myiasis results when species of necrophagous flies detect the scent of dead tissues associated with a wound or other health issues (like diabetic ulcers), lay eggs, and the feeding larvae infest the tissue. If left unchecked, the larvae may consume all of the necrotic tissues and then switch to feeding on adjacent live tissue. The disease can progress relatively quickly at that point, causing severe damage, even death, depending on the location of the initial infection. Facultative myiasis is a matter for medicocriminal entomology because of the factors (abuse) that lead to fly oviposition and the failure to detect or stop the parasitic condition (neglect). This form of myiasis is actually on the rise in the United States because of at least two factors: (1) an increase in the number of individuals diagnosed with diabetes and displaying diabetic ulcers on feet and toes, and (2) an increasingly aging population.

- Fly infestations are also associated with poor hygiene. For instance, failure to regularly change soiled diapers or clothing will lead to the attraction of some fly species that specialize on feces and urine. In most instances, the soiled diaper must be present long enough to be detected by flies living outdoors, for them to find a way inside, locate the diapers, lay eggs, and the eggs to eclose. In other words, fly infestation did not occur overnight when a child may have had an "accident" in his or her sleep. Instead, insect colonization of the individual was clearly an outcome of neglect since a parent, guardian, or caregiver did not provide appropriate supervision to prevent such poor hygiene practices. The situation can become more severe when a small child left alone, or an older individual unable to care for him- or herself, suffers from retroinvasion, in which the initial colonization of a soiled diaper over time leads to the fly maggots penetrating the anal or genital openings.

- Myiasis is not restricted to humans. Facultative myiasis is often an indication of neglect and abuse of pets, domesticated animals, and even wildlife. As might be imagined, when animals are living outdoors, it is much more difficult to establish whether fly infestations are caused by accidental wounds, wounds inflicted by other animals, or abuse or neglect. Typically, however, instances of neglect and abuse often go hand in hand with malnutrition and poor sanitary conditions. Forming a complete picture of the conditions that led to myiasis requires examining the living environment of the animal(s) in question, not just relying on entomological evidence.

Maggots on crack: Agents of toxicology

- If you are what you eat, then some maggots are toxicological receptacles. What types of drugs have been detected in fly larvae? Every conceivable compound, from cocaine, heroin, and amphetamines to prescription medications such as the psychoactive compounds used for learning and behavioral disorders and more severe conditions like bipolar disorder and schizophrenia has been extracted from fly larvae feeding on human remains. The importance of discovering such drugs in insect tissues is that some are retained in the insect body and may also influence aspects of insect development. The former implies that fly maggots can be used as surrogates to detect chemical compounds that were in the deceased in cases where insufficient tissues remain for direct toxicological analysis.

- Consumption of drugs or medications can alter fly development. For instance, cocaine has been shown to accelerate larval development. Heroin causes similar increases in developmental rates, yet extends the duration of stages within the puparia. The drug can even cause some species to be much larger than when heroin was not in their diet. This implies that heroin triggers fly larvae to overeat, essentially allowing them to ignore the satiety or fullness set point to continue eating. The effect of some drugs is just the opposite of cocaine and heroin, functioning to depress one or more stages of development. Eating is not inhibited; it simply takes longer to consume enough food to molt to the next stage of development, and high doses can be lethal to larvae or pupae. Regardless of whether an ingested compound is an accelerant or depressant, if it alters fly development in any way, it conceivably compromises the utility of that species for use in calculating the postmortem interval.

- The toxicological utility of fly larvae extends to human tissues exposed to explosives or bullet fragments. How so? When an explosive device is detonated, it leaves behind chemical residues on anything close by, which unfortunately includes humans and other animals. In instances of terrorism, bomb-makers tend to have a specific modus operandi for the chemical composition of their devices; if the residues can be located and analyzed, the terrorist organization responsible for the egregious act may be identified. A similar situation occurs with gunshots, in which the resulting residues have a defined chemical signature. Gunshot residues (GSRs) contain spheroid particles that are easily detectable to the trained eye. They also are composed of unique levels of three elements: barium (Ba), lead (Pb), and antimony (Sb). Explosive and gunshot residues accumulate on skin surfaces and also on mucus membranes, such as found in the eyes, nose, and mouth. But a badly decomposed body or one that has been consumed by insects and other creatures may leave behind no evidence of the residues. Here is where insects become useful to law enforcement once again. Insects, namely early colonizing flies, feed on surfaces where the residues accumulate. In situations in which foul play is suspected, the maggots can be ground up and analyzed for the chemical compounds that signify an explosive device was present, if not responsible for the death of the individual.

MUSHROOM FARMING (SELF-TEST)

Level 1: Knowledge/Comprehension

1. Define the following terms:
   - (a) carrion
   - (b) myiasis
   - (c) necrophagous
   - (d) postmortem interval
   - (e) stored product entomology
   - (f) defect action level

2. Several animals are attracted to the remains of a dead animal. Blow flies tend to be the first colonizers. What causes them to be so interested in arriving first?

3. What distinguishes the three branches of forensic entomology?

4. Why are necrophagous flies more useful to a homicide investigation than necrophagous beetles?

Level 2: Application/Analysis

1. Urban entomology and stored product entomology are both divisions of general entomology and forensic entomology. What determines when an insect matter falls under each umbrella?

2. Many different species of flies are drawn to carrion, creating intense competition. Explain how resource partitioning permits multiple species to use the same carcass.

3. What are the limitations in using fly larvae to detect gunshot residues on a corpse?

Level 3: Synthesis/Evaluation

1. Maggot masses that form on a corpse can be composed of thousands of individual fly larvae representing more than one species. The masses are known to generate internal heat that can exceed ambient air temperatures by several degrees. Explain the implications this would have on using necrophagous flies to calculate a postmortem interval.

2. If a corpse is discovered indoors and the deceased is believed to potentially have suffered from abuse, explain why it is imperative to determine whether facultative myiasis occurred before attempting to use fly evidence for estimating the time of death.

REFERENCES

Anderson, G. S. 2011. Comparison of decomposition rates and faunal colonization of carrion in indoor and outdoor environments. *Journal of Forensic Sciences* 56:136-142.

Benecke, M., and R. Lessig. 2001. Child neglect and forensic entomology. *Forensic Science International* 120:155-159.

Benecke, M., E. Josephi, and R. Zweihoff. 2004. Neglect of the elderly: Forensic entomology cases and considerations. *Forensic Science International* 146 (Suppl.): S195-S199.

Byrd, J. H., and J. L. Castner. 2010. Forensic Entomology: The Utility of Arthropods in Legal Investigations. CRC Press, Boca Raton, FL.

Campobasso, C. P., G. Di Vella, and F. Introna. 2001. Factors affecting decomposition and Diptera colonization. *Forensic Science International* 120:18-27.

Catts, E. P., and M. L. Goff. 1992. Forensic entomology in criminal investigations. *Annual Review of Entomology* 37:253-272.

Charabidze, D., B. Bourel, and D. Gosset. 2011. Larval-mass effect: Characterization of heat emission by necrophagous blowflies (Diptera: Calliphoridae) larval aggregates. *Forensic Science International* 211:61-66.

Gennard, D. 2012. Forensic Entomology: An Introduction. 2nd ed. Wiley-Blackwell Publishers, West Sussex, UK.

Goff, M. L., and W. D. Lord. 1994. Entomotoxicology: A new era for forensic investigation. *American Journal of Forensic Medicine and Pathology* 8:45-50.

Grassberger, M., and C. Reiter. 2001. Effect of temperature on *Lucilia sericata* (Diptera: Calliphoridae) development with special reference to the isomegalen- and isomorphen-diagram. *Forensic Science International* 120:32-36.

Greenberg, B. 1991. Flies as forensic indicators. Journal of Medical Entomology 28:565-577.

Greenberg, B., and J. C. Kunich. 2002. Entomology and the Law. Cambridge University Press, New York, NY.

Ireland, S., and B. Turner. 2006. The effects of larval crowding and food type on the size and development of the blowfly, *Calliphora vomitoria. Forensic Science International* 159:175-181.

Payne, J. A. 1965. A summer carrion study of the baby pig *Sus scrofa* Linnaeus. *Ecology* 46:592-602.

Richards, C. S., and M. H. Villet. 2008. Factors affecting accuracy and precision of thermal summation models of insect development used to estimate post-mortem intervals. *International Journal of Legal Medicine* 122:401-408.

Rivers, D. B. 2015. Heat production by necrophagous fly larvae: Implications for forensic entomology. *Annals of Forensic Research and Analysis* 2(1): 1013.

Rivers, D. B., and G. A. Dahlem. 2014. The Science of Forensic Entomology. Wiley-Blackwell Publishers, West Sussex, UK.

Rodriguez, W. C., and W. M. Bass. 1983. Insect activity and its relationship to decay rates of human cadavers in East Tennessee. *Journal of Forensic Sciences* 28:423-432.

Romero, A., M. E. Potter, D. A. Potter, and K. F. Hayes. 2007. Insecticide resistance in the bed bug: A factor in the pest's sudden resurgence. *Journal of Medical Entomology* 44:175-178.

Tomberlin, J. K., R. Mohr, M.E. Benbow, A.M. Tarone, and S. VanLaerhoven. 2011. A roadmap bridging basic and applied research in forensic entomology. *Annual Review of Entomology* 56:401-422.

VanLaerhoven, S. L., and G. S. Anderson. 1999. Insect succession on buried carrion in two biogeoclimatic zones of British Columbia. *Journal of Forensic Sciences* 44:32-43.

Villet, M. H., C. S. Richards, and J. M. Midgley. 2010. Contemporary precision, bias and accuracy of minimum post-mortem intervals estimated using development of carrion-feeding insects. In: Current Concepts in Forensic

Entomology (J. Amendt, C. P. Campobasso, M. L. Goff, and M. Grassberger, eds.), pp. 109–137. Springer, London, UK.

Voss, S. C., H. Spafford, and I. R. Dadour. 2009. Annual and seasonal patterns of insect succession on decomposing remains at two locations in Western Australia. *Forensic Science International* 193:26–36.

THE ENTOMOLOGIST BOOKSHELF (SUPPLEMENTAL READINGS)

Dent, B. B., S. L. Forbes, and B. H. Stuart. 2004. A review of human decomposition processes in soil. *Environmental Geology* 45:576–585.

Erzinçlioğlu, Z. 2000. Maggots, Murder and Men. Thomas Dunne Books, New York, NY.

Goff, M. L. 2000. A Fly for the Prosecution. Harvard University Press, Cambridge, MA.

Heinrich, B. 2013. Life Everlasting: The Animal Way of Death. Mariner Books, New York, NY.

James, M. T. 1947. Flies That Cause Myiasis in Man. USDA Miscellaneous Publication 631. United States Department of Agriculture, Washington, DC.

Maples, W. R. 1994. Dead Men Do Tell Tales. Broadway Books, New York, NY.

Norris, K. R. 1965. The bionomics of blowflies. *Annual Review of Entomology* 10:47–68.

Smith, K. G. V. 1986. A Manual of Forensic Entomology. Cornell University Press, Ithaca, NY.

ADDITIONAL RESOURCES

American Academy of Forensic Sciences
http://www.aafs.org/

American Board of Forensic Entomology
http://www.forensicentomologist.org/

Body Farm at the University of Tennessee
http://web.utk.edu/~fac/

Crime Scene Creatures
http://to.pbs.org/1dvbk1t

Forensic entomologist job description
http://www.crimesceneinvestigatoredu.org/forensic-entomologist/

Forensic entomology
http://www.forensic-ent.com/

Forensic entomology resources
http://www.forensic-entomology.com/

Identification keys for forensic flies
http://www.nku.edu/~dahlem/ForensicFlyKey/Homepage.htm

More than just blow flies and beetles
http://entomologytoday.org/2015/01/22/forensic-entomology-is-more-than-just-blow-flies-and-beetles/

North American Forensic Entomology Association
http://www.nafea.net/

16 Insect Mercenaries

Weapons for Human Warfare and National Security

Although the ways that humans have devised to inflict pain and misery on each other may seem limitless, they are rank amateurs in this enterprise compared to insects and their relatives.

Dr. May R. Berenbaum (2009)*
Professor of Entomology, University of Illinois

Insects generally have the upper hand in nearly every situation in which they come in contact with us. So the next statement may come as a complete surprise: humans victimize insects. The idea seems impossible if not absurd. True that! Yet throughout history, several species of insects haven fallen into the clutches of individuals (people that is) with less than forthright intentions. For what purpose, you might ask? To be recruited as soldiers. Actually "recruitment" is not quite a true reflection of the situation, as the insects do not volunteer. No, these critters are forcibly drafted with the intent of a one-way mission. The six-legged soldiers of fortune have been fashioned into weapons, bombs, and even bioterrorism agents, against an army that is not their enemy. A search of the historic record reveals stories of man using insects as weapons in conflicts with others, mostly relying on those that bite, sting, or otherwise evoke pain. Some species have even served as pawns in very imaginative yet diabolical torture methods. The details are too gruesome to tell, but for you, an exception will be made!

Of course, insect weapons with the most potent punch are those that cause disease through harboring pathogenic organisms, released during blood feeding on us. Fashioning insects into weapons is not only an idea of the past. In fact, insects are considered a very real threat to the national security of the United States and several European nations. Their value lies in the fact that they are easier, cheaper, and safer to work with than traditional disease-causing microorganisms. Hoorah for insects! Insects also represent three-pronged potential as bioterrorism agents, in that they can target humans directly or serve as vectors for disease transmission or as weapons that cripple agricultural systems. Chapter 16 examines these potential entoterrorism threats and also dives into the historical uses of insects as weapons of war, terror, and torture. If we stop with these topics, the future admittedly looks bleak. Fortunately, insect soldiers hold no allegiances to any nation or army; they can also be exploited to aid in the

---

*A telling and disturbing excerpt from Dr. Berenbaum's review of *Six-Legged Soldiers* by Jeffrey Lockwood (2009).

Figure 16.1. The Hatfield clan in 1897. The Hatfield family of West Virginia is famous for its long-standing feud with the McCoy clan of Kentucky. Violence between the families erupted several times over the years, including over land rights, a pig, and different allegiances during the American Civil War. Photo by the Iowa State Press in 1897 (http://bit.ly/1JKHQvR).

war *against* terror. We explore the role of insects in national security and discuss how insects can be used in surveillance, including for detecting materials used to make explosives or bioterrorist agents, being equipped with audio and video recording devices for short-term espionage, and being manipulated to serve as biosensors that detect weapons and even bodies. In reality, insects are the ultimate double agents.

Key Concepts
- Historical perspectives on entomological weaponry
- Insects as agents of terror
- Insects as bloodhounds: Sniffer systems
- Insect espionage: Cyborgs and surveillance
- An insect for an insect: Entomological counterterrorism

Historical perspectives on entomological weaponry

Humans seem to naturally seek out confrontation with each other. Most human conflicts, especially localized

skirmishes, are or have been over possessions: territories, boundaries, food, water, precious commodities, and even mates (http://hatfieldmccoycountry.com/feud/). In modern times, hostilities between groups and nations are centered more on ideological differences, especially those representing zealots or religious ideals. By no means does this imply that religious wars did not occur in the past. Of course they did, and frequently (figure 16.1). However, today's global environment is laced with factions, cells, or non-nation-affiliated organizations (commonly linked to terrorist groups) that appear to be continuously engaged in holy wars with all nonbelievers.

The point is that such disputes, whether examined from a historical perspective or simply by observing current events, often rise to the level of physical engagement. Escalating disagreements may lead to full-on war involving multiple participants. The forms of engagement, however, have not remained static over time. In practical terms this can be seen through the evolution of weaponry, from crude, simple low-efficiency devices to broad scale explosive munitions with the capacity to devastate large areas to the use of chemical and biological weapons designed to promote fear rather than carnage. That said, so-called **weapons of mass destruction** do encompass some forms of biological and chemical weaponry. The upper echelon of weapon evolution today exists as highly sophisticated smart weapons that can deliver their payload via **drones** with pinpoint accuracy, while being controlled by soldiers that are miles away from their targets. What is frightening is the very real possibility that smart technologies will someday be used in conjunction with biological and/or chemical weapons (figure 16.2).

Coinciding with the evolution of war has been a shift in attitude about who to target. At one time, battles enjoyed a type of gentlemen's agreement on the rules of engagement. For example, men, not women, served as soldiers,* and fighting was generally restricted to only the professionals (soldiers), often at prescribed locations and times of day. By World War II, a set of protocols existed (the Geneva convention)† that outlined proper etiquette with regard to the types of weaponry to use during war, how prisoners and wounded were to be treated, and promoted a hands-off mentality toward civilians. Obviously every army did not subscribe to the rulebook, but many nations at least made an attempt to comply. Modern warfare, however, does not really conform to any protocols. Why? Because the participants are not nations per se. Instead, radical religious groups are initiating wars on whomever they deem the enemy. From their vantage point, the Geneva convention does not apply to them. So the rules of war are theirs to write. This includes how to conduct a military campaign, which generally no longer conforms to conventional warfare tactics

Figure 16.2. A drone is launched from the amphibious dock landing ship USS *Tortuga* (LSD 46) for a scheduled missile exercise of the Singapore phase of Cooperation Afloat Readiness and Training (CARAT) 2008. US Navy photo by Cmdr. James Ridgway at http://bit.ly/1Jv7xOT.

and traditional weapons. Instead, a modern arsenal of weapons is employed that can be silent, insidious, and devastating.‡ We examine in detail some the new tactics of war in the next section. For now, suffice to say that insects are part of modern warfare.

Among these "new" weapons are ones that are considerably older than expected. Soldiers with six legs were used as weapons by nearly every ancient civilization. Initially insects were selected for their potent venoms. In some cases, their exuded toxins were used to coat arrows or spearheads, demonstrating that ancient peoples dabbled in rudimentary **pharmacology.** More common insect uses centered on designing bombs or grenades out of porcelain, clay, or ceramic, inside of which was inserted a small colony of bees or wasps or other suitable beasts. The device was then sealed with wax, clay, or mud, and flung toward the enemy. Upon contact with the walls of a fortress or entrenchment, or hard impact with the ground, the bomb would explode thereby releasing its contents. One can imagine that the bomb's inhabitants were quite agitated if not angry by then, and stung and bit at will. This was an affective technique for dislodging an entrenched enemy that, once exposed, could be attacked with other weapons quite easily. Of course the

---

*Exceptions abound, especially considering that some military forces have never discriminated based on gender and that, even since the founding of the United States, women have played prominent roles in key battles.

†The Geneva convention refers to four treaties and three protocols that establish humanitarian standards for treatment of prisoners and the wounded during times of war. The treaties were first established in 1864 and have been modified in subsequent years to address treatment of civilians and banning certain types of weapons, namely those that are chemically and biologically based. Details of the Geneva convention and each treaty can be found at http://bit.ly/1FnI5Ih.

---

‡To paraphrase Jeffrey Lockwood's (2009) description of modern entomological weapons.

problem with this method is that insect bombs could not be thrown from a long distance away from the enemy. The attackers were therefore nearby when the bees and wasps were released, which we have established meant the insects were quite agitated, and the angry insect mob cared not who they attacked. Undoubtedly compounding this problem for the bomb creators was the possibility that those who packed the insects were coated with alarm pheromones, which, as we learned in chapter 12, serve as an attractant for stinging Hymenoptera. Just squash a yellow jacket near the nest for a historic reenactment. On second thought, don't do that (http://bit.ly /1JMWxNr).

Sometimes the benefit of working with insects in the weapons industry is that they are just as effective whether released or not, because the mere sight of a multilegged creature evokes fear in some individuals. For others, the presence of insects can cause all sorts of irrational behaviors, often leading to a condition termed **delusional parasitosis**. As you might imagine, the powerful effect of stinging/biting insects on human behavior has been manipulated throughout history into methods of interrogation. In other words, insects have been used in torture techniques. Perhaps the most recognizable form is that depicted in a number of classic western genre movies in the United States, in which the settlers and soldiers are punished for intruding on Native American lands by being buried up to their necks near an ant mound. Given enough time, scouts from the colony would wander toward the detained humans, eventually sampling the skin with their mandibles. The preferred feeding areas were the moist mucosal membranes of the nose, mouth, and ears, and also the nutrient-rich eyeballs. To aid the process, a sugary mixture was often applied to exposed skin to ensure ants and other hungry creatures would be attracted. Death did not arrive quickly enough, and individuals subjected to this form of torture endured the very slow and incredibly painful process of being eaten alive.

The method described was not a Hollywood creation. No, it was an actual torture technique applied throughout Mesoamerica for centuries. Now, if the intent was merely to intimidate or interrogate the prisoners, then the individual might be tied naked to a tree or a pole near aggressive ants or in mosquito-infested areas. A few bites were usually more than sufficient to extract the informa-

tion desired. Unfortunately for the captured, the deal of exchanging information for freedom usually did not work out as hoped. One of the more diabolical methods of torture involved tying a captive naked to a raft, then force-feeding the prisoner milk and honey. The net effect of this gustatory concoction was induction of constipation with anal leakage. That alone would seem like torture enough. But no, the fluids would attract flies that in turn would oviposit near the anal opening. You can imagine what happened next; the eggs would hatch and the fly larvae ate the secretions and infested the anus. This is similar to **retroinvasion,** discussed in chapter 15. The pain would become so unbearable that the prisoner would reveal anything his or her captors wanted to know.

There would seem to be few insect-related torture techniques that could top one leading to anal discharge and infestation, but there certainly is at least one more that moves the cruelty meter to another level. What form of torture holds the distinction of unimaginable horror? The answer is actually a physical place rather than a technique: the infamous Bukhara Bug Pit of Zindon prison, located in what is now Uzbekistan (figure 16.3). Prisoners were lowered into the pit by rope, and then guards would pour scorpions, rats, sheep ticks, and insects on top of them. If the ruling emir was in an especially foul mood, he would have the animals starved for several days before releasing them into the pit, ensuring that they would attack the prisoners. The insects selected for this special purpose were actually a type of assassin bug (Hemiptera: Reduviidae), which produces salivary venoms that are injected into the prey. Salivary venom contains several potent digestive enzymes that work just as well on human tissues as they do on insect, so the effect was chemical digestion of muscle and soft tissues. If exposed to the venom for a long enough period, or a sufficient quantity was injected into a single individual, tissues would slough off the bone. Prisoners literally lost the capacity to use their limbs. They also endured unspeakable pain from bites, tissue loss, lingering effects of venoms, and subsequent infections. This was an experience that most individuals did not survive.

Employing insect weaponry has continued well into the last two centuries. Numerous examples of insect use in war can be cited, from medieval battles to the great world wars to even as recently as the Korean War and

Figure 16.3. Zindon prison at Bukhara, Uzbekistan, which was home to the infamous bug pit used to torture and punish captives. Photo by David Stanley at http://bit.ly/1JrHBzr.

Vietnam conflict. Two common threads tie the modern use of insect weapons together (table 16.1). The first is that war draws together large numbers of soldiers who in turn are more susceptible to insect-borne diseases because of the conditions they face: masses of individuals in close contact, stressed, and malnourished. In chapter 4, we discussed how typhus, yellow fever, and malaria caused more deaths during war than any battle-inflicted wounds until the discovery of the insecticide DDT just prior to World War II. The fact that insects harbor deadly pathogens did not go unnoticed. In fact, this knowledge—that insects vector disease and that insects as a whole are relatively easy to manipulate—is the second thread. A new class of weapons soon followed suit, in which vectors of potent diseases such as malaria, yellow fever, dengue fever, and plague, or those that could devastate important food crops, were raised en masse to purposely release into targeted enemy areas. Perhaps the most extreme example is the Japanese Imperial Army's covert chemical and biological warfare research and development during the World War II. Under the leadership of General Shiro Ishii, the Department of Epidemic Prevention and Water Purification began exploring Japan's potential to use weapons outlawed by the Geneva convention (figure 16.4).

The base of operations was in Manchuria, where the army's Unit 731 mass-produced millions of human fleas

(*Pulex irritans*), which is the insect that vectors bubonic plague. In turn, the fleas were infected with the bacterium *Yersinia pestis*, the causative agent of the plague. Within one year's time, the Manchurian facility was producing over five hundred million fleas harboring *Y. pestis*. The goal of the Japanese unit was to develop methods to effectively deliver living fleas to enemy targets. Thus, intense testing began in bomb development using Chinese villages and prisoners of war as test subjects. During the six years that the facility operated, nearly 250,000 men, women, and children—mostly Chinese, Korean, and Mongolians—died from the Japanese human experimentation phase of biological weapons development. Remarkably, none of the individuals involved with Unit 731 were tried during the war crimes tribunals conducted after the Allied victory. Instead, the scientists were given immunity in exchange for their data and the methods employed in raising the fleas and bacteria.

It is important to note that Japan was not the only nation involved in chemical and biological weapons development during World War II. The United States actively engaged in research using mosquitoes and fleas as disease delivery vectors, as did Germany and England. And the efforts were not restricted to just those insects that harbored human disease; accusations were flung from all sides that Germany, England, and the United States had used insects to attack crops. Actually, such accusations date back to the US Civil War, during which the Confederacy was convinced that Union forces were deliberately releasing insects to sabotage crops in the south. The point is that insect weapons have been used in the past and quite recently, and by multiple nations. Continued use by multiple users is a clear sign of effectiveness, which is a testament to the utility of insects even in human warfare.

Insects as agents of terror

Today's modern warfare relies heavily on **terrorism**. Terrorism literally is the use of terror tactics, most commonly through acts of violence, in the name of a religious, political, or other ideological purpose, toward chosen targets, with no regard for soldier or citizen status. The last point is the most evident difference from previous military campaigns and reflects a total disregard for the articles of the Geneva convention. Acts of violence are carried out using a broad array of weapons, including

Table 16.1. Insects suspected of or known to be used in military campaigns. *=insects suspected of being used but direct evidence is lacking. +=represent insects that were tested for potential use as biological weapons but actual deployment is not known. Information derived from Lockwood (2009).

Insects Used in War		
Common name	Order or Family	Military campaign
Bees and wasps	Presumed to be Apidae and Vespidae	Ancient Egypt, Maya and Aztec empires
Flies for torture	Calliphoridae	Prior to nineteenth Century
Harlequin bug	Pentatomidae	American Civil War*
Fleas	Pulicidae	WWII, Korean War
Mosquitoes	Culicidae	WWII*, Korean War*, Vietnam Conflict*
Filth flies	Calliphoridae	Korean War
Fruit flies	Tephritidae	Ecoterrorism in US
Caterpillars	Saturniidae	Desert Storm+

Figure 16.4. A building on the site of the Harbin bioweapon facility of Unit 731 located in Manchuria during World War II. The site was home to the Japanese Imperial Army's Biological Weapons Program, which was responsible for the deaths of more than 250,000 civilians as a result of biological weapons testing involving insects vectoring human disease. Photo by Masao Takezawa available at http://bit.ly/1syWQX2.

those classified as biological and chemical (figure 16.5). Use of such weapons represents an important distinction from any other era of human conflict,* especially since the United Nations banned the production and use of any type of biological weapon. The harsh reality is that there are likely more biological weapons in existence today than at any other time in history, and they also appear to be in the hands of groups ready and willing to put the weapons to use.

Biological weapons involve the use of living organisms and/or their products (i.e., toxins, venoms) by militant groups or nations to target other country, organization, or group. Traditionally, a more narrow definition had been used to describe biological weapons, limited to microorganisms like bacteria, viruses, or protozoan or other pathogens that specifically induce disease in humans or other animals. The 2001 anthrax attacks in the United States, labeled as Amerithrax by the Federal Bureau of

*While the Japanese did in fact develop biological weapons during World War II, they technically did not use them in actual military campaigns, and though other nations have been accused, evidence is lacking to prove that any did use them.

Figure 16.5. International symbols for weapons of mass destruction: from left to right, nuclear, biological, and chemical hazards or weapons warnings, respectively. Images created by Fastfission at http://bit.ly/1KK4dfu.

Investigation (FBI), solidified public perception that bioterrorism agents are deadly microbes. Of course public perception alone does not account for why microorganisms have been singled out for use in bioweapons development. What is the explanation? Quite simply, they possess the ideal traits for an agent to be used in biological terrorism:

1. They are easy to propagate and transmit.
2. They tax the infrastructure of the targeted area, city, or country.
3. They promote fear in the civilian population that in turn applies pressure on politicians to meet the demands of the terrorists.

Several disease-causing microorganisms meet these criteria through induction of potentially lethal diseases that can easily reach **epidemic** to **pandemic** status, which in turn overwhelms emergency responders and medical institutions. The mild outbreak of Ebola in the United States in 2014 demonstrates the fear that can sweep through a population in a short time. It also brought to light how ill-prepared the country is for dealing with one of the major candidates for biological terrorism.

A disease like smallpox represents yet another potential biological weapon that could quickly spread and tax infrastructure simply because vaccine supplies are so low throughout the world. You might ask, why is the vaccine stockpile low? Simple—smallpox has mostly been eradicated from Western nations. With no real natural threat of the disease, vaccinating the population has dissipated,

and in recent years, vaccine supplies are sufficient only for treating small, contained occurrences of the disease. Thus, reintroduction of smallpox to a large metropolis like New York City, Chicago, or Los Angeles could have immediate and devastating consequences. Now imagine that a new strain of smallpox is released, one that the current vaccine does not immunize against. Recent breakthroughs in recombinant DNA technologies and molecular biology make this a definite possibility. Newly created genetic tools permit manipulation of the genome of almost any organism, including those microorganisms feared to be the next agents of bioterrorism. In theory, enhancing pathogenic organisms is relatively simple, albeit expensive, making microbial pathogens even more attractive for deadly terrorist attacks. It is no wonder then that many nations are on high alert for terrorism involving microbial weapons.

Despite the doom and gloom discussion regarding bioterrorism agents, use of deadly microorganisms should not be viewed as inevitable. There are some severe limitations to working with such microbial weapons. For one, a terrorist cell or organization is just as susceptible to the microorganisms as the intended target. Without access to all the necessary safety equipment and facilities, there is absolutely no way to prevent exposure to the pathogens and suffering from their ill effects. This means that a terrorist organization must use sophisticated laboratory facilities like those found in research institutions equipped to work with biosafety level 3 and 4 organisms. In the United States, the Centers for Disease Control and

Prevention have established safety precautions for containing and working with organisms that are known to be or are potentially dangerous to humans (figure 16.6). Biosafety levels 3 and 4 categorize the most severe threats, including most microorganisms considered potential agents for biological terrorism (figure 16.7). The facilities required to contain such organisms generally are expensive to operate and nearly impossible to conceal from nations that monitor for biological weapons production sites. Also reducing the risk of attack by many (but not all) microbial weapons is the lack of effective delivery systems; even if a pathogen is highly virulent, if it cannot reach the target quickly, it offers little utility to a terrorist plot. In short, it is not as easy to develop biological weapons out of microbes as one might be led to believe from fictional suspense novels or depictions in espionage films. This does offer some relief but also inspires a question: What might the alternatives be to successful biological weapons development? Our answer lies in the multipurpose insect.

Insects are an excellent alternative to microorganisms in the development of biological weapons. Though often overlooked in discussions of biological weapons, animals with six legs have enormous potential for use in war or acts of terror. Obviously there is historic precedence for entomological weapons, as we have discussed earlier. However, entomological weapons, especially in the form of **entomological terrorism** or **entoterrorism**, overcome the limitations of microbial weapons and actually provide a broader array of targets. In terms of bioweapons development, thousands of insects can be raised cheaply and quickly—certainly in comparison to pathogenic microorganisms—in a relatively small space (say, in a basement or garage of a house) and do not require sophisticated rearing or containment facilities. Equally important is the fact that many adult insects can fly, so there is no need for an elaborate delivery system for bioweapons to reach their target. Instead, insects can disperse from the release point with no need for human intervention. Entomological weapons also offer a broad spectrum of targets. For example, insects can be used or manipulated in entoterrorism in a number of ways.

1. Target human or other animal populations directly
2. Serve as vectors of disease targeting animals and plants
3. Affect food production or agricultural economy

Figure 16.6. Individuals working in positive pressure suits in a biosafety level 4 facility. Such protection is needed to work with most virulent human pathogens and thus represents a significant obstacle to a terrorist organization attempting to developing biological weapons using deadly microorganisms. Image created by National Institute of Allergy and Infectious Diseases, http://bit.ly/1quiWrx.

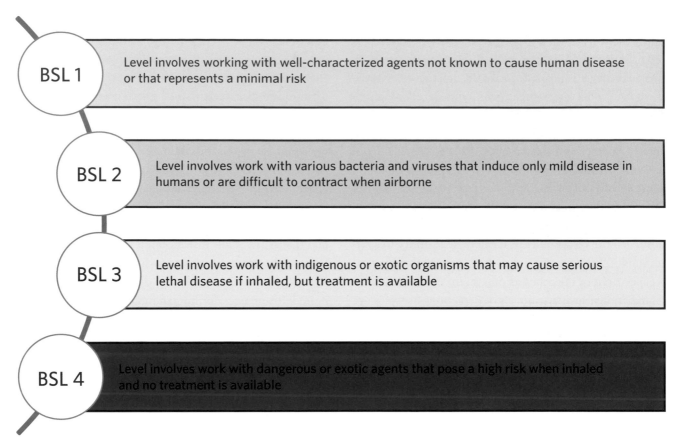

Figure 16.7. Biosafety level precautions when working with biological agents that pose a potential health risk to humans or the environment. The safety levels were developed by the Centers for Disease Control and Prevention.

The point is that insect weapons are not restricted to disease transmission and can actually tax infrastructure through spreading disease, diminishing food supply, and weakening the economy. A quick glance at the above list reveals that each category is not mutually exclusive and that a vector for plant disease could easily fall under the heading of affecting agriculture, otherwise known as **agroterrrorism**, or a disease vector may also directly attack human or animal populations. That said, each represents a distinctive target or means for using insect weapons.

Direct attack or assault by entomological weapons implies that the insects used bite, sting, or produce noxious or toxic secretions. It is the aggressive behavior of the insects and/or the toxins they produce that create terror and havoc in the targeted population (figure 16.8). Bee bombs used in premodern warfare and torture techniques involving ants were among the precursors to

today's entomological weapons. The most effective insects used in this fashion tend to be social Hymenoptera and true bugs (Hemiptera) that produce nondiscriminate toxins (i.e., do not require binding to a cell receptor to elicit effects) or venoms that, when injected into a human or other animal, produce very painful and long-lasting effects. Pain often results from lysing of plasma membranes and inflammation triggered by histamine release. Remember, the goal of terrorist acts is not necessarily to induce carnage but fear. Intense pain that does not dissipate quickly produces long-term memories that achieve the end goal. By contrast, using insects as vectors of disease is essentially the same as simply developing bioweapons out of pathogenic microorganisms. The difference of course is that the insect serves as an effective delivery system and targets more broadly, including livestock and crops, not just human populations. This method still suffers from the health risks and expense

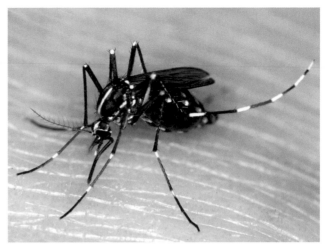

Figure 16.8. An adult Asian giant hornet, *Vespa mandarinia* (Hymenoptera: Vespidae), produces a potent venom that, when combined with the aggressive group attack behavior of the nest, represents a potential entomological weapon for direct assault. Photo by Yasunori Koide at http://bit.ly/1KK74VE.

Figure 16.9. An adult female Asian tiger mosquito, *Aedes albopictus* (Diptera: Culicidae), is a potential choice as an entomological weapon to deliver several insect-borne diseases to populations in the United States and in other western nations. Photo by Susan Ellis, Bugwood.org.

discussed earlier, but insect vectors overcome some of the drawbacks as well.

Mosquitoes serve as an excellent example of the efficiency of using insects as vectors of disease. Several species of *Aedes* (Diptera: Culicidae) mosquitoes can easily be raised in a small laboratory or well-humidified room (figure 16.9). After the adult females blood feed—which by the way does not require a host but instead can occur by providing warm blood in a thin-sleeved latex material (e.g., a condom or glove)—they lay eggs, collected on paper towels. With a large colony of mosquitoes, hundreds of eggs can easily cover a one-inch-by-one-inch square of paper towel. Placed in a shirt pocket, the insect vectors will go undetected while traveling through an airport. If the eggs are in turn dropped into freshwater, an **inoculative release** of the mosquitoes occurs, potentially infiltrating the new region with a deadly disease. As can be seen, this process is relatively simply to accomplish without endangering the individual making the insect delivery.

Perhaps the greatest threat of insects to national security is associated with agriculture. Insects have a long history of functioning as pests of crops, livestock, and stored food products. It thus is a small step to move from an accidental or opportunistic pest of agriculture to one that is deliberately released as a weapon of war or ter-

rorism. In this respect, many species of insects are well suited to be agents of agroterrorism. Agroterrorism is the deliberate introduction of animal or plant pathogens, or pests that directly target cropping systems, livestock, or food held in storage after harvest. Before we focus on the entomological side of agroterrorism, the question to be addressed is, why focus on agriculture instead of human pathogens for terrorist acts? In the United States, though farming as an occupation represents less than 2% of the national workforce, more than 15% of Americans are engaged in jobs associated with food or fiber production; a well-placed agroterrorism agent could deliver a severe blow to the US economy.

More concerning is the way that agriculture is practiced in the North America: **monoculture** dominates cropping systems and livestock production and finishing occurs at very few facilities. For example, over two hundred million acres of farmland in the United States are planted in only four crops: corn, hay, soybeans, and wheat. Similarly, more than 90% of the pork produced in the country is derived from fewer than one hundred producers. Again, just a few agricultural pests or disease vectors could conceivably cripple specific aspects of food production. The effect would be a greatly reduced food supply, a diminished economy, and widespread panic and fear. It is less than comforting that protecting our food supply from potential terrorist threats falls predominantly in the lap of the United States Department of Agriculture, an agency that has been grossly underfunded to carry out its primary domestic functions of ensuring food safety and quality and working to improve food production. The take-home message is that the United States, like most other nations, is not prepared for a major attack on agriculture.

Insects are ideal candidates to use as agents in agroterrorism. Our discussions from chapter 4 already laid out the toll that just a small number of species inflict on agriculture annually through their destructive feeding tendencies and/or their ability to transmit pathogens to crops and livestock. A potential agroterrorism agent really does not require extraordinary or unusual life history characteristics. In fact, the process could be as simple as introducing an **exotic** or **invasive species** into an agricultural district and letting the species become established. Depending on the monitoring or surveillance tactics used for a given cropping system, the introduced insect may go undetected for months to years. Exotic species can wreak havoc in a short time because there are no natural enemies like predators or parasites to keep the population in check, at least not initially. Even once the insect is discovered and then determined to be a pest, an effective management strategy (see chapter 14 for details of insect control and management) can take a long time to develop.

The brown marmorated stink bug, *Halyomorpha halys* (Hemiptera: Pentatomidae) and emerald ash borer, *Agrilus planipennis* (Coleoptera: Buprestidae) represent exotic pests that have proven very difficult to manage and contain even a decade after their accidental introductions into the United States (figure 16.10). Now imagine the agricultural pest is deliberately introduced into a region with the intent of maximizing destruction. Just a few steps can ensure success. For example, rather than releasing small numbers (termed inoculative release) of a given agroterrorism agent, such as occurs with accidental introductions, an **inundative release** would be a more effective approach. Inundative release relies on mass release of thousands of individuals into an area so that the insect pest quickly establishes and begins to immediately cause severe damage. The downside is that the entoterrorists (aka, the insects) will be

Figure 16.10. The recently introduced emerald ash borer, *Agrilus planipennis* (Coleoptera: Buprestidae), has quickly become an invasive pest species causing millions of dollars in damage to ash trees annually. Photo by the Pennsylvania Department of Conservation and Natural Resources–Forestry, Bugwood.org.

discovered sooner than would an inoculative release, but by that point, most of the damage will have already occurred.

Another way to ensure success is to manipulate the genetic status of the biological weapon. Ideally, the insect would be selected through a breeding program or possibly rely on genetic knockout techniques to confer insecticide resistance. This approach is based on the notion that insecticides are the most likely and powerful tools to combat agroterrorism agents. Insecticide resistance essential neutralizes the only rapid counterespionage weapons available to a susceptible nation. From this perspective, some insect species that are indigenous to a targeted region are also well suited to serve as biological agents because they demonstrate insecticide resistance and **cross-resistance** to a wide range of biocides and have evolved into some of the most economically important agriculture pests. Augmenting the utility of native pests for purposes of agroterrorism is that their discovery would not be considered out of the ordinary, and thus they may not even be recognized as a biological weapon at all.

Quick check
What makes insects so useful as biological weapons used for terror?

Insects as bloodhounds: Sniffer systems

Insects are naturally gifted at detecting chemicals in the environment. They rely on odors derived from all sorts of living and dead things, such as those emanating from flowers and decaying carcasses, as well as from live animals, to find food. The scent of the opposite sex helps them locate mates. Even the chemical signature of their hive or other abode is used like an insect version of Google Maps* to navigate their way home. In fact, chemical perception is the underlying mechanism by which insects interact within the environment and communicate, regardless of whether information is shared with **conspecifics** or **allospecifics**.

Detecting chemicals in the environment relies on the acute olfactory system that is characteristic of nearly all insects. Not only is insect olfaction superior to that of most any other animal, but sensory hairs that perceive chemicals also cover the body of an insect. Okay, maybe a bit overzealous—insects are not literally covered in sensory receptors. However they do possess hundreds to thousands of such structures positioned in multiple locations on the body, primarily on antennae, mouthparts, legs, footpads, and between the head and thorax. They can smell from just about anywhere on the body! When this keen sense of olfaction is coupled with the ability to train certain insect species to complete a task, you then have the makings of a **sniffer system**. What is a sniffer system? In the context of insects or other animals, sniffer systems are essentially living entities that can be used for odorant detection. Equip the beast with some sort of tracking device, and you now have the means of following an animal's movement as it is released to search for

---

*Google Maps is one of many apps for smart phones, tablets, and computers to obtain directions for travel from one location to another, or to simply visualize a map of a particular destination.

the odor of interest. Many insects are also useful for chemical detection in the form of **biosensors** or through **chemical trapping**. At this point, we likely have created more questions than have been answered. So next we explore sniffer systems and biosensors, and how they are used specifically with regard to national security.

Several animal models have been tested for potential as sniffer systems, including vertebrate and nonvertebrate animals. The key to the potential or success of a given animal in odorant detection is the application; in other words, what needs to be detected directly influences selection of a sniffer system. For example, canines have very sensitive and acute olfaction that makes them ideal for tracking a missing person through a wooded area based on odorant and gas detection, or determining if an accelerant has been used to start a building fire. They also have an excellent capacity to be trained to detect an odor, which, as we discuss a little later, many insects do as well. However, there are limitations to using dogs in this capacity. Canines are expensive to maintain, the training process is long, there is a limited window for how long they can be used and under what conditions (dogs perform poorly when it is hot and humid, or during heavy precipitation), a handler is required even after training, other sensory stimulation can distract from the task, and some locations are too dangerous to send a valuable dog into. The latter is especially true when searching for unexploded munitions like landmines. As a consequence, alternative sniffer systems have been explored, in particular, ones involving insects, with the intent of overcoming the problems encountered with vertebrate systems.

How or why are insects better choices than dogs for use in sniffer systems? "Better" probably does not best describe the differences between sniffer systems based on insects and canines. Insects provide alternatives based on differences in their life history traits that allow them to be used in ways that other organisms cannot. For one, insects can be trained in a fraction of the time that a dog can be. A honey bee, for instance, can be trained to associate food with an odor (say, that of unexploded land mine) within minutes to hours, as opposed to the multiple weeks needed for conditioning a canine. Thousands of individuals can be trained simultaneously, and the overall cost is minuscule compared with the cost of using a vertebrate animal. Perhaps most important, flying insects can be sent into hazardous regions that land-based creatures cannot. If in turn the flying biosensor is equipped with optical or GPS devices, their movements can be monitored and any behavioral changes detected. What would the latter reveal?

Let's use the honey bee again to illustrate the importance of behavioral monitoring. Bees can be conditioned to detect volatile compounds (TNT, TATP, 4-DNT, and others) emitted from landmines or other explosives that have not detonated. When released into the target area as a free-moving system, the bees begin normal foraging behavior (figure 16.11). If during flight they encounter the odorants used for laboratory conditioning, a momentary pause or just a hiccup in flight pattern occurs. To the naked eye, or when it occurs far removed from human interlopers, this would go undetected. However, with the aid of optical sensors strapped to the backs (dorsal surface) of the bees, an electrical signal is generated that records the change and sends it to transponders wherever the investigative team is located. The information provided includes GPS coordinates of the bee's location, which is also presumed to be the site of the unexploded landmine. Again, this technique allows insects to enter regions far too dangerous for any walking animal. The system also provides instantaneous feedback so that the site of interest can be immediately examined and, if necessary, the munitions removed before encountered by an unsuspecting human or other animal.

Bug bytes

Bees find explosives

https://www.youtube.com/watch?v=_T7d0bze4kM

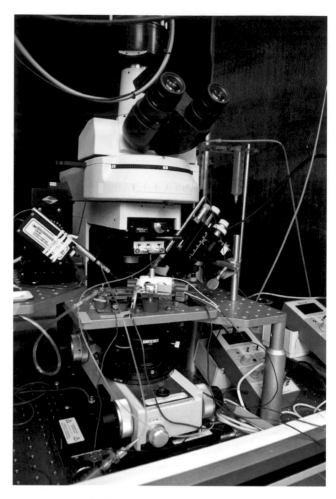

Figure 16.11. The basic design of a biosensor is to integrate a living entity, which could be a whole organism like an insect, or cells, or cellular products (enzymes, antibodies) with an electrical transducer that in turn permits an output signal to be generated when exposed to samples of interest. Limitless combinations can be used for the integration, which in turn means a wide range of abiotic and biotic samples can be tested. Photo by the US Food and Drug Administration at http://bit.ly/206Xhlr.

Throughout the discussion thus far, we have referred to conditioning or training insects. What exactly does it mean to "condition" an insect? There are two general methods for training an insect. One is **operant conditioning**, in which a reward (or positive reinforcer) or punishment (negative reinforcer) is used in association with a particular behavior. A classic example is that of larvae of the vinegar (fruit) fly, *Drosophila melanogaster*, which can be trained to associate an odorant like amyl acetate (weak smell of bananas) with electric shock. When they are exposed to a different odor like benzaldehyde (cherries),

no punishment is applied. After sufficient conditioning, when the larvae are exposed to both odors simultaneously, they migrate to the one not associated with negative reinforcement. Canine training usually relies on operant conditioning via positive reinforcement.

A second form of training is **classical or Pavlovian conditioning**. The basic principle involves the use of an unconditioned stimulus—one that does not result in a dramatic behavioral response—with a conditioned stimulus that causes a significant change in behavior. This type of conditioning is used to condition honey bees to search for munitions, as described above. Bees are trained to associate their food (unconditioned stimulus) with a volatile chemical compound normally emitted from an unexploded munition (conditioned stimulus). Once training is complete, an individual is released into the natural environment to seek out the odor that it has been conditioned to relate to food. Again, using tracking devices on the trained forager permits discovery of a potential site of interest.

Certain insects, namely social Hymenoptera, are especially suited for the task of free-moving or roaming sniffer systems. Social Hymenoptera possess a highly refined system of chemical communication, equipped with a broad range of chemosensory receptors. They also demonstrate short- and long-term memory, and are longer-lived than most other species. As a consequence, honey bees or wasps will retain the training for a long period of time (relative to other insects), making the investment in developing the sniffer system worthwhile. And since they rely on a haplo-diploid mechanism of sex determination (the idea of **arrhenotoky** discussed in chapter 9), nearly all individuals in the colony are females and essentially clones of one another. So once the ideal training conditions have been developed for a particular odorant of interest, the process should basically work for any individual from that hive.

Other insects have been used as sniffer systems, including parasitic wasps, fruit flies, and even moths, and their roles go beyond just explosives detection. Some are being developed for locating buried or concealed bodies, relying on volatile organic compounds emitted from human cadavers during the decay process. Others are used as biological sensors, or biosensors, for detecting weapons, explosives, or illicit drugs, particularly in buildings like airports. A biosensor is an analytical device that integrates living organisms or biological materials into a physiochemical detector (or transducing system). The goal is to measure some feature of a particular environment, such as the presence of a chemical, by relying on the biological properties of the organisms or its products. Does this mean that flying insects are released in airports to chase after baggage with drugs? No. The insect is restrained in a device so that the behavioral response can be observed immediately if the odorant of interest is detected. The whole insect is not even used in some in-

stances. Instead, isolated cells modified to detect specific chemical odorants are contained within boxes positioned in various locations throughout a building. If the desired chemical is recognized, the cell generates an electrical signal, which alarms officials of a possible threat so appropriate action can be taken. Obviously the benefits to using biosensors over sniffer systems are that insects do not have to be maintained or conditioned for odor recognition tasks. The major disadvantage is that isolated cells are very difficult to maintain outside a laboratory setting, so functionality of in vitro biosensors is short-lived.

The simplest method for chemical detection using insects is chemical trapping. The process is almost entirely passive. An insect is released into the designated area of interest, then, after a period of foraging, the individual is captured and the body is analyzed for trapped chemicals. The technique relies on the idea that chemical particles become attached or stuck to the bodies of insects during encounters with living or nonliving objects. In some instances, the hairs covering the body demonstrate electrostatic properties, facilitating the trapping of common particles like pollen, pollutants, and even chemical residues. In the applied world of national security, these properties are useful for monitoring chemicals associated with drugs, explosives, and biological or chemical weapons. For example, honey bees have been used in these roles by once again taking advantage of their natural tendencies to forage for food. Prior to release, bees are equipped with a tracking device and stamped (not literally) with a barcode. Upon returning to the hive, the bee must pass through a detection system that evaluates any chemical signatures on the forager's body. If a chemical residue of interest is detected, then the barcode scan will reveal the identity of a specific bee, so that, in turn, that individual's daily travels can be retraced. Viola, the source of the chemical threat can be revealed. As can be seen from most of the examples provided for odorant detection, honey bees play a major role in national security. Perhaps recruitment into covert operations and espionage is the real bases for colony collapse of bees.

Bug bytes
Air Force bugbots
https://www.youtube.com/watch?v=z78mgfKprdg

Insect espionage: Cyborgs and surveillance

Reality can be far more intriguing than fiction. Take spy movies for example. The technologies depicted in films portraying espionage and covert operations dabble slightly in true science but mostly dive head-deep into fiction. It makes for great action-packed entertainment, but at the end of the day, the audience knows it's not real. Now contrast this to the ideas of conducting covert surveillance operations with cyborg insect species or using specimens with neural circuitry, both controlled remotely by humans. Are these examples just science fiction as well? Not all! Cyborg insects have been used for many years in surveillance operations targeting organized crime and militant organizations, as part of search and rescue missions, and also in efforts to locate explosives and hazardous materials (like those used to generate biological weapons). To avoid detection, the miniature robots, termed **micro air** or **land vehicles** (**MAV** or **MLV**), are designed to look like flying or walking insects. To achieve this desired effect, the micro spies are about the size of medium- to large-sized insects. This size permits attachment of high-tech audio and video surveillance equipment that provides instantaneous feedback to the team of investigators. Of course with this advanced technology comes a hefty price tag, which is one of the major limitations to using cyborg technology. A lack of external power to permit continuous operation is another drawback to more common usage of MAVs or MLVs. A natural remedy to these limitations is to use live insects.

Early use of insects as spies relied on attaching surveillance equipment directly to the bodies. Though an intriguing idea, the approach is not practical with most species of insects, largely because the equipment weight overburdens the six-legged spies. Of course continued advances in technology are resulting in increasingly smaller devices that can do more than ever before. Simply look at what smart phones can do today, and at a fraction of the size of the versions of cell phones from just ten years ago. And those phones could function only as phones. So the problem of equipment size is likely to soon be rectified (figure 16.12). That said, one solution is already being developed: piezoelectric generators attached to the wings. The generators actually capture heat energy released by an insect during aerobic metabolism, such as occurs with muscle movement, and then returns the energy to flight muscles in the form of an electric output. The net effect is increased energy for flight, which means the potential to extend the duration of surveillance flights by an insect and also increase the payload that can be carried. This technology is still in the development stages but it has been tested successfully with at least one species of beetle.

If energy-capturing electrical generators are not cool enough technology for you, then how about biological circuits inserted into insect neurons? What sounds completely like something for science fiction television is

Figure 16.12. Evolution of mobile or cell phones from the beginnings of non-land lines until ~2006. Since that time, cell phones are moving in the direction of "bigger is better" as they take on many of the functions of laptop computers and cameras. Improvements in mobile phone technologies are making the concept of insect tracking devices for surveillance more feasible. Photo by Anders at http://bit.ly/1IgKcLy.

actually quite feasible. In fact, it has already been done, and on multiple insect species. Microdissections on any size insect are relatively easy when using the right microscope. Insects also have a small number of neurons extending from the frontal ganglion to innervate other tissues, making it easy to intercept and manipulate messages related to **central motor programs** with implanted neural circuits. Central motor programs control the movements of appendages, like the legs and wings. You might ask, how can this be applied to using insects as spies? Good question! The reason for developing such technology is to be able to manipulate an insect's movements. As you might imagine, one of the weaknesses with trusting insects with high-tech surveillance equipment on their bodies is that they might not come back. However, equipped with neural circuits implanted in the ventral nerve cord and associated ganglia, leg and wing movement can be controlled remotely. Not only does this allow the insect to be recovered once the espionage or surveillance tasks are complete, but it also permits directing the spy to move exactly where needed. There is an almost endless number of uses for surveillance technologies relying on controllable insects.

Bug bytes

Remote-controlled cockroaches

https://www.youtube.com/watch?v=adkJfDaR_FI

An insect for an insect: Entomological counterterrorism

An overlooked topic among insect roles in national security is that of **entomological counterterrorism**, meaning the use of insects to combat entomological terrorism agents. An insect for an insect, if you will. This discussion is more conceptual than focused on how insects are actually used in this capacity, because entomological counterterrorism is theoretical. That said, it is based on concepts from pest management used to deal with insects of agricultural importance. In particular, the use of natural enemies to reduce pest populations has a great deal of potential to combat agents of agroterrorism.

For example, imagine that an entomological attack has occurred toward one of the highly susceptible monoculture crops discussed earlier in the chapter. Let's say that corn has been targeted with a major insect pest. Several responses could be pursued, provided that the insect weapon has even been detected and that it is susceptible to common biocides, which it likely is not. If this was simply a pest outbreak under normal agricultural conditions, then one avenue that might be considered is the use of natural enemies. One type of prime importance is insect parasitoids, most commonly in the form of parasitic wasps. Parasitoids are more selective in host range than predators seeking prey and, as a consequence, are fairly good at host location. In other words, parasitic wasps would be good choices for locating the agroterrorism agent. What value would this have for countering the effect of the introduced pest? At least two benefits could arise from using parasitoids in entomological counterterrorism: (1) the wasps would parasitize the hosts, thereby helping reduce the pest populations; and (2) if wasps are equipped with tracking devices, the precise location of the agroterrorism agent can be revealed, so that more invasive control measures could be implemented. Is the latter even possible? Well, certainly this would be a challenge, because most parasitic wasps are very small. In fact the majority of parasitic Hymenoptera are less than two millimeters in length as adults. However, at least one wasp, *Microplitis croceipes* (Hymenoptera: Braconidae) has been trained successfully as a sniffer system equipped with tracking devices.

There certainly are limitations to using insect parasitoids. For one, though they will parasitize certain insect pests, the reduction in pest populations will not be instantaneous (e.g., some parasitized species continue to eat for several days). Parasitism also does not achieve "control" but instead yields reduction in numbers so that a pest population is lowered by 10% to 30%. Those numbers may seem low, but that is more than enough to reduce a potential pest below **economic injury levels**. Still, a more dramatic reduction is needed when facing an agroterrorism agent, especially if an inundative release has been used to introduce the pest. A particularly important limitation is that natural enemies do not exist for exotic or invasive species used as agroterrorism agents. Thus, a highly effective insect weapon would be an exotic species that is also insecticide resistant. In that scenario, an immediate control strategy is not available, permitting the crop pest to thrive unmolested.

Homegrown entomological weapons

Militant groups living in some far-off land are the only ones who perform terrorist acts—or at least that is what we like to believe. It gives us some level of comfort convincing ourselves that terrorism originates in the Middle East or parts of Africa. The logic makes sense on one level, since such locations are known to be home to several terrorist organizations. Perhaps even during earlier discussions on the use of insects as weapons of war and agents of terror, you formed a mental image of non-US citizens fashioning insects into weapons. Again, to a degree this is understandable. However, the reality is that the United States has been a big-time player in the development of biological weapons in the form of insects. For how long? The answer is not entirely certain, but perhaps

since the American Civil War (1861–1865). The Confederacy accused the Union Army of releasing the harlequin bug, *Murgantia histrionica* (Hemiptera: Pentatomidae) into the South with the goal of disrupting agriculture. This insect attacks a wide range of food crops and, during the time of the war, wreaked havoc on cash crops and the food supply of southern states. Most likely *M. histrionica* arrived via Mexico rather than from a northern introduction, although whether this pest was aided on its journey by Union forces was never established. What this example does reveal is that the idea of using insects as weapons was well appreciated in the United States by the middle of the nineteenth century. Following World War I, the Geneva Protocol (properly called the Protocol for Prohibition of the Use in War of Asphyxiating, Poisonous or other Gases, and of Bacteriological Methods of Warfare) was ratified in 1925 by most of the world powers in an attempt to thwart the growing concern over biological weapons.

Interestingly, insects were not identified as a weapons threat, despite the numerous historic examples available to the framers of the Geneva Protocol. The United States did not ratify the treaty until 1975, largely because of the country's practice of isolationism at the time. In the intervening years, between the introduction of the Geneva Protocol and its ratification by the United States, the country become actively involved in biological weapons development, including those relying on insects. At the onset of World War II, biological weapons development was considered a necessity to keep up with other nations, especially Japan. President Roosevelt proclaimed that the United States would never go on the offensive with chemical or biological weapons, but would retaliate in kind if such weapons were deployed against the country. By 1942, three facilities in the United States were dedicated to exploring biological weapons. Most prominent of these research programs was the one

---

It should be obvious from the limitations identified for agroterrorism that entomological counterterrorism is not appropriate for insects transmitting disease or directly assaulting humans. Insects like parasitic wasps simply do not operate quickly enough or provide massive enough reductions in pest numbers for use in protecting human life.

CHAPTER REVIEW

- Historical perspectives on entomological weaponry
  - Modern warfare does not really conform to any protocols. Why? Because the participants are not

nations per se. Instead, radical religious groups are initiating wars on whomever they deem the enemy. From their vantage point, the Geneva convention does not apply to them. So the rules of war are theirs to write. This includes how to conduct a military campaign, which generally no longer conforms to conventional warfare tactics and traditional weapons. A modern arsenal of weapons is employed that includes biological weapons, many which have been fashioned out of insects.

headquartered at Fort Detrick in Frederick, Maryland. It was here that entomological weapons development reached perhaps the most advanced level in the world. The United States explored the possibility of weaponization of several human and livestock diseases, including insect-borne diseases such as bubonic plague (fleas), typhus (lice), yellow fever (mosquitoes), dengue fever (mosquitoes), Rift Valley fever (mosquitoes), and Chikungunya fever (mosquitoes).

Research at the facility also focused on development of protocols for mass-rearing the insect vectors, maximizing infectivity of these insects, and designing delivery systems for getting the infected insects to the target enemy populations. The last even resulted in field tests of "bug" bomb delivery systems using noninfected insects over parts of the United States. Such initiatives had memorable names like Operation Big Itch (release of Oriental rat fleas), and Operation Big Buzz, Operation May Day, and Operation Drop Kick, which all involved test releases of hundreds of thousands of live mosquitoes over Georgia and Florida. No doubt the citizenry of these southern states would have been thrilled to be used as guinea pigs for blood-feeding mosquitoes, if they had had any clue what was occurring in the skies above them. By the end of the 1950s, the United States was prepared to engage in biological warfare at the highest level, and insects were clearly integral components of the military strategy.

The United States bioweapons arsenal was not strictly focused on human disease. Several agents of agricultural significance were developed. Again, insects were examined for potential roles in these campaigns, including being used as mechanical vectors of plant diseases (i.e., rusts, blasts) and to directly attack cropping systems. The most significant entomological agent for agricultural attack was the Colorado potato beetle, *Leptinotarsa decemlin-* *eata* (Coleoptera: Chrysomelidae). Even by the late 1940s, this generalist beetle was already showing signs of insecticide resistance, which would make slowing its onslaught on several crops a difficult task. There is no direct evidence demonstrating that the United States actually used any of the agricultural weapons or, for that matter, disease vectors, during any military campaign. That said, certainly the country's enemies made accusations during World War II, the Korean War, and the Vietnam conflict. The majority of military documents relevant to the US bioweapons program remain classified at this time and may never become available for public consumption. As a consequence, the exact nature of this country's involvement in entomological weapons development may forever remain a secret. However, what can be said is that a biological weapons arsenal has not been restricted to just our enemies. Several have been homegrown, much closer than most of us ever imagined!

■ Soldiers with six legs were used as weapons by nearly every ancient civilization. Initially insects were selected for their potent venoms. In some cases, their exuded toxins were used to coat arrows or spearheads. More common insect uses centered on designing bombs or grenades out of porcelain, clay, or ceramic. Upon contact with the walls of a fortress or entrenchment, or hard impact with the ground, the bomb would explode, thereby releasing its contents. One can imagine that the bomb's inhabitants were quite agitated if not angry by then, and stung and bit at will.

■ The presence of insects can cause all sorts of irrational behaviors, often leading to a condition termed delusional parasitosis. As you might imagine, the powerful effect of stinging/biting insects on human behavior has been manipulated throughout history into methods of interrogation. In other words, insects have been used in torture techniques. Perhaps the most recognizable form is that depicted in a number of classic western genre movies in the United States, in which the settlers and soldiers are punished for intruding on Native American lands by being buried up to their necks near an ant mound. To aid the process, a sugary mixture was often applied to exposed skin to ensure ants and other hungry creatures would be attracted. Now, if the intent was merely to intimidate or interrogate the prisoners, then the individual might be tied naked to a tree or a pole near aggressive ants or in mosquito-infested areas. A few bites were usually more than sufficient to extract the information desired. One of the more diabolical methods of torture involved tying a captive naked to a raft, then force-feeding the prisoner milk and honey. The net effect of this gustatory concoction was induction of constipation with anal leakage. The fluids would attract flies that in turn would oviposit near the anal opening. The pain from feeding fly larvae would become so unbearable that the prisoner would reveal anything his or her captors wanted to know. The infamous Bukhara Bug Pit of Zindon prison, located in what is now Uzbekistan was also used to torture prisoners. Prisoners were lowered into the pit by rope, and then guards would pour scorpions, rats, sheep ticks, and insects on top of them. The insects selected for this special purpose were actually a type of assassin bug, which produces salivary venoms that are injected into the prey. Salivary venom contains several potent digestive enzymes that work just as well on human tissues as they do on insect, so the effect was chemical digestion of muscle and soft tissues.

- Employing insect weaponry has continued well into the last two centuries. Numerous examples of insect use in war can be cited, from medieval battles to the great world wars to even as recently as the Korean War and Vietnam conflict. Two common threads tie the modern use of insect weapons together. The first is that war draws together large numbers of soldiers who in turn are more susceptible to insect-borne diseases because of the conditions they face: masses of individuals in close contact, stressed, and malnourished. Several insects that transmitted disease took advantage of these conditions during every military campaign prior to World War II to infect the masses of soldiers. The fact that insects harbor deadly pathogens did not go unnoticed. In fact, this knowledge—that insects vector disease and that insects as a whole are relatively easy to manipulate—is the second thread.

- A new class of weapons soon followed suit, in which vectors of potent diseases such as malaria, yellow fever, dengue fever, and plague, or those that could devastate important food crops, were raised en masse to purposely release into targeted enemy areas. Perhaps the most extreme example is the Japanese Imperial Army's covert chemical and biological warfare research and development during the World War II. The goal of the Japanese unit was to develop methods to effectively deliver living fleas to enemy targets. Thus, intense testing began in bomb development using Chinese villages and prisoners of war as test subjects. During the six years that the facility operated, nearly 250,000 men, women and children—mostly Chinese, Korean, and Mongolians—died from the Japanese human experimentation phase of biological weapons development. It is important to note that Japan was not the only nation involved in chemical and biological weapons development during World War II.

Insects as agents of terror

- Biological weapons involve the use of living organisms and/or their products by militant groups or nations to target other countries, organizations, or groups. Traditionally, a more narrow definition had been used to describe biological weapons, limited to microorganisms like bacteria, viruses, or protozoan or other pathogens that specifically induce disease in humans or other animals. Though often over-looked in discussions of biological weapons, animals with six legs have enormous potential for use in war or acts of terror. Obviously there is historic precedent for entomological weapons, as we have discussed earlier. However, entomological weapons, especially in the form of entomological terrorism or entoterrorism, overcome the limitations of microbial weapons and actually provide a broader array of targets. In terms of bioweapons development, thousands of insects can be raised cheaply and quickly—certainly in comparison to pathogenic microorganisms—in a relatively small space and do not require sophisticated rearing or containment facilities.

- Entomological weapons offer a broad spectrum of targets. Insects can be used or manipulated in entoterrorism to directly target human or other animal populations; serve as vectors of a disease targeting animals and plants; and affect food production or agricultural economy. The point is that insect weapons are not restricted to disease transmission and can actually tax infrastructure through spreading disease, diminishing food supply, and weakening the economy.

- Direct attack or assault by entomological weapons implies that the insects used bite, sting, or produce noxious or toxic secretions. It is the aggressive behavior of the insects and/or the toxins they produce that create terror and havoc in the targeted population. Bee bombs used in premodern warfare and torture techniques involving ants were among the precursors to today's entomological weapons. The most effective insects used in this fashion tend to be social Hymenoptera and true bugs that produce nondiscriminate toxins or venoms that, when injected into a human or other animal, produce very painful and long-lasting effects. Pain often results from lysing of plasma membranes and inflammation triggered by histamine release.

- Using insects as vectors of disease is essentially the same as simply developing bioweapons out of pathogenic microorganisms. The difference of course is that the insect serves as an effective delivery system and targets more broadly, including livestock and crops, not just human populations. This method still suffers from the health risks and expense discussed earlier, but insect vectors overcome some of the drawbacks as well. Mosquitoes serve as an excellent example to show the efficiency of using insects as vectors of disease.

- Perhaps the greatest threat of insects to national security is associated with agriculture. Insects have a long history of functioning as pests of crops, livestock, and stored food products. It thus is a small step to move from an accidental or opportunistic pest of agriculture to one that is deliberately released as a weapon of war or terrorism. In this respect, many species of insects are well suited to be agents of agroterrorism. Agroterrorism is the deliberate introduction of animal or plant pathogens, or pests that directly target cropping systems, livestock, or food held in storage after harvest.

Insects as bloodhounds: Sniffer systems

- Insects are naturally gifted at detecting chemicals in the environment. They rely on odors derived from all sorts of living and dead things, such as those emanating from flowers and decaying carcasses, as well as from live animals, to find food. Not only is insect olfaction superior to that of most any other animal, but insects possess hundreds to thousands of sensory receptors positioned in multiple locations on the body. Couple this keen sense of olfaction with the ability of certain insects to be trained to complete a task and you then have the makings of a sniffer system. In the context of insects or other animals, sniffer systems are essentially living entities that can be used for odorant detection. Equip the beast with some sort of tracking device, and you now have the means of following an animal's movement as it is released to search for the odor of interest.

- Insects can be trained in a fraction of the time that a dog can be. A honey bee, for instance, can be trained to associate food with an odor (say, that of unexploded land mine) within minutes to hours, as opposed to the multiple weeks needed for conditioning a canine. Thousands of individuals can be trained simultaneously, and the overall cost is minuscule compared with the cost of using a vertebrate animal. Perhaps most important, flying insects can be sent into hazardous regions that land-based creatures cannot.

- Certain insects, namely social Hymenoptera, are especially suited for the task of free-moving or roaming sniffer systems. Social Hymenoptera possess a highly refined system of chemical communication, equipped with a broad range of chemosensory receptors. They also demonstrate short- and long-term memory, and are longer-lived than most other species. As a consequence, honey bees or wasps will retain the training for a long period of time, making the investment in developing the sniffer system worthwhile.

- Some insects are used as biological sensors, or biosensors, for detecting weapons, explosives, or illicit drugs, particularly in buildings like airports. A biosensor is an analytical device that integrates living organisms or biological materials into a physiochemical detector. The goal is to measure some feature of a particular environment, such as the presence of a chemical, by relying on the biological properties of the organisms or its products. The insect is restrained in a device so that the behavioral response can be observed immediately if the odorant of interest is detected. The whole insect is not even used in some instances. Instead, isolated cells modified to detect specific chemical odorants are contained within boxes positioned in various locations throughout a building. If the desired chemical is recognized, the cell generates an electrical signal, which alarms officials of a possible threat so appropriate action can be taken.

- The simplest method for chemical detection using insects is chemical trapping. The process is almost entirely passive. An insect is released into the designated area of interest, then, after a period of foraging, the individual is captured and the body is analyzed for trapped chemicals. The technique relies on the idea that chemical particles become attached or stuck to the bodies of insects during encounters with living or nonliving objects. In some instances, the hairs covering the body demonstrate electrostatic properties, facilitating the trapping of common particles like pollen, pollutants, and even chemical residues. In the applied world of national security, these properties are useful for monitoring chemicals associated with drugs, explosives, and biological or chemical weapons.

Insect espionage: Cyborgs and surveillance

- Cyborg insects have been used for many years in surveillance operations targeting organized crime and militant organizations, as part of search and rescue missions, and also in efforts to locate explosives and hazardous materials. To avoid detection, the miniature robots, termed micro air or land vehicles, are designed to look like flying or walking insects. To achieve this desired effect, the micro spies are about the size of medium- to large-sized insects. This size permits attachment of high-tech audio and video surveillance equipment that provides instantaneous feedback to the team of investigators. Of course with this advanced technology comes a hefty price tag, which is one of the major limitations to using cyborg technology. A lack of external power to permit continuous operation is another drawback to more common usage of MAVs or MLVs.

- A natural remedy to the limitations of cyborgs is to use live insects. Early use of insects as spies relied on attaching surveillance equipment directly to the bodies. Though an intriguing idea, the approach is not practical with most species of insects, largely because the equipment weight overburdens the six-legged spies. The problem of equipment size is likely soon to be rectified. That said, one solution is already being developed: piezoelectric generators attached to the wings. The generators actually capture heat energy released by an insect during aerobic metabolism, such as occurs with muscle movement, and then returns the energy to flight muscles in the form of an electric output. The net effect is increased energy for flight, which means the potential to extend the duration of surveillance flights by an insect and also increase the payload that can be carried.

- The use of biological or neural circuits allows a human operator to be able to manipulate an insect's movements: neural circuits implanted in the ventral nerve cord and associated ganglia allow leg and wing movement to be controlled remotely. Not only does this allow the insect to be recovered once the espionage or surveillance tasks are complete, but it also permits directing the spy to move exactly where needed.

- **An insect for an insect: Entomological counterterrorism**
  - An overlooked topic among insect roles in national security is that of entomological counterterrorism, meaning the use of insects to combat entomological terrorism agents. This discussion is more conceptual than focused on how insects are actually used in this capacity, because entomological counterterrorism is theoretical. That said, it is based on concepts from pest management used to deal with insects of agricultural importance. In particular, the use of natural enemies to reduce pest populations has a great deal of potential to combat agents of agroterrorism.
  - One avenue that might be considered is the use of natural enemies. One type of prime importance is insect parasitoids, most commonly in the form of parasitic wasps. Parasitoids are more selective in host range than predators seeking prey and, as a consequence, are fairly good at host location. In other words, parasitic wasps would be good choices for locating the agroterrorism agent. Using parasitoids in entomological counterterrorism would lead to parasitism of the hosts and thereby help reduce the pest populations; and, if the wasps are equipped with tracking devices, the precise location of the agroterrorism agent can be revealed, so that more invasive control measures could be implemented.
  - A particularly important limitation is that natural enemies do not exist for exotic or invasive species used as agroterrorism agents. Entomological counterterrorism is also not appropriate for insects transmitting disease or directly assaulting humans.

MUSHROOM FARMING (SELF-TEST)

Level 1: Knowledge/Comprehension

1. Define the following terms:
   - (a) sniffer system
   - (b) biosensor
   - (c) entoterrorism
   - (d) inoculative release
   - (e) entomological counterterrorism
   - (f) micro air vehicle

2. What are some of the biological characteristics of insects that make them desirable for use as biological weapons?

3. What is the difference between a sniffer system, a biosensor, and chemical trapping?

4. Describe the ways that insects have been used historically as weapons of war, torture, and terror.

5. Provide examples of how insects can be used in modern forms of terrorism.

6. What are the advantages and disadvantages of relying on insects for surveillance?

Level 2: Application/Analysis

1. Explain how insects like honey bees are better choices for odorant detection than canines, and under what circumstances they are used.

2. The Geneva Protocol of 1925 banned the development and use of chemical and biological weapons. Is there any evidence that the treaty did or did not work to achieve this desired goal?

3. If tasked with the job of producing the optimized insect agent for agroterrorism targeting the United States, what biological features of the entomological weapon should be selected?

Level 3: Synthesis/Evaluation

1. Explain the steps that would be required to train an insect like a bee or wasp for a role in odorant detection of a concealed human corpse.

2. Describe how insects relying on arrhenotokous parthenogenesis are more useful for operant conditioning than those that depend only on sexual reproduction.

REFERENCES

Aktakka, E. E., H. Kim, and K. Najafi. 2011. Energy scavenging from insect flight. *Journal of Micromechanics and Microengineering* 21:95–116.

Centers for Disease Control and Prevention, Public Health Service, and National Institutes of Health (eds.). 2010. Biosafety in Microbiological and Biomedical Laboratories. 5th ed. US Department of Health and Human Services, Washington, DC.

Chalk, P. 2004. Hitting America's Soft Underbelly: The Potential Threat of Deliberate Biological Attacks against the U.S. Agricultural and Food Industry. National Defense Research Institute, Santa Monica, CA.

Floreano, D., J.-C. Zufferey, M. V. Srinivasan, and C. Ellington. 2010. Flying Insects and Robots. Springer, New York, NY.

Frederickx, C., F. J. Verheggen, and E. Haubruge. 2011. Biosensors in forensic sciences. *Biotechnology, Agronomy, Society and Environment* 15:449–458.

Habib, M. K. 2007. Controlled biological and biomimetic systems for landmine detection. *Biosensors and Bioelectronics* 23:1–18.

Hinkle, N. C. 2011. Ekbom syndrome: A delusional condition of "bugs in the skin." *Current Psychology Reports* 13:178–186.

King, T. L., F. M. Horine, K. C. Daly, and B. H. Smith. 2004. Explosives detection with hard-wired moths. *IEEE Transactions of Instrumentation and Measurement* 53:1113–1118.

Kirby, R. 2005. Using the flea as a weapon. *Army Chemical Review* (July-December): 3035.

Koblentz, G. D. 2011. Living Weapons: Biological Warfare and International Security. Cornell University Press, New York, NY.

Lockwood, J. A. 2009. Six-Legged Soldiers: Using Insects as Weapons of War. Oxford University Press, New York, NY.

Lockwood, J. A. 2012. Insects as weapons of war, terror, and torture. *Annual Review of Entomology* 57:205–227.

Marshall, B., C. G. Warr, and M. de Bruyne. 2010. Detection of volatile indicators of illicit substances by the olfactory receptors of *Drosophila melanogaster*. *Chemical Senses* 35:613–625.

Monke, J. 2007. Agroterrorism: Threats and preparedness. Congressional Research Service Report for Congress Order Code RL 32521, March 12. www.fas.org/sgp/crs/terror/RL32521.pdf.

Monthei, D., S. Mueller, J. Lockwood, and M. Debboun. 2010. Entomological terrorism: A tactic in asymmetrical warfare. *Army Medical Department Journal* (April-June). http://www.cs.amedd.army.mil/dasqaDocuments.apex?type=1.

Rains, G. C., J. K. Tomberlin, and D. Kulasiri. 2008. Using insect sniffing devices for detection. *Trends in Biotechnology* 26:288–294.

Rapasky, K. S., J. A. Shaw, R. Scheppele, C. Melton, J. L. Carsten, and L. H. Spangler. 2006. Optical detection of honeybees by use of wing-beat modulation of scattered laser light for locating explosives and land mines. *Applied Optics* 45:1839–1843.

Rivers, D. B., and G. A. Dahlem. 2014. The Science of Forensic Entomology. Wiley-Blackwell Publishers, West Sussex, UK.

Rotz, L. D., A. S. Khan, S. R. Lillibridge, S. M. Ostroff, and J. M. Hughes. 2002. Public health assessment of potential biological terrorism agents. *Emerging Infectious Diseases* 8:225–230.

Sakar, M. 2010. Bio-terrorism on six-legs: Insect vectors are the major threat to global health security. http://www.webmedcentral.com/article_view/1282.

Tabachnick, W. J., W. J. Harvey, J. J. Becnel, G. G. Clark, R. C. Connelly, J. F. Day, P. J. Linser, and K. J. Linthicum. 2011. Countering a bioterrorist introduction of pathogen-infected mosquitoes through mosquito control. *Journal of the American Mosquito Control Association* 27:165–167.

United Nations Office of Disarmament Affairs. 2011. The Biological Weapons Convention. http://www.un.org/disarmament/WMD/Bio/.

Zufferey, J.-C. 2008. Bio-Inspired Flying Robots. EPFL Press, Lausanne, Switzerland.

THE ENTOMOLOGIST BOOKSHELF (SUPPLEMENTAL READINGS)

Garrett, B. C. 1996. The Colorado potato beetle goes to war. *Chemical Weapons Convention Bulletin* 33:1–2.

Horn, F. P., and R. G. Breeze. 1999. Agriculture and food safety. *Annals of the New York Academy of Sciences* 894:9–17.

Levin, D. B., and G. Valadares De Amorim. 2003. Potential for aerosol dissemination of biological weapons: Lessons from biological control of insects. *Biosecurity and Bioterrorism: Biodefense Strategy, Practice and Science* 1:37–42.

Rose, W. H. 1981. An evaluation of the entomological warfare as a potential danger to the United States and European NATO nations. http://www.thesmokinggun.com/archive/mosquito1.html.

ADDITIONAL RESOURCES

Anthrax investigation in the United States
http://1.usa.gov/1IwgGTB

Biotoxins: bioweapons
http://bit.ly/1M4hpSb

Bug pit of Bukhara
http://abt.cm/1UtdYWl

Did Nazis study insects for use in biological warfare?
http://bit.ly/1M4hlla

Insects could be used as bioweapons
http://abcnews.go.com/Technology/Story?id=99552&page=1

Lockwood, J. A. 2007. Bug bombs: Why our next terrorist
    attack could come on six legs
http://bit.ly/1KwLlmC

National Institute of Justice: Agroterrorism: Why we are not
    ready
https://www.ncjrs.gov/pdffiles1/nij/214752.pdf

Office of Public Health Preparedness and Response
http://www.cdc.gov/phpr/index.htm.

17 Invasive and Elusive

New Insect Threats to the Human Condition

Extinction by habitat destruction is like death in an automobile accident: easy to see and diagnose. Extinction by the invasion of exotic species is like death by disease: gradual, insidious, requiring scientific methods to diagnose.

Dr. Edward O. Wilson (1997)*
Professor Emeritus and Honorary Curator of Entomology, Harvard University

In an era dominated by almost daily advances in science and technology, how is it even remotely possible that new insect threats to the human condition are occurring? After all, molecular biology, genetic engineering, and chemical manipulation have provided an arsenal of previously untapped weapons to control and manage insect pests. Technological advances have yielded surveillance and remote sensing capabilities that allow us to monitor not only human but also insect activities. Right? Not totally. Yes, we are better equipped than ever to combat the challenges associated with six-legged neighbors. However, living organisms are not static creatures, and insects, like other organisms, are in a continual process of evolution. In other words, they adapt and change, sometimes gradually and in other cases relatively rapidly. What once worked effectively to manage pest populations can nearly overnight be rendered ineffective toward critters with high genetic variability and rapid reproductive rates. Couple these biological traits with modern realities like globalization of trade and long-term climatic changes, and new issues arise with insects. How so?

Insects are notorious for hitching a ride on man-made or natural objects. As a consequence, the introduction of nonindigenous species—so-called invasive or exotic species—into new regions of the world creates pest problems that did not exist previously. As global trade increases, new opportunities are created for insects to be dispersed and hence become problematic. Opportunities also abound for insect invasions following natural disasters, as competing species and natural enemies often do not recover from fire, hurricanes, tornadoes, or other severe climatic or natural events the way many insect species are able to. Global warming represents a complete unknown in terms of how animals have responded and will respond, although one would be safe to predict that insects will adapt more quickly than man will. Evidence already exists that several species are taking advantage of changing weather patterns, maximizing pest opportunities that were limited before or achieving the status of pest in regions in

---

*From the foreword to *Strangers in Paradise: Impact and Management of Nonindigenous Species in Florida,* by D. Simberloff, D. C. Schmitz, and T. C. Brown.

which they were relatively benign. Chapter 17 examines new insect threats to the human condition, deciphering not only the biology of the insects and their effects on man, but also exploring the factors that have contributed to their achieving pest status. A discussion also focuses on how human intervention has accelerated or augmented the pest status of some species, and what can be learned to prevent a repeat in history. Global warming falls under the umbrella of human intervention and is examined in terms of how insect populations are changing in response to climatic change and what this means to the functioning of ecosystems.

Key Concepts

+ How can there be new threats today?
+ Same old story. Accidental introduction
+ Effects of human interference: Nonaccidental introductions
+ Insect activity following natural disasters
+ Global warming, climate change, and insects

How can there be new threats today?

The world we live in is dynamic and actively changing. Very few places on Earth look exactly as they did twenty or even ten years ago. The neighborhood you grew up in probably little resembles today what it looked like when you started elementary school. For me, I grew up in rural Indiana in a town that had a population just above forty thousand around 1975. A few months ago, I returned to visit family only to discover a population explosion. The town (and surrounding suburbs) now claims upward of one hundred thousand residents; entire neighborhoods exist where cornfields dotted the landscape; even new major highways have been built that in turn promoted development of strip malls, restaurants, gas stations, and every other type of infrastructure conceivable to support the expansion of the city. My childhood home is nearly unrecognizable and no longer resembles John Mellencamp's* homage to small town America (http://bit.ly/1wYNUKq). Rural Indiana is but a microcosm of the entire nation and, really, most regions of the world. What this type of growth

---

*John J. Mellencamp is an American-born singer, songwriter, musician, and artist best known for his heartland-themed music that emphasizes traditional instrumentation.

represents more than anything else is an increase in the human population as a whole, although to say "increase" is downplaying reality. It is more correct to say that the human race has undergone massive population growth.

Estimates of the human population in 1950 were 2.5 billion people. In just forty years that number doubled, and by the end of the twentieth century, six billion humans occupied the planet. In some parts of the world, an uncontrolled, nearly exponential rise in the human population has been experienced. On present course, the human population is expected to reach eight to twelve billion by the close of the twenty-first century (figure 17.1). Suffice to say, "massive" is the right adjective to describe the growth that has occurred. Perhaps this human geography lesson seems out of place to you in an entomology textbook. You at least are questioning what any of this information has to do with insects. It is even entirely possible that you are not impressed or concerned by the numbers relating actual growth of the human species with anticipated growth fifty to eighty years from now. After all, in chapter 1 we made comparisons between insect and human populations, and ours is but a blip in comparison to those creatures with six legs. Undeniably true. But humans are far and away more destructive to ecosystems than insects, regardless of the fact that insects dominate in terms of species diversity and abundance. In fact, human activity is the major contributing factor leading to new insect threats to our own populations.

Over the past several decades, insects have become pests in regions where previously they were not. In other cases, the timing or severity of pest status has changed. Why have these changes occurred? Much of the reason is tied to human population growth. An immediate link between the two seemingly unrelated events may not be obvious, but there is indeed a causal relationship. For one, the rapid rise in the human population is at, if not exceeding, the **carrying capacity** of the Earth's ecosystems. Land, water, and all other natural resources are being manipulated, in some cases depleted, at an alarming rate to satisfy the growing demands of an expanding human existence. For example, forested areas in tropical zones have been cleared to raise food and provide homes for the increased population. Not only are harsh methods

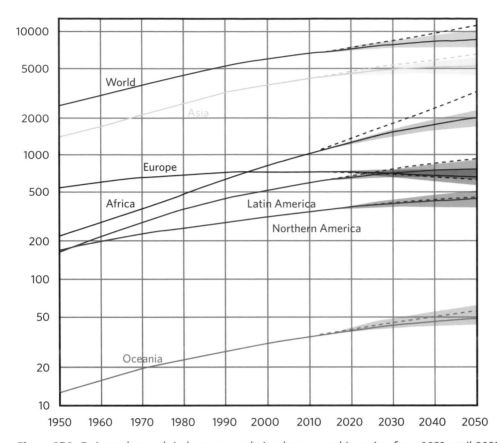

Figure 17.1. Estimated growth in human population by geographic region from 1950 until 2050. The x-axis is a logarithmic scale representing millions of people. Figure created by Conscious at http://bit.ly/1OwBFeY.

(e.g., **slash-and-burn**,* brute force) used to remove trees and vegetation, but the lands are now dedicated to cultivation of nonnatural agricultural and livestock herding, often including the introduction of nonindigenous crops to the region.

The results of modifying existing ecosystems for artificial functions are potentially devastating and irreversible: taxing freshwater supplies, altering associated marine environments, losing biodiversity, releasing carbon dioxide from storage sinks, destroying other natural resources, and creating new niches for insects. The

last example means that species that were not previously pests now have opportunity; ones that achieved pest status infrequently or almost never became severe threats are regularly evoking damage beyond established **economic thresholds** or injury levels; and exotic species arrive to take advantage of the introduced agriculture. An **exotic** or **invasive** insect species is not native to a particular region and, once introduced, begins to spread or increase its range, causing damage to the environment and/or to humans. As we discussed in chapter 14, nonindigenous species can be some of the most difficult insects to manage once firmly entrenched in a region.

Globalization of the world's marketplaces and climate change linked to **global warming** are factors contributing to the rise of new insect threats. Both topics can be relatively controversial depending on your political or even religious ideologies, and each is directly linked to human population growth. Though the twenty-first

---

*Slash-and-burn is a technique used to cut down trees and other plants on a plot of land and then setting it and any remaining vegetation a blaze. The ash and debris is then used as nutrients worked into the soil, so that crops can be grown or livestock maintained on the clear plot of land. After a short period of time, the land is abandoned so that succession occurs to reforest the area. The entire process from clearing to succession is also termed **shifting cultivation**.

Figure 17.2. Repeat glacier photography is a technique used to monitor changes in glacier appearance over time. Historic photos of glaciers are used as a point of reference, so that new photos can be taken at the same place and time of year for comparison. Photo of Iceberg Lake by GlacierNPS (http://bit.ly/1W3IF5q).

century is still in its infancy, it is clear that this period of human history can be characterized by globalization of economies. From an entomological point of view, this equates to new trade partners and the potential for insects to travel with crops, or on or in freight containers, ships, planes, or any other conveyance exchanging goods between countries. Introduction of exotic species has occurred throughout history, but generally on a small scale; in the past, few invasive species were accidentally introduced at any given time. Enhanced trade, simultaneous with many new global partners, has the potential to create widespread accidental releases of new insect species into the United States, or from North America to other nations. Part of the fuel behind seeking new global partners is that nations need resources to feed, clothe, and house their burgeoning populations. The nuclear arms agreement in 2015 between the United States and Iran—two nations locked in ideological and religious conflict for more than three decades—has led to wide-scale discussions of new economic opportunities for US businesses in the middle eastern nation. Again, the result is quite likely to be the introduction of exotic species, some of which will establish as econom-

ically or medically important invasive insects. Remember from our discussions of entomological terrorism in chapter 16 the difficulties in detection and control once such species beachhead in the United States.

The climate is changing. Although not everyone is in agreement that this is true, especially with regard to global warming, some solid pieces of evidence cannot be argued against, at least not intelligently. For example, the surface temperature of the Earth has elevated by 0.6°C since the beginning of the twentieth century. While that increase seems like it could be viewed as insignificant or modest, it has contributed to the melting of centuries-old sea ice and glaciers. Significant glacial receding is occurring in New Zealand, Africa, India, and at the polar caps (figure 17.2). A simple comparison of photos taken twenty years ago with those from today shows that melting of glacial ice has occurred. That evidence is indisputable. What can be contested are reasons or causes for the temperature increases leading to the melt. Massive clearing of forested lands in the tropics and destruction of coral reefs along heavily populated regions is known to release carbon dioxide, one of the gases that facilitate the so-called **greenhouse effect**, into the atmosphere, which in turn is believed to contribute to global warming.

Later in this chapter, we delve further into features that contribute to global warming and the counterarguments that this phenomenon is actually not occurring. To begin, we focus on, what effect would climate change caused by warming temperatures have on insect populations? Several entomological changes are anticipated.

1. Expanded range or distribution of several insect species.
2. Decreased survivorship of some species.
3. Increased opportunities for other species as ecological niches are vacated and new habitats created.
4. Modified life cycles, such as shifting from a single generation (**monovoltine**) each year to multiple generations (**univoltine**), or averting dormancy altogether.

Certainly many other alterations to insect life histories can occur and will not be fully realized until or if global warming continues to manifest its effects on climatic conditions. Regardless of the extent of life history changes, what are the implications for the human condition? Such alterations mean new agricultural pest situations, including new insect pests; extended range for disease vectors; and possibly more or new insect species invading human habitation, in part seeking food, since many plant species and other food sources will decline with global changes in climate. Suffice to say, human influence is leading to wonderful opportunities for insects to annoy, attack, or threaten us. In the coming sections, we explore in more detail some of impending new threats from insects toward the human condition, attempt to decipher the causes, and also discuss whether there are means to prevent the scenarios from playing out or to deal with the insects once pest status is realized.

Quick check

How is human population growth increasing opportunities for insects?

Same old story: Accidental introduction

The United States is known worldwide as the land of opportunity. This ideal or dream has inspired millions of people from across the globe to seek out opportunities or entirely new lives by migrating to North America. In fact, the very identity of this country is derived from the millions of immigrants who have made the United States their home. The largest period of human transference to this country occurred from 1830 to 1914, during which over thirty million Europeans migrated to the United States Peak migration came in 1907, with just over 1.2 million individuals arriving during that one year alone. Throughout history the United States has embraced all people, but the reality is that uncontrolled migration is not sustainable for any nation. Numerous attempts have been made to contain the flow of immigrants into the country, including even today, with rising concern over illegal migration from Mexico into the southwestern portion of the United States. Most efforts at enforcing legal restrictions to immigration or even employing physical barriers have yielded only modest success in curtailing the human influx. As human populations grow exponentially in many third world and developing nations, the pace of attempted immigration will only increase.

Insect immigrations—meaning accidental introductions—into the United States parallel closely the path of human migration. This implies a strong link between the origins of invasive species and mass movement of people from one region of the world to another. Why does this occur? Human immigrants generally do not travel alone, as numerous nonnative or alien species of plants, animals, and microorganisms—whether intentionally or accidentally—join the two-legged migrants on their journeys. Insects travel in food, luggage, and other goods carried by human migrants. In the nineteenth century, the majority of immigrants, and their alien travel companions, came from European nations. Coincidentally, two major insect pests, the European corn borer, *Ostrinia nubilalis* (Lepidoptera: Crambidae) and the gypsy moth, *Lymantria dispar* (Lepidoptera: Erebidae)* were introduced to the United states during the same period of time. By 2010, the tide of human immigration had shifted away from Europe (figure 17.3).

---

*The European Gypsy moth is believed to have been imported to the United States on purpose but was accidentally released in Massachusetts in 1869. However, undoubtedly other introductions occurred as a result of mass European immigration during this same window of time.

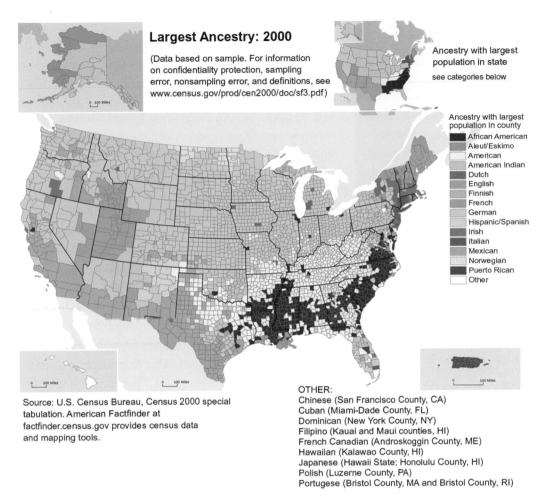

Largest Ancestry: 2000

(Data based on sample. For information on confidentiality protection, sampling error, nonsampling error, and definitions, see www.census.gov/prod/cen2000/doc/sf3.pdf)

Ancestry with largest population in state

see categories below

Ancestry with largest population in county

- African American
- Aleut/Eskimo
- American
- American Indian
- Dutch
- English
- Finnish
- French
- German
- Hispanic/Spanish
- Irish
- Italian
- Mexican
- Norwegian
- Puerto Rican
- Other

Source: U.S. Census Bureau, Census 2000 special tabulation. American Factfinder at factfinder.census.gov provides census data and mapping tools.

OTHER:
Chinese (San Francisco County, CA)
Cuban (Miami-Dade County, FL)
Dominican (New York County, NY)
Filipino (Kauai and Maui counties, HI)
French Canadian (Androskoggin County, ME)
Hawaiian (Kalawao County, HI)
Japanese (Hawaii State; Honolulu County, HI)
Polish (Luzerne County, PA)
Portugese (Bristol County, MA and Bristol County, RI)

Figure 17.3. Distribution of immigrants from different regions of the world throughout the United States. Many instances of invasive species introductions are correlated with human migrations into a particular geographic region of the United States. Image created by United States Census Bureau (http://www.census.gov/prod/2004pubs/c2kbr-35.pdf).

Instead, most migrations stemmed from parts of southeast Asia, several eastern and western Caribbean nations, and Mexico. Correspondingly, many of the most recent accidentally introduced insect species that have become invasive species in the United States are derived from these same regions of the world.

For example, the brown marmorated stink bug, *Halyomorpha halys* (Hemiptera: Pentatomidae); Asian longhorn beetle, *Anoplophora glabripennis* (Coleoptera: Cerambycidae); and emerald ash borer, *Agrilus planipennis* (Coleoptera: Buprestidae) represent recent insect immigrants from Asia, while the sugarcane lace bug, *Leptodictya tabida* (Hemiptera: Tingidae), and West Indian cane weevil, *Metamasius hemipterus* (Coleoptera: Curculionidae) arrived in the United States fewer than twenty

years ago from the Caribbean (figure 17.4). The point is that many introductions of insects into the United States are directly tied to human immigration. As a consequence, the threat of new invasive species into the United States is ever present as long as the nation permits sizeable migration and individuals are arriving from regions of the world with insects that can potentially thrive in areas at or near the ports of entry. Isolationism, or essentially the practice of closing our borders, is not an option as it once was in the early part of the twentieth century. No, globalization, especially in terms of national economy, is the norm today, so the borders of the United States must be more 'open' than ever before.

The trend toward globalization in all aspects of trade and our economy is another major factor contributing to

Figure 17.4. An adult female Asian longhorn beetle, *Anoplophora glabripennis* (Coleoptera: Cerambycidae) chewing on a red oak. Photo by Dean Morewood, Health Canada, Bugwood.org.

accidental introductions of insects. In reality, globalization means more than simply an increase in the number of trade partners for a given country. Corporations in the United States have increasingly adopted an international focus, in which **marketing globalization** is the business model. Marketing globalization combines marketing and selling goods with an integrated global economy. In essence, the business is nationless in terms of consumers, targeting its products to meet the specific needs of citizens of many nations, beyond the country where the company is physically home. Such business models have the potential to enormously expand the range of export for a particular company. The United States is home to literally thousands of such corporations. Similarly, the country is now targeted by thousands of businesses not based in North America. As a consequence, import and export of goods from the United States is at an all-time high and expected to continue to rise. What does this mean in terms of invasive insects? Exactly what you would guess: an anticipated sharp increase in the number of exotic species entering the United States. Importation of goods, along with the associated shipping

materials and transport vehicles (e.g., ships, planes) is the most common means for accidental introductions of invasive insect species into North America, or any nation. And it is not new. The rise of the global marketplace simply magnifies what has occurred throughout history. For example, the silverleaf whitefly, *Bemisia tabaci* (Hemiptera: Aleyrodidae), was introduced from India, most likely on tropical flowers or Christmas poinsettias. It is now considered invasive throughout most of the United States (figure 17.5).

A similar story occurred with the brown marmorated stink bug, *Halyomorpha halys* (Hemiptera: Pentatomidae), which entered the United States via Allentown, Pennsylvania, in 1998, apparently stowed away in packing crates from either Japan or China. The insect is now firmly established as an invasive pest of a wide range of crops and is currently spreading its range westward and south in the United States. Introduced species are not always potential agricultural pests. The Asian tiger mosquito, *Aedes albopictus* (Diptera: Culicidae) is an important vector of several human pathogens and arrived in used tires shipped from Japan (figure 17.6). The adults of this species are ferocious blood-feeders, actively seek out hosts during the day and night, and live in close proximity to human habitation.

Equally threatening to the health of humans and others animals is the red imported fire ant, *Solenopsis invicta* (Hymenoptera: Formicidae), which has brought havoc to nearly every living organism it has encountered since its importation from South America in the early part of the twentieth C (figure 17.7). It is clearly an immigrant that we could do without. These examples represent just a small fraction of the thousands of nonindigenous insects that have been accidentally introduced. Table 17.1 lists of some of the most economically or medically important invasive species to be introduced into the United States.

Now that we have an understanding of how accidental introductions of insects occur, it is time to examine why all the concern. After all, importation of a handful of exotic insects seems insignificant to a nation with an abundant supply. Right? Actually it is not so much the numbers of species that get accidentally brought into the country that raises the concern, although certainly the odds of trouble increase as the number of accidental re-

Figure 17.5. Damage to melons by feeding activity of the silverleaf whitefly, *Bemisia tabaci* (Hemiptera: Aleyrodidae). The insect is an invasive species that was accidentally introduced from India. Photo by David Riley, University of Georgia, Bugwood.org.

Figure 17.6. An adult female Asian tiger mosquito, *Aedes albopictus* (Diptera: Culicidae), blood-feeding on a human host. This mosquito is a medically important invasive species accidentally introduced into the United States. Photo by James Gathany at phil.cdc.gov/phil.

Figure 17.7. Distribution of the red imported fire ant, *Solenopsis invicta* (Hymenoptera: Formicidae) in the United States as of 2006. The ant was accidentally introduced into the United States via Mobile, Alabama, and now resides in all southern states from coast to coast. Image by Strongbad82 at http://bit.ly/1FNMd29.

leases happen more frequently with more species. The issue is *which* species are introduced. Not every organism that travels to the United States is equipped with the life history traits needed to thrive, even survive, in the region of importation. As a consequence, the vast majority of accidental introductions either die out in a relatively short period of time or establish as minor players in local ecosystems. Winter conditions in the northern states prevent species derived from tropical environments from

becoming established, and the extreme heat of the south and arid conditions found in the southwestern regions of the United States are equally effective barriers to exotic insects from regions with much different climatic conditions. The point is that for an exotic insect to become invasive, it needs to be introduced into a region in which its life history characteristics are compatible with the climatic conditions. Unfortunately, this pairing occurs far too frequently in the United States.

Once established, invasive insect species cause damage to agriculture, human habitation, materials or goods

Table 17.1. Common exotic insects that have established as invasive species in the United States. Information derived from the US Department of Agriculture's list of invasive insect species (http://www.invasivespeciesinfo.gov/animals/main.shtml).

Common name	Scientific name	Order	Family
Africanized honey bee	Apis mellifera scutellata	Hymenoptera	Apidae
Asian long-horned beetle	Anoplophora glabripennis	Coleoptera	Cerambycidae
Asian tiger mosquito	Aedes albopictus	Diptera	Culicidae
Cactus moth	Cactoblastis cactorum	Lepidoptera	Pyralidae
Emerald ash borer	Agrilus planipennis	Coleoptera	Buprestidae
European corn borer	Ostrinia nubilalis	Lepidoptera	Crambidae
European gypsy moth	Lymantria dispar	Lepidoptera	Erebidae
Formosan subterranean termite	Coptotermes formonsanus	Isoptera	Rhinotermitidae
Hemlock woolly adelgid	Adelges tsugae	Hemiptera	Adelgidae
Khapra beetle	Trogoderma granarium	Coleoptera	Dermistidae
Kudzu bug	Megacopta cribraria	Hemiptera	Plataspididae
Mediterranean fruit fly	Ceratitis capitata	Diptera	Tephritidae
Mexican fruit fly	Anastrepha ludens	Diptera	Tephritidae
Oriental fruit fly	Bactrocera dorsalis	Diptera	Tephritidae
Pink bollworm	Pectinophora gossypiella	Lepidoptera	Gelechiidae
Pink hibiscus mealybug	Maconellicoccus hirsutus	Hemiptera	Pseudococcoidae
Red imported fire ant	Solenopsis invicta	Hymenoptera	Formicidae
Silverleaf whitefly	Bemisia tabaci	Hemiptera	Aleyrodidae
Sirex woodwasp	Sirex noctilio	Hymenoptera	Siricidae

of importance to humans, or cause disruption to the immediate environment. The most obvious effect of invasive species is alteration of **biodiversity**. In this context, biodiversity refers to all the different organisms that exist in a particular ecosystem influenced directly by the invasive insect species of interest. One possible outcome for an alien insect species is that it reduces or outcompetes other species in the environment, potentially displacing indigenous species. However, invasive species may also alter trophic interactions within an environment in such a way that populations of indigenous species now flourish. This in turn directly or indirectly influences the population dynamics of other species, not just insects. What is set in motion is a cascade of outcomes that changes the delicate balance of a particular ecosystem. The potential effects of an invasive species invasion are almost limitless. That said, the effects are categorized based on relatively predictable outcomes. For example, an alien insect will generate either direct or indirect effects on its new environment. Direct effects include predation, parasitism, or herbivory by the invasive species. In contrast, indirect effects are more commonly manifested as interactions with organisms in which the alien insect competes for resources or otherwise influences, but not directly, the life cycle of another species. In practical terms, alien or invasive species exert their influence through

1. Predation/parasitism of native species of insects and other organisms (direct)
2. Competition for resources (direct or indirect)
3. Triggering a loss of pollinator species (direct or indirect)
4. Herbivory (direct)
5. Disease transmission (direct)
6. Replacement of species in niches or habitat (direct or indirect)

This is just a partial list of some of the effects that invasive species can evoke on an ecosystem once they become established. Any of the examples given reflect mechanisms for disruption or loss of biodiversity from a given ecosystem. They are also means for reaching pest status. Now consider that if the avenues that promote accidental introductions are expanded, namely in the forms of enhanced global importations of goods and people, the likelihood that multiple invasive species will be introduced simultaneously, or nearly so, will increase, and in turn magnify the disruptions to ecosystems and the complex species communities and interactions that occur within them. Ultimately, the effects on the human condition could be enormous.

The unfortunate reality is that this nation is not really prepared to deal with an increase in alien species arrivals in the United States. The reasons are basically the same as those discussed in chapter 16 when examining entomological agents for agroterrorism: a lack of resources and funding to develop the methods of detection (surveillance) and control of potential insect threats. In the ideal world, detection of exotic insects occurs before the infested goods enter the country (at the point of origin). Currently, if discovered at a point of entry, the insect

containing materials are confiscated, and either placed in quarantine or destroyed. This approach has been effective at reducing some alien species from being distributed in the United States. However, at least as many species escape detection, and the introduced species become unwelcome insect citizens of North American ecosystems.

Effects of human interference: Nonaccidental introductions

Exotic insects are not always opportunistic in their path to finding the United States. In many instances, they have been deliberately introduced. Nonaccidental release of alien species is generally for one of two reasons: (1) as part of an act of terror or war in which the exotic species is being used as an entomological weapon, or (2) the insect is being introduced as part of a biological control program targeting an invasive organism (the pest is not always an insect) from the same region as the control agent (i.e., the alien insect). The concept of insects as weapons and agents of terror was discussed in great detail in chapter 16. It is important to note that the same effects as discussed earlier for accidentally introduced exotic species apply to those deliberately introduced. Perhaps the major difference is that with agents used in entomological terrorism, the species employed are selected based on the characteristics that essentially guarantee success; that is, the life history traits are ideally suited for the climate and region released (figure 17.8).

By contrast, insects used for biological control are introduced with good intentions. The practice is based on the idea that invasive pests, especially those that are herbivorous on agriculturally important crops, initially thrive upon introduction to a new region because there is a lack of natural enemies to keep the insect population in check. Rather than simply rely on a management strategy in which chemical biocides are used—which creates a whole series of health and environmental issues—natural enemies in the form of predators and parasites are released to suppress the pest population. Generally such natural agents of control, termed **biological control** agents, are collected from the native land of the invasive species causing problems, raised en masse in a laboratory, and then released into the problem areas in the United States. The initial application of the alien species is

Figure 17.8. Color variants of the Asian lady beetle, *Harmonia axyridis* (Coleoptera: Coccinellidae). The beetle was deliberately introduced by the USDA as a biological control agent. Photo by Louis Tedders, USDA Agricultural Research Service, Bugwood.org.

usually in the form of an **inundative release**, in which a large number of the predators or parasites are released to lower the pest population. In many instances, additional or **augmentative releases** are needed to supplement the biological control agent's population to help it drive the pest below economic thresholds or injury levels and to aid in establishing a permanent population of the natural enemy.

Biological control deliberately introduces exotic insects to combat other exotic species. Does the strategy work? More than two thousand species of exotic insects and other arthropods have been deliberately introduced in nearly two hundred nations. The result has been a permanent reduction in 165 invasive species. So on one hand, the answer is yes, introduction of exotic species to control invasive species has worked. At least in some scenarios, biological control has been a success. Often, however, the outcomes are mixed: some reductions in pest numbers occur, but not to the extent that the invasive species is driven below pest status. That fact in itself is not reason to abandon biological control strategies. The issue that arises from release of exotic species is that they potentially pose the same problems as accidentally introduced insects. When using predators and parasites, there is always the risk that biological control agents will expand their range. In other words, the natural enemies do not restrict their activities to our intended targets and instead follow their own desires: they prey upon native species. In some cases, the nontarget species were endangered and on the verge of extinction. In others, the biological control agents actually attacked other exotic species that were introduced for control efforts toward a different pest. The reality is that there are now more exotic species in the United States released deliberately as biological control agents than there are accidentally introduced invasive species.

What is truly surprising is that, in the past, very little risk assessment has been required prior to the introduction of an exotic biological control agent. Essentially a good number of the attempts using exotic agents have relied on trial and error rather than solid empirical testing. Thus, when an insect agent goes rogue, everyone is surprised! That approach is thankfully being abandoned. A full risk assessment is now mandated in the United States, including isolation in quarantine before any exotic insects can be used in small field trials. Obviously the hope is that we have learned from past mistakes and now will minimize future threats from alien species deliberately introduced. Another option that is being examined is to rely more on native species for biological control.

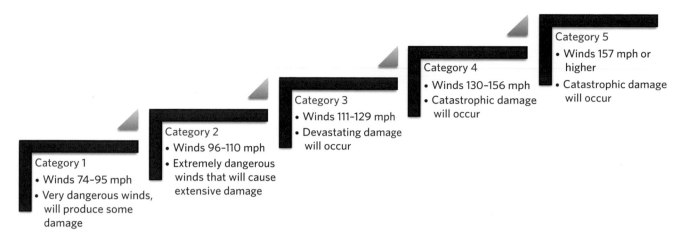

Figure 17.9. Saffir-Simpson Hurricane Wind Scale. The scale is based on sustained wind speed. Hurricanes reaching category 3 are considered major hurricanes because of the potential of loss of life and significant property damage. Further details are available at http://www.nhc.noaa.gov/aboutsshws.php.

While this option is appealing for many reasons, it is still limited in that indigenous insects often display no or only limited interest in an exotic species to use as prey or a host.

Beyond the text

What characteristics of exotic biological control agents should be selected to avoid negative effects on native species?

Insect activity following natural disasters

Natural disasters occur in every geographic region of the United States. Hazards in the form of earthquakes, landslides, forest fires, floods, and severe weather patterns (e.g., tornadoes, hurricanes, and blizzards) threaten human lives and cause huge economic losses annually. The events occur frequently, depending on climatic conditions, geography, and differences in topography. For instance, the National Oceanic and Atmospheric Administration (http://www.noaa.gov/), the agency in the United States that monitors weather patterns in the region, predicted that in 2015, six to ten named storms* would occur in the portion of the Atlantic Ocean affect-

ing the United States, of which four would develop into hurricanes, and one would become a major hurricane threat, meaning classified as at least category 3 status (figure 17.9). The predictions were not far off (eleven named storms, four that became hurricanes, and two of those developed into major hurricanes) and were consistent with the thirty-year average for the region, in which more than ten named storms typically occur, with four developing into hurricanes. Most storms do not make landfall, but when they do, significant wind, water, and flood damage occurs.

By contrast, the west coast of the United States rarely experiences hurricanes (large tropical cyclones), largely because strong **wind shear** (gradients in wind speed and/or direction within the atmosphere) coming from the north Pacific prevents tropical storms from intensifying into full-blown hurricanes. Parts of California more than make up for their lack of climatic natural disasters through a torrent of major and minor earthquakes and forest fires. Earthquakes actually occur throughout North America but have been most severe along fault lines (e.g., San Andreas) in California and Nevada. The west coast is also susceptible to tsunamis, large tidal waves that develop as a result of major offshore earthquakes (figure 17.10). Periods of extreme and long-term drought plague several regions from the Pacific coast to the Great Plains, which in turn is the major contributing factor to large scale forest fires.

Between the coasts, fires and tornadoes are also common natural disasters that occur during warmer

---

*The practice of naming storms is done more for convenience to weather reporting agencies, who communicate with each about storm systems that may last as long as a week. In the Atlantic Ocean, a tropical storm receives a name once winds are sustained at 34 knots or 39 miles per hour.

Figure 17.10. High-level clouds showing wind shear, with changes in wind speed and direction. Photo by Fir0002 at http://bit.ly/1KhShSu.

months, and blizzards occur quite commonly in December through February in the northern Midwest (lake-effect snows) and northeastern states (commonly resulting from nor'easters) (figure 17.11). This list of natural hazards is not exhaustive, but it does serve to show that such events are not rare for North America. What they all have in common is that each tends to affect a large geographic area all at once or in a very short period of time. The effects of such disasters can be enormous. They can also serve as great opportunities—that is, if you are small and possess six legs.

Natural disasters for humans are opportunities for insects. Not for all species, of course; some insects suffer the consequences of being in the path of a natural disaster. Flooding is more deadly for nonflying species than it is for those with wings and, certainly, for humans. The

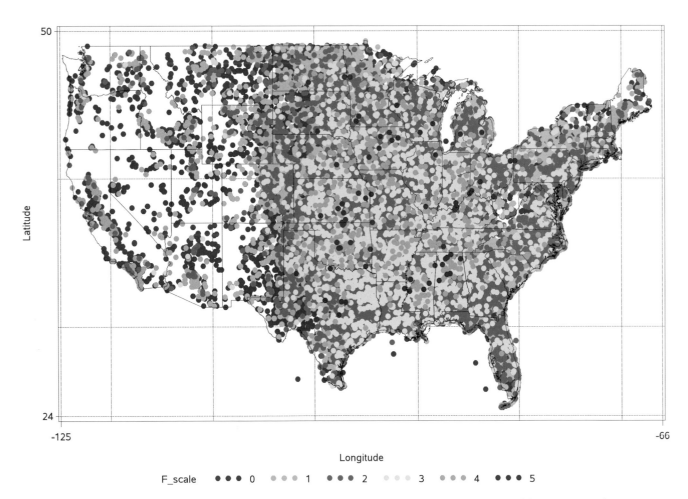

Figure 17.11. Occurrence of tornadoes in the United States from 1950 to 2013. The most intense and frequent tornado activity occurred in the eastern Great Plains through the Midwest. Image by Tertius51 at http://bit.ly/1YdHnXw.

same can be said for fire, volcanic eruptions, and most any other natural disaster. Landslides and avalanches are exceptions since, for the most part, they occur unpredictably. Natural disaster preparedness is one area where humans—who, if they heed warning signs and alerts for predictable natural hazards, can escape immediate harm—have a clear advantage over the class Insecta. Insects, by contrast, do not get text alerts from the National Weather Service warning of impending danger. Arguably an exception is with blizzards and avalanches, since insects are on the same frequency with Mother Nature and avoid being active during winter altogether.

Some species take advantage of human misery during times of natural disaster. They do so specifically by using the event for dispersal opportunities or to transmit disease. Some may even augment the effect of the natural disaster while it is occurring. For instance, during Hurricane Katrina, the floodwall system in New Orleans failed, causing extensive flooding throughout the city. New Orleans sits below sea level at the Gulf Coast, with the city designed, essentially, like a soup bowl. The floodwall system was built to keep floodwaters from flowing down into the city (figure 17.12). During the hurricane in 2004, the failure of the levees was believed to have an insect-related cause. Much speculation has been directed toward the Formosan subterranean termite, *Coptotermes formosanus* (Isoptera: Rhinotermitidae) as a culprit in weakening portions of the floodwalls through their normal feeding behavior. The seams of the levees were (and still are) composed of sugarcane waste, which is known to attract the termites. More than 70% of the seams examined showed some level of termite damage, which at the very least compromised the integrity of the floodwalls. The reality is that this threat still exists today, so that another major hurricane to that region could result in similar devastating flooding.

Climate-induced disasters can dramatically alter insect populations within a given region. Strong winds associated with cyclones, hurricanes, tropical storms, nor'easters, and even tornadoes are capable of transmitting small-bodied arthropods, whether or not they possess wings, from one location to another. Numerous reports from the seventeenth and eighteenth centuries have suggested that hurricanes were directly responsible for colonization of many Caribbean Islands with insects

Figure 17.12. Hurricane Katrina as a category 5 storm, making it one of the strongest storms ever to strike the United States. The devastating effects of this hurricane were augmented by opportunistic insects. Photo by Jeff Schmaltz, MODIS Rapid Response Team, NASA/GSFC (http://visibleearth.nasa.gov/view_rec.php?id=7938).

that in turn established themselves as pests of sugarcane, people, and other island inhabitants. Wind has never been reported to induce localized extinction of a species, but the same is not true for the flooding that occurs with many of these major storms. Heavy precipitation, storm surges, and intense wave action are often associated with cyclones, hurricanes, tropical storms, and nor'easters, leading to flooding. As mentioned earlier, if you're an insect that cannot fly, flooding is bad news, especially for subterranean species.

Hurricanes are well known to have changed the pest status of many insects in regions of the Caribbean as well as along the Atlantic coast. As expected, some pest species have been driven (or drowned) from islands and other regions. And of course, with new vacancies in ecological niches, other species move in to fill the void, potentially becoming pests for the very first time. Floodwaters are also instrumental in redistributing organisms from one location to another. Whether this occurs by wind or water, the movement of insects in this manner represents mechanical dispersal, which is aided by natural disasters. In the case of flooding, most species of insects can float because the exoskeleton has hydrophobic properties, and they have a very small mass. As a consequence, insect bodies do not break the surface tension of water, and thus many can walk or float on top of water. If the insect species can survive the threat of drowning or physical damage during the flooding event, the waters will transport any such species to a new location. Once the floodwaters recede, the insects begin to establish themselves in a new region. If the insect is already indigenous to the area, then little change in the ecosystem may be expected. However, if they are in a new region or possibly even a new habitat, the introduced insect is now considered an exotic that may well become a problem species—invasive by achieving pest status. All the problems with invasive insect species discussed earlier in the chapter also apply to insects introduced by natural disasters. Fly Spot 17.1 discusses how flooding from storms has actually been an asset for the red imported fire ant, *Solenopsis invicta,* spreading its range throughout the southern United States.

In the wake of Hurricane Katrina, the Internet was abuzz with concerns over new insect infestations caused by the cleanup process. The apparent threat was actually not a threat at all, but nonetheless, a story circulated that downed trees from Louisiana were chopped into mulch and then sold throughout the country. One supposed consumer reported online that Formosan termites were present in the mulch purchased in a northern state. This in turn caused an Internet sensation as the story spread. Warnings were posted to avoid all mulch from the region. Several individuals were panicked because they had already spread the "infested" mulch in flowerbeds and around trees in their yards.

Was the story true? No. The process of mulching hardwood is incredibly destructive to anything passing through the blades. The likelihood that a colony of termites could survive such harsh treatment is very slim. However, even if they could, mulch undergoes further treatment and inspection to ensure that live insects are not present before being shipped to retailers. The same processes were used with the mulch generated from felled trees resulting from Hurricane Katrina. Insects transported in mulch from one location to another would represent a form of accidental introduction. Now, you may want to challenge this account because you have actually witnessed termites and ants in bags of mulch purchased at a local home improvement store, an infestation that is quite likely. Insects commonly enter mulch (as well as other products) after it has been delivered but before the product has been sold. The same is true for when truckloads of mulch are stored on-site. Such large accumulations of shredded wood product are highly attractive to indigenous wood-infesting insects. However, a link between termites in mulch and Hurricane Katrina should simply be marked down as an urban myth propagated on the web.

The greatest opportunities awaiting insects in the face of any type of natural disaster are associated with disease transmission. Insects are not deliberately seeking to exploit humans during times of misery. Instead, the way humans respond to disaster creates opportunities for several insect species that transmit disease. What types of opportunities are suddenly available? Essentially, the same types that occur during times of war, for both soldiers and refugees. For example, people gather together in large groupings, often being confined together while seeking shelter, food, and water. Most forms of natural hazards or disasters will temporarily knock out electricity and, coupled with flooding, can also make drinking water scarce and proper sanitation impossible. Dealing with the natural disaster itself and then the subsequent aftermath is highly stressful, which in turn can compromise sleep and proper eating, and lower immune responses.

Each of these factors alone can increase the susceptibility of individuals to disease. When multiple factors are at play all at once for a large cluster of people, opportunities abound for insects to take advantage. Many of the species most frequently encountered after natural disasters are blood-feeders like mosquitoes and lice, the very same ones mentioned in chapter 4 as causing problems during times of war. Mosquitoes are especially problematic because flooding creates new breeding sites. Even after waters recede, fresh or brackish water remaining in

A deadly combination: fire ants and hurricanes

The red imported fire ant, *Solenopsis invicta* (order Hymenoptera, family Formicidae) is one of the nastiest invasive species to have ever made North America home. When encountering humans, or really any living creature, the ant usually engages in battle mode. Workers and soldiers are fully equipped for physical battle. Each packs potent venom that when injected into a human victim evokes intense pain, followed by swelling and pustule formation. Now multiply this image by ten or even one hundred. Why? Fire ants attack as a group, relying on chemical signals (pheromones) to recruit other members of the colony to join in pursuit of the prey. The resulting pustules last for several days, leaking interstitial fluid when bumped, irritated by rubbing of clothes, or movement during sleep. Actually, it is probably more correct to say attempted sleep because the presence of several pustules makes it nearly impossible to rest comfortably. The aggressive ants are also ferocious biters, inserting their mandibles repeatedly into the "enemy." Like most species of ants, salivary venom—which is actually a weak acid—is injected during the attack. So for the unfortunate victim of fire ant aggression, venom is dispensed from both ends of the attacker! As if things could not get worse, in regions in which *S. invicta* thrives, extreme severe weather in the form of hurricanes is the norm. The hurricane season in the Caribbean and along the southeastern United States generally occurs from June to November. This also happens to be the peak period of fire ant activity. Historically, natural disasters have been credited with decimating plant and animal populations in the paths of fire, tornadoes, hurricanes, or volcanic eruptions. Hurricanes in the Caribbean have brought relief from vicious ant species that tormented white European settlers attempting to colonize island nations like Barbados, Jamaica, and Hispaniola (composed of Haiti and the Dominican Republic). Of course, many species recover quickly or even depend on natural disturbances for dispersal. The red imported fire ants fall into this latter category. When flooding occurs following a hurricane, fire ants evacuate their colonies to form floating masses on the water surface. The ants can float because their exoskeleton is hydrophobic (repels water) and their mass is too small to break the surface tension of water. Workers will actually form mats or balls, where they link together using their mandibles. If time permits, the queen and larvae and pupae from the nursery are inserted into the center of the ball. The water current sends the ant masses to new locations. Ants will cling to anything in the water that will help them float. This can include any type of natural or artificial debris. Once the water recedes, the fire ants are no longer anywhere near their old colony. As consequence, they begin the process of building new colonies, wherever the mat or ball has landed. In most cases, the new colony is built in the precise location that the debris came to rest. The destructive forces of flood waters and hurricane force winds will drive the debris anywhere, including inside of homes and other buildings. Thus, the fire ants form new colonies in rural and urban locations, expanding the colony and their range. In essence, hurricanes and floodwaters actually augment the destructive capabilities of this invasive species. At the moment, about the only thing keeping the red important fire ant in check in terms of expansion is adverse climatic conditions of winter. But that force of nature may not be a foe for much longer with the way this species operates!

containers, cars, debris, or in natural or artificial drainage sites is still sufficient to serve as oviposition sites for mosquitoes and as places for larval development to be completed. Development is temperature-dependent, but in most cases a mass emergence of adult mosquitoes can occur in as few as seven to ten days. Assuming that the infrastructure has not been restored to any area devastated by a hurricane, the mass of hungry mosquitoes will easily find blood meals within the shelters housing displaced people. Such conditions favor transmission of malaria, dengue fever, encephalitis, and even yellow fever. Surely not in North America? It is easy to forget that as recently as the early 1900s, parts of the United States were plagued by malaria and yellow fever. Adding to the threat of diseases like malaria is the emergence of drug-resistant pathogens. Thus, even with early intervention following a disaster by national and world relief organizations, it is possible that many individuals may still succumb to diseases transmitted by insect vectors.

Bug bytes
Volcanic eruptions and insects
http://bit.ly/1KdAUph

Global warming, climate change, and insects

Global warming is one of the most divisive and polarizing subjects in the United States today. If you are unsure about the political allegiances of a neighbor or friend, simply make some proclamation regarding your view on the topic. No doubt you will know in short order their stance. Some people believe that global warming is not a real phenomenon, but instead is simply a topic dreamed up by liberal-minded politicians or scientists trying to save trees. All this talk of climate change is nothing more than propaganda or glorified scare tactics designed to guilt people into regression from their overindulgent use of resources. Maybe. Others are at the extreme opposite pole, advocating that humans need to eliminate our entire **carbon footprint**, which of course is only possible if we move quickly toward extinction. Maybe that's a bit of an exaggeration, but it's not too far off from statements I have personally overheard. The point is that most people have very strong opinions when it comes to global warming. Perhaps most individuals have views that lie

somewhere in the middle. Of course, they likely do not express such ideas in an open forum; moderation with regard to global warming seems to be unfashionable, especially among academics. Here, rather than go on a soapbox one way or another, we briefly explore what is meant by global warming, examine the major factors considered responsible for the condition, and then examine how insects are responding or will respond to changing climatic conditions. The last topic is especially important to investigate from the standpoint of how altered insect populations will affect the human condition.

Our exploration of how climate change is likely to affect insect populations needs to begin with an examination of what global warming is. Global warming, and the associated changes in the climate that are expected to accompany it, is the idea that the temperature of the Earth's climatic system is increasing. The general concerns about rising temperatures are related to the changes in climatic conditions that will occur and the subsequent effects on the Earth's ecosystems. We will examine those aspects in a moment. But first, we must also consider one of the biggest concerns in the eyes of many: that is, that we, as a species, are responsible for the phenomenon, which also represents where the majority of the controversy exists.

Most scientists agree that global warming is largely the result of an accumulation of greenhouse gases in the atmosphere. Gases such as water vapor, carbon dioxide, methane, nitrous oxide, and ozone trap radiation within the **thermal infrared wavelength** range in the atmosphere. This in turn elevates atmospheric temperatures. It is easy to see what can occur over time: the more gases that accumulate in the atmosphere, the higher the temperature elevation that occurs from radiation absorption. The planet's climatic system is designed to handle the accumulations of greenhouse gases released from natural processes only; those are not a problem. The real problems are believed to arise when artificial processes, namely those instigated by humans, cause gases to be produced and released. As a result, higher than normal greenhouse gas levels become available in the atmosphere. The so-called greenhouse effect, or atmospheric warming, then has the potential to become magnified. Advocates of global warming indicate that is exactly what is occurring. In reality, reliable data concerning average atmospheric and surface temperatures is available from only about the last hundred years. So at the moment, it is difficult to fully assess how much of the temperature change we are currently experiencing is caused by human activity and how much by natural variability. Complicating the debate on global warming is the fact that controlled experiments really cannot be performed to show cause-and-effect relationships when it comes to global climate changes.

Despite inherent weaknesses in arguments over whether global warming is real or not, the surface temperature of the Earth has warmed by 0.6°C over the last century. While this may seem like a small change in the grand scheme of things, the effects of increased temperatures vary from region to region. The planet has already experience reductions in the amounts of permafrost and sea ice as well as rapid melting of glaciers that have existed for centuries. If such warming trends continue, regardless of cause, what are some of the anticipated changes to the climate and living organisms?

In terms of climatic changes, unpredictability will be the norm. Storm and weather patterns are expected to be intense or extreme, such that heavy rains or snowfall will be common precipitation occurrences, drought will be more frequent, and wave intensity will increase along large bodies of water. Let's pause for a second to underscore that: yes, increased snow accumulation is predicted to be a possible outcome from global warming. In recent years, naysayers point to particularly cold, snowy winters as evidence that global warming is bunk. In reality, such winters are consistent with climate change predictions. Other expected outcomes include expansion of deserts in tropical regions and rises in sea levels. It is important to again state that these are predictions, not necessarily certainties. However, substantial data has been analyzed to create the models from which global warming predictions have been made (figure 17.13).

How will these climatic changes affect life on Earth? Attempting to address the implications of global warming on all life forms would be an enormous task, so only the expected outcomes on insects at the hands of global warming will be examined here. Four broad areas of change are anticipated.

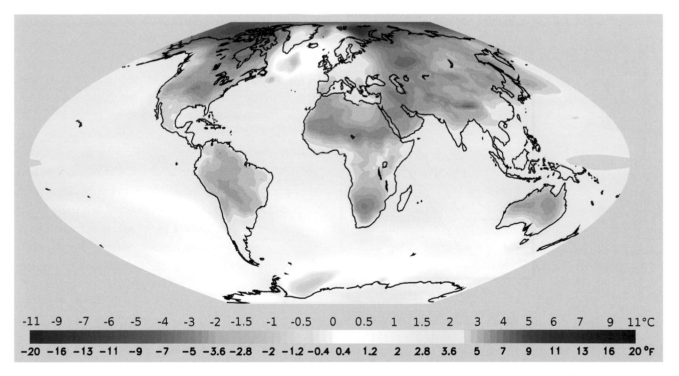

| -11 | -9 | -7 | -6 | -5 | -4 | -3 | -2 | -1.5 | -1 | -0.5 | 0 | 0.5 | 1 | 1.5 | 2 | 3 | 4 | 5 | 6 | 7 | 9 | 11°C |
| -20 | -16 | -13 | -11 | -9 | -7 | -5 | -3.6 | -2.8 | -2 | -1.2 | -0.4 | 0.4 | 1.2 | 2 | 2.8 | 3.6 | 5 | 7 | 9 | 11 | 13 | 16 | 20 °F |

Figure 17.13. Projected changes in surface temperatures from the late twentieth century until the mid-twenty-first century based on increases in greenhouse gases and aerosols in the Earth's atmosphere. Any climatic temperature changes will affect ecothermic animals, especially insects. Image created by Author NOAA Geophysical Fluid Dynamics Laboratory (GFDL) (http://bit.ly/1KhVdyF).

1. Changes in insect physiology
2. Changes in species distribution
3. Changes in insect phenology
4. Adaptation

Insects are the most diverse and abundant animals in terrestrial ecosystems, which should make them the most susceptible to the effects of global warming. And they are, but mostly because they are **poikilothermic**. As a result, any changes, up or down, in ambient temperatures will have a direct effect on the physiology of terrestrial insects. Aquatic species will also be affected, but to a lesser extent, since the heat capacity of water buffers them from rapid climatic changes. Insect respiration, metabolic rate, all aspects of development, and locomotion are all temperature-dependent. So even a modest elevation of 0.6°C in surface or air temperatures will result in faster rates of food acquisition and processing, decreased length of developmental duration, and, conceivably, the ability for locomotion at times that it was not possible previously (e.g., within a twenty-four-hour cycle, or even seasonally). Really, any aspect of insect development that depends on **degree-days**—thermal energy in the environment—is vulnerable to climate change.

All species, however, will not be equally susceptible to global warming. For example, tropical insects will be affected more negatively than any other types of insects. Why? Insects living near the equator tend to be very sensitive to temperature changes and are already living near their temperature optima. Thus, elevations in ambient temperatures would move them closer to the upper end of their **zone of temperature tolerance** and consequently into thermal or proteotaxic stress (figure 17.14). In contrast, species living at higher latitudes generally reside in conditions cooler than their optimum temperature. They thus could tolerate some increases in overall environmental temperatures. In fact, initial increases from climatic change would enhance their development for the reasons mentioned earlier. Any shift in insect developmental patterns has the potential to alter the status of a given species from benign to a pest, from minor to major pest, or even possibly drive a species toward extinction.

A warming environment will induce shifts in insect populations. The most likely movements or migrations will be away from the equator to higher elevations or toward the poles in latitude. Of course such changes in insect populations will be species- and region-dependent.

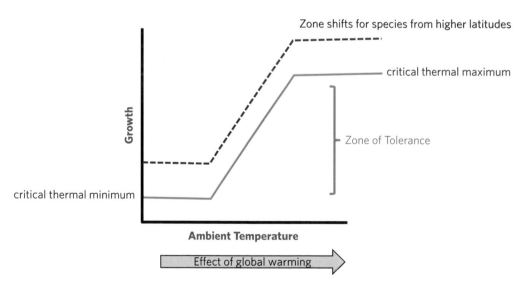

Figure 17.14. The relationship between rising environmental temperatures caused by climate change and insect growth. Insects are poikilotherms, so their body temperature changes directly with ambient conditions. Species residing at higher latitudes are likely best suited to adapt to increases in atmospheric temperatures; acclimatization is anticipated to lead to shifts in such insects' zones of tolerance over time. In comparison, tropical species will adapt poorly since they already live at near optimal temperatures. Global warming will drive them toward their critical thermal maxima.

For instance, insects that develop at rapid rates, those that do not enter diapause (non-diapausing species), or species that diapause but do not require low temperatures for the induction of dormancy are predicted to expand their ranges in response to global warming. In other words, they will thrive as surface temperatures elevate. This also makes them prime candidates for assuming new relationships with human populations; undoubtedly many will become pests of agriculture or the human condition for the first time, while others will shift from rare or occasional pests to serious threats. In contrast, insect species that display prolonged growth periods or that depend on low temperatures to induce the onset of diapause will be at a clear disadvantage from climate change. At the very least, such species should experience a retraction in range. They are also expected to encounter increased competition for resources as other species expand their distribution, which could ultimately lead to localized extinction. The fossil record indicates that such shifts in insect populations have occurred previously during past climatic changes.

Related to changes in physiology and population distributions will be concurrent life cycle or phenological disturbances. This will be especially true for life cycle events linked to degree-days, such as the onset or termination of dormancy and the synchronization of development with other species. The latter is especially critical for insect species in mutualistic relationships. Perhaps the most crucial example is that of flowering plants and their insect pollinators. We have already discussed the effects that warmer climatic conditions can have on development of poikilothermic animals like insects. If corresponding developmental changes do not also occur with mutualistic partners, then their life cycles will no longer be in synchrony. The result, for flowering plants, is that the timing of key insect species visiting flowers and serving as pollinators will be severely altered. The obvious outcome would be diminished sexual reproduction in flowering plants. Some mathematical models predict reductions in floral resources between 15% and 50% as a result of shifts in insect phenology. The effect would be felt throughout entire ecosystems, creating cascading effects that would alter the populations of most living organisms in that system. In terms of the human condition, reductions in pollination equate to reduced food production. How severe the effect would be on human populations depends on the extent of climate changes in key agricultural zones across the globe, the level of cooperation between nations, and whether our populations are still growing at the rate they are today.

The final category of change is adaptation. Insects are masters at adapting. Thus, they will be among the best of all organisms, and certainly among animals, at responding to environmental change resulting from global warming. However, not all will be equally suited to the task of evolving to ensure survival. Climate change will serve as a selection force that will favor species with short generation times. Consequently, adaptive traits in populations with short generation times will be selected for much faster than in longer-lived species. Species that require months to more than a year to complete their reproductive cycle are more likely to suffer retractions in populations and even move toward species extinction locally or more globally. The successful adapters will also be drawn to human habitation seeking food, especially since flowering plants as a whole are expected to offer fewer resources. Potentially, ecosystem changes resulting from global warming could generate more insect species as pests than has heretofore been realized.

CHAPTER REVIEW

How can there be new threats today?

- Over the past several decades, insects have become pests in regions where previously they were not. In other cases, the timing or severity of pest status has changed. Why have these changes occurred? Much of the reason is tied to human population growth. The rapid rise in the human population is at, if not exceeding, the carrying capacity of the Earth's ecosystems. Land, water, and all other natural resources are being manipulated, in some cases depleted, at an alarming rate to satisfy the growing demands of an expanding human existence. Such changes alter existing ecosystems, creating opportunities for insects that either did not exist before or providing them the means to go unchecked.

- Globalization of the world's marketplaces and climate change linked to global warming are factors contributing to the rise of new insect threats. Both topics can be relatively controversial depending on your political or even religious ideologies, and each is directly linked to human population growth. From an entomological point of view, this equates to new trade partners and the potential for insects to travel with crops, or on or in freight containers, ships, planes, or any other conveyance exchanging goods between countries.

- The climate is changing. Although not everyone is in agreement that this is true, especially with regard to global warming, some solid pieces of evidence cannot be argued against. Climate change is anticipated to lead to new insect threats, such as through expanded range or distribution of several insect species; decreased survivorship of some species; concomitant increased opportunities for other species as ecological niches are vacated and new habitats created; and modified life cycles, such as shifting from a single generation each year to multiple generations, or averting dormancy altogether. Any of these changes has the potential to create new pest opportunities for some species or allow other insects to elevate in stature in terms of the damage caused to the human condition.

✦ Same old story: Accidental introduction

- Insect immigrations—meaning accidental introductions—into the United States parallel closely the path of human migration. This implies a strong link between the origins of invasive species and mass movement of people from one region of the world to another. Why does this occur? Human immigrants generally do not travel alone, as numerous nonnative or alien species of plants, animals, and microorganisms—whether intentionally or accidentally—join the two-legged migrants on their journeys. Insects travel in food, luggage, and other goods carried by human migrants. The point is that many insect introductions into the United States are directly tied to human immigration. As a consequence, the threat of new invasive species into the United States is ever-present as long as the nation permits sizeable migration and individuals are arriving

from regions of the world with insects that can potentially thrive in areas at or near their ports of entry.

- Import and export of goods from the United States is at an all-time high and expected to continue to rise. What does this mean in terms of invasive insects? Exactly what you would guess: an anticipated sharp increase in the number of exotic species entering the United States. Importation of goods, along with the associated shipping materials and transport vehicles is the most common means for accidental introductions of invasive insect species into North America, or any nation. And it is not new. The rise of the global marketplace simply magnifies what has occurred throughout history.

- It is not so much the numbers of species that get accidentally brought into the country that raises the concern, although certainly the odds of trouble increase as the number of accidental releases happen more frequently with more species. The issue is *which* species are introduced. Not every organism that travels to the United States is equipped with the life history traits needed to thrive, even survive, in the region of importation. As a consequence, the vast majority of accidental introductions either die out in a relatively short period of time or establish as minor players in local ecosystems. Winter conditions in the northern states prevent species derived from tropical environments from becoming established, and the extreme heat of the south and arid conditions found in the southwestern regions of the United States are equally effective barriers to exotic insects from regions with much different climatic conditions. The point is that for an exotic insect to become invasive, it needs to be introduced into a region in which its life history characteristics are compatible with the climatic conditions.

- Once established, invasive insect species cause damage to agriculture, human habitation, materials or goods of importance to humans, or cause disruption to the immediate environment. The most obvious effect of invasive species is alteration of biodiversity. In this context, biodiversity refers to all the different organisms that exist in a particular ecosystem influenced directly by the invasive insect species of interest. One possible outcome for an alien insect species is that it reduces or outcompetes other species in the environment, potentially displacing indigenous species. However, invasive species may also alter trophic interactions within an environment in such a way that populations of indigenous species now flourish. This in turn directly or indirectly influences the population dynamics of other species, not just insects.

Effects of human interference: Nonaccidental introductions

- Exotic insects are not always opportunistic in their path to finding the United States. In many instances, they have been deliberately introduced. Nonaccidental release of alien species is generally for one of two reasons: (1) as part of an act of terror or war in which the exotic species is being used as an entomological weapon, or (2) the insect is being introduced as part of a biological control program targeting an invasive organism from the same region as the control agent (i.e., the alien insect).

- Biological control deliberately introduces exotic insects to combat other exotic species. More than two thousand species of exotic insects and other arthropods have been deliberately introduced in nearly two hundred nations. The result has been a permanent reduction in 165 invasive species. At least in some scenarios, biological control has been a success. Often, however, the outcomes are mixed. That fact in itself is not reason to abandon biological control strategies. The issue that arises from release of exotic species is that they potentially pose the same problems as accidentally introduced insects. When using predators and parasites, there is always the risk that biological control agents will expand their range. In other words, the natural enemies do not restrict their activities to our intended targets and instead follow their own desires.

- What is truly surprising is that, in the past, very little risk assessment has been required prior to the introduction of an exotic biological control agent. Essentially a good number of the attempts using exotic agents have relied on trial and error rather than solid empirical testing. Thus, when an insect agent goes rogue, everyone is surprised! That approach is thankfully being abandoned.

Insect activity following natural disasters

- Natural disasters occur in every geographic region of the United States. Hazards in the form of earthquakes, landslides, forest fires, floods and severe weather patterns threaten human lives and cause huge economic losses annually. The events occur frequently, depending on climatic conditions, geography, and differences in topography. The effects of such disasters can be enormous. They can also serve as great opportunities—that is, if you are small and possess six legs.

- Natural disasters for humans are opportunities for insects. Not for all species, of course; some insects suffer the consequences of being in the path of a natural disaster. Flooding is more deadly for nonflying species than it is for those with wings and, certainly, for humans. The same can be said for fire, volcanic eruptions, and most any other natural disaster. Landslides and avalanches are exceptions since, for the most part, they occur in an unpredictable fashion. Natural disaster preparedness is one area where humans—who, if they heed warning signs and alerts for predictable natural hazards, can escape immediate harm—have a clear advantage over the class Insecta. Insects, by contrast, do not get text alerts from the National Weather Service warning of impending danger. Some species take advantage of human misery during times of natural disaster. They do so specifically by using the event for dispersal opportunities or to transmit disease. Some may even augment the effect of the natural disaster while it is occurring.

- Climate-induced disasters can dramatically alter insect populations within a given region. Strong winds associated with cyclones, hurricanes, tropical storms, nor'easters, and even tornadoes are capable of transmitting small-bodied arthropods, whether or not they possess wings, from one location to another. Wind has never been reported to induce localized extinction of a species, but the same is not true for the flooding that occurs with many of these major storms. Heavy precipitation, storm surges, and intense wave action are often associated with cyclones, hurricanes, tropical storms, and nor'easters, leading to flooding. Hurricanes are well known to have changed the pest status of many insects, again in regions of the Caribbean as well as along the Atlantic coast. As expected, some pest species have been driven from islands and other regions. And of course, with new vacancies in ecological niches, other species move in to fill the void, potentially becoming pests for the very first time. Floodwaters are also instrumental in redistributing organisms from one location to another.

- The greatest opportunities awaiting insects in the face of any type of natural disaster are associated with disease transmission. Insects are not deliberately seeking to exploit humans during times of misery. Instead, the way humans respond to disaster creates opportunities for several insect species that transmit disease. For example, people gather together in large groupings, often being confined together while seeking shelter, food, and water. Most forms of natural hazards or disasters will temporarily knock out electricity and, coupled with flooding, can also make drinking water scarce and proper sanitation impossible. Dealing with the natural disaster itself and then the subsequent aftermath is highly stressful, which in turn can compromise sleep and proper eating, and lower immune responses. Each of these factors alone can increase the susceptibility of individuals to disease. When multiple factors are at play all at once for a large cluster of people, opportunities abound for insects to take advantage.

Global warming, climate change, and insects

- Global warming is one of the most divisive and polarizing subjects in the United States today. If you are unsure about the political allegiances of a neighbor or friend, simply make some proclamation regarding your view on the topic. No doubt you will know in short order their stance. Some people believe that global warming is not a real phenomenon, but instead is simply a topic dreamed up by liberal minded politicians or scientists trying to save trees. All this talk of climate change is nothing more than propaganda or glorified scare tactics designed to guilt people into regression from their overindulgent use of resources. Maybe. Others are at the extreme opposite pole, advocating that humans need to eliminate our entire carbon footprint, which of course is only possible if we move quickly toward extinction. The point is that most people have very strong opinions when it comes to global warming. Perhaps most individuals have views that lie somewhere in the middle.

- Global warming, and the associated changes in the climate that are expected to accompany it, is the idea that the temperature of the Earth's climatic system is increasing. The general concerns about rising temperatures are related to the changes in climatic conditions that will occur and the subsequent effects on the Earth's ecosystems. The biggest concern in the eyes of many is that humans are responsible for the phenomenon, which also represents where the majority of the controversy exists. Most scientists agree that global warming is largely the result of an accumulation of greenhouse gases in the atmosphere. Gases such as water vapor, carbon dioxide, methane, nitrous oxide, and ozone trap radiation within the thermal infrared wavelength range in the atmosphere. This in turn elevates atmospheric temperatures. The more gases that accumulate in the atmosphere, the higher the temperature elevation that occurs from radiation absorption. The planet's climatic system is designed to handle the accumulations of greenhouse gases released from natural processes only; those are not a problem. The real problems are believed to arise when artificial processes, namely those instigated by humans, cause gases to be produced and released.

- Insects are the most diverse and abundant animals in terrestrial ecosystems, which should make them the most susceptible to the effects of global warming. And they are, but mostly because they are poikilothermic. As a result, any changes, up or down, in ambient temperatures will have a direct effect on the physiology of terrestrial insects. Aquatic species will also be affected, but to a lesser extent, since the heat capacity of water buffers them from rapid climatic changes. Insect respiration, metabolic rate, all aspects of development, and locomotion are all temperature-dependent. So even a modest elevation of 0.6°C in surface or air temperatures will result in faster rates of food acquisition and processing, decreased length of developmental duration, and, conceivably, the ability for locomotion at times that it was not possible previously. Really, any aspect of insect development that depends on degree-days is vulnerable to climate change.

- A warming environment will induce shifts in insect populations. The most likely movements or migrations will be away from the equator to higher elevations or toward the poles in latitude. Of course such changes in insect populations will be species- and region-dependent. For instance, insects that develop at rapid rates, those that do not enter diapause, or species that diapause but do not require low temperatures for the induction of dormancy are predicted to expand their ranges in response to global warming. In other words, they will thrive as surface temperatures elevate. This also makes them prime candidates for assuming new relationships with human populations; undoubtedly many will become pests of agriculture or the human condition for the first time, while others will shift from rare or occasional pests to serious threats. In contrast, insect species that display prolonged growth periods or that depend on low temperatures to induce the onset of diapause will be at a clear disadvantage from climate change.

- Related to changes in physiology and population distributions will be concurrent life cycle or phenological disturbances. This will be especially true for life cycle events linked to degree-days, such as the onset or termination of dormancy and the synchronization of development with other species. The latter is especially critical for insect species in mutualistic relationships. Perhaps the most crucial example is that of flowering plants and their insect pollinators. If corresponding developmental changes do not also occur with mutualistic partners, then their life cycles will no longer be in synchrony.

- Insects are masters at adapting. Thus, they will be among the best of all organisms, and certainly among animals, at responding to environmental change resulting from global warming. However, not all will be equally suited to the task of evolving to ensure survival. Climate change will serve as a selection force that will favor species with short generation times. Consequently, adaptive traits in populations with short generation times will be selected for much faster than in longer-lived species. Species that require months to more than a year to complete their reproductive cycle are more likely to suffer retractions in populations and even move toward species extinction locally or more globally. The successful adapters will also be drawn to human habitation seeking food, especially since flowering plants as a whole are expected to offer fewer resources.

MUSHROOM FARMING (SELF-TEST)

Level 1: Knowledge/Comprehension

1. Define the following terms:
   - (a) invasive species
   - (b) mechanical dispersal
   - (c) greenhouse effect
   - (d) global warming
   - (e) insect phenology
   - (f) biodiversity

2. What is the difference between an exotic and invasive species?

3. How can you distinguish the difference between direct and indirect effects of invasive species?

4. Provide examples of nonaccidental introductions of invasive species into a region.

5. Explain the relationship between natural disasters and potential new threats from insects.

6. Describe some of the anticipated changes in insect populations caused by global warming.

Level 2: Application/Analysis

1. Explain why increases in the human population and globalization of trade are factors that may well lead to new problems with insects.

2. Discuss how the use of biological control agents to suppress populations of a pest species can at times actually cause more harm than good.

3. Why are insects considered among the most vulnerable of species to the effects of global warming?

Level 3: Synthesis/Evaluation

1. Assuming that climate change in the form of atmospheric warming is or does occur over your lifetime, what would be indicators that insects in the United States are responding? In other words, how can you tell that insects have been affected?

2. Speculate on what type of natural disaster would potentially have the greatest effect on humans from an entomological perspective. Provide examples to support your answer.

REFERENCES

Colunga-Garcia, M., R. A. Haack, and A. O. Adelaja. 2009. Freight transport and the potential for invasions of exotic insects in urban and periurban forests of the United States. *Journal of Economic Entomology* 102:237-246.

Deutsch, C. A., J. J. Tewksbury, R. B. Huey, K. S. Sheldon, C. K. Ghalambor, D. C. Haak, and P. R. Martin. 2008. Impacts of climate warming on terrestrial ectotherms across latitude. *Proceedings of the National Academy of Sciences (USA)* 105:6668-6672.

Edwards, B. 1794. The History, Civil and Commercial, of the British Colonies in the West Indies. 2nd ed. John Stockdale, Piccadilly, UK.

Gullan, P. J., and P. S. Cranston. 2014. The Insects: An Outline of Entomology. 5th ed. Wiley-Blackwell, West Sussex, UK.

Holzapfel, E. P., and J. C. Harrell. 1968. Transoceanic dispersal studies of insects. *Pacific Insects* 10:115-153.

Hughes, L. 2000. Biological consequences of global warming: Is the signal already apparent? *Trends in Ecology and Evolution* 15:56-61.

Kenis, M., M.-A. Auger-Rozenberg, A. Roques, L. Timms, C. Péré, M. J. W. Cock, J. Settle, S. Augustin, and C. Lopez-Vaamonde. 2008. Ecological effects of invasive alien insects. *Biological Invasions.* doi:10.1007/s10530-008-9318-y.

Levine, J. M., and C. M. D'Antonio. 2003. Forecasting biological invasions with increasing international trade. *Conservation Biology* 17:322-326.

Louda, S. M., D. Kendall, J. Connor, and D. Simberloff. 1997. Ecological effects of an insect introduced for the biological control of weeds. *Science* 277:1088-1090.

Mattson, W. J. 1997. Exotic insects in North American forests: Ecological systems forever altered. http:www.invasive.org/symposium/mattson.html.

McCullough, D. G., R. A. Werner, and D. Neumann. 1998. Fire and insects in northern and boreal forests ecosystems of North America. *Annual Reviews of Entomology* 43:107-127.

Memmott, J., P. G. Craze, N. M. Waser, and M. V. Price. 2007. Global warming and the disruption of plant-pollinator interactions. *Ecology Letters* 10:1-8.

Menéndez, R. 2007. How are insects responding to global warming? *Tijdschrift voor Entomologie* 150:355-365.

Nunes, L. F., S. R. Leather, and F. C. Rego. 2000. Effects of fire on insects and other invertebrates: A review with particular reference to fire indicator species. *Silva Lusitana* 8:15-32.

Pearson, D. E., and R. M. Callaway. 2003. Indirect effects of host-specific biological control agents. *Trends in Ecology and Evolution* 18:456-461.

Pimentel, D. (ed.). 2002. Biological Invasions: Economic and Environmental Costs of Alien Plants, Animal and Microbe Species. CRC Press, Boca Raton, FL.

Ruiz, G. M., and J. T. Carlton. 2003. Invasive Species: Vectors and Management Strategies. Island Press, Washington, DC.

Simberloff, D., D. C. Schmitz, and T. C. Brown. 1997. Strangers in Paradise: Impact and Management of Nonindigenous Species in Florida. Island Press, Washington, DC.

van Lentern, J. C., J. Bales, F. Bigler, H. M. T. Hokkanen, and A. J. M. Loomans. 2006. Assessing risk of releasing exotic biological control agents of arthropod pests. *Annual Review of Entomology* 51:609-634.

Vinson, S. B. 1997. Invasion of the red imported fire ant (Hymenoptera: Formicidae): Spread, biology and impact. *American Entomologist* 43:23–39.

Waring, S. C., and B. J. Brown. 2005. The threat of communicable diseases following natural disasters: A public health response. *Disaster Management and Response* 3:41–47.

THE ENTOMOLOGIST BOOKSHELF (SUPPLEMENTAL READINGS)

Archer, D. 2011. Global Warming: Understanding the Forecast. 2nd ed. John Wiley and Sons, Sussex, UK.

Klein, N. 2015. This Changes Everything: Capitalism vs. Climate. Simon and Schuster, New York, NY.

Louda, S. M., D. Kendall, J. Connor, and D. Simberloff. 1997. Ecological effects of an insect introduced for the biological control of weeds. *Science* 277:1088–1090.

Simberloff, D. 2013. Invasive Species: What Everyone Needs to Know. Oxford University Press, New York, NY.

Waters, T. 2008. The Persistence of Subsistence Agriculture: Life beneath the Level of the Marketplace. Lexington Books, Lanham, MD.

Williamson, M. 1996. Biological Invasions. Chapman and Hall, London, UK.

ADDITIONAL RESOURCES

Could ants solve global warming?
http://dailym.ai/1LElucT

Crops, beetles and carbon dioxide: Will global warming bring more voracious insects to farms?
http://bit.ly/1PFTZjQ

Fire ants and Hurricane Katrina
http://bit.ly/1JO9Zgn

Global warming could trigger insect population boom
http://bit.ly/1N3ou5J

How do volcanoes affect plants and animals?
http://bit.ly/1idp8Rq

Human population growth
http://www.prb.org/Publications/Lesson-Plans/HumanPopulation/PopulationGrowth.aspx

Hurricanes and mosquitoes
http://bit.ly/1UgeOJm

Interception of pineapple pest
http://1.usa.gov/1imdC6k

Invasive and exotic insects
http://www.invasive.org/species/insects.cfm

National invasive species information center
http://www.invasivespeciesinfo.gov/index.shtml

Natural disasters in North America
http://www.sfu.ca/geog/geog351fall06/group06/

Rebirth of an island after volcanic eruption
http://bit.ly/1JqLANG

Termites and hurricane Katrina
http://bit.ly/1K445rz

Appendix

One question often asked by students interested in insects is, Where can I get insect stuff? And by stuff, I mean just about anything insect-related, from collecting equipment to fashion to, yes, ones that can be eaten.

Included here are sources of insects (live, preserved, or stored by other means) and materials for working with insects (for collecting and rearing, and field guides and other study materials needed for working with insects). Also included are vendors of all sorts of insect-related items, ranging from books, artwork, and jewelry to a whole host of odds and ends with connections to "bugs." This list is by no means exhaustive, as a quick examination using any search engine will reveal. Sorting through the pages and pages of useful (and some not-so-useful) insect-related sources and vendors can take days. The sources given here are meant to help you get started. Keep in mind that almost any major land grant university is home to an entomology department and cooperative extension agency filled with individuals who can provide a wealth of information regarding insects. The same is true of federal and local government agencies, such as the US Department of Agriculture, state Department of Natural Resources and Agriculture Divisions, and natural history museums.

Sources of Insects

A-1 Unique Insect Control
5504 Sperry Drive
Citrus Heights, CA 95621
(916) 961-7945
www.a-1unique.com

ARBICO (Arizona Biological Control, Inc.)
PO Box 4247 CRB
Tucson, AZ 85738
(800) 827-2847
http://www.usit.net/BICONET

Associates Insectary
PO Box 969
Santa Paula, CA 93060
(805) 933-1301
www.associatesinsectary.com

Beneficial Insectary, Inc.
9664 Tanqueray Court
Redding, CA 96003
(800) 477-3715
www.insectary.com

Benzon Research
(717) 258-1183
http://www.benzonresearch.com

Berkshire Biological Supply Company
264 Main Street
Westhampton, MA 01060
(413) 527-3932

Biocontrol Network
5116 Williamsburg Road
Brentwood, TN 37027
(800) 441-2847
www.biconet.com

Bloomington Drosophila Stock Center at Indiana University
Department of Biology
Indiana University
1001 East Third Street
Bloomington, IN 47405-7005
(812) 855-2577
http://flystocks.bio.indiana.edu/

Carolina Biological Supply Company Biology/Science Materials
PO Box 6010
Burlington, NC 27216
(800) 334-5551
http://www.carolina.com

Connecticut Valley Biological Supply Co., Inc.
PO Box 326, 82 Valley Road
Southampton, MA 01073
(800) 628-7748

Delta Education, Inc.
PO Box 3000
Nashua, NH 03061-3000
(800) 442-5444

EcoSolutions, Inc.
2948 Landmark Way
Palm Harbor, FL 34684
(727) 787-3669
www.ecosolutionsbeneficials.com

Entomology Solutions, LLC
1355 Bardstown Road, Suite 118
Louisville, KY 40204
(502) 384-8953
www.bugsbehavingbadly.com

Feeder Source
PO Box 2677
Cleveland, GA 30528
http://www.feedersource.com

Fluker's Cricket Farm, Inc.
1333 Plantation Road
Port Allen, LA 70767
(800) 735-8537
http://www.flukerfarms.com/

Forked Tree Ranch
8347 Farm to Market Road
Bonners Ferry, ID 83805
(208) 267-2632
http://www.forkedtreeranch.com/

Green Spot, Ltd.
93 Priest Road
Nottingham, NH 03290
(603) 942-8925
www.greenmethods.com

Hydro-Gardens, HGI Worldwide Inc.
PO Box 25845
Colorado Springs, CO 80936
(888) 693-0578
www.hydro-gardens.com

Insect Lore Products
PO Box 1535
Shafter, CA 93263
http://www.insectlore.com/

KLM BioScientific
8888 Clairemont Mesa Boulevard
San Diego, CA 92123
(858) 571-5562
http://labsuppliesusa.com/

Koppert Biological Systems, Inc.
28465 Beverly Road
Romulus, MI 48174
(734) 641-3763
www.koppertonline.com

Louisiana Biological Products, Inc.
(504) 235-1607
ssackett@usa.net

Mulberry Farms
3920 Gird Road
Fallbrook, CA 92028
(760) 731-6088
http://www.mulberryfarms.com/

Nature's Control
PO Box 35
Medford, OR 97501
(541) 245-6033
www.naturescontrol.com

Peaceful Valley Farm Supply
PO Box 2209
Grass Valley, CA 95945
(888) 784-1722
www.groworganic.com

Rincon-Vitova Insectaries, Inc.
PO Box 1555
Ventura, CA 93002-1555
(800) 248-2847
www.rinconvitova.com

Tip Top Bio-Control
PO Box 7614
Westlake Village, CA 91359
(800) 525-0004
www.tiptopbiocontrol.com

UC San Diego Drosophila Stock Center
University of California, San Diego
9500 Gilman Drive #0116
La Jolla, CA 92093-0116
(858) 246-0350
https://stockcenter.ucsd.edu/info/welcome.php

Wards Biological Supply Company
PO Box 92912
Rochester, NY 14692-9012
(800) 962-2660
https://www.wardsci.com

Worm's Way Inc.
7850 North State Road 37
Bloomington, IN 47404
(800) 274-9676
www.wormsway.com

Sources of Insect Materials

Amazon, Inc.
http://www.amazon.com

American Entomological Society
1900 Benjamin Franklin Parkway
Philadelphia, PA 19103-1195
(215) 561-3978
http://darwin.ansp.org/hosted/aes/

American Museum of Natural History
Central Park West at 79th Street
New York, NY 10024-5192
(212) 769-5100
http://www.amnh.org/

Berkshire Biological Supply Company
264 Main Street
Westhampton, MA 01060
(413) 527-3932

BioQuip Products
17803 LaSalle Avenue
Gardena, CA 90248-3602
(310) 324-0620, FAX (310) 324-7931.
http://www.bioquip.com

BioServ
One Eighth Street, Suite 1
Frenchtown, NJ 08825
(800) 996-9908
http://www.insectrearing.com

Carolina Biological Supply Company Biology/Science Materials
PO Box 6010
Burlington, NC 27216
(800) 334-5551
http://www.carolina.com

Connecticut Valley Biological Supply Co., Inc.
PO Box 326, 82 Valley Road
Southampton, MA 01073
(800) 628-7748
http://www.connecticutvalleybiological.com/

Delta Education, Inc.
PO Box 3000
Nashua, NH 03061-3000
(800) 442-5444

Entomological Society of America
9301 Annapolis Road
Lanham, MD 20706-3115
(301) 731-4535
http://www.entosoc.org

John W. Hock Company
PO Box 12852
Gainsville, FL 32604
(352) 378-3209
http://www.johnwhock.com

Lane Science Equipment Corporation
225 West 34th Street, Suite 1412
New York, NY 10122-1496
(212) 563-0663
http://www.lanescience.com

Mississippi Entomological Museum
Mississippi State University
100 Old Highway 12
Box 9775
Mississippi State, MS 39762-9775
(662) 325-2990
http://mississippientomologicalmuseum.org.msstate.edu
/index.html#.VftWzs6G7V0

O. Orkin Insect Zoo Smithsonian Natural History Museum
Tenth Street and Constitution Avenue NW
Washington, DC 20560
(202) 633-1000
http://www.mnh.si.edu/education/exhibitions
/insectzoo.html

Orkin: BUGS! A Closer Look
Public Relations Department, Orkin Pest Control
2170 Piedmont Road NE
Atlanta, GA 30324

Sante Traps
1118 Slashes Road
Lexington, KY 40502
(859) 268-9534
http://www.santetraps.com

Sigma Scientific LLC
(352) 505-5793
http://www.sigma-sci.com

Sonoran Arthropod Studies, Inc.
PO Box 5624
Tucson, AZ 85703
(520) 883-3945
http://www.sasionline.org

Southland Products, Inc.
201 Stuart Island Road
Lake Village, AR 71643
(870) 265-3747
http://www.tecinfo.com/~southland/

Wards Biological Supply Company
PO Box 92912
Rochester, NY 14692-9012
(800) 962-2660
https://www.wardsci.com

Sources for Insect-Related Paraphernalia

Bug Under Glass
http://www.bugunderglass.com/

Entomology Illustration Archive
Department of Entomology
Smithsonian Institute
PO Box 37012, MRC 165
Washington, DC 20013-7012
(202) 633-0982
http://entomology.si.edu/IllustrationArchives.htm

Hexapoda
Insects/Art/Jewelry
http://www.thehexapodacollection.com/

How to Make an Insect Collection
http://www.extension.entm.purdue.edu/401Book/default
.php?page=field_identification

Insect Art
http://www.insectartonline.com/

Insect Identifications and Keys
BugGuide
http://www.bugguide.net/node/view/15740

Household and Structural Insect Pests
http://www.insectid.ento.vt.edu/insect-id/identify-pests
/index.html

Insect Identification
http://www.insectidentification.org/insect-key.asp

Identifying Nature (UK forum)
http://www.nhm.ac.uk/take-part/identify-nature.html

Insect Identification Key
http://www.knowyourinsects.org/

Insect Images and Identification
http://ento.psu.edu/public/insect-images

Introduction to Identification of Adult Arthropods
http://www.entnemdept.ifas.ufl.edu/choate/insectid.pdf

Keys of Insects to Order
http://www.earthlife.net/INSECTS/orders-key.html

Keys to Adult Insects with Wings
http://www.amnh.org/learn/biodiversity_counts/ident
_help/Text_Keys/arthropod_keyA.htm

Keys to Insects in the Southeastern United States
http://entnemdept.ufl.edu/choate/insecpdf.htm

Keys to Mosquito Identification
http://fmel.ifas.ufl.edu/key/

Pest Insect Identification Guide
http://www.pestworld.org/pest-guide/

Tree of Life: Insecta
http://tolweb.org/tree?group=Insecta&contgroup
=Hexapoda

What's That Bug?
http://www.whatsthatbug.com/

Insect Images
Center for Invasive Species and Ecosystem Health
4601 Research Way
Administrative Building, Room 113

PO Box 748
Tifton, GA 31793
(229) 386-3298
http://www.insectimages.org/

Insect Models

Store for Knowledge
PO Box 501
Stephenville, TX 76401
http://www.storeforknowledge.com/Insect-Replicas
-C209.aspx

Insect Repellant Clothing

Bugbaffler
(800) 662-8411
http://www.bugbaffler.com/

Insects Are Food Recipes

http://www.insectsarefood.com/recipes.html

Insect T-shirts

Café Express
(877) 809-1659
http://www.cafepress.com/+insect+t-shirts

Zazzle.com
(888) 892-9953
http://www.zazzle.com/insect+tshirts

Pheromone/Christopher Marley Studio

PO Box 4451
Salem, OR 97302
(503) 990-8132
http://www.pheromonegallery.com

The Real Insect Company

2860 Spring Street, Suite 3
Redwood City, CA 94063
(650) 261-9888
http://www.realinsect.net/

Think Geek—Edible Insects

http://www.thinkgeek.com/product/1716/?rkgid=1961107519
&cpg—ogcmcl&device=c&matchtype=b&network=s

Glossary

abdomen (or **opisthosoma**): the posterior-most body region of arthropods.

abiogenesis: the idea that living organisms can form from inanimate objects. Also referred to as spontaneous generation.

accessory glands: glands associated with both the male and female reproductive systems, producing a wide range of substances from hormones to adhesives to venom.

acron: the primitive or simple head of ancestral insect species.

aedeagus (or **intromittent organ**): the organ of male insects used for sperm deposition.

aesthetic injury level (AIL): an arbitrary threshold level at which pest status is determined subjectively by an individual. Pest status is achieved subjectively, not through empirical data.

aestivation: a period of suppressed activity in response to high temperatures or arid conditions.

agamogenesis: reproduction that does not involve male gametes.

agroterrorism: the deliberate introduction of animal or plant pathogens, or pests that directly target cropping systems, livestock, or food held in storage after harvest.

allele: a different form of the same gene.

allelochemicals: semiochemicals used for interspecific communication.

allomones: chemical substances produced and released by an individual of one species that is harmful to the receiver.

allospecific: an individual of a different species.

ametabolous metamorphosis: a form of insect development in which juveniles gradually become more adult-like by getting larger in size with each molt. Few others changes occur.

anaphylaxis: an acute systemic allergic reaction that can lead to severe complications or even death.

anthropomorphism: the practice of assigning human purpose and characteristics to other animals as a means to explain what we observe.

anus: opening of rectum to outside of body.

apitherapy: the practice of using bees or bee products for medicinal purposes.

apneumones: chemical signals emitted from a nonliving object that trigger a response in the recipient.

apodemes: the inward points of the exoskeleton that serve as sites for muscle attachment.

apolysis: physical separation of the cuticle from the epidermis.

apomictic: without meiosis.

aposematic coloration: warning coloration, usually in the form of a combination of orange, red, yellow, and black, to signal that an insect bites, stings, or is otherwise noxious or toxic.

apterous: wingless (adult insects).

arrhenotoky: a type of parthenogenesis in which males are produced from unfertilized eggs and consequently possess half (haploid) the chromosome content of the mother (diploid).

articulation: a joint or moveable point of attachment.

augmentative release: generally associated with biological control efforts in which periodic introductions of a biological control agent are released to supplement the existing population.

autointoxication: self-susceptibility to toxins produced by an insect.

automictic: changes that occur after meiosis; in other words post-gamete production.

autotomy: the release of an appendage at a point of weakness during an attack.

axon: a region of neuron that transmit messages to other cells.

Batesian mimicry: when an otherwise harmless insect evolves to mimic one that is harmful if eaten or messed with.

bilateral symmetry: body arrangement of animals that display a mirrored image or handedness along the longitudinal axis of body.

binomial system of nomenclature: system for classifying plants and animals developed by the Swedish naturalist Carolus Linnaeus.

biodiversity: all organisms found in a particular ecosystem.

biogeoclimatic zone: a geographical area with a relatively uniform macroclimate, characterized by a mosaic of vegetation, soils, and animal life reflecting that climate.

bioinformatics: an interdisciplinary field that develops methods and software for analyzing biological data.

biological control: a strategy for managing pest populations in which natural enemies (e.g., predators, parasites, pathogens) are used to reduce the pest status of a particular organism.

bioluminescence: the production of light through biochemical reactions.

biosensor/biological sensor: an analytical device that integrates living organisms or biological materials into a physiochemical detector.

blastoderm: the one-cell-thick layer formed early during embryonic development.

bursa copulatrix: the genital chamber of most insects.

campaniform sensilla: sensory organs that function in mechanoreception.

cantharidin: a terpenoid compound secreted by blister beetles for defense.

carbon footprint: the total amount of greenhouse gases produced by an activity, product, population, or individual.

carnivorous: animals that consume other animals for sustenance.

carrion: the remains of a dead animal (or corpse).

carrying capacity: the maximum load of a species population that the environment can sustain indefinitely.

central motor programs: programs that control the movements of appendages, like the legs and wings.

cephalon: the head or anterior region of arthropods.

cephalothorax: an anterior region of some arthropods in which the head and thorax are fused together.

cercus (pl., **cerci**): a sensory appendage extending from the abdomen of some arthropods.

chelicerae: pre-oral appendages of arthropods belonging to the subphylum Chelicerata.

chemical digestion: the breakdown of food molecules via the action of enzymes and/or the chemical environment in which the food resides.

chemical trapping: a technique that relies on chemical particles becoming attached or stuck to the bodies of insects during encounters with living or nonliving objects.

chilling injury: damage resulting from exposure to low temperatures above freezing.

chordotonal sensillum: a sensory organ that functions in vibration detection.

chorion: the outer covering or membrane of an insect egg.

circadian clock (also **internal** or **biological clock**): an internal timer in the form of neuronal cells that respond to environmental signals to initiate or inhibit cellular events.

circadian rhythm: a physiological or behavioral change that occurs once approximately every twenty-four hours.

claspers: a generic reference to a structure used by males to grab a female during courting and/or copulation.

classical (Pavlovian) conditioning: the use of an unconditioned stimulus—one that does not result in a dramatic behavioral response—with another stimulus (conditioned) that causes a significant change in behavior, to train or condition an animal to associate one with the other.

clutch: a batch of eggs all laid at the same time by a female insect.

coeloconic sensillum: the sensory organs that function in thermoreception.

coelomoducts: the excretory tubules found in some invertebrates.

cold hardiness: acquired resistance to adverse effects of low temperature exposure, usually as part of diapause.

cold shock: a response to a rapid drop in temperature.

colony collapse disorder: a condition in which honey bee adults disappear in large numbers from the hive, never to return; literally leads to the collapse of the entire hive.

common oviduct: connects the lateral oviducts to the genital chamber.

complete metamorphosis: a type of development in insects in which a pupal stage is used to transform from immature to an adult.

compound eyes: the image-forming eyes of arthropods.

conspecific: an individual of the same species.

contact chemoreceptors: receptors that perceive chemical signals that make physical contact, used in gustation and olfaction.

cranium: the insect head.

critical thermal maxima: temperature at or above that which is lethal to a particular insect species depending on length of exposure.

critical thermal minima: temperature at or below that which is lethal to a particular insect species depending on length of exposure.

cross-resistance: development of resistance to a chemical, usually an insecticide, as a result of becoming resistant to a different chemical.

cryoprotectants: compounds synthesized to function like antifreeze, preventing insect fluids from forming ice crystals.

crypsis: body coloration that allows an insect to blend into its surroundings.

cryptic species: a species considered to be anatomically indistinguishable from another closely related species.

cryptonephridial arrangement: a condition in which Malpighian tubules wrap around portions of midgut or hindgut as a means to rapidly extract water from a liquid diet.

cuticle: the nonliving layers of exoskeleton.

cuticular sensillum (pl., **sensilla**): the basic sense organ that extends through the exoskeleton.

debride: to clean a wound by removing dead, injured, and infected tissues.

defect action level (DAL): levels established by the US Food and Drug Administration regulating the acceptable amount of contamination in food by insects, microorganisms, and other materials that are unavoidable during the process of harvesting or processing foods and that pose no health risk to humans.

degranulation: the process of granulocytes releasing granules onto parasites to attract other hemocytes to respond.

degree-days: the thermal energy available in the environment.

delusional parasitosis: an irrational or uncontrollable fear of insects that can render an individual physiologically and emotionally debilitated.

dendricles: cuticular teeth-like projections found in the proventriculus of some insects.

dendrite: the region of a neuron that receives an incoming stimulus.

determinate growth: displayed by insects that have a fixed number of instars during their life cycle.

detritivores: organisms that consume organic matter decaying from other organisms.

deuterotoky: a form of parthenogenesis in which an unfertilized egg can become either a male or female.

diapause: a dynamic physiological state of dormancy commonly associated with winter and enhanced cold hardiness.

dichotomous keys: identification keys in which external features are presented in contrasting couplets. The features in the first couplet are examined, and if relevant to the insect specimen being examined, then instructions lead to which feature to compare next.

dicondylic: mandibles attached to the posterior portion of the head capsule at two points of articulation.

diploid: a condition in which a cell or organism has two copies of each chromosome.

direct insemination: transfer of spermatozoa between insects in which the mating pair make direct contact with each other.

diverticuli (s., **diverticulum**): blind end pouches or sacs extending from the esophagus.

DNA barcoding: a method of insect identification in which short, unique sequences of DNA for a given insect species can be used to identify any developmental stage to the exact genus and species.

drone: a male honey bee, or an unmanned aerial vehicle.

ecdysis: the removal of the old skin or exuvia during molting to unveil the new exoskeleton.

ecdysone/ecdysteroids: a complex of hormones produced by the prothoracic gland that stimulates epidermis to initiate a molt.

economic injury level (EIL): the lowest population density of an insect pest that will cause economic damage to a crop or commodity, or the amount of pest injury that justifies some type of control strategy to be applied.

economic (or action) threshold: the population density of an insect below the economic injury level that is used for making a decision as to whether artificial control should be applied.

ectotherms: animals that cannot regulate body temperature directly or at all.

egg eclosion: hatching of an insect egg so that the first instar juvenile emerges into the environment.

ejaculatory duct: extends through the male aedeagus to the gonopore, permitting sperm to leave male body.

embryonic development: the period of insect development that occurs within the egg.

encapsulation: hemocyte response to parasitic insects and nematodes in which several layers of cells surround the invaders to contain and destroy the parasites.

endocuticle: innermost layer of procuticle that is digested during a molt.

endopterygotes: these insects form wings internally from imaginal discs during pupal development of complete metamorphosis.

endotherm: animals that can regulate their body temperature regardless of ambient conditions.

entomological counterterrorism: the use of insects to combat entomological terrorism agents.

entomological terrorism (entoterrorism): the use of insects or their products in acts of terrorism.

entomology: the study of insects and the closely related terrestrial arthropods.

entomophagy: the practice of eating insects as part of the normal or typical diet.

entomophobia: fear of insects.

environmental stewardship: the practice of maintaining a healthy environment through personal actions.

environmental tokens: abiotic cues in the environment, such as photoperiod, humidity, or temperature, that evoke behavioral or physiological responses in insects.

ephemeral: temporary, or lasting for a short period of time, such as resources that do not last, like an animal carcass.

epicuticle: the outermost layer of cuticle.

epidemic: a disease reaches levels of infection that were unexpected, perhaps even new to a particular location.

epidermis: the living tissue layer that makes up the exoskeleton of arthropods.

essential nutrients: nutrients that must be obtained in the diet because they cannot be synthesized in the body.

excreta: waste material released or eliminated from the alimentary canal of an insect, usually via the anus, which may contain undigested or partially digested foodstuffs, urine, microorganisms, and water.

exocrine gland: glands that have ducts that release secretions to the outside of the body, often in the form of pheromones.

exocuticle: the outermost layer of procuticle that is shed during ecdysis.

exopterygote: insects in which the wings develop externally and gradually during juvenile development.

exoskeleton: the outer covering or skin of an arthropod used for muscle attachment.

extant: living organisms, as opposed to those that are extinct or no longer living.

exuvia: exocuticle shed during molting.

facultative myiasis: infestation of living tissue by necrophagous flies that ordinarily restrict feeding to dead tissues.

fecundity: reproductive output of a female.

ferculum: the fork-like projection of Collembola used for jumping.

file: the hard edge of insect wing used to make sound.

filter chamber: arrangement of hindgut in close contact with the midgut to remove water from a liquid diet, usually covered by a sheath to conceal the arrangement.

fitness: overall reproductive success of an individual based on ability to produce viable offspring that in turn are reproductively successful.

fixed action pattern: a form of innate behavior in which the overall behavior occurs in a series of sequential steps, each with a threshold requirement or internal readiness before a response can be initiated.

foregut (stomodeum): anteriormost portion of the gut tube.

formicophilia: a fetish in which sexual arousal is achieved from insects walking or biting a person, most often on the genitals.

frass: material eliminated from the digestive tube via the anus, generally composed of undigested and nonabsorbed food materials, metabolic wastes, and water.

frontal ganglion: ganglion located in the cranium that functions in regulating appetite and satiety.

ganglia (s., ganglion): functional tissue masses composed of large numbers of neurons and glial cells, including all of their processes.

gastric cecae (s., cecum): finger-like projections at the anterior end of the midgut that function in food absorption and synthesis of peritrophic membrane.

general adaptation syndrome: better known as the general stress response, a series of cellular, physiological, and behavioral responses to environmental stressors.

glial cell: a support cell in the nervous system that primarily functions to insulate axons.

global warming: warming of the Earth's atmosphere and surface caused by accumulation of greenhouse gases and subsequent thermal radiation in the atmosphere.

glossa: tongue-like structure in the mouth of insects that is covered with taste receptors and can be used for lapping up liquids.

gonads: the primary reproductive organs of animal, which produce gametes.

gonopore: the outside opening to the reproductive system of male and female insects.

gravid: the condition of a female possessing mature eggs ready for oviposition.

greenhouse effect: the warming of the Earth's surface caused by radiation from the atmosphere.

gustation: the process of tasting, typically involving sensory neurons via taste receptors or taste buds.

gynogenesis (pseudogamy): a form of parthenogenesis in which sperm makes contact with an egg to activate mitosis, does not exchange genetic material.

habituation: a simple form of learned behavior in which an insect stops responding to a stimulus that previously elicited a reaction.

haploid: a cell or organism with half the chromosome number of the diploid state.

heat shock: response to a rapid elevation in temperature.

heat shock proteins (HSPs): a large family of stress proteins synthesized in response to a large number of seasonal and aseasonal stressors.

hematophagous (hematophore): an organism, usually an animal, that feeds on the blood of a living animal, which by definition is a host-parasite relationship.

hemimetabolous (incomplete) metamorphosis: a type of development in which an insect gradually becomes more adult-like with each molt, including the formation of wings from external buds.

hemocoel: the body cavity of an animal in which circulating fluid is not contained within vascular tissues and thus bathes internal organs.

hemocyte: the circulating cells in hemolymph responsible for several defensive functions, wound healing, and precipitation.

hemolymph: the circulating fluid of insects that performs both the functions of circulation (hemo) as well as body defense (lymph) and bathes all internal organs of insect body.

heraldry: a form of symbolism in which a coat of arms or simply the image of an animal is used to adorn metal armor for the purpose of identification.

herbivorous: an organism (herbivore) that feeds on plants or plant material for sustenance.

hermaphroditism: an individual that possess both male and female reproductive organs, at least with respect to gonads. Simultaneous hermaphrodites possess both sets of reproductive structures, while sequential hermaphrodites begin as one sex and then become the other.

heterothermy: the production of internal heat in poikilothermic animals.

hindgut (proctodeum): posteriormost portion of the gut tube.

holometabolous (complete) metamorphosis: a form of insect development in which juveniles and adults are completely dissimilar in appearance and often in habitat, and a pupal stage is used to transform from larva to adult.

homeostasis: the tendency of a system, particularly a physiological system, to maintain stability owing to the normal or healthy function of its components.

homozygosity: the condition of a diploid organism in which both alleles of a gene are identical.

hydrofuge hairs: water-repellant hairs located on the bodies of insects, which serve a variety of functions.

hyperphagia: elevated appetite.

hypoglossa: the tongue of insects.

humancentric: concerned for or focused on humans or mankind.

ileum: anterior portion of hindgut.

imaginal discs: undifferentiated or stem cell tissue that develop into adult structures during pupal development.

imago: an adult insect.

indeterminate growth: displayed by insects that have a variable number of instars during their life cycle.

indirect insemination: transfer of spermatozoa between insects without making direct contact with each other.

innate behavior: behaviors that an animal is genetically preprogrammed to do; no experience or learning is required to perform the behavior.

inoculative release: deliberate or intentional introduction of small numbers of an insect species with the goal of the species reproducing and establishing within that area.

insect growth regulators: hormone analogs that mimic that action of insect hormones critical to molting, thereby disrupting various aspects of growth.

insect phylogeny: the study of evolutionary relationships among insects.

insecticide: biocidal chemical compounds used specifically to reduce insect populations when reaching pest status to humans.

instar: a developmental stage of insects, most often applied to juvenile stages.

instrumental learning: a form of learning that depends on an insect using memory from past experiences to modify or even improve performance, when subjected to an identical stimulus.

integrated pest management: an approach to managing insect pest populations in which an understanding of the pest's biology is used to develop a comprehensive strategy to reduce pest status.

intromittant organ: the male aedeagus or penis.

inundative release: the deliberate introduction of large numbers of an insect with the intent of that species causing an immediate effect in the area of release.

invasive (exotic) species: an organism accidentally or purposively introduced into a new region, typically in a new country, that becomes a pest because of a lack of natural controls in the new region.

Johnston's organ: a mechanoreceptor located on the antennae of adult insects.

juvenile hormones: a complex of hormones that stimulates epidermis to produce a juvenile cuticle during molting.

kairomone: a type of allelochemical by which the emitter may or may not be harmed by the chemical released but the receiver usually benefits.

kinesis: a form of innate behavior in which an insect responds to a stimulus by random motion.

labium: the lower lip of an insect's mouth.

labrum: the upper lip of an insect's mouth.

larva (pl., **larvae**): immature stage of development, when an insect ecloses from an egg to begin a period of feeding.

larval competition: battles for nutrients or survival between allospecifics and/or conspecifics for a resource that they coinhabit.

larval mimicry: a type of mimicry that occurs in a few families (e.g., Papilionidae, Saturniidae) of Lepidoptera, in which images of the caterpillar are evident along the margins of the mesothoracic (fore) wings.

latent learning: learning that occurs with no apparent reward or punishment.

lateral oviduct: connects ovaries to common oviduct.

lysozymes: digestive enzymes that digest microorganisms or old cells.

maggot mass: large feeding assemblages of necrophagous flies feeding on carrion, also known as larval aggregations.

maggot therapy: treatment of wounds or lesions on humans with sterile, necrophagous fly larvae as a means for debriding and cleaning the wound.

Malpighian tubules: a series of tubes extending from the hindgut that function in excretion and osmoregulation.

mandibles: teeth-like structures that serve as mouthparts for some arthropods.

marbling: mosaic pattern of skin discoloration resulting from formation of sulfhemoglobin in capillaries near the skin surface.

marketing globalization: a business model that combines marketing and selling goods within an integrated global economy.

maxillae: accessory jaws to the mandibles.

mechanical digestion/mastication: the physical breakdown of food molecules as a result of mouthpart manipulation, muscular contractions, or other physical distortion.

medicocriminal entomology: the branch of forensic entomology focused on the use of insect evidence in criminal investigations, especially those associated with violent acts.

melanization: tanning or pigment deposition in the exoskeleton.

micro air or land vehicles: cyborg robots designed to mimic the appearance and movement of insects while engaged in surveillance operations.

micropyle: the opening or canal on egg that sperm enters for fertilization.

midgut (mesenteron): the middle portion of the gut tube, dedicated to chemical digestion and absorption.

mimesis: body coloration that allows insects to appear as a natural object in the environment.

molting: the process by which arthropods and other ecdysozoans grow by shedding the old cuticle and replacing it with a new, larger one.

monocondylic: mandibles with a single point of articulation to the posterior portion of the head capsule.

monoculture: the practice of growing one row crop alone in a field during a season.

monophyly: grouping an ancestral species with all of its descendants.

monovoltine: having only one generation a year for a particular species.

Müllerian mimicry: an evolutionary adaptation in which multiple species, each being unpalatable or possessing defense capabilities, evolve very similar aposematic coloration.

mushroom bodies: a region of the protocerebrum that functions in storing short- and long-term memories, particularly those associated with olfactory cues; also called the corpora pedunculata.

mycetocyte: a specialized cell in the hindgut that harbors endosymbiotic organisms.

myiasis: the invasion or infestation of living or necrotic tissue of a host by flies during the larval stages.

naiad: the juvenile stage of some hemimetabolous insects in which the immatures develop in aquatic environments but the adults are terrestrial.

natural enemies: organisms that function as regulators of plant and animal populations through their behaviors as predators, parasites, or herbivores.

nauplius larvae: the juvenile stage of the majority of crustaceans, in which head appendages are used for locomotion.

necrophagous: feeding on dead organisms, usually in reference to consuming dead animals.

necrosis: physiological changes that occur in cells and tissues after death.

Neoptera: the group of adult insects capable of folding their wings flat against the body at rest.

neuron: a nerve cell that specifically receives stimuli and passes the information on to the central nervous system for interpretation and response.

nodule formation: hemocyte response to large numbers of microorganisms, in which the foreign invaders are walled off by cell layers.

nonessential nutrient: a nutrient that can be synthesized from precursor materials obtained in the diet, and thus is not required to be in the food consumed.

nymph: a juvenile insect resulting from hemimetabolous development. Generally the term is reserved for terrestrial species.

ocellus (pl., **ocelli**): a simple eye, a non-image-forming, light-sensing structure located on the head of arthropods.

ommatidium: functional units of the compound eye, which contain photoreceptors and light sensory units covered by a cornea.

ontogeny: the developmental history of an organism.

ootheca: protective coating surrounding the developing embryos of some hemimetabolous insects.

operant conditioning: conditioning in which a reward (positive reinforcer) or punishment (negative reinforcer) is used in association with a particular behavior.

osmolyte: solute dissolved in a fluid that contributes to the osmotic pressure of that liquid.

outbreeding: when related individuals mate with nonrelatives (as opposed to inbreeding).

ovaries: the primary reproductive organs of a female, which produces oocytes that mature into eggs.

ovariole: a tubular sheath in an ovary that serves as the site of oocyte production and maturation.

ovicidal: the behavior of killing eggs.

oviparity: a mechanism of oviposition in which an egg with a chorion is deposited to the outside of the female's body.

oviposition: the process of egg laying.

ovipositor: egg-laying structure of female insects.

ovotestis: a structure that produce both eggs and sperm.

ovoviviparity: an egg with a chorion hatches inside the mother, giving appearance of live birth.

ovulation: release of eggs from the ovaries.

Palaeoptera: the group of adult insects that cannot fold the wings against the body at rest.

palps: accessory mouthpart structures that function in mechanical and chemical perception.

pandemic: an epidemic disease that spreads to other countries or continents.

parasitism: the process of one organism (parasite) feeding exclusively on an individual (host) to complete a particular stage of development. The host is usually not killed during the association.

parasitoidism: a specialized condition of parasitism in which the host is always killed as a result of parasitism.

parthenogenesis: a form of asexual reproduction in which unfertilized eggs develop into new individuals.

patchy: irregular or unpredictable in occurrence, as with carrion in the natural environment.

pedipalps: a second pair of appendages on the cephalothorax of chelicerates, often considered analogous to the mandibles of insects.

peritrophic membrane: a noncellular membranous tube that lines the midgut.

pest management: a practice of managing insect populations in which the goal is to reduce the effects of the pests, not necessarily to control or kill them.

pesticides: biocidal chemical compounds used to control any type of living organism deemed a pest to humans.

phagocytosis: the engulfing of cells, debris, or chemicals by hemocytes.

pharmacology: the study of drugs, their sources, mode of action, and properties.

pheromones: chemicals produced in exocrine glands and released to the outside of the body to promote communication between members of the same species.

photoperiod: the amount of daylight in relation to darkness in a twenty-four-hour period.

photoreception: the sensory perception of visual or light-related stimuli.

phylogeny: the development or evolution of a particular group of organisms.

phytophagous: organisms that consume plant or vegetative materials as a food source.

pleopods: swimming legs or swimmerets found on the abdomen of crustaceans.

pleural region: the lateral side of the thorax of winged insects.

poikilothermic: a condition in which body temperature varies with ambient conditions.

polyembryony: a form of asexual reproduction in which a single egg divides to produce two or more individuals.

polynomial nomenclature: a method for naming animals that consisted of using a generic name to label a particular species and a second name that was meant to be descriptive of the species.

positive feedback: a type of feedback in which the response generated provokes further stimulation, resulting in an enhanced response that continues in a feedback loop until the desired outcome is achieved or additional stimulation is no longer possible.

postembryonic development: the period of insect growth that occurs after egg eclosion.

postmortem interval: an estimation of the time since death for a human.

predation: the process of one organism (predator) consuming an animal (prey) as food.

preprogrammed motor program: an innate behavior in which an internal readiness must be achieved before the behavior can be manifested.

primer pheromones: pheromones that do not cause or expect an immediate reaction by the receiver.

procuticle: the first layer of cuticle produced during a molt.

propolis: a substance derived by bees from tree sap and other plant materials, used in bee hives for sealing cracks in the hive.

proprioceptors: mechanoreceptors that detect change in body position or movement of an appendage or some other body part.

prothoracicotropic hormone (PTTH): a neural hormone that stimulates prothoracic gland to produce and release ecdysone.

proventriculus: a muscular valve positioned at the posterior end of the foregut.

pulvillus (pl., **pulvilli**): the soft cushion-like pad located between terminal claws in Diptera.

puparium: the hardened outer covering of fly pupae that is formed by the last larval skin undergoing apolysis (release from the epidermis) but not ecdysis (actual shedding of the old exoskeleton).

pygidium: the posterior body region or shield of some arthropods.

pyrethrum: an insecticidal compound derived naturally from *Chrysanthemum* flowers.

quiescence: a physiological state in which metabolic activity is lowered during unfavorable conditions and resumes to normal immediately upon return to favorable conditions.

rapid cold hardening: the process of quickly acquiring resistance to the adverse effects of low temperatures from a brief exposure to nonlethal low temperatures before exposure to more extreme conditions.

rapid heat hardening: the process during which a brief exposure to a nonlethal high temperature stimulates the production of several HSPs, which in turn protect intracellular proteins from a subsequent lethal aseasonal elevated temperature change.

rectum: the posterior portion of the hindgut.

reflex (pl., **reflexes**): innate response to a stimulus that results in an immediate reaction, often to avoid a perceived threat.

reflex bleeding: release of hemolymph in response to attack by a predator.

regressive evolution: the process by which an organism loses a trait over evolutionary time.

releaser pheromones: pheromones that demand an immediate behavioral response in the recipient.

resource partitioning: the processes by which two or more organisms competing for the same limited resources coexist by using the resource differently.

retroinvasion: the condition that occurs when fly larvae that were accidentally ingested pass through the entire alimentary canal and then migrate back into the anus to feed.

reverse peristalsis: vomiting.

saprophagous: an organism (a saprophage) that feeds on dead or decaying plant, animal, or other organism material for sustenance.

satiety: the state or sensation of being full after consuming a meal.

scent glands: specialized exocrine glands that produce volatile sex pheromones released into the environment.

scientific method: a method for testing questions using an approach centered on formulating hypotheses, making observations from carefully designed experiments, refining questions, and narrowing possible explanations for observed phenomena.

sclerite: an external plate on the exoskeleton that functions like protective armor.

sclerotization: the process of hardening the exoskeleton.

scraper: the comb-like edge of insect wing used to make sound.

seminal vesicle: functionally stores sperm in males until ejaculation.

semiochemical: chemicals used for communication that modify behavior of recipients.

sensitization: a simple form of learned behavior in which a repeated stimulus evokes an increasingly progressive response.

sequestration: acquiring compounds from a food source that are stored in the body for use in defense.

sericin: adhesive or glue-like material produced by silkworms and composed of protein that holds silk together during the process of forming a cocoon.

sericulture: the practice of breeding silk moths for the production of raw silk.

setae: hair-like structures commonly referred to as bristles.

shellac: the resinous secretion derived from lac-producing scale insects that functions as an adhesive.

slash-and-burn: a technique used to clear land for agricultural purposes by cutting down vegetation and setting it on fire. The ash is used as nutrients once worked back into the soil.

sniffer systems: living entities that can be used for odorant detection.

spatial partitioning: the physical separation of competing organisms that use the same resource.

species: the largest group of organisms that are capable of interbreeding and producing viable offspring.

sperm competition: the phenomenon associated with males competing to fertilize eggs of females that can mate more than once and store sperm until fertilization.

spermathecae: the sperm storage organ in females.

spermathecal gland: an accessory gland that provides nutriment to keep sperm in the spermathecae alive.

spermatophore: a gelatinous package produced by male insects to protect sperm from desiccation.

spiracles: the external openings to the tracheal system of insects, usually paired on thoracic and abdominal segments.

stadium: synonymous with **instar**; a stage of insect development.

startle response: the temporary surprise or startle of a potential predator so that an escape response can be generated, caused by use of sudden visual displays or noise by prey.

sternum: the ventral surface of the thorax and abdomen.

stilting: the behavior of male Madagascar hissing cockroaches of lifting the body or just the abdomen up and down to ward off other males.

stored product entomology: the branch of forensic entomology focused on insect infestation of foods and food products and the legal disputes that arise from the presence of insects, their body parts, or evidence of insect activity in foodstuffs.

stressors: stimuli that cause an organism to deviate from the homeostatic condition for long enough that normal physiology is altered.

stridulation: the sound production of crickets and grasshoppers produced by rubbing wings together.

stylus (pl., **styli**)**:** sensory appendage extending from the abdomen of some arthropods.

subesophageal ganglion: part of the central nervous system of insects, representing three fused ganglion that connect to the ventral nerve cord and have nerves that innervate several organs and muscles in the thorax.

sublimation: a physical phase change of matter, commonly in reference to transformation of liquid to a solid.

supercooling point: the temperature at which ice crystals begin to form in body fluids.

supraesophageal ganglion: the portion of the central nervous system that is considered the true brain. It represents the fusion of three large ganglionic centers, the protocerebrum, deutocerebrum, and tritocerebrum.

suture: an external groove in the exoskeleton.

swarming: behavior of fly masses of male insects attempting to attract females based on visual display.

symbiont: an organism living in association with another organism, in which the outcome may be of mutual benefit, parasitic, or the host is neither harmed nor benefits from the interactions.

symports: membrane proteins that function as amino acid receptors/transporters into the cell.

synanthropic: an association or dependence on humans or human activity.

synaptic junction: the space between neurons in which terminal membranes of axon are facing the dendritic membranes of another neuron. Neurotransmitters are released into the synapsis to bind to postsynaptic membranes (dendrites).

syncytium: a tissue layer in which multiple nuclei are present in an apparent single cell.

synomones: chemical substance that when released generate a response in the receiver that is also beneficial to the emitter

tagmata: distinct regions of the insect body that represent specific functions.

tarsi: the last true segments of insect legs.

taxis: a form of innate behavior in which the insect changes body position either toward or away from the stimulus.

taxonomy: the branch of science dealing with describing, identifying, naming, and classifying of organisms.

tenaculum: the hook-like structure on the sternum of Collembola into which the ferculum is inserted to allow jumping.

tentorium: apodemes in the head that form an internal skeleton used for muscle attachment.

tergum: the dorsal surface of the thorax and abdomen.

terrorism: the use of terror, frequently through acts of violence, to instill fear in a group of people.

thanosis: playing dead as a defensive strategy.

thelytoky: a form of parthenogenesis in which females develop from unfertilized eggs but remain diploid.

thermal hysteresis: the phenomenon in which antifreeze proteins attach to ice that has already formed in body fluids, inhibiting further growth of the ice crystals.

thermal infrared wavelengths: electromagnetic radiation or invisible radiant energy.

thorax: the middle region of insect body, to which legs and wings are attached.

totemism: the practice of using wooden or other carved structures, known as totems, adorned with images of living creatures to represent ancestors, tribe, or clan.

trachea: a series of tubes located internally that compose the majority of the insect ventilator system.

traumatic insemination: the insertion of the male's aedeagus through the body wall of the female to ejaculate.

trichoid sensillum: a sensory organ used for mechanoreception in insects.

tympanum: membranes used to make and receive sound.

univoltine: insects having multiple generations in one year.

urban entomology: the branch of forensic entomology focused on insect invasions of human habitation and the legal disputes resulting from such insects.

vagina: the genital chamber in insects, possessing a narrow vulva.

vas deferens: tubing that transports sperm from the gonads to the seminal vesicles.

ventral nerve cord: large, paired nerve connectives that link all the ganglionic centers of the insect together.

viviparity: an egg without an eggshell hatches in mother, giving the appearance of live birth.

vulva: the opening of the genital chamber to the outside of the body.

wax: yellowish or brown lipid-rich secretions from honey bees used for the construction of honeycomb in the hive.

weapons of mass destruction: nuclear, radiological, chemical, biological, or other weapons that can kill or bring significant harm to a large number of humans or cause great damage to man-made structures, natural structures, or the environment.

widow making: behavior of female insects that kill mate after copulation.

wind shear: gradients in wind speed or direction within the atmosphere.

xylophagous: organisms that consume wood as a food source.

zone of temperature tolerance: the range of temperature in which an insect species can complete development without deleterious consequences and survive indefinitely.

zoology: the branch of biology focused on the study of animals.

Index

Page numbers followed by f refer to figures and t to tables. Page numbers in bold refer to terms in the glossary.